# INTRODUCTORY
# DIFFERENTIAL EQUATIONS

---

## *From Linearity to Chaos*

---

Eric J. Kostelich

Arizona State University


Dieter Armbruster

Arizona State University

 **ADDISON-WESLEY**

An imprint of Addison Wesley Longman, Inc.

Reading, Massachusetts • Menlo Park, California • New York • Harlow, England
Don Mills, Ontario • Sydney • Mexico City • Madrid • Amsterdam

SPONSORING EDITOR *Laurie Rosatone*
DEVELOPMENT EDITOR *Marianne Lepp*
SENIOR PRODUCTION SUPERVISOR *Karen Wernholm*
PRODUCTION SERVICES *Diane Freed*
MARKETING MANAGER *Michelle Babinec*
SENIOR MANUFACTURING MANAGER *Roy Logan*
PREPRESS BUYER *Caroline Fell*
SR. TECHNICAL ART SPECIALIST *Joseph Vetere*
TEXT DESIGNER *Ron Kosciak, Dragonfly Design*
COMPOSITOR *TechBooks, Fairfax, VA*
ILLUSTRATOR *Horizon Design*
ART COORDINATOR *Sandra Selfridge*
COVER DESIGNER *Barbara T. Atkinson*
COVER PHOTOS *Jaguar © All Stock/Tim Davis/PNI; Kayaker © All Stock/Tim Davis/PNI; Pendulum © PhotoEdit/Tony Freeman/PNI; and Antelope © Aurora/Jose Azel/PNI*

Library of Congress Cataloging-in-Publication Data

Kostelich, Eric John.
    Introductory differential equations / Eric J. Kostelich, Dieter
Armbruster.
      p.    cm.
    ISBN 0-201-76549-7
    1. Differential equations.   I. Armbruster, Dieter.   II. Title.
QA372.K8438   1996
515′.352–dc20                         96-23882
                                      CIP

1 2 3 4 5 6 7 8 9 10-DOC-99989796

# PREFACE

Differential equations are the heart of mathematical modeling, because they allow scientists to make predictions about a wide variety of natural phenomena. The goal of this book is to introduce the student to the principal ideas in ordinary differential equations, of which prediction is one of several important problems. We pay particular attention to the following issues:

- the qualitative behavior of solutions and its relation to the behavior of a corresponding physical system;

- the sensitivity of solutions to initial conditions and parameters and how it affects scientists' ability to make specific predictions about a process from mathematical models;

- the notion of linearity in differential equations and other areas of mathematics.

The ordinary differential equations course is often the last formal mathematics course for many students in the physical sciences and engineering. Therefore, we believe that it is important to stress the connection between the basic mathematical theory and its relevance to contemporary scientific practice.

## APPROACH AND FEATURES

The text is appropriate for a one-semester or one- to two-quarter course at the sophomore level for students in mathematics, engineering, and the physical and biological sciences who have had one full year of calculus. The book emphasizes the theory of linear ordinary differential equations, but also includes a self-contained introduction to various nonlinear equations of current scientific interest. The topics and pedagogical approach in the book are organized according to the principal scientific ideas, as opposed to a more conventional ordering in terms of "solution methods" for various differential equations.

This book, which grew out of a curriculum reform project that began at Arizona State University in January 1993, moves away from the teaching of solution methods for their own sake. Rather, we have tried to emphasize that solution methods are a means to a more important end, that is, to understand what a given mathematical model implies about an underlying physical process. Thus, Chapters 1 through 8 discuss, respectively, qualitative analysis of first-order equations, exponential growth, continuous dependence on initial conditions, linearity as an abstract mathematical concept, linear oscillators, qualitative analysis of first-order systems, linear systems, and nonlinear equations, including a brief introduction to bifurcation theory. The only exceptions to this organizational rule are Chapters 9 and 10, which consider numerical methods and Laplace transforms, respectively. Even in these chapters, considerable emphasis has been placed on interpreting the qualitative behavior of the solutions.

## EXERCISES: ANALYTICAL, GRAPHICAL, AND FLEXIBLE

Most of the exercises in this text have multiple parts. Often, we ask the student to derive a solution, then plot it and interpret it in various ways. The questions are designed to get the student to think about what the solution means. There are fewer "drill" problems and more exercises that investigate a given equation in detail. For this reason, instructors will find that many exercises in this book require more analysis and interpretation from the student than those in traditional texts.

Many of the exercises ask students to explain some aspect of the problem in a sentence or paragraph. We believe that it is important to help students to develop their writing ability and to explain the implications of mathematical results in ways that are comprehensible to nonmathematicians.

We have written the exercises to give the instructor maximum flexibility in using them as teaching tools. Most of the exercises can be assigned as traditional homework problems. Likewise, many exercises can easily be adapted into worksheets for students to complete as part of an in-class computer laboratory or problem session. With slight modifications, some of the exercises can be made into small projects with two- to three-page writeups. Others can be assigned as part of a group project, chapter project, or take-home exam. Due to this flexibility, there is no restrictive labeling of exercises as "computer projects," "chapter exercises," and the like. We encourage instructors to read the exercises carefully and to decide in advance how much detail they want their students to provide when written explanations are required.

# APPROPRIATE USE OF TECHNOLOGY

Powerful and relatively inexpensive personal computers have become valuable scientific and instructional tools. We have found them particularly helpful for the following purposes:

1. for graphing direction fields and phase portraits;

2. for illustrating basic numerical methods;

3. for investigating the effects of changes in initial conditions and parameters on solutions.

While the use of computer-supported symbolic and numerical calculations relieves students of tedious calculations, it also increases the importance of solid theoretical understanding of differential equations. For instance, questions about the existence and uniqueness of solutions become much more important when a computer determines a solution numerically. Many of the exercises are designed to develop students' theoretical understanding.

We believe that an important part of students' scientific training is learning how to use a computer appropriately. Our advice to students is that if a homework exercise can be solved readily with a pencil and paper, then the solution should be written up in the traditional way. On the other hand, if a student finds that the required calculations are too lengthy or tedious to do by hand, then he or she should use a software package that is appropriate for the task. Students (and instructors) are expected to use their best judgment on whether computer assistance is desirable for a particular problem.

The text is designed to accommodate a variety of technological tools. It is not tied to any specific hardware or software platform, and it gives instructors considerable leeway to determine the extent to which technology is used in the course. As mentioned above, however, the exercises often ask students to plot a solution and comment on some aspect of the graph. At a minimum, therefore, students should have ready access to equipment that can plot functions over a specified interval.

Some possibilities for technological tools include:

• The Maple symbolic algebra package. The student edition of Maple contains all the necessary commands for the computations required in this text. A brief supplement to this text is available for Release 4 of Maple V. The supplement's organization parallels that of the main text and briefly outlines those commands that are most directly relevant for the exercises in the corresponding chapter of the main text.

• The *Mathematica* symbolic algebra package. A supplement analogous to the Maple supplement is available.

- Other packages, such as Macsyma, Derive, and MATLAB, are also suitable for use with the text.

- A graphing calculator. At a minimum, students should have access to graphing calculators, as homework exercises routinely ask students to graph solutions. (In addition, many graphing calculators are able to plot direction fields for scalar first-order equations.) The text has been used successfully in classes where only graphing calculators were available.

Although symbolic algebra programs are excellent technological aids, we strongly advise instructors to use a minimum number of commands. By doing so, the learning curve for students is shortened, and more importantly, students can focus on the mathematical ideas instead of the computer. The Maple and *Mathematica* supplements show how only a handful of commands can be used effectively for many of the exercises. The instructor's solutions manual also includes some sample computer labs, which can be used in class or assigned as homework.

## CONTENT AND ORGANIZATION: A DIFFERENT PHILOSOPHY

### Chapter 0: A Survey of the Main Ideas

In our experience, most students who enroll in differential equations courses have little idea of the nature of the subject matter that will be covered. Chapter 0, whose inclusion is a unique feature of the text, presents a brief overview of some of the principal scientific ideas that are studied with the aid of differential equations, namely, the concepts of exponential growth, oscillatory motion, and the dependence of solutions on initial conditions.

The material in Chapter 0 can be used as the basis of a first lecture, which can be a survey of the course. Alternatively, instructors may wish to focus on one or two examples from Sections 0.1 and 0.2, then assign these sections for reading. The material in Section 0.3, which can be postponed until later in the course, makes a nice one- to two-lecture introduction to nonlinear differential equations; we have found it useful when there was not enough time to cover the material in Chapter 8 on nonlinear equations in detail.

### Chapter 1: Fundamental Ideas

This chapter discusses the basic ideas of first-order differential equations. The presentation is inspired by various calculus reform projects, insofar as we focus first on the geometrical ideas (direction fields), second on numerical approximations to solutions (Euler's method), and finally on an analytical approach (the separation of variables). One goal of the chapter is to emphasize to students that most differential equations do have solutions, that the qualitative behavior of solutions can be discerned in simple cases, and that solutions can be approximated accurately by numerical means, even though convenient formulas for the solutions often cannot be found. For this reason, the separation of variables method is postponed to the end of the chapter.

## Chapter 2: Exponential Growth

This chapter emphasizes applications from continuous compounding of interest, Newton's law of cooling, and chemical mixing. We suggest that the instructor choose two applications (e.g., financial modeling and Newton's law of cooling) to illustrate the basic points. The first three sections can be covered fairly rapidly if students have had a solid introduction to the exponential growth equation in their calculus courses.

Section 2.4 is unique to textbooks at this level, as it discusses the predictive value of differential equations and some of the limitations that imperfect data present in mathematical modeling. The section introduces the notions of absolute and relative error and the effects of errors in initial conditions and parameters on the predictions of simple mathematical models. The material in this section can be augmented nicely with computer labs or graphing calculators.

## Chapter 3: Existence and Uniqueness of Solutions

Chapter 3 discusses the existence and uniqueness theory for first-order equations. Although the material in the first two sections includes the standard theorems, the third section, on Gronwall's inequality, is not usually found in textbooks at this level. Our goal is to help students understand how Gronwall's inequality gives a priori bounds on the rate of separation between two initial conditions without solving the differential equation. Although the bounds are often crude, the continuous dependence on initial conditions is a key idea. The fact that solutions vary continuously with initial conditions under mild hypotheses gives useful predictive value to mathematical models of many natural phenomena. (The material in Section 3.3 requires somewhat more mathematical maturity than the other sections, and some instructors may wish to skim this material.)

## Chapter 4: Linearity

The purpose of Chapter 4 is to acquaint students with the notion of linearity as an abstract mathematical concept. It is simple enough to show students how to recognize linear differential equations. However, this chapter has a more ambitious goal, namely, to show that linearity is a useful abstraction that appears in many places in higher mathematics.

Our decision to include a separate chapter on linearity grew out of our experiences in giving research talks on nonlinear dynamics (our research specialty) to varied audiences of physicists and engineers. Although the word *nonlinear* is much in vogue, we have found that many workers in the engineering and physical sciences do not have a good understanding of how linearity (or the lack of it) affects the nature of solutions of differential equations. This chapter illustrates the notion of linearity in a variety of contexts and explains how linearity and the superposition principle are intertwined.

## Chapter 5: Oscillatory Motion

This chapter presents the essential material on linear second-order equations with constant coefficients. In contrast to many textbooks, which present a variety of solution methods for nonhomogeneous linear equations with constant coefficients in comparable chapters, we emphasize the method of undetermined coefficients because it is a nice illustration of linearity as well as being computationally straightforward.

The latter half of the chapter focuses primarily on linear second-order systems with periodic forcing functions. We emphasize periodic forcing, because the basic ideas have wide technological application. Section 5.6 discusses beating solutions in detail, paying particular attention to the nature of the solution as the frequency of the forcing function approaches the frequency of the homogeneous solution. Resonant solutions are shown to arise in the limit as the two frequencies become equal.

One important theme in the chapter, which as far as we are aware is unique to this textbook, is the effect of errors in parameters in LRC circuit equations. Parameter errors are important in electrical engineering applications, as circuit components often vary appreciably from their rated values. The chapter also includes an in-depth discussion of the parameter estimation problem for linear second-order equations with constant coefficients. Computer labs or graphing calculators are helpful supplements for these sections.

## Chapter 6: Systems of First-Order Equations

The presentation in this chapter parallels that in Chapter 1. We introduce the main ideas graphically in the form of direction fields and phase portraits. The chapter concludes with a discussion of Euler's method for first-order systems.

The goal of this chapter is to highlight the main ideas geometrically before introducing analytical methods for linear equations. In our experience, students find the pictures intuitive and memorable. In particular, it is easier to motivate the role of eigenvalues and eigenvectors in the solution of linear equations (discussed in Chapter 7) when students can relate the analysis to a graphical representation.

## Chapter 7: Linear Systems

This chapter presents the basic analytical methods for linear first-order systems with constant coefficients. The first four sections contain the basic material. One underlying theme in this chapter is the question of sensitivity to changes in initial conditions and how it is related to the eigenvalues of the matrix that defines the system. The remaining sections discuss the fundamental matrix, $3 \times 3$ and $4 \times 4$ systems, and the variation of parameters for linear systems; they can be omitted if the instructor desires.

## Chapter 8: Nonlinear Differential Equations

Our goal in this chapter is to present a self-contained introduction to some nonlinear differential equations of contemporary scientific interest, as well as the notions of bifurcations and chaotic dynamics. Although many introductory textbooks now mention the idea of chaos, few of them contain much in-depth discussion.

    The material has been arranged so that the instructor can select one or more sections and skip the others. However, the sections can be covered sequentially so that instructors who have more time can give their students a good introduction to these subjects at an appropriate level.

    Each of the first three sections discusses a "case study" of a nonlinear differential equation. The equations in succeeding sections exhibit increasingly complex behavior. Section 8.5, on linearization, is a nice follow-up to Chapters 4 and 7. Sections 8.6 and 8.7 give a short overview of bifurcation theory, and Section 8.8 discusses chaotic dynamics. Section 0.3 can be used with Section 8.8 to introduce students to the fact that deterministic systems are not necessarily predictable.

## Chapter 9: Numerical Methods

Our focus is on three classical numerical methods: Heun's method, Taylor series methods, and the fourth-order Runge-Kutta method. One important feature of the chapter, which is not often found in conventional textbooks at this level, is a self-contained introduction to linear difference equations. One of our goals is to explain that numerical methods lead to dynamical systems whose solutions may not necessarily mimic those of the related differential equation. The inclusion of linear difference equations lets us treat the notion of numerical stability in some depth.

## Chapter 10: Laplace Transforms

The material in Chapter 10 on the Laplace transform is classical and self-contained. This chapter can be covered directly after Chapter 5. In contrast to many textbooks, whose primary focus is the "plug and chug" aspect, the exercises in this chapter emphasize the qualitative behavior of solutions as well as their derivation, particularly in response to discontinuous forcing functions.

## A Suggested Syllabus

The heart of the subject matter is covered in Chapters 1–7. We recommend that these chapters be covered sequentially, but some variations are possible depending on the interests of the instructor. The following chart illustrates the logical interdependence of the chapters.

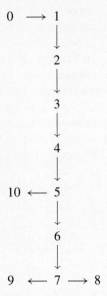

A typical one-semester course might consist of a brief introduction, using portions of Sections 0.1 and 0.2, followed by Chapters 1 and 2 in their entirety, though only one or two of the applications in Section 2.3 need to be covered. Sections 3.1 and 3.2 are important; if time is running short, Section 3.3 can be omitted without loss of continuity. Chapter 4 should be covered in its entirety, as well as the first six sections of Chapter 5. We recommend that all of Chapter 6 be covered as well as the first four sections of Chapter 7.

Aside from this core material, there are many options, depending on the interests of the instructor and the class. The last three sections of Chapter 5, the remainder of Chapter 7, most of Chapter 10, most of Chapter 9, or several sections in Chapter 8 can be used as self-contained topics; typically, each of these options can be covered in three to five lectures. Additional coverage of the method of undetermined coefficients and the method of variation of parameters for second-order equations is given in the appendices.

## ACKNOWLEDGMENTS

This book would not have been possible without the assistance of the Addison Wesley Longman Publishing Company. We particularly want to thank Laurie Rosatone, whose enthusiasm got the project underway and sustained it to its completion. Marianne Lepp provided invaluable editorial assistance. Karen Wernholm and her staff ran the entire production process smoothly and on schedule. Michelle Babinec coordinated the initial marketing and class testing efforts.

Wanzie McAuley from Northeast Tennessee Community College taught from an early version of the manuscript and suggested many improvements to the presentation and the figures. Glenn Ledder (University of Nebraska), Kathy Shay (Middlesex County College), and Carolyn Tucker (Westminster College) class tested the preliminary edition. Their suggestions and criticisms were very helpful. Paul Lorczak carefully checked the manuscript and the exercises for accuracy; any remaining errors are the sole responsibility of the authors.

In addition, the following reviewers provided valuable comments:

| | |
|---|---|
| Edgar Chandler | Scottsdale Community College |
| Rebecca Hill | Rochester Institute of Technology |
| Dar-Veig Ho | Georgia Institute of Technology |
| Michael Kirby | Colorado State University |
| Renate McLaughlin | University of Michigan, Flint |
| Vince Panico | University of the Pacific |
| Thomas T. Read | Western Washington University |
| Zwi Reznik | Fresno City College |
| Tim Sauer | George Mason University |
| Asok K. Sen | Indiana University/Purdue University at Indianapolis |
| Steven Wilson | Johnson County Community College |
| Ali Zakeri | California State University |

The National Science Foundation provided matching funds in 1992 for the purchase of a computer laboratory, which was the first such facility at Arizona State University devoted full time to instruction in undergraduate mathematics courses. Additional financial support for the computer laboratory was provided by the following entities at Arizona State University: the Vice President's office, the Consortium for Instructional Innovation, and the Department of Mathematics. Special thanks go to Christian Ringhofer, Tom Trotter, and Marlene Salvato of the ASU Department of Mathematics for their assistance in class scheduling and other matters related to our curriculum development efforts.

*Eric J. Kostelich and Dieter Armbruster*
*Tempe, Arizona*

# CONTENTS

# *Chapter 0*

## INTRODUCTION

This introductory chapter serves as a preview and appetizer for the whole course. Differential equations are a surprisingly powerful tool for solving many kinds of problems. There is a vast literature on the subject, which is still an active area of research more than three centuries after the initial investigations of Newton and Leibniz. The sections in this chapter consider three topics that are both central to the mathematical content of the subject and of fundamental scientific importance:

1. exponential growth and its implications,
2. simple oscillatory motion, and
3. the predictability of natural systems and the effects of errors in measurement.

The goal of this chapter is to give you an overview of these questions and to pique your curiosity about some of the details. Therefore, you should not worry about understanding every step in the mathematical exposition. Instead, you should read it casually to get a feel for the subject as a whole. Sections 0.1 and 0.2 are overviews of two classical applications of differential equations, namely, exponential growth and oscillatory motion, that are covered in detail later in the book. Section 0.3 introduces a more challenging subject, chaotic dynamics and unpredictability in difference equations, that is an active area of research in contemporary mathematics.

## 0.1 A LOTTERY JACKPOT

Let us start by finding out how to become independently wealthy and retire to do mathematics just for fun. Many states run a lottery, which is marketed as a chance to strike

1

it rich. The Arizona Lottery offers you two options if you hit the jackpot. If you win $1 million, then you can receive the money in 20 annual installments of $50,000 each, or you can receive a single cash payment of slightly less than $500,000. Assuming that the Arizona Lottery is still in business for the next 20 years and ignoring unforeseen economic influences (such as taxes) that may change things dramatically in that time, we can ask which payment option is the bettter choice from a mathematical point of view.

Assume that the jackpot is $1 million and that your choice is between 20 annual payments of $50,000 and a single cash payment of $490,000. (For simplicity we will neglect taxes, which in practice is unwise.) Here are some of the questions that you might ask as a lucky winner:

- Are you likely to be alive and healthy enough in 20 years to enjoy the money?

- What kind of car will you be able to buy in 20 years with $50,000? (In other words, how will inflation erode the value of your annual cash payments over time?)

- Assuming that you invest a single cash payment of $490,000, how much money can you withdraw annually and still have your principal untouched? How much money can you withdraw each year and have the value of your investment remain constant after taking inflation into account?

- How much money can you withdraw annually and have spent everything, interest and principal, after 20 years?

- Suppose you take the 20 annual installments, of which you spend half and invest half each year. How much money do you have after 20 years?

- Assuming that you take the 20 installments and invest all the money in a mutual fund, what rate of return must you receive to have as much money in 20 years as you would receive by investing a single $490,000 cash payment?

Of course, you can ask many similar kinds of questions. Although we have not done any mathematics so far, we have already identified one important point. There is not a clear-cut answer to the question of whether it is better to take the jackpot as a payback over 20 years or as a single (but smaller) cash payment. The answer depends heavily on your personal circumstances, your willingness to make long-term investments, global economic trends, and so on.

These kinds of questions are typical of those that are addressed in applied mathematics. We are not interested in "the answer" in the sense of a single formula or number; rather, we are interested in the behavior of a process as a function of other variables. For example, at a 10 percent annual rate of return, an initial investment of $490,000 grows into $3.62 million after 20 years. This is fine as it far as it goes, but how can we be assured of a 10 percent annual return? Instead, we are interested in knowing how an investment will grow over 20 years under different scenarios. For example, a conservative investment might yield only a 6 percent annual rate of return, while more risky investments, such as junk bonds, might yield a 15 percent annual rate of return. Typically, such results are best represented in a graph like the one in Fig. 0.1. Depending on the annual rate of return, an initial investment of $490,000 can grow to between $1 million and $10 million after 20 years.

With all these reservations about the applicability of mathematical modeling in mind, let us try to answer the question concerning which lottery jackpot option is better. To do this, we first have to understand how the graph in Fig. 0.1 is generated.

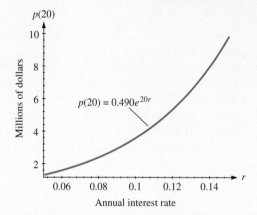

**FIGURE 0.1**  The value of an initial investment of $490,000 after 20 years as a function of the annual rate of return, $r$.

We assume that the rate at which the investment grows is proportional to the amount of money (the *principal*) that is invested at any given instant. This assumption can be expressed in words as

rate of change of principal $\propto$ current principal,

and it leads to the following differential equation for $p(t)$, the principal at time $t$:

$$\frac{dp}{dt} = rp(t). \tag{0.1}$$

The rate of change of the principal is the derivative $dp/dt$. The constant of proportionality is $r$, otherwise known as the *interest rate* or the *rate of return*. Equations like Eq. (0.1) are called *differential equations*, because they express a relation between the derivative of a function (here $p$) and the current value of the function and other variables.

Let us examine some of the consequences of Eq. (0.1). If you put the money under your mattress, then you receive no return on the funds, that is, $r = 0$. Hence

$dp/dt = 0$—the principal does not change. In this case, $p(t) = p_0$, where $p_0$ is a constant (the initial size of the principal). The constant function $p(t) = p_0$ solves Eq. (0.1), because the left-hand side is 0 when you substitute in the constant function, and the right-hand side is 0 because $r = 0$.

Suppose instead that you invest in a spectacularly successful mutual fund that gives you an annual rate of return of 100 percent (i.e., $r = 1$). In this case, Eq. (0.1) says that $p(t)$ is a function whose derivative is itself. The exponential function has this property:

$$\frac{d}{dt}(ce^t) = ce^t, \tag{0.2}$$

where $c$ is any constant. Thus, the function $p(t) = p_0 e^t$ solves Eq. (0.1) in the case $r = 1$. Here, $p_0$ is a constant equal to the initial value of the principal, $p(0)$.

Closer investigation reveals a way to solve Eq. (0.1) for other values of $r$. Equation (0.1) says that $p$ is a function whose derivative is a constant multiple of itself. In fact, the general solution of Eq. (0.1) is

$$p(t) = p_0 e^{rt}.$$

For convenience, we express $p$ in units of millions of dollars. If we set $p_0 = 0.490$ (why?) and $t = 20$, we get an expression for $p(20)$ as a function of the rate of return, $r$:

$$p(20) = 0.490 e^{20r}. \tag{0.3}$$

Figure 0.1 is a plot of the resulting function for values of $r$ between 5 percent and 15 percent.

Next let us see what happens if we choose to take 20 annual payments of $50,000. To make a fair comparison with the previous scenario, we assume that every installment of $50,000 is invested right away and that no withdrawals are made for a 20-year period. Under these assumptions, the principal still grows according to Eq. (0.1), except that we also add $50,000 every year. We can describe this situation in words as

rate of change of principal = rate of return + size of annual deposits,

which translates into the differential equation

$$\frac{dp}{dt} = rp + 0.05, \tag{0.4}$$

where $p$ is expressed in millions of dollars and $p(0) = 0$.[1] The condition $p(0) = 0$ reflects the fact that we start initially with no money.

---

[1] The model is approximate because it assumes that the money is paid out at a constant rate throughout the year for a total of $50,000. In practice, lottery winners receive a single payment of $50,000 once a year. However, the difference is unimportant for this illustration. A detailed discussion of discrete versus continuous compounding of funds is given in the next chapter.

In Chapter 2, we show that the solution of Eq. (0.4) is

$$p(t) = ce^{rt} - \frac{0.05}{r},\tag{0.5}$$

where $c$ is a constant of integration. If $p(0) = 0$, then

$$p(t) = \left(\frac{0.05}{r}\right)(e^{rt} - 1).\tag{0.6}$$

Figure 0.2 shows the graphs of Eqs. (0.6) and (0.3) as a function of $r$ when $t = 20$. If the rate of return is less than about 8.3 percent annually, then the 20-payment option produces a greater principal at the end of 20 years. However, if the rate of return is greater than 8.3 percent, then the single cash payment grows to a larger amount. Under the assumptions above, you are better off taking the jackpot as one lump sum if you are able to invest the money at an annual rate of return that is at least 8.3 percent.

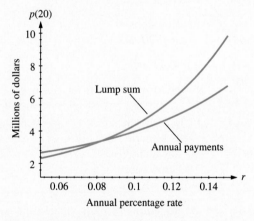

**FIGURE 0.2**   Comparison plot of the amount of money after 20 years as a function of the annual rate of return for a lump-sum investment (Eq. 0.3) and annual payments (Eq. 0.6).

The point of our discussion is this: To analyze these financial scenarios, we are less interested in finding a formula for a solution of the differential equation as in understanding the consequences of the outcomes based on various assumptions. To keep the problem simple, we have ignored taxes, assumed a constant rate of return, and made no withdrawals for 20 years. The differential equations above can be modified to reflect assumptions about all of these other factors. The algebra will be more complicated, but the strategy of the investigation is the same.

The previous calculations also show that some constants in the problem are more important than others. Let us consider this question in more detail. So far, we have assumed that the one-time cash payment is $490,000. Suppose instead that it is less, say 10 percent less, that is, $441,000. How does this affect the size of the principal after 20 years, assuming that the money earns a constant rate of return and we make no withdrawals?

Figure 0.3 shows the principal after 20 years as a function of time for the case $p_0 = \$441,000$ (bottom curve) and $p_0 = \$490,000$ (top curve). Here we have assumed an annual rate of return of 10 percent. The initial difference between the two curves corresponds to $49,000. The difference grows to $133,000 after 10 years and to $362,000 after 20 years.

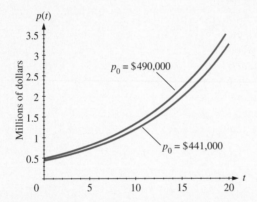

**FIGURE 0.3**    The difference in principal as a function of time assuming two different initial values.

One way to think about the difference between two quantities is to compare it to the size of one of them. Suppose, for instance, that you are shortchanged by $1 in the course of a financial transaction. If the transaction involves the purchase of a house for $100,000, you probably will not worry about the error. On the other hand, a $1 error in the purchase of a loaf of bread will send you back to the cashier to get the rest of your change. Of course, $1 is a trivial amount in the context of a $100,000 transaction, but it is not trivial compared to the price of a loaf of bread that costs $1.60.

We define the *relative error* in a nonzero quantity or (the *relative difference* between two nonzero quantities) as

$$\epsilon_r = \left| \frac{\text{true value} - \text{approximate value}}{\text{true value}} \right|. \tag{0.7}$$

Thus, a $1 error corresponds to a relative error of 0.001 percent in a $100,000 financial transaction but a relative error of 62.5 percent in the purchase of a $1.60 loaf of bread.

We can think of the difference in the amount of money received from the lottery in a similar way. If the initial size of the investment is $490,000, then the amount of money after $t$ years is $490{,}000e^{rt}$ when interest is compounded continuously at an annual rate $r$. A similar equation applies if instead we receive only $441,000 from the lottery. The relative difference in the projected amount of cash available after $t$ years is

$$\epsilon_r = \left| \frac{490{,}000e^{rt} - 441{,}000e^{rt}}{490{,}000e^{rt}} \right|, \tag{0.8}$$

which remains constant at 10 percent.

In contrast, a 10 percent difference in the relative size of the annual rate of return leads to much bigger discrepancies in the size of the principal over time. Suppose that instead of a 10 percent annual rate of return, we earn 11 percent. Figure 0.4(a) shows the growth of an initial principal of $490,000 under both assumptions. After 20 years, the 10 percent annual return yields $3.62 million, but an 11 percent annual return yields $4.42 million—about 22 percent more money. Figure 0.4(b) is a plot of the relative difference between the two curves shown in Fig. 0.4(a). In contrast to the previous example, the relative difference does not remain constant if we assume a difference in the rate of return.

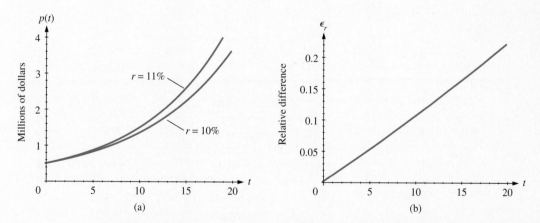

FIGURE 0.4   (a) The value of an initial $490,000 investment as a function of time with an annual rate of return of 10 percent (bottom curve) and 11 percent (top curve). (b) The relative difference between the two curves.

## *Exercises*

**1.** In 1995, the average cost of a new car was about $16,000. What will the average cost of a new car be in 20 years, assuming that the cost increases by 4 percent per year?

**2.** Assume that you win the $1 million lottery jackpot and opt for payment in 20 annual installments. You decide to spend half of your payment every year and invest the other half at an 8 percent annual rate of return. How much money will you have after 20 years?

**3.** Discuss how taxes might affect your decision to take a 20-year annuity versus a one-time payment if you win a $1 million lottery.

**4.** This exercise discusses the concept of a relative error and its interpretation.

(a) Consider Eq. (0.8). What happens as $t$ increases? Extend the graphs in Fig. 0.4 for a longer time.

(b) What does your result in (a) tell you about the relationship between the two investment scenarios for large $t$?

(c) Often we do not know which of two expressions represents the true value and which is the approximation. Assume that a projection for the investment is made with an annual rate of return of 11 percent but that the actual rate is 10 percent. Determine the relative error $\epsilon_r$ in this case.

(d) What happens to $\epsilon_r$ in (c) as $t$ increases?

(e) What does your result in (d) tell you about the relationship between the approximate and the true value for large $t$? Compare this to the discussion in (a) and (b).

## 0.2   OSCILLATORY MOTION

Many processes in nature are oscillatory, including the rotation of the earth about its axis, the motion of the earth around the sun, biological rhythms such as heartbeats and hormonal secretions, the swinging of the pendulum of a clock, and so on. One of the simplest prototypes for an oscillatory model is a mass hanging from a spring that is attached to some immovable object. For the purpose of illustration, let us imagine a person (the mass) attached to a bungee cord (the spring) that is anchored to a bridge. The bungee jumper leaps from the bridge, bounces up and down many times with a steadily decreasing amplitude, and eventually comes to rest if the process continues long enough (Fig. 0.5). The person responsible for the safety of the bungee jumper must keep the following questions in mind:

1. How strong must the bungee cord be?
2. What are the extreme positions that the jumper is likely to attain? (The jumper should not splash into the water or crash into the bridge on the way up.)

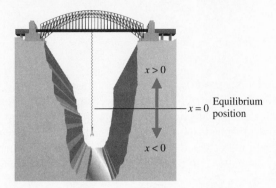

FIGURE 0.5   Schematic drawing of a bungee jumper from a bridge.

A mathematically minded bungee jumper might also like to know the following:

3. What is the maximal speed that can be attained?
4. How long will it take to go up and down once?

The answers to some of these questions can be estimated by observing some jumps. It is also possible to formulate a mathematical model for the bungee cord and the jumper that is based on simple physical arguments. For the purpose of this discussion, we will assume that the bungee cord acts like a simple spring. (This assumption is not completely realistic, but it is adequate for this illustration.) In Chapter 5, we show that one model for a spring acting on a mass $m$ is

$$ma = -kx. \tag{0.9}$$

Equation (0.9) says that the net force of the spring on the mass is proportional to the amount by which the spring is stretched or compressed from its natural length. (The proportionality constant $k$ is called the spring constant.) Because the acceleration is the second derivative of the position $x$, Eq. (0.9) is equivalent to the differential equation

$$\frac{d^2x}{dt^2} = -\left(\frac{k}{m}\right)x. \tag{0.10}$$

Let us step back and contemplate what Eq. (0.10) tells us. The function $x(t)$ gives the position of the bungee jumper at any time $t$; the second derivative of the function is a constant multiple of the function itself. A moment's thought suggests that $x$ is a sine or cosine function, because

$$(\sin \omega t)'' = -\omega^2 \sin \omega t \quad \text{and} \quad (\cos \omega t)'' = -\omega^2 \cos \omega t$$

for any constant $\omega$.

A complete description of the position as a function of time requires that we know both the initial position and the initial velocity of the bungee jumper. (For instance, there are slightly different outcomes, depending on whether the bungee jumper simply topples off the bridge or jumps upward while leaping away from the bridge.) The length of the bungee cord also is important; we let $x = 0$ correspond to the imaginary point at which the bungee jumper would eventually come to rest if we waited for friction to completely damp the motion. It can be shown (see Chapter 5) that if the jumper simply topples off the bridge (corresponding to an initial velocity of 0), then

$$x(t) = x_0 \cos \left( \sqrt{\frac{k}{m}} t \right), \tag{0.11}$$

where $x_0$ is a constant that depends on the length of the cord and the weight of the jumper.

Equation (0.11) can be used to predict the motion of the bungee jumper if we know the mass of the jumper and the spring constant and length of the cord. For example, Fig. 0.6 shows the position of an 80-kg bungee jumper, relative to the rest point, on a bungee cord with a spring constant of $k = 25$ N/m. We can now try to answer some of our initial questions.

1'. Our considerations above do not address the issue of the strength of the cord. This is a different kind of problem that is not directly related to the question of how the bungee jumper moves.

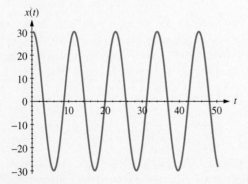

**FIGURE 0.6**    A plot of Eq. (0.11) for $m = 80$, $k = 25$, and $x_0 = 30$.

2′. Equation (0.11) can be used to determine the extreme displacements of the bungee jumper, which are $x_0$ units on either side of the imaginary rest point. However, we have not discussed the question of how to determine the rest point; more details are given in Chapter 5. (Note also that Eq. (0.11) relies on our assumption that the bungee cord acts like a simple spring, which is only approximately correct. A different model would be required for a safety analysis for a real bungee cord.)

3′. The velocity of the motion is given by $x'(t)$, so we can determine the maximum speed of the bungee jumper.

4′. The time required to go up and down once can easily be read from the graph in Fig. 0.6. The cycle time clearly depends on the period of the cosine function, which in this case is $T = 2\pi \sqrt{m/k}$. Given the constants above, $T = 11.2$ s.

The model described above ignores physical factors such as air resistance and internal friction in the cord. To make our model more realistic, we can include a term that represents friction, a force that acts to damp the oscillations. In practice, as time goes on, the amplitude of the bungee jumper decreases; eventually, the bungee jumper ceases to bounce up and down. You may have noticed that a spring or a rubber band that is repeatedly extended and compressed becomes warm. This effect results from internal friction, which converts some of the kinetic energy of the spring into heat.

Theories of frictional motion tend to be quite complicated. For our purposes, we can make the crude approximation that the forces of friction (including air resistance) are proportional to the velocity and oppose the motion. This assumption can be written mathematically as

$$F_{\text{friction}} = -cv, \tag{0.12}$$

where $c$ is the *coefficient of friction*. The minus sign indicates that the frictional force acts in a direction opposite the direction of the motion. It can be shown (see Chapter 5) that the force of friction leads to the model

$$\frac{d^2x}{dt^2} = -\left(\frac{k}{m}\right)x - \left(\frac{c}{m}\right)\frac{dx}{dt}. \tag{0.13}$$

The solutions of Eq. (0.13) are discussed in detail in Chapter 5. Figure 0.7 shows three solutions for different values of the coefficient of friction $c$. If $c$ is not too large, then Eq. (0.13) yields a plausible model for the motion of the bungee jumper. The solution still oscillates, but its amplitude gets smaller with time.

If $c$ is sufficiently large, then no oscillations occur at all. This situation arises in an automobile shock absorber, which consists of a spring in a viscous liquid. A good shock absorber does not oscillate (or at least quickly damps any oscillations), lest continued bouncing cause the driver to lose control of the car.

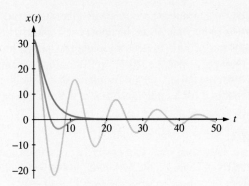

**FIGURE 0.7**   Three solutions for Eq. (0.13) corresponding to different choices of the coefficient of friction, $c$.

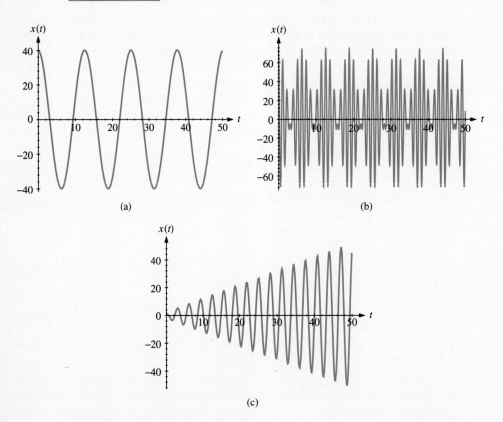

**FIGURE 0.8**   Three representative solutions of Eq. (0.14):  (a) a simple periodic solution; (b) a beating solution; (c) a resonant solution.

One remarkable fact is that disparate physical systems sometimes can be modeled accurately with similar (or even identical) differential equations. For instance, Eq. (0.13) is a good model of certain simple electrical circuits, even though an electrical circuit has no physical resemblance to a mass-spring system.

Oscillating electrical circuits of various kinds have important technological uses. One class of models of considerable scientific interest has the form

$$\frac{d^2x}{dt^2} = -\left(\frac{k}{m}\right)x + A\sin\omega t. \tag{0.14}$$

The term $A\sin\omega t$ might represent a time varying potential that is applied to a circuit. (Household alternating current is a good example of such a potential.)

Figures 0.8(a)–(c) are graphs of some representative solutions of Eq. (0.14). The qualitative nature of the solution depends on the values of $k$, $m$, and $\omega$. Figures 0.8(a)–(b) show two periodic solutions; one of them is a simple periodic function, and the other is a more complicated periodic function called a *beating* solution. Figure 0.8(c) shows an oscillating solution that is not periodic; its amplitude grows with time. This kind of function is called a *resonant* solution. Beating and resonant solutions are discussed in detail in Chapter 5.

## Exercises

**1.** Suppose you throw a baseball straight up into the air. In the simplest case, we ignore air resistance and suppose that gravity is the only force acting on the ball. A better model takes into account the effect of air resistance.

(a) Suppose you want to predict the maximum height of the ball above the ground as a function of its initial velocity. Is a model that ignores air resistance more likely to overestimate or underestimate the maximum height? Explain.

(b) Suppose that the force of air resistance is proportional to the current velocity of the ball. Write a differential equation that relates the acceleration on the ball as a function of its current velocity.

(c) Suppose instead that the force of air resistance is proportional to the square of the velocity of the ball. Write the corresponding differential equation that relates the acceleration on the ball to its present velocity.

**2.** Let

$$x(t) = \sin\tfrac{1}{2}t + 2\cos\tfrac{1}{2}t - 4\sin 3t.$$

(a) What is the period of the term $-4\sin 3t$?

(b) What is the period of the remaining two terms?

(c) What is the period of $x$?

(d) Plot $x(t)$ together with each of the three terms that comprise it. Check that the plot is consistent with your answer in (c).

**3.** Equation (0.9) implies that the force exerted by a spring is proportional to the amount by which it is stretched or compressed from its natural length. In particular, it implies that a spring "pushes back" when it is compressed. How good a model is Eq. (0.9) of a real bungee cord? Discuss ways in which the equation is and is not realistic.

## 0.3  Determinism and Unpredictability

One of the principal goals of mathematical modeling is prediction. For instance, it is possible to formulate more sophisticated models of bungee jumping that can be used to predict the maximum amplitude of the motion as a function of the weight of the bungee jumper and other variables. Predictions can also be of a general nature; for instance, a financial planner might want to know the future value of an investment under various assumptions.

### Random and Deterministic Models

It has long been realized that certain physical systems cannot be predicted in the usual sense. One example in physics involves the dispersion of a drop of red dye in a clear liquid. On one hand, you can predict that in the long term, the dye will be distributed uniformly in the liquid. Conversely, it is impossible to say where any particular molecule of dye will be after two minutes. A drop of dye consists of a large number of interacting molecules, and it is impossible to enumerate all the forces that act on each one of them. Nevertheless, it is possible to give a statistical description of the forces and say something about their average size and direction.

These considerations lead to a distinction between a probabilistic model and a deterministic model. Probabilistic models contain terms whose values are random variables; that is, the values of such terms vary randomly within a certain interval. The solutions of probabilistic models can be used to answer questions about the likelihood of particular events or the average value of some quantity of interest. In contrast, a deterministic model can be used to make specific predictions, such as the position and velocity of a particle at a given instant in the future.

Mathematical models consisting of differential equations are deterministic insofar as they usually do not contain variables or parameters whose values are random.[2] The existence and uniqueness theorem, discussed in Chapter 3, guarantees that most differential equations of practical interest have unique solutions. If a process is governed by a particular differential equation, and if the initial state is known, then it is possible in principle to make specific predictions about the state of the process at any time in the future. For instance, if you deposit a certain amount of money in a bank account for which interest is compounded continuously at a known rate, then the amount of money in the account at any time in the future is given precisely by the solution of Eq. (0.1).

In most cases, however, it is not possible to measure the state of the process with complete certainty, and no differential equation is likely to be a completely accurate

---

[2]There is such a thing as a stochastic differential equation, which contains random terms. A discussion of this more specialized topic is beyond the scope of this text.

model. For instance, the radioactive decay of an unstable isotope can be modeled to high accuracy by Eq. (0.1). However, it is impossible to know the size of a radioactive sample down to the last atom, and there is also some uncertainty in measuring the half-life (which determines the rate constant $r$). These uncertainties lead to discrepancies between the predicted behavior (in this case, the precise mass of the remaining isotope at any time in the future) and the actual observation.

An important question in any given mathematical model is whether small uncertainties in the initial state grow into large uncertainties when one needs to make predictions about the future state of a system. If a differential equation is *linear*, then it is possible to give a general theory concerning the structure and behavior of solutions. (We discuss what is meant by a linear differential equation in Chapter 4.) In particular, it is often possible to quantify the rate at which initial uncertainties in the initial state grow with time if the mathematical model is a linear differential equation.

However, most of the differential equations that arise in practice are not linear. Examples include models for weather prediction, the spread of infectious diseases, the flow of liquid through a pipe, and many others. (Indeed, some mathematicians have quipped that nonlinear differential equations are as ubiquitous as non-elephant animals.) There is no general theory that characterizes the solutions of nonlinear equations. In fact, the solutions of very simple nonlinear equations can be extraordinarily complicated, as we now illustrate in the context of a simple biological model.

## The Logistic Map

One of the simplest ecological systems is a seasonally breeding population, such as a mosquito population in a cold climate. In such a case, the insects live and breed during the brief warmth of summer but die at the first frost; successive generations do not overlap. A biologist might be interested in the population, $p_n$, of the insects in year $n$. An ecologist might be interested to know how the population fluctuates from year to year.

In this example, the data consist of the sequence of numbers $\{p_1, p_2, \ldots\}$, where $p_n$ is the total population in the $n$th generation. If we suppose that this year's insect population is the sole source of eggs for the next generation, then we can consider a model of the form

$$p_{n+1} = f(p_n). \tag{0.15}$$

Equation (0.15) is an example of a *difference equation*. It expresses the population in the next generation (generation $n + 1$) as a function of the population in the current generation. There is a close mathematical relationship between difference equations and differential equations. We consider Eq. (0.15) instead of a differential equation because it is easier to conduct numerical experiments.

A specific formula for $p_{n+1}$ depends on the assumptions one makes about the behavior of the population. Let us suppose that if the population of the $n$th generation is large, then competition for food and other resources leads to a low reproductive rate, and so the population in the $(n + 1)$st generation is small. However, if the current population is small, then it grows at nearly an exponential rate. Obviously, if the population ever becomes extinct, that is, $p_n = 0$, then it must remain extinct ($p_k = 0$ for $k > n$).

One model that reflects these assumptions is the *logistic map*, given by the difference equation

$$p_{n+1} = p_n(a - bp_n), \tag{0.16}$$

where $a$ and $b$ are constants (also called *parameters*) that depend on the population. Although many factors affect population growth, Eq. (0.16) incorporates some plausible assumptions that might hold at least approximately for a real population.

For convenience, we take the largest possible population as 1 and rescale Eq. (0.16) by the change of variable $x = bp/a$. (In other words, $x_n$ expresses the population of the $n$th generation as a fraction of the maximum possible population.) This rescaling leads to the difference equation

$$x_{n+1} = ax_n(1 - x_n). \tag{0.17}$$

Notice that if $x_n \geq 1$, then $x_{n+1} \leq 0$; thus the population dies out if it ever reaches a certain threshold.

We can describe the behavior of the population graphically by plotting $x_{n+1}$ as a function of $x_n$. The parabola in Fig. 0.9 is the graph of Eq. (0.17) when $a = 2.5$. Let the initial population be $x_0$. The population of the first generation, $x_1$, can be read from the graph in Fig. 0.9 as shown. For example, if $x_0 = 0.95$ then $x_1$ is approximately 0.119.

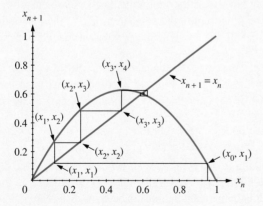

**FIGURE 0.9**    A staircase diagram for the orbit of $x_0 = 0.95$ when $a = 2.5$.

To find $x_2$, we repeat (*iterate*) the procedure using $x_1$ as the input, and so on. The sequence $x_0, x_1, x_2, \ldots$ is called the *orbit* of $x_0$. For example, if $x_0 = 1$, then the orbit is $1, 0, 0, \ldots$; the population crashes and becomes extinct. If $x_0 = 0.95$ and $a = 2.5$, then the orbit is given approximately by $0.95, 0.119, 0.262, 0.483, 0.624, \ldots$.

Figure 0.9 shows a *staircase diagram* of the orbit of $x_0 = 0.95$ when $a = 2.5$. Given $x_0$, we can read $x_1$ from the vertical axis. The value $x_2$ can be found by starting at the point $(x_0, x_1)$ and moving horizontally to the point $(x_1, x_1)$ on the diagonal, then up to the point $(x_1, x_2)$ on the parabola. A similar procedure can be followed to obtain every subsequent point on the orbit.

It is straightforward to write a computer program to generate staircase diagrams. In this way, we can investigate the behavior of different initial conditions and different values of $a$ in Eq. (0.17). We make the following observations:

1. The maximum value that can be attained by the population occurs at $x_{n+1}$ whenever $x_n = 1/2$. In this case, $x_{n+1} = a/4$.

2. The logistic map makes sense as a population model only if $0 \le x_n \le 1$. Hence we make the restriction $0 \le a \le 4$.

3. There are two distinguished points that represent constant populations. One of them is $x_0 = 0$, corresponding to an extinct population. (Clearly, if $x_0 = 0$, then $x_n = 0$ for every positive $n$.) The other is the point on the parabola where $x_{n+1} = x_n$, given by the nonzero root of the equation

$$x = ax(1 - x),$$

namely, $x = x_{\text{eq}} = 1 - 1/a$. Points that correspond to constant populations are called *equilibrium points*; they are often important in the analysis of difference equations.

4. Figure 0.9 suggests that most orbits starting from points other than $x_{\text{eq}}$ eventually approach $x_{\text{eq}}$. All orbits that start close to $x_0 = 0$ move away from 0. We say that 0 is an *unstable equilibrium* because points that start close to 0 do not stay close to 0 (unless, of course, they start exactly on 0). The point $x_{\text{eq}}$ is an example of a *stable equilibrium:* Orbits that start close to $x_{\text{eq}}$ continue to get closer. This observation implies that a population governed by Eq. (0.17) with $a = 2.5$ oscillates with gradually decreasing amplitudes and eventually approaches a value given by $x_{\text{eq}}$.

5. If $a$ is increased to 3.3, an interesting change occurs, as illustrated in Fig. 0.10. The equilibrium $x_{\text{eq}}$ remains, but it is no longer stable; orbits that start nearby do not remain nearby for long. Instead, the orbits appear to alternate between two points, as illustrated by the rectangle in the staircase diagram. These two points, $p_1$ and $p_2$, lie on the diagonal and map to each other; that is, the orbit of $p_1$ is $p_1, p_2, p_1, p_2, \ldots$. We call this orbit a *period-2* orbit: If the initial population is $p_1$, then the population in subsequent generations oscillates between the two values. Moreover, the period-2

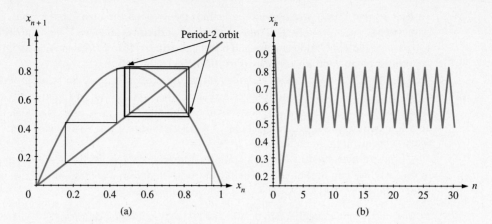

**FIGURE 0.10**    (a) A staircase diagram for Eq. (0.17) with $a = 3.3$. The orbit approaches a stable period-2 orbit. (b) A plot showing $x_n$ as a function of $n$.

orbit is *stable* because points near $p_1$ and $p_2$ approach the orbit. Figure 0.10(b) shows a plot of $p_n$ as a function of $n$; it illustrates the period-2 behavior quite clearly.

6. Typical orbits become more complicated when $a = 3.45$. They now approach a period 4-orbit, as illustrated by the staircase diagram in Fig. 0.11. The arrows show a stable period-4 orbit. (The period-2 orbit is no longer stable.)

7. As $a$ is increased further, stable periodic orbits of period 8, 16, 32, . . . appear. In fact, it can be shown that there is an infinite sequence of such stable periodic orbits, each

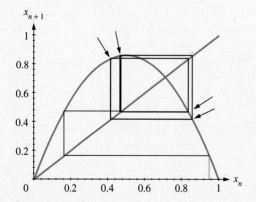

**FIGURE 0.11**    Staircase diagram for $a = 3.45$. The arrows indicate a period-4 orbit.

with a period twice as large as its predecessor. It can also be shown that when $a$ is between approximately 3.57 and 4, there are periodic orbits of all possible periods (almost all of which are unstable). The logistic map is an example of a simple equation with very complicated behavior.

8. When $a = 4$, the orbit of $x_0$ does not follow any apparent pattern. Figure 0.12 shows some corresponding graphs of a representative orbit. We say that the orbit is *chaotic*. The staircase diagram almost completely covers the interval from 0 to 1 as we add more iterations, as shown in Fig. 0.12(c).

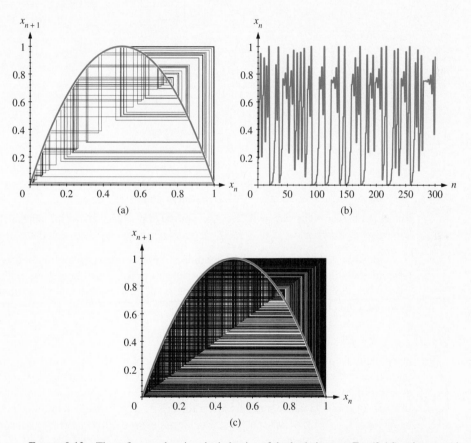

**FIGURE 0.12**   Three figures showing the behavior of the logistic map, Eq. (0.16), when $a = 4$: (a) Staircase diagram showing the first 100 points in the orbit of $x_0 = 0.95$; (b) a plot of $x_n$ versus $n$; (c) staircase diagram showing the first 1000 points in the orbit.

## Sensitive Dependence on Initial Conditions

One of the most important properties of chaotic orbits is their sensitive dependence on initial conditions. A chaotic system is sensitive to small initial errors. For example, let $a = 4$ and consider the orbits of the logistic map for $x_0 = 0.95$ and $\hat{x}_0 = 0.95001$. Although $|x_0 - \hat{x}_0|$ is small, the difference $|x_n - \hat{x}_n|$ does not remain small. Figure 0.13 illustrates the phenomenon. When $0 < n < 12$, the two curves are indistinguishable because the corresponding points on the orbit are close together. However, when $n > 12$, the orbits appear to diverge.

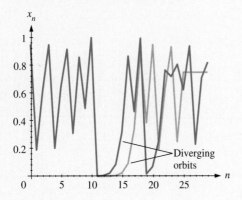

**FIGURE 0.13**    The orbits of $x_0 = 0.95$ and $\hat{x}_0 = 0.95001$ as a function of $n$ for $a = 4$.

Sensitive dependence on initial conditions implies that detailed long-term predictions of the state of the system are impossible to make, even though the underlying system is completely deterministic (i.e., there are no random terms in the system). For example, given the logistic map

$$x_{n+1} = 4x_n(1 - x_n) \tag{0.18}$$

and the initial condition $x_0 = 0.1$, we can accurately predict the values of, say, $x_5$ and $x_{10}$, but not $x_{50}$. (See Exercise 4.) As we mentioned above, no measurement of a physical system is ever completely accurate. The error in $x_0$ given above represents an uncertainty of 1 part in 1000—a rather small error, particularly in a biological situation. The chaotic behavior of Eq. (0.18) means that the size of the population cannot be predicted accurately beyond a few generations.

The prevalence of chaotic behavior in a variety of settings has been generally appreciated only in the past two decades. (One important reason has been the proliferation of personal computers, which have allowed scientists to experiment with mathematical models to a greater extent than ever before.) Chaos has been found in simple physical models (some of which are discussed in Chapter 8), epidemiological models of disease, weather prediction models, and economic models. The presence of chaos leads to an ironic situation in which a completely deterministic process can defy accurate prediction!

## Exercises

**1.** Plot a staircase diagram and calculate the first five points in the orbit of $x_0 = 0.3$ for each map in the following list:

(a) The *tent map*, given by

$$x_{n+1} = \begin{cases} 2x_n, & 0 \le x_n \le 1/2; \\ 2(1 - x_n), & 1/2 \le x_n \le 1. \end{cases}$$

(b) $x_{n+1} = \cos \pi x_n$.

(c) $x_{n+1} = \sin \pi x_n$.

**2.** Determine all of the fixed points for each difference equation in the previous problem. Which ones are stable and which are unstable?

**3.** Equation (0.4) models the situation in which a lottery jackpot is paid out at the rate of $50,000 per year and the entire proceeds are invested in an account in which interest is compounded continuously at an annual rate $r$. However, Eq. (0.4) is not a completely accurate model because it assumes that the $50,000 is paid out continuously (imagine receiving 50,000/365 dollars every day of the year). In practice, such a lottery jackpot is paid out as a single payment of $50,000 at the start of each year.

We can model this situation as follows. If you invest your first payment of $50,000 on January 1 at an annual interest rate of 10 percent, then at midnight on December 31 you will have $50,000e^{0.1}$ dollars. At 12:01 A.M. on January 1, you will have $50,000e^{0.1} + 50,000$ dollars as you receive the next payment. Derive a difference equation that

gives the amount of money you have at 12:01 A.M. on January 1 of year $n + 1$ as a function of the amount of money at the same time in year $n$. (Initially, $x_0 = 50,000$.) Draw a staircase diagram for $x_0, x_1, \ldots, x_5$.

**4.** Let $x_{n+1} = 3.99x_n(1 - x_n)$. This difference equation gives a completely unambiguous rule for generating a list of numbers, yet it is difficult to estimate $x_{50}$ given $x_0$. (A similar statement holds for Eq. (0.18), but a numerical demonstration of this fact is easier for the value 3.99.) Demonstrate this claim as follows.

(a) Let $x_0 = 0.1$. Using a programmable hand calculator, compute and store $x_1, x_2, \ldots, x_{50}$.

(b) Using a personal computer, write a program or use a spreadsheet to recompute the same quantities.

(c) Compare the results that you obtain for $x_1$ through $x_{10}$. Are they about the same?

(d) Next compare the results for $x_{11}$ through $x_{20}$. Do any discrepancies start to appear?

(e) What do you observe for $x_{21}$ through $x_{50}$?

(f) Suggest a reason why the answers in (a) and (b) differ. (Note: It is important to use two different kinds of computers for this problem. You can take this observation as a hint.)

(g) How might you attempt to determine the "correct" value of $x_{50}$?

# *Chapter 1*

---

# FUNDAMENTAL IDEAS

---

This introductory chapter introduces the basic ideas that underlie the study of ordinary differential equations. The purpose of this chapter is to explore the following fundamental questions:

- What is a differential equation?
- What is meant by an explicit solution of a differential equation?
- Can we describe the general characteristics of a solution even without an explicit formula for it?
- How can solutions be approximated numerically?
- How can explicit solutions be derived, at least for one class of differential equations?

We motivate the ideas in three ways. First, we explain a graphical device called a *direction field* that gives a geometrical interpretation of the relationship between a differential equation and its solution. Second, we describe a simple method that can be used to find numerical approximations of the solutions. Third, we explain an analytical procedure for finding solutions for a class of differential equations often encountered in applications.

## 1.1 WHAT IS A DIFFERENTIAL EQUATION?

### A Motivating Example

Consider the problem of forecasting the future size of the population in some country. Census data can be used to determine birth and death rates. For instance, we might find that among every 1000 people alive today, on the average there will be 25 births and 20 deaths in the next year. We might also find that, on the average, 3 people out of every 1000 emigrate every year and that a total of 100,000 people immigrate. If the population today is 50 million, what will be the population 10 years from now if present trends continue?

Notice the following aspects of the problem.

- Most of the relevant information describes the *rate* at which the population changes. We speak of birth, death, and immigration and emigration rates, which are expressed as people per year. If $p(t)$ represents the population at time $t$, then the stated information tells you only about $dp/dt$. The subject of differential equations is the study of equations that involve the derivatives of an unknown function.

- To forecast the population 10 years from now, we must know the size of the population today. (Identical demographic trends lead to different predictions of future population if the initial population is 100 million instead of 50 million.) The present population is an example of an *initial condition*, so called because we must know the initial size of the population to forecast the future size.

- Suppose $t = 0$ corresponds to the present day and $t = 10$ corresponds to 10 years from now. We want to know $p(10)$. If we have a formula for $p(t)$ in terms of $t$, then we can predict the future population at any future time, assuming that present trends continue. However, we have only indirect information about $p(t)$; if we know $dp/dt$ and $p(0)$, can we determine $p(t)$? The answer is a qualified *yes*, at least in most cases of practical interest. In most cases, the value of $p(t)$ can be estimated by numerical means. In some cases, an explicit formula for $p(t)$ can be derived analytically.

In the present example, we can write

$$\frac{dp}{dt} = \text{birth rate} - \text{death rate} + \text{immigration rate} - \text{emigration rate.} \qquad (1.1)$$

Births and immigration increase the population; therefore, they are positive terms in Eq. (1.1). Deaths and emigration decrease the population; therefore, they are negative terms in Eq. (1.1).

If we assume 25 births and 20 deaths per 1000 people per year on the average, then

$$\text{birth rate} = 0.025p$$

and

$$\text{death rate} = 0.020p,$$

where $p$ is the present population. Similarly,

$$\text{emigration rate} = 0.003p$$

if 3 out of every 1000 people emigrate each year. The immigration rate is assumed to be constant; here,

$$\text{immigration rate} = 100,000.$$

In many countries, the immigration rate is fixed by law and is independent of the size of the current population. This assumption is implicit in the present example. Thus, a differential equation that models the population is

$$\frac{dp}{dt} = 0.025p(t) - 0.020p(t) + 100{,}000 - 0.003p(t). \qquad (1.2)$$

Notice that each term in Eq. (1.2) has units of people per year. A *solution* of Eq. (1.2) is a function $p(t)$ whose derivative satisfies the differential equation.

## Notation and Terminology

We say that $t$ is the *independent variable* in Eq. (1.2). The terminology reflects the idea that the passage of time is independent of any trends in population. Conversely, $p$ is the *dependent* variable, because the population clearly depends on time. A differential equation like Eq. (1.2) is *first order* because the highest-order derivative in the equation is $dp/dt$. An equation that involves $d^2p/dt^2$ as well as $dp/dt$ is *second order*. An *ordinary* differential equation expresses the derivatives of one or more functions with respect to a *single* independent variable (typically interpreted as time). A *partial* differential equation involves the derivatives of an unknown function with respect to two or more independent variables.

Various notations are used to denote the dependent and independent variables in differential equations. The derivative $dp/dt$ may be expressed as $p'$, $\dot{p}$, or $p'(t)$. Often, we write $p$ instead of $p(t)$.

✦ *Example 1*   The equations

$$\frac{dp}{dt} = p(t) + 3t \qquad \text{and} \qquad \dot{p} = p^2$$

are first-order ordinary differential equations. In each case, the independent variable is $t$ and the dependent variable is $p$. Notice, however, that the independent variable is not stated explicitly in the equation on the right.

The equation

$$p'' + p' + p = 0$$

is a second-order ordinary differential equation. As above, the independent variable is $t$ and the dependent variable is $p$.

The equation

$$\frac{\partial u}{\partial t} = \frac{\partial^2 u}{\partial x^2}$$

is a second-order partial differential equation. The dependent variable is $u$, and the independent variables are $t$ and $x$. Partial differential equations often are used to express rates of change that depend on position (denoted by $x$) as well as time.     ∎

## Some Applications and Limitations of the Model

The equation

$$\frac{dp}{dt} = 0.025p(t) - 0.020p(t) + 100{,}000 - 0.003p(t) \qquad (1.3)$$

is an example of a mathematical model of population growth. Questions that can be answered by Eq. (1.3) include the following:

1. If present trends continue and the current population is 50 million, what will be the population 1 year from now? 10 years from now? 20 years from now?

2. If present trends continue, what will be the population 10 years from now if the present population is 51 million? 60 million? 100 million?

   Equation (1.3) also has some limitations. A partial list includes the following:

- Populations are discrete (it makes no sense to talk about half a person). However, $p(t)$ is not an integer for most values of $t$. At best, $p(t)$ is a continuous approximation of a discrete, time-varying population.

- Equation (1.3) was derived by using average values of birth and death rates. Although it may be a reasonable model for population growth in a country as a whole, Eq. (1.3) probably is not applicable to every city and town, in part because small subsets of the population may not have the same characteristics as the population as a whole. (For instance, the birth rate is negligible in a retirement community where most of the residents are over age 55.)

- Similarly, Eq. (1.3) does not capture population trends on a week-by-week or month-by-month basis, in part because birth rates tend to fluctuate over the course of a year.

## The Definition of a Differential Equation

Equation (1.3) is a simple model for population growth. More generally, we can express $dp/dt$ as a more complicated function of $p$ and $t$. The following definition summarizes the main ideas.

**Definition 1.1**   *A first-order ordinary differential equation is a relation of the form*

$$\frac{dp}{dt} = f(t, p) \qquad (1.4)$$

*in which the rate of change of p with respect to t is a function of t and of p itself. We call p the* dependent variable *and t the* independent variable. *Equation (1.4) is called* first order *because it expresses a functional relationship between dp/dt, p, and t. Equation (1.4) is an* ordinary *differential equation because all derivatives are taken with respect to a single independent variable. An* initial condition *is a specification of the value of p at some initial $t_0$; we write $p(t_0) = p_0$. Equation (1.4) together with the initial condition is called an* initial value problem. *A* solution *is a differentiable function p for which $p(t_0) = p_0$ and for which Eq. (1.4) is satisfied for an interval of t that includes $t_0$.*

The term "dependent variable," although standard, is something of a misnomer. The dependent variable actually is a differentiable function of $t$.

✦ *Example 2*   The relation

$$x' + \sin x = \cos t$$

is a first-order differential equation, because it involves only the first derivative of $x$. Here, $x$ is the dependent variable and $t$ is the independent variable. We can rewrite the equation as

$$x' = \cos t - \sin x,$$

which has the same form as Eq. (1.4), where $f(t, x) = \cos t - \sin x$.                                    ∎

✦ *Example 3*   The relation

$$x'' + x = 1 \tag{1.5}$$

is a second-order ordinary differential equation, because it involves the second derivative of the function $x$. The dependent variable is $x$; the independent variable is not stated explicitly. In most cases, we take the independent variable as $t$, because we are interested in time-varying processes. (Definition 1.1 covers only first-order differential equations. However, an analogous definition for second-order differential equations can be formulated easily.)

It is straightforward to verify by substitution that $x(t) = 1 + 2 \cos t$ is a solution of Eq. (1.5). We have

$$x'(t) = -2 \sin t \qquad \text{and} \qquad x''(t) = -2 \cos t,$$

so

$$x'' + x = (-2 \cos t) + (1 + 2 \cos t) = 1.$$                                    ∎

## The Exponential Functions

Exponential functions arise frequently as the solutions of differential equations. The function

$$y(t) = Ce^{kt} \tag{1.6}$$

solves the first-order differential equation

$$\frac{dy}{dt} = ky, \tag{1.7}$$

because

$$y'(t) = kCe^{kt} = ky(t). \tag{1.8}$$

We call Eq. (1.7) the *exponential growth equation*. Equation (1.6) is a *family of solutions* of the exponential growth equation, because the equality (1.8) holds for every constant $C$. Equation (1.6) often is called the *general solution* of Eq. (1.7), because $C$ can be chosen to satisfy any initial condition of the form $y(t_0) = y_0$.

✦ *Example 4*   Suppose

$$x' = 2x \tag{1.9}$$

with $x(1) = 3$. The general solution of Eq. (1.9) is

$$x(t) = Ce^{2t}. \tag{1.10}$$

The initial condition implies $x(1) = Ce^2 = 3$. Therefore, $C = 3e^{-2}$. The solution of the initial value problem is

$$x(t) = 3e^{-2}e^{2t} = 3e^{2t-2}. \qquad\blacksquare$$

Example 4 illustrates the following fact:

*A general solution of a differential equation contains one or more arbitrary constants. A solution of an initial value problem does not.*

The solution of a first-order initial value problem like the one in Example 4 can be derived from the general solution by substituting the initial condition and solving for the arbitrary constant.

✦ *Example 5*   The function

$$x(t) = Ca^t \tag{1.11}$$

is a general solution of the differential equation

$$\frac{dx}{dt} = (\ln a)x,$$

because

$$x'(t) = (\ln a)Ca^t = (\ln a)x.$$

If the initial condition is $x(0) = x_0$, then

$$x(t) = x_0 a^t.$$

For instance, the solution of the initial value problem

$$y' = (\ln 2)y, \qquad y(0) = 3$$

is

$$y(t) = 3(2^t).$$                                                                        ∎

## How Many Initial Conditions Are Needed?

Our previous discussion suggests that the general solution of the exponential growth equation has at least one arbitrary constant, whose value is determined uniquely if an initial value is specified. A natural question concerns the number of initial conditions that are needed to completely determine a solution. To motivate the answer, consider the following familiar examples from calculus.

✦ *Example 6*   You drive a car at a constant velocity of 100 km/h. The position $x(t)$ of the car at time $t$ is governed by the first-order differential equation $x' = 100$, which can be integrated to yield

$$x(t) = 100t + C,$$

where $C$ is a constant of integration. The value of $C$ is determined if the initial position of the car is specified. Here only one initial condition is needed.                 ∎

✦ *Example 7*   A particle falls in a vacuum. If the acceleration due to gravity is constant, then the position $x(t)$ of the particle is governed by the second-order differential equation

$$x'' = -g.$$

The general solution, which can be found by two integrations, is

$$x(t) = -\tfrac{1}{2}gt^2 + c_1 t + c_2,$$

where $c_1$ and $c_2$ are arbitrary constants of integration. The values of $c_1$ and $c_2$ are determined by the initial position and velocity of the particle. Here two initial conditions are needed. (Notice that $g$ is *not* an arbitrary constant but assumes a fixed value in this model.)                                                                              ∎

These two examples illustrate the following general result:

- The general solution of a first-order differential equation contains one arbitrary constant. One initial condition suffices to determine the value of the constant.
- The general solution of a second-order differential equation contains two arbitrary constants. Two initial conditions are needed to determine the value of the constants.

## The Concept of Proportionality

If we say that $x$ is proportional to $y$, then $x = ky$ by definition. Here $k$ is the *constant of proportionality*. Similarly, we say that $x$ is *inversely* proportional to $y$ if $x = k/y$. The notion of proportionality is fundamental in mathematics, and it is used frequently in the construction of differential equation models. The value of the proportionality constant must be determined from other sources. For instance, a sequence of measurements or other independent information may be required.

✦ *Example 8*   Suppose a population of bacteria in a petri dish grows at a rate that is proportional to its present size. Then

$$p' = kp.$$

The statement of the problem defines a differential equation because the statement gives a relation between the rate of population growth ($p'$) and the present size of the population ($p$). Because one is proportional to the other, we introduce a constant of proportionality, $k$. If the population of bacteria is increasing, then $k > 0$. If the population is decreasing, then $k < 0$.

Additional information is needed to specify the exact value of $k$. For instance, if we know that the population grows at a rate of 1000 bacteria per minute when there are 1 million bacteria in the dish, then $k = 0.001/\text{min}$. The units of $k$ are inverse time, because the units of $p'$ are bacteria per unit time.   ■

All terms that appear in a given differential equation must have consistent units. The following examples show how changes in the units of measurement affect the factors of proportionality.

✦ *Example 9*   Suppose that time is measured in hours instead of minutes in Example 8. Then

$$k = \frac{0.001}{\text{min}} = \frac{0.001}{\text{min}} \times \frac{60 \text{ min}}{\text{h}} = \frac{0.06}{\text{h}}.$$

Therefore, $p' = 0.06p$ if time is measured in hours.   ■

✦ *Example 10*    The rate at which liquid drains from a tank is proportional to the square root of the height $h$ of the liquid inside. Therefore, $h' = kh^{1/2}$. Suppose $h$ is measured in meters (m) and time in minutes. If the height decreases at the rate of 0.2 m/min when $h = 1$, and if $k$ is constant, then

$$h' = -0.2\sqrt{h}, \tag{1.12}$$

where $k = -0.2$ m$^{1/2}$/min. (Why is $k$ negative?)

Suppose $h$ is measured in centimeters (cm) instead. Then

$$k = \frac{-0.2 \text{ m}^{1/2}}{\text{min}} \times \sqrt{\frac{100 \text{ cm}}{1 \text{ m}}} = \frac{-2 \text{ cm}^{1/2}}{\text{min}}.$$

Thus Eq. (1.12) may be written equivalently as

$$h' = -2\sqrt{h}$$

when $h$ is measured in centimeters.                                           ∎

## Exercises

**1.** For each of the following equations, identify the independent variable, the dependent variable, and the order.

(a) $dp/dt + p = 0$         (b) $p'' + \sin p = \cos t$

(c) $p'' + \sin t = -p$       (d) $\dot{x} + x^2 = 0$

(e) $t' = \sqrt{t}$             (f) $(t'')^2 + t = p$

(g) $x' = \tan x$

**2.** The equation $p' = p$ can be used to model a rapidly growing population. There is an apparent difficulty with the units on the right-hand side of the equation, because $p$ does not have units of time. What is wrong?

**3.** Price inflation can be modeled by the exponential growth equation $dp/dt = kp$ if the inflation rate is constant.

(a) If the price of a basket of goods and services that is $100 today increases to $103 one year from now, determine $k$.

(b) Determine the constant of proportionality if time is measured in months instead of years.

**4.** Suppose the growth of bacteria in a petri dish is governed by the *logistic equation*

$$\frac{dp}{dt} = p(a - bp), \tag{1.13}$$

where $a$ and $b$ are constants.

(a) If $t$ is measured in days, what are the units of $a$ and $b$?

(b) Suppose that the numerical values of $a$ and $b$ are $a = 1/2$ and $b = 2$ when $t$ is measured in days. Write an equivalent expression for Eq. (1.13) when $t$ is measured in hours.

**5.** Consider Eq. (1.2).

(a) Show by substitution that

$$p(t) = 50,000,000\left(e^{0.002t} - 1\right)$$
$$+ Ae^{0.002t}$$

is a solution, where $A$ is a constant.

(b) Explain what $A$ signifies.

(c) If the current population is 50 million, and if the population grows at a rate given by Eq. (1.2), what will the population be in 20 years?

(d) How does your answer in (c) change if the current population is 100 million? Is the population in 20 years twice as large? Explain.

**6.** For each of the following equations, find the value(s) of $a$ for which the function $x(t) = e^{at}$ is a solution.

(a) $x' + x = 0$       (b) $x' + 2x = 0$

(c) $x'' = x$           (d) $x'' + 3x' + 2x = 0$

(Note: The equation that defines the appropriate values of $a$ is called the *characteristic equation*.)

**7.** For which value(s) of $\omega$ is the function $x(t) = \sin \omega t$ a solution of the differential equation $x'' = -4x$?

**8.** Suppose that $x(t) = Ae^{2t}$ satisfies the relation $x' = kx$. What conclusions, if any, can be drawn about the values of $A$ and $k$?

**9.** Determine which functions (1)–(5) solve equations (a)–(d). (Note: Some equations may have more than one solution, and others may have none.)

(a) $x' + x = 3t$       (b) $x' + x^2 = 0$

(c) $x' - x^2 = 1$       (d) $x' - x = t^2$

(1) $x(t) = 1/(t+1)$

(2) $x(t) = \tan(t + \pi/4)$

(3) $x(t) = 3e^t - t^2 - 2t - 2$

(4) $x(t) = 3t - 3 + 4e^{-t}$

(5) $x(t) = (t+1)^2 + 4e^t + 1$

**10.** Consider a point particle moving along a line. Let $x(t)$ be the position of the particle at time $t$. For each of the following scenarios, write a differential equation that is consistent with the given information. Identify your equation as first order or second order.

(a) The particle moves with constant velocity.

(b) The particle moves with constant acceleration.

(c) The velocity of the particle is proportional to its position.

(d) The acceleration of the particle is proportional to its position.

(e) The acceleration of the particle is proportional to its velocity.

(f) The velocity of the particle is inversely proportional to its position.

(g) The acceleration of the particle is proportional to the square of its velocity.

**11.** Find the solution of the differential equation $x' = 1.4x$ that satisfies the initial conditions in the list below.

(a) $x(0) = -1$

(b) $x(-1) = 1$

(c) $x(1) = -1$

**12.** Initial values specified at *different* values of $t_0$ sometimes yield the *same* solution curve. Distinct initial values specified at the *same* value of $t_0$ generally yield *different* solution curves. This exercise illustrates the general principle with a specific example. Let $p' = p$.

(a) Find the solution that satisfies the initial condition $p(0) = 1$.

(b) Find the solution that satisfies the initial condition $p(1) = e$.

(c) Find the solution that satisfies the initial condition $p(0) = 1/2$.

(d) Compare your answers in (a)–(c). Which are the same? Which are different?

(e) Show that, for the equation $p' = p$, the initial condition $p(0) = c_1$ and the initial condition $p(0) = c_2$ lead to different solutions if $c_1 \neq c_2$.

## 1.2  QUALITATIVE ANALYSIS OF SOLUTIONS

The goal of this section is to introduce you to methods that can be used to interpret the qualitative behavior of solutions of differential equations. Qualitative analysis answers questions like the following:

- Does the solution go to infinity, to zero, or to some other constant value as time goes on?

- Is the solution periodic? If so, what is the period?

- Does the solution reach 0 only in the limit as $t \to \infty$, or is there some specific value, such as $t = 100$, for which the solution equals 0?

- Does the solution eventually approach a more familiar function? For instance, does the solution "converge" to a straight line whose slope can be estimated?

In many cases, the above questions can be answered *even if it is difficult or impossible to find an analytical formula for the solution of the differential equation.* Qualitative analysis often provides important insights about the properties of the solutions that may not be obvious from analytical formulas.

### Explicit Solutions Cannot Always Be Found

The equation

$$\frac{dx}{dt} = kx \tag{1.14}$$

describes a function in terms of its derivative; that is, $x$ is a function whose derivative is a constant multiple of $x$. A typical calculus exercise asks you to find the tangent line given a formula for a function. In differential equations, the situation is reversed. A differential equation like (1.14) gives $x'(t)$, and hence the slope of the tangent line, at any point along the graph of the function $x(t)$, which is unknown.

How easy is it to determine $x(t)$ given information about $x'(t)$? In some cases, $x(t)$ can be found readily by integration. For example, the general solution of the differential equation

$$\frac{dx}{dt} = t^2 + \sin t$$

is

$$x(t) = \tfrac{1}{3}t^3 - \cos t + C.$$

The solution is easy to find because the derivative $x'$ is expressed solely in terms of $t$. In contrast, it is not obvious what the solution of

$$\frac{dx}{dt} = t^2 + \sin x \qquad (1.15)$$

might be. (The derivative of the function must be the sine of the current value of the function, with a factor of $t^2$ left over.) *More often than not, no explicit solution of a given differential equation can be found.* That is, it is not possible to find a formula for the solution $x(t)$ consisting of a finite sum of polynomial, exponential, trigonometric, and logarithmic functions.[1]

In summary, it may or may not be possible to express solutions of a given differential equation explicitly as functions of the independent variable. Even if an explicit formula for the solution can be found, the formula may be complex and difficult to interpret.

## The Direction Field

As stated above, a differential equation provides information about the derivatives of a differentiable function. The curve in Fig. 1.1(a) is the graph of a function $x(t)$. Figure 1.1(b) shows a sequence of six tangent line approximations to the same curve. Although the graph of $x(t)$ is not drawn in Fig. 1.1(b), the tangent lines—which provide information only about $x'(t)$—provide an accurate outline of the graph of $x(t)$.

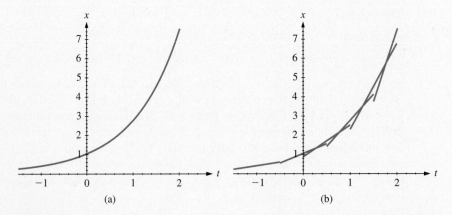

**FIGURE 1.1**    (a) The graph of a function. (b) A sequence of six tangent line approximations to the function in (a).

---

[1]The polynomial, exponential, trigonometric, and logarithmic functions often are called the *elementary functions.*

This observation is the basis for the use of a graphical device, called a *direction field*, that permits a graphical analysis of the solutions of first-order ordinary differential equations. Every first-order ordinary differential equation can be written in the general form

$$\frac{dp}{dt} = f(t, p). \tag{1.16}$$

The slope of the solution curve at every point $(t, p)$ in the plane is given by $f(t, p)$. A direction field consists of a large number of line segments, called *slope lines*, which are drawn on a grid. Equation (1.16) is used to evaluate $dp/dt$ at each grid point; the slope of the corresponding line segment (slope line) is $f(t, p)$.

Figure 1.2(a) shows the direction field corresponding to the equation

$$\frac{dp}{dt} = p. \tag{1.17}$$

The slope lines are generated on a grid in the following manner. Consider, for example the point $(t, p) = (0, 1)$. Equation (1.17) implies that $dp/dt = 1$ at $(0, 1)$. Thus, the slope line at $(0, 1)$ has unit slope. An identical slope line is drawn at $(0.2, 1)$, because $dp/dt = 1$ there also. In fact, the slope lines are identical at every grid point of the form $(t, 1)$.

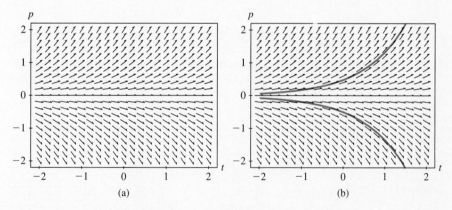

(a)                                                        (b)

**FIGURE 1.2**   (a) The direction field corresponding to Eq. (1.17). (b) The same direction field with two representative solution curves.

Similarly, the slope lines are horizontal along the $t$ axis, because $dp/dt = 0$ when $p = 0$. The slope lines have slope $-1$ along the line $p = -1$. In general, the slope lines point up when $p > 0$ and point down when $p < 0$. The slopes become steeper as $|p|$ becomes larger, because $|dp/dt| = |p|$.

The slope lines indicate the general behavior of the solutions of the differential equation, as illustrated in Fig. 1.2(b). The direction field provides the following graphical interpretation of solutions:

> *A solution of a differential equation is a function whose graph is consistent with the direction field.*

The tangent line of the graph of any solution must coincide with the slope lines at every point on the curve.

## Plotting Direction Fields

### The Grid Method

Computers are very useful tools for plotting the direction field associated with a particular differential equation, but they are not essential. It is straightforward to sketch direction fields by careful inspection of the differential equation. (Graph paper is quite helpful.) As an example, consider the differential equation

$$x' = t - x. \tag{1.18}$$

The independent variable is $t$, which by convention is plotted on the horizontal axis.

One way to sketch the direction field is to draw the slope lines on a grid of points. Figure 1.3(a) shows the direction field that is produced on a grid that is spaced at intervals of 0.5 along each axis. For example, at $(t, x) = (0, 1)$, we have $x' = -1$. The slope line drawn through $(0, 1)$ is a segment of slope $-1$. The slope line drawn through $(1/2, 1)$ is a segment of slope $-1/2$, and so on. The software used to draw the direction fields in this

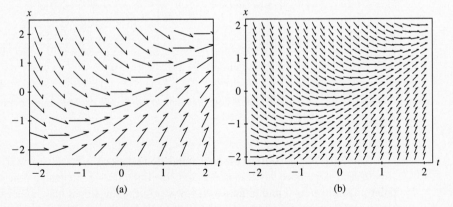

**FIGURE 1.3**    Direction field for $x' = t - x$. (a) coarse grid; (b) fine grid.

text draws the slope lines as arrows that point in the direction of positive $t$. The arrows in Fig. 1.3(b) illustrate the use of a grid that is spaced at intervals of 0.2 along each axis.

### The Method of Isoclines

An *isocline* is a curve along which all the slope lines in the direction field have the same slope. The *method of isoclines* is an alternative procedure for drawing direction fields, wherein the slope lines are drawn along isoclines corresponding to a particular value of the derivative $x'$.

In the case of Eq. (1.18), the slope line has slope $C$ when $x' = t - x = C$. Thus, the isoclines are lines of the form $x = t - C$; that is, the isoclines are straight lines of unit slope. For example, the line $x = t$ is an isocline along which the slope lines are horizontal, corresponding to the case $C = 0$. Figure 1.4 shows three isoclines; from left to right, they correspond to lines along which $x' = -1$, $x' = 0$, and $x' = 1$.

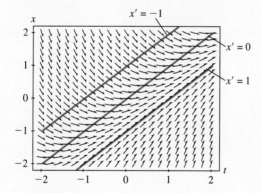

**FIGURE 1.4**   The method of isoclines, illustrated here for Eq. (1.18).

Isoclines are *not* the same thing as solution curves. Isoclines are merely a device to help you sketch the direction field more quickly. In the case of Eq. (1.18), however, the isocline corresponding to $x' = 1$ happens to be a solution curve, as you can easily check by substitution. This situation is merely a coincidence, but it does identify one solution of Eq. (1.18).

## An Example of Qualitative Analysis

The direction field permits a qualitative analysis of the solution that can answer many of the questions outlined at the beginning of this section, as the following examples illustrate.

✦ *Example 1*    The direction field in Fig. 1.2(b), corresponding to the differential equation

$$\frac{dp}{dt} = p,$$

illustrates the following facts about the solutions:

- The solution increases with time if the initial condition is positive, that is, if $p(t_0) > 0$. Here, $t_0$ can be any arbitrary starting time.
- The solution always decreases with time if $p(t_0) < 0$.
- Solutions cannot change sign.
- The zero function is a solution corresponding to initial conditions of the form $p(t_0) = 0$. The zero function is the only constant solution.
- Solutions other than the zero function never approach a constant value; that is, they never level off.                                                                                  ∎

✦ *Example 2*    Figure 1.5(a) shows the graph of a function that *cannot* be a solution of the differential equation

$$\frac{dp}{dt} = 2 - p. \tag{1.19}$$

Equation (1.19) implies that $p' < 0$ if $p > 2$. The derivative of the function in Fig. 1.5(a) is not consistent with Eq. (1.19), because the derivative sometimes is positive when $p > 2$ and sometimes is negative when $p < 2$.

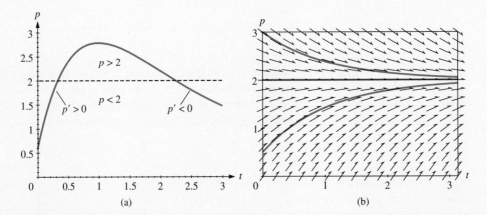

**FIGURE 1.5**    (a) An implausible solution of $p' = 2 - p$. (b) The corresponding direction field with three solution curves.

Figure 1.5(b) shows the direction field corresponding to Eq. (1.19), together with three solution curves. The qualitative behavior of the solutions can be summarized as follows:

- Solution curves increase if $p < 2$ and decrease if $p > 2$.
- Solution curves level off as they approach the line $p = 2$.
- The constant function $p(t) = 2$ is a constant solution of Eq. (1.19).
- No solution curve can cross the line $p = 2$. (See Exercise 9.)
- Solutions of Eq. (1.19) cannot oscillate periodically. The slope of a periodically oscillating function changes sign, because the function alternately increases and decreases. However, in the case of Eq. (1.19), the slope of a solution curve cannot change sign unless the solution curve crosses the line $p = 2$, which is impossible.                                                                ∎

## Qualitative Analysis and Simple Population Models

Qualitative analysis often can determine the extent to which a given differential equation is a plausible model for some phenomenon of interest. The following examples illustrate how qualitative analysis can assess whether simple population models are plausible.

✦ *Example 3*    The exponential growth equation

$$p' = kp \qquad (1.20)$$

models a population whose rate of growth is proportional to its present size. Suppose that, if the current population is $p_0$, then the population after $t$ hours is $p_0 e^{2t}$. Then the population is described by Eq. (1.20) with $k = 2$, at least initially. However, the solutions of Eq. (1.20) grow forever: $p(t) \to \infty$ as $t \to \infty$. Obviously, any real population of bacteria on a petri dish eventually reaches some limiting value, because the amount of space and the amount of nutrients are finite.

In summary, Eq. (1.20) may be a reasonable model for population growth as long as the size of the population does not get too large. Once the population is large enough, other factors act to limit its growth. Therefore, Eq. (1.20) is not an accurate model of population growth if the population is very large.                                                              ∎

✦ *Example 4*    Consider the extent to which the equation

$$p' = 2 - p \qquad (1.21)$$

is a plausible model of a population. The direction field, shown in Fig. 1.5(b), shows that solution curves eventually approach the constant value 2. Equation (1.21) reflects the idea that $p = 2$ is an *equilibrium* for the population, often called the *carrying capacity*. (The carrying capacity is the maximum sustainable population.) Values of $p$ greater than 2 are not sustainable in this example, because $p' < 0$ if $p > 2$, reflecting the idea that competition for food and habitat causes the population to decline. If the population is less than the carrying capacity, then there are sufficient resources to allow the population to grow back. In this respect, Eq. (1.21) may be a plausible model for the growth of certain populations.

However, there are limits to which Eq. (1.21) reflects the behavior of a real population. For instance, Eq. (1.21) implies that $p' > 0$ when $p = 0$. Insofar as an extinct population cannot grow, Eq. (1.21) is not a realistic model if $p = 0$.                    ∎

## A Comparison Theorem

Sometimes, it is possible to compare the qualitative behavior of the solutions of two differential equations by a direct comparison of the differential equations themselves. Consider the following equations:

$$x_1' = x_1, \tag{1.22}$$

$$x_2' = x_2 + 1. \tag{1.23}$$

Let $x_1(t)$ and $x_2(t)$ be the solutions of Eqs. (1.22) and (1.23), respectively, that satisfy the initial condition $x(0) = 1/4$. It is not necessary to derive explicit formulas for $x_1$ and $x_2$ to prove that the graphs of the two functions *cannot* resemble the sketch in Fig. 1.6(a).

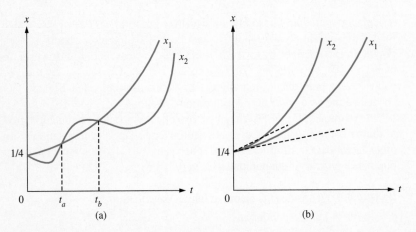

**FIGURE 1.6**    (a) Two functions that cannot be solutions of $x_1' = x_1$ and $x_2' = x_2 + 1$. (b) Two functions that are consistent with the two equations. See text for details.

Here are two reasons why:

- Figure 1.6(a) suggests that the tangent line to $x_1$ has positive slope and the tangent line to $x_2$ has negative slope when $t = 0$. Both functions are equal (in this case, $x_1(0) = x_2(0) = 1/4$), but the graphs contradict Eqs. (1.22)–(1.23), which imply that $x_1'(0) = 1/4$ and $x_2'(0) = 5/4$.
- More generally, at any point where $x_1(t) = x_2(t)$, Eq. (1.23) implies that the function $x_2$ increases more swiftly than $x_1$. Thus, the curves cannot cross as indicated at $t = t_b$.

Instead, the solution curves must resemble the sketch shown in Fig. 1.6(b). The two solutions are equal at $t = 0$, but $x_1$ increases more slowly than $x_2$, which is consistent with Eqs. (1.22)–(1.23).

Figure 1.7 shows the direction fields corresponding to Eqs. (1.22)–(1.23), together with solution curves that satisfy the initial condition $x(0) = 1/4$. At any fixed value of $x$, the slope lines for Eq. (1.23) are steeper than those for Eq. (1.22).

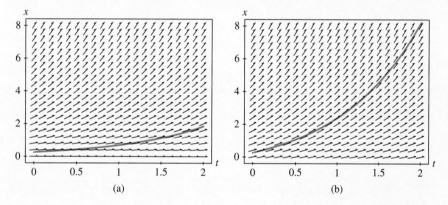

**FIGURE 1.7**   (a) The direction field for $x_1' = x_1$. (b) The direction field for $x_2' = x_2 + 1$.

The above discussion illustrates the following general result.

**Theorem 1.2   _The Comparison Theorem_**   _Let_ $x_1' = f(t, x_1)$ _and_ $x_2' = g(t, x_2)$, _where_ $f$ _and_ $g$ _are continuous and_ $x_1(t_0) = x_2(t_0) = x_0$. _Let_ $R$ _be a rectangle containing the initial point_ $(t_0, x_0)$, _and suppose that_ $f(t, x) < g(t, x)$ _for any point_ $(t, x)$ _in the interior of_ $R$. _Then_ $x_1(t) < x_2(t)$ _for all values of_ $t > t_0$ _for which the solution curves remain in_ $R$.

The *interior* of $R$ consists of the area enclosed by $R$, but it excludes the points on the line segments comprising the boundary of $R$.

◆ *Example 5*   Let

$$x_1' = x_1, \qquad x_2' = \sqrt{x_2},$$

with the initial condition $x_1(0) = x_2(0) = 1$. Let $R_1$ be a rectangle whose lower left-hand corner is $(t_0, x_0) = (0, 1)$, as depicted in Fig. 1.8. If $(t, x)$ is a point in the interior of $R_1$, then $f(t, x) = x > \sqrt{x} = g(t, x)$. Therefore, $x_1(t) > x_2(t)$ when the solution curves enter $R_1$.

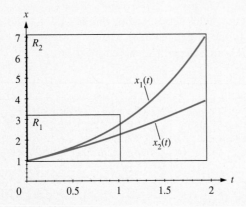

**FIGURE 1.8**   Two rectangles that are appropriate for comparing the solutions in Example 5.

There is nothing special about the choice of $R_1$ in Fig. 1.8. The results are equally valid in a bigger rectangle, such as $R_2$, also shown in Fig. 1.8. In fact, the hypotheses of Theorem 1.2 are valid for any rectangle whose lower left-hand corner is at $(0, 1)$; in particular, it holds for $R_2$ no matter how far $R_2$ is extended upward and to the right. Therefore, we may conclude that $x_1(t) > x_2(t)$ for all $t > 0$, assuming that $x_1$ and $x_2$ are defined for all $t > 0$.

Notice that the initial point $(0, 1)$ is on the boundary of $R_1$ (and $R_2$). We may still conclude that $x_1(t) > x_2(t)$ for $t > 0$ even though $f(0, 1) = g(0, 1)$, because the initial point is not in the interior of $R_1$ (or $R_2$). ■

## *Exercises*

**1.** Match the direction fields in Fig. 1.9 with the differential equations in the following list.

  (i) $x' = 2 - x$      (ii) $x' = 2 - \frac{1}{4}x^2$

  (iii) $x' = \sin x$      (iv) $x' = \sin t$

  (v) $x' = x(2 - x)$

**2.** Match the solution curves in Fig. 1.10 with the differential equations in the following list.

  (i) $x' = \sin t$      (ii) $x' = x - 2$

  (iii) $x' = 1 - x$      (iv) $x' = t - 2$

  (v) $x' = \sin x$

**3.** For each of the direction fields in Fig. 1.11, sketch a solution curve corresponding to the initial condition $x(0) = 1$ and a solution curve corresponding to $x(0) = 0$. In each case, summarize what happens as $t$ becomes large. (This corresponds to the eventual behavior of the system for the given initial condition.)

**4.** Consider Fig. 1.4.

  (a) What is the equation of the rightmost isocline shown in the figure?

  (b) Show by substitution that your answer in (a) is a solution of Eq. (1.18).

**5.** Let $p' = t - 2p$.

  (a) Describe the behavior of solutions as $t$ gets large.

  (b) Show by substitution that $p(t) = At + B$ is a solution for appropriate values of $A$ and $B$. (What are they?)

**6.** Draw the direction field for each of the differential equations in the following list. In each case,

  • identify the isoclines, even if you use the grid method to draw the direction field, and

  • discuss what the direction field implies about the qualitative behavior of the solutions. (Use Examples 1 and 2 as a guide.)

  (a) $p' = \sqrt{p}$      (b) $p' = p(2 - p)$

  (c) $p' = tp^2$      (d) $p' = t + \sin p$

**7.** Draw a direction field for the differential equation $p' = p$. Next, compare it to the direction field corresponding to each of the following equations. Discuss how the indicated insertion of the constant 2 changes the qualitative behavior of the solutions.

  (a) $p' = 2p$

  (b) $p' = -2p$

  (c) $p' = p - 2$

**8.** Draw a direction field for the differential equation $p' = \sin p$ and discuss the qualitative behavior of the solutions. Then, for each equation in the following list, draw the corresponding direction field and discuss how the indicated insertion of the constant 2 changes the qualitative behavior of the solutions.

  (a) $p' = 2 \sin p$      (b) $p' = \sin 2p$

  (c) $p' = \sin(p + 2)$      (d) $p' = 2 + \sin p$

**9.** Example 2 considers the equation

$$p' = 2 - p,$$

where $p = 2$ is an equilibrium solution, indicated by the dashed line in Fig. 1.12. The figure shows two ways in which a solution curve might cross the line $p = 2$. Argue that both sketches contradict the differential equation.

**10.** Discuss the difficulties involved in plotting a direction field for the differential equation $x' = \sin(1/x)$ near $x = 0$. What, if anything, can you say about the qualitative behavior of the solution that satisfies the initial condition $x(0) = C$, where $C$ is a small positive constant?

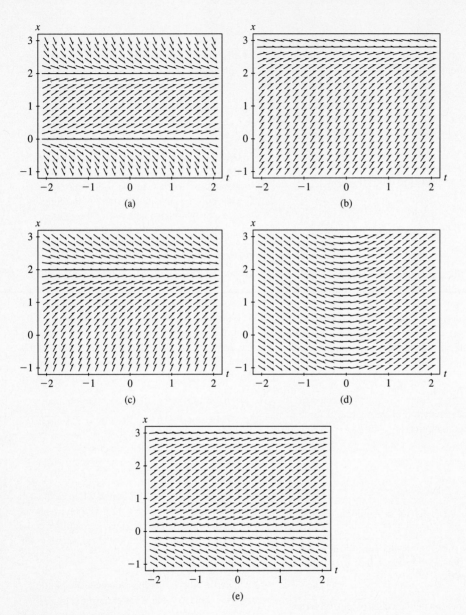

**Figure 1.9**    For Exercise 1.

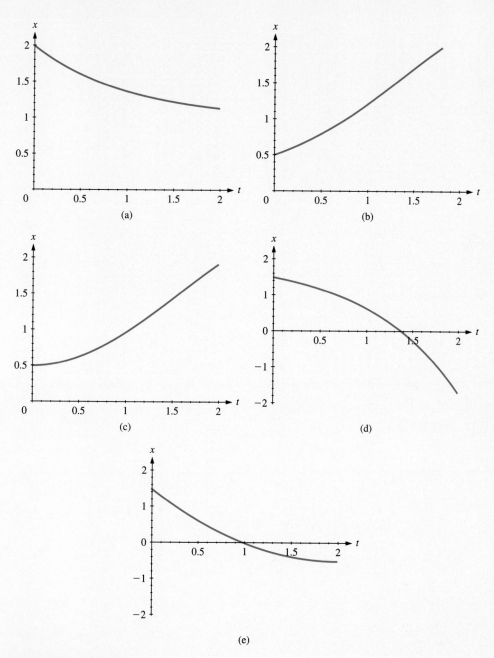

FIGURE 1.10 For Exercise 2.

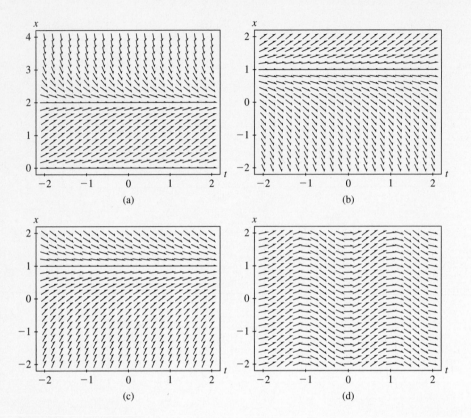

**FIGURE 1.11**    For Exercise 3.

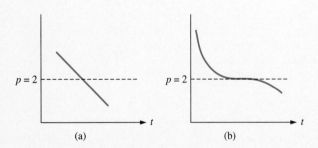

**FIGURE 1.12**    Two ways in which a curve can cross the line $p = 2$.

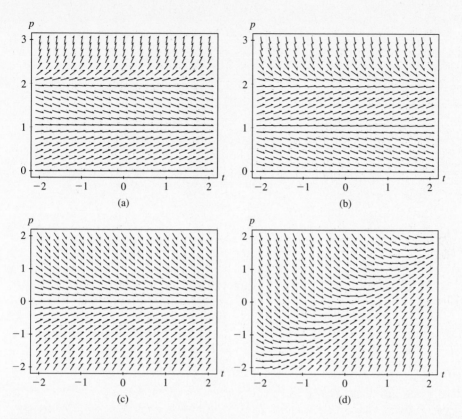

**FIGURE 1.13**    For Exercise 11.

---

**11.** Which of the direction fields in Fig. 1.13 might describe the growth of a population in a biologically plausible way? If the model is unrealistic, then discuss specific features that make it unrepresentative of a real population. If the model is realistic under certain circumstances, then explain what they are.

**12.** Each of the direction fields in Fig. 1.14 models the growth of a population. Explain why the direction field in Fig. 1.14(a) describes a population that will probably become extinct. (Hint: Every population model is an idealization that assumes constant living conditions at all times. In nature, small changes in living conditions are inevitable and often occur unpredictably.) Explain why Fig. 1.14(b) might be a more realistic model for a given population. In addition, suppose that you are responsible for developing a management plan to conserve the population modeled in Fig. 1.14(b). What strategies might you pursue, assuming that the model is reasonably accurate?

**13.** Consider the equations $x' = x^p$ and $x' = x^q$, where $p$ and $q$ are positive numbers. Let $x_p(t)$ and $x_q(t)$ denote the respective solutions, where $x_p(0) = x_q(0) = 1$. Suppose you want $x_p(t) > x_q(t)$ for $t > 0$. How must $p$ and $q$ be related?

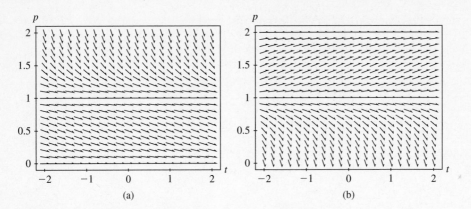

(a)

(b)

**FIGURE 1.14**    For Exercise 12.

**14.** Repeat Exercise 13 under the assumption that $p$ and $q$ are negative.

**15.** Let $R$ be the rectangle for which $0 \leq t \leq 4$ and $0 \leq x \leq 1$. Let

$$x_1' = x_1, \qquad x_2' = \sqrt{x_2}.$$

(a) Let $x_1$ and $x_2$ be the solutions that satisfy the initial conditions $x_1(0) = x_2(0) = 1/4$. For an appropriate set of values of $t$, does the comparison theorem predict $x_1(t) < x_2(t)$, $x_1(t) > x_2(t)$, or neither as long as both solution curves remain in $R$? Justify your answer.

(b) Show by substitution that $x_1(t) = e^t/4$ and $x_2(t) = (t+1)^2/4$ are solutions of their respective initial value problems.

(c) Show that $x_1(2) < x_2(2)$ but $x_1(3) > x_2(3)$. Does this finding contradict the comparison theorem? Explain.

**16.** Let

$$x_1' = x_1, \qquad x_2' = x_2^2,$$

and consider the solutions that satisfy the initial condition $x_1(0) = x_2(0) = 1$.

(a) Does the comparison theorem predict that $x_1(t) > x_2(t)$, $x_1(t) < x_2(t)$, or neither for $t > 0$?

(b) Show by substitution that $x_2(t) = 1/(1-t)$ solves the indicated initial value problem. What is $x_1(t)$?

(c) For what values of $t$ is the comparison theorem valid?

**17.** Is it possible for a solution of the equation $x' = \tan x$ to pass through the line $x = \pi/2$ in the manner suggested in Fig. 1.15? Explain.

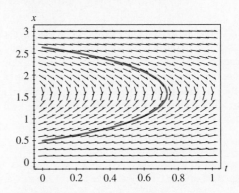

**FIGURE 1.15**    For Exercise 17.

# 1.3 AUTONOMOUS EQUATIONS AND EQUILIBRIUM SOLUTIONS

## What Is an Autonomous Process?

An *autonomous* differential equation is one for which the derivatives of the dependent variable $x$ depend only on the current value of $x$. More precisely, a first-order differential equation is autonomous if it can be written in the form

$$x' = f(x) \tag{1.24}$$

for a suitable function $f$. (In contrast, a general first-order differential equation is written as $x' = f(t, x)$. See Definition 1.1.) A first-order differential equation that is not autonomous is *nonautonomous*.

✦ *Example 1*　Figure 1.16 shows a plot of two populations as a function of time. The top curve shows $p(t)$, corresponding to the initial condition $p(0) = 0.3$. The bottom curve shows $q(t)$, corresponding to $q(0) = 0.2$. Assuming that both curves correspond to solutions of a first-order differential equation, let us ask whether they are consistent with an autonomous growth process.

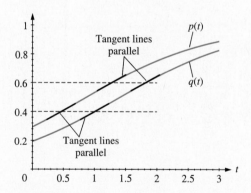

**FIGURE 1.16**　Two populations that are consistent with a first-order autonomous differential equation.

The dashed lines illustrate two representative cases. The population $p$ reaches 0.4 when $t \approx 0.5$, and the population $q$ reaches 0.4 when $t \approx 1$. Although the two populations reach 0.4 at different times $t$, their *rate* of growth is the same when the populations are 0.4. In other words, the tangent line to $p(t)$ at $(0.5, 0.4)$ and the tangent line to $q(t)$ at $(1, 0.4)$ are parallel.

Similarly, $p(t) = 0.6$ when $t \approx 1.3$, and $q(t) = 0.6$ when $t \approx 1.8$, and the tangent line to $p(t)$ at $(1.3, 0.6)$ is parallel to the tangent line to $q(t)$ at $(1.8, 0.6)$. The situation is analogous at every value of the population that is shown in Fig.1.16; that is, the tangent lines are parallel where $p = q = 0.3$, $p = q = 0.5$, and so on.

The solution curves are consistent with an autonomous differential equation: The rate of growth of the population depends only on its current size. In this situation, the growth rate of the population can be determined in principle simply from a census. Factors such as the time of year do not affect an autonomous population growth process.                                                                                  ■

◆ *Example 2*    Figure 1.17 shows two populations, $p$ and $q$, as a function of time. As in the previous example, $p(0) = 0.3$ and $q(0) = 0.2$. The two curves are not consistent with a first-order autonomous differential equation because the tangent lines are not parallel when the populations reach equal values. For instance, when $0.7 < t < 0.9$, $q(t)$ is nearly constant at a value of approximately 0.38. Thus, $q' \approx 0$ in this interval. In contrast, when the population described by $p(t)$ (top curve) reaches 0.38, the rate of growth $p'$ is clearly positive. Thus the rate of change depends on $t$ as well as the size of the population, so the underlying first-order differential equation is nonautonomous. In this example, a census and a calendar are required to determine the rate of population change.

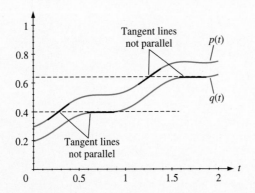

**FIGURE 1.17**    Two populations that grow according to a nonautonomous first-order differential equation.

_____                                                                                  ■

◆ *Example 3*    The isoclines in the direction field corresponding to an autonomous differential equation are horizontal lines, because $x'$ depends only on $x$, not on $t$. (See Exercise 10.)                                                                                  ■

## Equilibrium Solutions and Stability

We frequently speak of equilibrium processes in nature. In the simplest case, an equilibrium is a state in which the system does not change. An *equilibrium solution* of an autonomous first-order differential equation is a solution that is a constant function.

**Definition 1.3**   *Let $x' = f(x)$ be an autonomous first-order differential equation. An* equilibrium solution *is a constant function $x(t) = C$ such that $f(C) = 0$.*

   A constant function does not change, so its time derivative is zero. The concept of an equilibrium solution applies only to autonomous differential equations.

✦ *Example 4*   Suppose that the rate at which a glass of water warms or cools is proportional to the current temperature difference between the water and the room. That is, if $T(t)$ is the temperature of the water at time $t$ and if the room temperature $R$ is held constant, then

$$\frac{dT}{dt} = k(T(t) - R).$$

If the water initially is at room temperature, then no heat flows between the water and the room, and the water temperature remains constant. Thus, $T(t) = R$ is an equilibrium solution.                                                                              ■

✦ *Example 5*   The zero function is an equilibrium solution of the differential equation $x' = \sin x$, but it is not a solution of $x' = \sin t$.                                              ■

   Equilibrium solutions are often of interest, because they are easy to analyze qualitatively and because they often provide important information about the behavior of a system. For example, suppose that a population of birds has been constant for a long time (an equilibrium state) but pesticide contamination kills some of the birds. If the contamination is cleaned up, will the population grow back to its previous level?

   This kind of question concerns the *stability* of the equilibrium state. Suppose a system is initially at equilibrium and is perturbed in some way. (A perturbation is a disturbance.) Will the system return to the equilibrium? If the answer is *yes*, then we say that the equilibrium is *stable*. (This is one of many notions of stability in mathematics.) We can give an informal definition of stability as follows:

   *Let $p = p_0$ be an equilibrium solution of the differential equation $p' = f(p)$. The equilibrium solution is stable if solution curves starting at initial values near $p_0$ return to $p_0$ as $t \to \infty$. Otherwise, the equilibrium is unstable.*

According to this criterion, a stable equilibrium is one for which all solution curves that start sufficiently close to $p_0$ eventually return to $p_0$. In the population of birds mentioned above, the equilibrium state is stable if the population grows back to its previous level. The concept of stability refers only to small perturbations, however, as the next example shows.

✦ *Example 6*    The population of some species of plants and animals does not recover if the number of individuals in a given area drops below a certain threshold. For instance, sexual reproduction is inhibited if individuals are too widely scattered. (Pollen is less likely to fertilize the flowers of distant plants, and animals are less likely to encounter a suitable mate.)

Suppose that a population can recover provided that the number of individuals is not reduced below half of the carrying capacity. A simple model of this situation is given by the equation

$$p' = p(p - 1)(2 - p). \tag{1.25}$$

In this example, the carrying capacity of the population is 2. Values of $p$ larger than 2 are not sustainable, because if $p > 2$, then $p' < 0$, reflecting the influence of population pressures, as the direction field in Fig. 1.18 shows. If $1 < p < 2$, then $p' > 0$, reflecting the recovery of the population. Thus, $p = 2$ is a stable equilibrium.

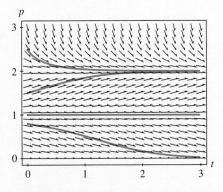

**FIGURE 1.18**    The direction field corresponding to Eq. (1.25). The curves are the graphs of some representative solutions.

On the other hand, if $p$ drops below 1, then $p' < 0$, reflecting the eventual extinction of the population. Thus, $p = 1$ is an unstable equilibrium. Figure 1.18 shows the direction field corresponding to Eq. (1.25), together with some representative solution curves.

Equation (1.25) implies that the population can survive indefinitely at $p = 1$ but that the population does not grow. Any reduction below the threshold $p = 1$ results in eventual extinction.

This example also illustrates what we mean about small perturbations. For instance, $p = 2$ is a stable equilibrium, but the population recovers only if the perturbation does not reduce $p$ to a value that is less than or equal to 1. If $p$ is reduced from 2 to a value that is less than 1, then the perturbation is so large that $p$ no longer returns to 2.    ■

## Phase Diagrams

Direction fields often are time consuming to draw, and information about the stability of the equilibrium solutions sometimes is difficult to discern. In the case of autonomous equations of the form

$$p' = f(p),$$

all of the information that is needed to analyze the qualitative behavior of the solutions can be extracted from a plot of $f(p)$ versus $p$. Such a plot often is called the *phase diagram*.

Figure 1.19(a) shows the phase diagram for the exponential growth equation $p' = p/2$, and Fig. 1.19(b) shows the corresponding direction field. The graphs represent two different ways of looking at the qualitative properties of the solutions. Notice the following features of Fig. 1.19:

- The dependent variable $p$ is plotted on the vertical axis in the direction field, Fig. 1.19(b). However, $p$ is plotted on the horizontal axis in the phase diagram, Fig. 1.19(a).

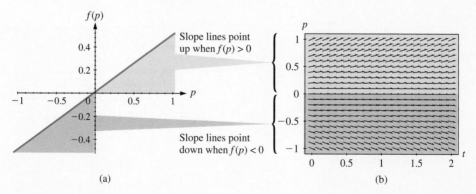

(a)                                                (b)

**FIGURE 1.19**    (a) A phase diagram for $p' = p/2$. (b) The corresponding direction field.

- The phase diagram does *not* contain explicit information about the independent variable $t$.

- The slope lines in the direction field point up for those values of $p$ for which $f(p) > 0$ and point down when $f(p) < 0$. In Fig. 1.19, $p' = f(p) > 0$ when $p > 0$ and $p' = f(p) < 0$ when $p < 0$. Hence, the slope lines in the direction field point up for positive $p$ and down for negative $p$.

- As $|p|$ becomes larger, so does $|f(p)|$. The slope lines in the corresponding direction field become steeper as $|p|$ increases.

- The constant function $p(t) = 0$ is an equilibrium solution. The existence of an equilibrium solution is apparent from the phase diagram, where the graph of $f(p)$ crosses the $p$ axis at $p = 0$.

In the phase diagram, an equilibrium solution corresponds to the point where the graph of $f(p)$ crosses the $p$ axis. For this reason, we often refer to an equilibrium solution as an "equilibrium point."

The equilibrium solution $p = 0$ is unstable. To see this, observe that $p' < 0$ whenever $p < 0$. Thus, a solution starting at a negative value of $p$ continues to decrease (away from $p = 0$). Similarly, $p' > 0$ whenever $p > 0$, which implies that a solution starting at a positive value of $p$ continues to increase (away from $p = 0$). The corresponding direction field in Fig. 1.19(b) shows that solution curves move away from the line $p = 0$. The phase diagram reflects this fact, because $f(p) > 0$ to the right of the point $p = 0$ and $f(p) < 0$ to the left of the point $p = 0$.

Figure 1.20(a) shows the phase diagram for the exponential decay equation $p' = -p/2$. The direction field in Fig. 1.20(b) shows that the equilibrium solution $p = 0$ is stable, because $p' > 0$ whenever $p < 0$, and $p' < 0$ whenever $p > 0$. The stability

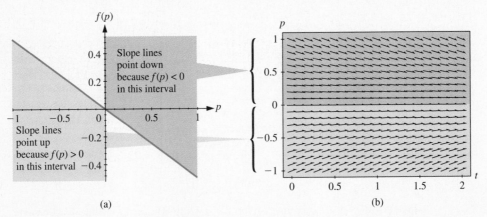

**FIGURE 1.20**    (a) A phase diagram for $p' = -p/2$. (b) The corresponding direction field.

of the equilibrium solution is apparent from the phase diagram, where $f(p) < 0$ to the right of the point $p = 0$ and $f(p) > 0$ to the left of the point $p = 0$.

The phase diagrams in Figs. 1.19(a)–1.20(a) illustrate the following general result.

**Theorem 1.4**   *Let $p' = f(p)$ for some differentiable function $f$, and suppose that $p(t) = p_0$ is an equilibrium solution. If*

$$\left.\frac{df}{dp}\right|_{p=p_0} > 0,$$

*then $p = p_0$ is an unstable equilibrium solution. If*

$$\left.\frac{df}{dp}\right|_{p=p_0} < 0,$$

*then $p = p_0$ is a stable equilibrium solution.*

✦ *Example 7*   Figure 1.21(a) shows the phase diagram corresponding to the equation

$$p' = f(p) = p(p - 1)(2 - p), \tag{1.26}$$

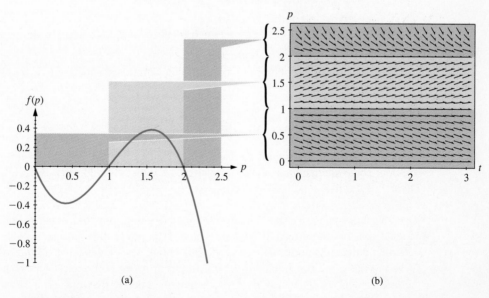

(a)                                                    (b)

**FIGURE 1.21**   (a) The phase diagram corresponding to $p' = p(p - 1)(2 - p)$. (b) The corresponding direction field.

discussed in Example 6. Figure 1.21(b) shows the corresponding direction field. The product rule implies that

$$\frac{df}{dp} = (p - 1)(2 - p) + p(2 - p) - p(p - 1).$$

The equilibrium solutions are $p(t) = 0$, $p(t) = 1$, and $p(t) = 2$. In the phase diagram, the graph of $f(p)$ crosses the $p$ axis at $p = 0$, 1, and 2. Therefore, we often say that 0, 1, and 2 are the equilibrium points of Eq. (1.26). Theorem 1.4 can be used to determine the stability of the equilibrium solutions as follows:

- $\left. \dfrac{df}{dp} \right|_{p=0} = -2$, so 0 is stable;

- $\left. \dfrac{df}{dp} \right|_{p=1} = 1$, so 1 is unstable;

- $\left. \dfrac{df}{dp} \right|_{p=2} = -2$, so 2 is stable.                                               ∎

◆ *Example 8*  Let $p' = f(p)$ and suppose that $p_0$ is an equilibrium solution. Theorem 1.4 cannot be used to determine the stability of $p_0$ if

$$\left. \frac{df}{dp} \right|_{p=p_0} = 0.$$

For instance, $p = 0$ is an equilibrium solution of the differential equation $p' = f(p) = p^3$. However,

$$\left. \frac{df}{dp} \right|_{p=0} = 3p^2 |_{p=0} = 0.$$

Therefore Theorem 1.4 cannot be used to determine the stability of the equilibrium solution; some alternative analysis is required. A direction field often is useful when Theorem 1.4 fails.                                                                          ∎

The phase diagram provides one additional piece of information about the behavior of solutions that the direction field does not. The phase diagram shows where the inflection points of solutions are located. Inflection points indicate where the solution curve reaches a relative minimum or maximum in its rate of change with respect to the independent variable $t$. If $p' = f(p)$, then the relative maxima and minima of $f$ correspond to the relative maxima and minima of $p'$—but such points are precisely the inflection points in the graph of $p(t)$, as the following example shows.

✦ *Example 9*   Let

$$p' = f(p) = p(p - 1)(2 - p). \tag{1.27}$$

Assuming that $p'' = d^2p/dt^2$ exists, the chain rule implies that

$$\frac{d^2p}{dt^2} = \left(\frac{d}{dp} f(p)\right)\left(\frac{dp}{dt}\right)$$

$$= [(p - 1)(2 - p) + p(2 - p) - p(p - 1)]\frac{dp}{dt}.$$

Inflection points may occur wherever $p'' = 0$. Candidate points include the equilibrium solutions as well as the points where the term in brackets is 0. The equilibrium solutions have no inflection points. (Why?) The term in brackets is 0 when $p = 1 \pm 1/\sqrt{3}$. The solution is not constant in this case, and the graph of $p'$ has a relative extremum at $p = 1 - 1/\sqrt{3}$ and $p = 1 + 1/\sqrt{3}$ (see Fig. 1.21(a)). Therefore, the solution curve has an inflection point wherever it passes through the values $p = 1 - 1/\sqrt{3}$ and $p = 1 + 1/\sqrt{3}$.

The top curve in Fig. 1.22 shows the solution curve for the initial condition $p_0 = 1.15$. The concavity of the solution curve changes when the solution crosses the line $p = 1 + 1/\sqrt{3} \approx 1.577$, indicated by the top dashed line. The bottom curve shows the solution curve for the initial condition $p_0 = 0.8$. The inflection point is located where the solution curve crosses the line $p = 1 - 1/\sqrt{3} \approx 0.423$, indicated by the bottom dashed line.

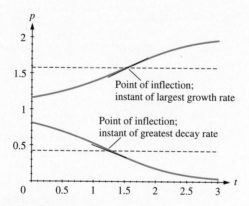

**FIGURE 1.22**   Two solutions of $p' = p(p - 1)(2 - p)$. The dashed lines indicate the location of the inflection point in each solution. The short solid line segments indicate the tangent line to each inflection point.

We can interpret the meaning of the inflection points in the following way. Suppose that Eq. (1.27) is a model for the growth of some population. If $p_0$ is slightly greater than 1, then the population growth accelerates until the population reaches $p = 1 + 1/\sqrt{3}$. When the population passes this value, the rate of growth begins to level off. Eventually, the population reaches a limiting value of $p = 2$, which is a stable equilibrium solution.

Conversely, if $p_0$ is slightly less than 1, then the population begins to die off. The rate of decline begins to level off when the population drops below $1 - 1/\sqrt{3}$. Eventually, the population becomes extinct.                                               ∎

## *Exercises*

**1.** Most populations in nature grow at a rate that depends on the time of year as well as the current size of the population. Can the rate of growth of such populations be described by an autonomous first-order differential equation? Explain.

**2.** Assume that the curve shown in Fig. 1.23 is the solution of a first-order differential equation. Determine whether the corresponding differential equation is autonomous. Give specific reasons to justify your answer.

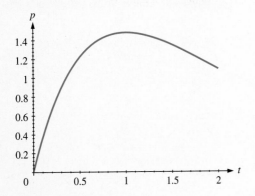

**FIGURE 1.23**   For Exercise 2.

**3.** Suppose the rate at which a kettle of hot water cools depends on the current temperature of the water. Determine whether each scenario in the following list is an autonomous process. Give specific reasons to justify your answer.

(a) At any instant, you can determine the rate at which the water cools merely by measuring the current temperature of the water.

(b) At any instant, the rate at which the water cools depends on how long it has been cooling as well as the current temperature of the water.

**4.** If the rate at which a sum of money (the *principal*) earns interest is proportional to the size of the principal, then the factor of proportionality by definition is the *interest rate*. Determine whether each scenario in the following list is an autonomous process. Give specific reasons to justify your answer.

(a) A bank certificate of deposit earns interest at a rate proportional to the size of the principal.

(b) As in (a), but a higher rate of interest is paid after the deposit is held for six months or more.

(c) As in (a), but the interest rate is a function of interest rates in the bond market.

(d) As in (a), but the interest rate depends on the size of the principal.

**5.** Suppose the rate at which a kettle of hot water cools is proportional to the difference between the current temperature of the water and the room temperature. For each of the following scenarios determine whether it is an autonomous process. Give specific reasons to justify your answer.

(a) The room temperature is held constant as the water cools.

(b) The room temperature fluctuates because of an external thermostatic control that is independent of the temperature of the water in the kettle.

(c) The room is heavily insulated, and the temperature of the room increases as the kettle cools. Suppose the total change in the room temperature is proportional to the total amount of heat lost by the water in the kettle.

**6.** For each of the following scenarios concerning a single population, sketch a direction field for an autonomous differential equation that is consistent with the given information. Draw in a solution curve to indicate what happens after a long time. Consider what happens for a variety of initial population sizes as you sketch your direction field.

(a) The population increases at a rate proportional to its present size.

(b) The population decreases at a rate proportional to its present size.

(c) If the population initially is sufficiently small, then it eventually becomes extinct. If the initial population is sufficiently large, then it eventually reaches a constant (positive) level.

**7.** Suppose an object falls toward the earth from a great height. For each of the following scenarios, determine whether it is an autonomous process. Give specific reasons to justify your answer.

(a) The acceleration of the object is proportional to the length of time that it has fallen.

(b) The acceleration of the object is proportional to its height above the ground.

(c) The acceleration of the object is proportional to its present velocity.

(d) The acceleration of the object is proportional to its present velocity and to its height above the ground.

**8.** Find all equilibrium solutions for the following differential equations and determine whether each is stable or unstable. Use graphical analysis or Theorem 1.4 as appropriate.

(a) $x' = x^2 - 1$           (b) $x' = x^2$

(c) $x' = \sin 2x$           (d) $x' = (\sin x) - 1$

(e) $x' = x - x^2$

**9.** We say that a differential equation is *homogeneous* if the zero function is an equilibrium solution. Identify the homogeneous equations in the previous problem.

**10.** Which of the direction fields in Fig. 1.24 corresponds to an autonomous first-order differential equation? Explain.

**11.** Figure 1.25 shows the direction field of a first-order ordinary differential equation.

(a) Determine whether 0 is a stable equilibrium.

(b) Sketch the corresponding phase diagram.

(c) If $x' = f(x)$ and you knew a formula for $f(x)$, could you use Theorem 1.4 to determine the stability of the equilibrium at 0? Explain.

**12.** The *logistic equation* refers to a class of simple population models that incorporates the idea that the population grows rapidly when the population is small but that the rate of growth drops as the population reaches the carrying capacity. (The carrying capacity is the maximum sustainable population.) The equation

$$p' = p(1 - p)$$

is one example of a logistic equation.

(a) What is the carrying capacity of the population in this example?

(b) Discuss the extent to which the equation is a plausible model of a real population.

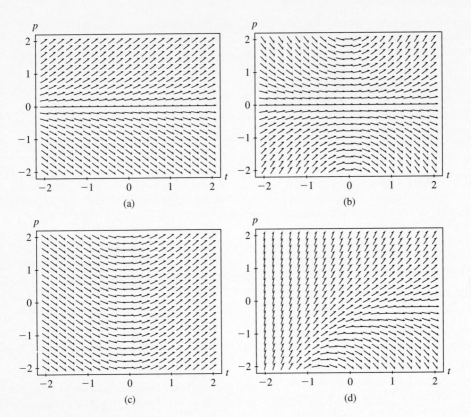

**FIGURE 1.24**    For Exercise 10.

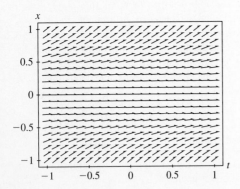

**FIGURE 1.25**    For Exercise 11.

**13.** Consider the logistic equation given by $p' = p(1 - p)$. Discuss the qualitative behavior of the solutions of this equation, including any equilibrium solutions and their stability. Next, consider each equation in the following list. Discuss how the indicated insertion of the constant 2 affects the qualitative behavior of the solutions. (In other words, compare and contrast the qualitative behavior of the solutions in the list below to the qualitative behavior of the solutions of the equation $p' = p(1 - p)$.)

(a) $p' = 2p(1 - p)$

(b) $p' = p(2 - p)$

(c) $p' = p(1 - 2p)$

**14.** Give an example of a first-order autonomous differential equation whose equilibria satisfy the following criteria.

(a) An unstable equilibrium at 1 and a stable equilibrium at 2

(b) A stable equilibrium at 1 and an unstable equilibrium at 2

(c) An unstable equilibrium at 1 and an unstable equilibrium at 2

(d) A stable equilibrium at 1 and a stable equilibrium at 2

**15.** Find all equilibrium solutions of the differential equation

$$\frac{dx}{dt} = \sin\left(\frac{2\pi}{x}\right)$$

and classify each as stable or unstable.

**16.** Suppose $dp/dt = f(p)$, where the corresponding phase diagram is shown in Fig. 1.26 Which of the following statements concerning $p(t)$ are true and which are false? Justify your answer.

(a) If $p(0) = 1$, then $p(t) = 1$ for all $t$.

(b) $p(t) < 0$ if $t > 1$.

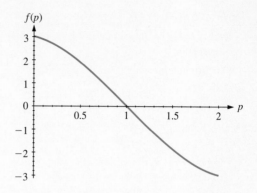

**FIGURE 1.26**   The graph of $f(p)$ for Exercise 16.

(c) $\lim_{t \to \infty} p(t) = 1$ whenever $p(0) > 0$.

(d) $p$ has an inflection point at $p = 1$.

(e) $p$ is undefined for $t < 0$.

(f) $p$ is increasing for $t < 1$.

**17.** Suppose $dp/dt = f(p)$, where the corresponding phase diagram is shown in Fig. 1.27. Which of the following statements concerning $p(t)$ are true and which are false? Justify your answer.

(a) $p(t) = 1$ for all $t$.

(b) $p(t)$ is a periodic function.

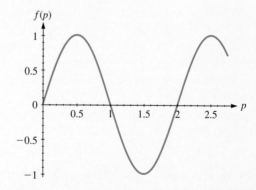

**FIGURE 1.27**   The graph of $f(p)$ for Exercise 17.

(c) If $p(2) = 2$ then $p(t) = 2$ for all $t$.

(d) 1 is a stable equilibrium, and 2 is an unstable equilibrium.

(e) If $p(0)$ is sufficiently large, then

$$\lim_{t \to \infty} p(t) = \infty.$$

**18.** Let $f$ be a continuous function and let $p' = f(p)$.

(a) In your own words, describe the intermediate value theorem. (Look it up in any calculus text if you don't remember what the term "intermediate value" refers to.)

(b) Suppose you know that $p' < 0$ if $p < 2$ and that $p' > 0$ if $p > 3$. Use your statement in (a) to show that the differential equation must have at least one equilibrium solution. Where must an equilibrium solution be located?

**19.** The logistic equation can be written more generally as

$$\frac{dp}{dt} = ap - bp^2, \qquad (1.28)$$

where $a$ and $b$ are positive constants.

(a) Graph $f(p)$ for some representative values of $a$ and $b$, and discuss how variations in $a$ and $b$ affect the graph of $f$.

(b) Find the carrying capacity of the population in terms of $a$ and $b$.

(c) Determine the equilibrium solutions and whether they are stable or unstable.

(d) Find an expression for $p''$ in terms of $a$, $b$, and $p$.

(e) Do solution curves have an inflection point? If so, determine the inflection point in terms of $a$ and $b$.

**20.** Prove or disprove: The solutions of the differential equation $x' = \cos x$ do not oscillate. Contrast the qualitative behavior of the solutions to that of $x' = \cos t$.

**21.** Prove the following theorem: *A first-order, autonomous ordinary differential equation of a single variable cannot have a periodically oscillating solution.* (Hint: Suppose there were an oscillating solution. Draw the direction field. What is wrong with the picture?)

**22.** Prove the following theorem: *Every solution of a first-order, autonomous ordinary differential equation must do exactly one of the following as $t \to \infty$: (a) the solution goes to $+\infty$; (b) the solution goes to $-\infty$; (c) the solution approaches a constant value.* Use direction fields to develop your argument.

**23.** Suppose $dp/dt = f(p)$, where $f$ is a continuously differentiable function.

(a) Find an expression for $p''(t)$.

(b) If $f$ is always increasing as a function of $p$, can solutions of the differential equation have an inflection point? Explain.

(c) Suppose $f$ changes sign at $p = a$. Is there necessarily an equilibrium solution at $a$? Explain why or why not.

(d) If $f$ has a relative maximum at $p = b$, what if anything can you say about solutions $p(t)$ at $p = b$? (For example, does $p$ have a relative maximum at $b$? An inflection point?)

**24.** Let $x' = f(x)$ and suppose that $x = 0$ is an equilibrium solution. Theorem 1.4 does not apply if

$$\left.\frac{df}{dx}\right|_{x=0} = 0.$$

(a) Show that $x = 0$ is a stable equilibrium for $x' = f(x) = -x^3$. (Hint: See Example 8.)

(b) Let $x' = f(x) = x^2$. Use an argument similar to the one in part (a) to show that 0 is an unstable equilibrium solution.

(c) Argue that 0 is an unstable equilibrium for the equation

$$x' = x^{2n},$$

where $n$ is a positive integer.

(d) Argue that 0 is an unstable equilibrium for the equation

$$x' = x^{2n+1},$$

where $n$ is a positive integer.

(e) Argue that 0 is a stable equilibrium for the equation

$$x' = -x^{2n+1},$$

where $n$ is a positive integer.

(f) Suppose $x_{eq}$ is an equilibrium solution of $x' = f(x)$, where

$$\left. \frac{df}{dx} \right|_{x=x_{eq}} = 0.$$

State a conjecture regarding the values of

$$\left. \frac{d^2 f}{dx^2} \right|_{x=x_{eq}}, \left. \frac{d^3 f}{dx^3} \right|_{x=x_{eq}}, \ldots$$

that lets you check whether $x_{eq}$ is a stable equilibrium solution, given that

$$\left. \frac{df}{dx} \right|_{x=x_{eq}} = 0.$$

**25.** The spruce budworm is a forest pest. Let $p(t)$ denote the budworm population at time $t$, and suppose that

$$\frac{dp}{dt} = ap - sp^2 - R(p). \qquad (1.29)$$

The constants $a$ and $s$, which are positive, describe the growth of the budworm population in the absence

of predation. The function $R(p)$ represents the effect of predation by birds. Suppose

$$R(p) = \frac{p^2}{1 + p^2}.$$

(a) Plot $R(p)$ as a function of $p$. How can you interpret the rate at which budworms are eaten as a function of the current budworm population? (In other words, what is the predation rate if $p$ is small? Large?)

(b) Let $a = 0.5$ and $s = 0.1$. Plot $a - sp$ and $R(p)/p$ as a function of $p$ on the same axes. What do the intersection points represent?

(c) Repeat (b) for $a = 0.6$, $s = 0.1$.

(d) Show that, as $a$ increases from 0.5 to 0.6, there is a value of $a$ at which Eq. (1.29) has exactly two nonzero equilibrium solutions.

(e) Show that, depending on the values of $a$ and $s$, there can be one, two, or three nonzero equilibrium solutions Eq. (1.29).

(f) Give an example of a value of $a$ and a value of $s$ for which

- Eq. (1.29) has exactly one nonzero equilibrium solution;
- Eq. (1.29) has exactly two nonzero equilibrium solutions; and
- Eq (1.29) has exactly three nonzero equilibrium solutions.

(g) Discuss the stability of each equilibrium in (f). Draw an appropriate phase diagram.

(h) Discuss the biological meaning of each of the three scenarios discussed in (f).

# 1.4  EULER'S METHOD

The first-order differential equation $x' = f(t, x)$ describes the function $x(t)$ in terms of its derivative. The previous two sections show how graphical interpretation can provide useful information about the qualitative behavior of $x(t)$ without finding an explicit formula for $x(t)$. In this section, we consider one way to find a quantitative approximation of the solution. The resulting numerical scheme is called *Euler's method* after Leonhard Euler, who first used it almost three centuries ago.[2]

From a graphical point of view, a solution of a first-order differential equation is a function whose graph is consistent with the direction field. That is, the graph of the solution must be tangent to the slope lines at every point. Euler's method approximates the solution by a sequence of points that follows the slope lines.

We begin with the differential equation

$$\frac{dx}{dt} = f(t, x) \tag{1.30}$$

and the initial condition $x(t_0) = x_0$. Euler's method approximates the solution at $t_1, t_2, \ldots$, where the $t$'s are equally spaced. That is, $t_1 = t_0 + h$, $t_2 = t_1 + h$, and so on. The increment $h$ is called the *time step*. We denote by $x_i$ the approximated value of the solution $x(t_i)$.

The basic idea behind Euler's method is to approximate the solution at each step by the slope line. Consider Fig. 1.28. We start at the initial point $(t_0, x_0)$. The straight line segment is the tangent line to the solution curve at $(t_0, x_0)$. Its slope is $x'(t_0)$, which by Eq. (1.30) is equal to

$$x'(t_0) = f(t_0, x_0).$$

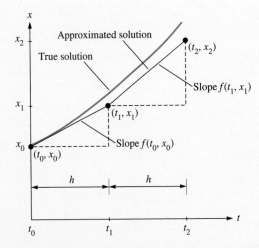

**FIGURE 1.28**    Schematic illustration of Euler's method.

---

[2]Euler is pronounced "oiler."

Therefore, Euler's method yields

$$x_1 = x_0 + hx'(t_0)$$
$$= x_0 + hf(t_0, x_0).$$

We proceed in a similar way to obtain an estimate of $x_2$:

$$x_2 = x_1 + hx_1'$$
$$= x_1 + hf(t_1, x_1).$$

Euler's method generates the sequence

$$x_{n+1} = x_n + hf(t_n, x_n). \tag{1.31}$$

Figure 1.28 illustrates the idea for the first two steps.

The step size $h$ determines how many times Eq. (1.31) must be iterated (repeated) to follow the solution from $t_0$ to some prespecified time $t$. (For instance, if $t_0 = 0$ and we want to approximate $x(1)$, then eight iterations of Eq. (1.31) are needed if $h = 1/8$.) The initial condition $(t_0, x_0)$ also must be specified. The following examples illustrate the use of Euler's method.

✦ *Example 1* Let

$$\frac{dx}{dt} = \frac{x}{2} \tag{1.32}$$

with $x(0) = 1$. Euler's method can be used to generate numerical approximations of $x(1)$ as follows.

Suppose first that $h = 1$. Then $x_1$ is an estimate of $x(1)$, because $t_1 = t_0 + h = 1$. Euler's method yields

$$x_1 = x_0 + 1 \times f(t_0, x_0) = 1 + \tfrac{1}{2} = 1.5.$$

Therefore, a crude estimate is $x(1) \approx 1.5$. (Equation (1.32) is autonomous, so $t$ is not used explicitly in evaluating $f$.)

Now suppose that the step size is $h = 1/2$. Then two iterations of Euler's method are needed to approximate $x(1)$. The first iteration gives an approximation to the solution at $t = 1/2$, and the second iteration gives an approximation at $t = 1$, as follows:

$$x_1 = x_0 + \tfrac{1}{2}f(t_0, x_0)$$
$$= x_0 + \tfrac{1}{2}f(0, 1)$$
$$= 1 + \left(\tfrac{1}{2}\right)\left(\tfrac{1}{2}\right)$$
$$= 1.25 \qquad \text{an estimate of } x\left(\tfrac{1}{2}\right)$$

and

$$x_2 = x_1 + \tfrac{1}{2} f(t_1, x_1)$$
$$= x_1 + \tfrac{1}{2} f\left(\tfrac{1}{2}, 1.25\right)$$
$$= 1.25 + \left(\tfrac{1}{2}\right)(0.625)$$
$$= 1.5625 \qquad \text{an estimate of } x(1).$$

Therefore, if the step size is $h = 1/2$, then Euler's method yields the estimate $x(1) \approx 1.5625$, which is slightly larger than the initial estimate above.

We proceed similarly if the step size is $h = 1/4$. Then $t_1 = 1/4$, $t_2 = 1/2$, $t_3 = 3/4$, and $t_4 = 1$, so four steps of Euler's method are needed to approximate $x(1)$, as follows:

$$x_1 = x_0 + \tfrac{1}{4} f(0, x_0) = 1 + \tfrac{1}{4}(0.5) = 1.125,$$
$$x_2 = x_1 + \tfrac{1}{4} f\left(\tfrac{1}{4}, x_1\right) = 1.125 + \tfrac{1}{4}(0.5625) = 1.2656,$$
$$x_3 = x_2 + \tfrac{1}{4} f\left(\tfrac{1}{2}, x_2\right) = 1.2656 + \tfrac{1}{4}(0.6328) = 1.4238,$$
$$x_4 = x_3 + \tfrac{1}{4} f\left(\tfrac{3}{4}, x_3\right) = 1.4238 + \tfrac{1}{4}(0.7119) = 1.6018.$$

Hence, if $h = 1/4$, then Euler's method yields the estimate $x(1) \approx 1.6018$, where we have rounded the computations to four decimal places.

If $h = 1/n$, then $n$ steps are required to approximate $x(1)$. Figure 1.29 shows the approximations to the solution of Eq. (1.32) using step sizes of $1/2$, $1/4$, and $1/16$.

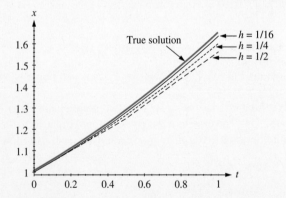

**FIGURE 1.29**   Numerical solution of Eq. (1.32) generated by Euler's method for step sizes $1/2$, $1/4$, and $1/16$. The top curve is a plot of the true solution.

Table 1.1 shows the approximations of $x(1)$ that are obtained for various values of $h$. The exact solution of Eq. (1.32) is $x(t) = e^{t/2}$, so

$$x(1) = e^{1/2} = 1.6487\ldots .$$

The figure and the table suggest that the approximations become more accurate as the step size gets smaller.

| $h$ | $\hat{x}(1)$ | $|\hat{x}(1) - x(1)|$ |
|---|---|---|
| 1/2 | 1.5625 | 0.0862 |
| 1/4 | 1.6018 | 0.0469 |
| 1/8 | 1.6242 | 0.0245 |
| 1/16 | 1.6362 | 0.0125 |
| 1/32 | 1.6424 | 0.0063 |

TABLE 1.1   Table of approximations of $x(1)$, where $x$ is the solution of $x' = x/2$ with the initial value $x(0) = 1$ and $\hat{x}$ is the numerical approximation.    ■

✦ *Example 2*   Use Euler's method with a step size of 1/4 to approximate the solution of the initial value problem

$$x' = t - x, \qquad x(0) = 0.5 \tag{1.33}$$

at $t = 1$.

*Solution*   Let $h = 1/4$ and $f(t, x) = t - x$. Then $t_1 = 1/4$, $t_2 = 1/2$, $t_3 = 3/4$, and $t_4 = 1$. Hence four steps of Euler's method are needed to approximate $x(1)$. We proceed as in the previous example to obtain the sequence

$$x_0 = 0.5,$$
$$x_1 = x_0 + \tfrac{1}{4} f(0, x_0)$$
$$= 0.5 + \tfrac{1}{4}(-0.5)$$
$$= 0.375,$$
$$x_2 = x_1 + \tfrac{1}{4} f\left(\tfrac{1}{4}, x_1\right)$$
$$= 0.375 + \tfrac{1}{4}(-0.125)$$
$$= 0.3438,$$

$$x_3 = x_2 + \tfrac{1}{4} f\left(\tfrac{1}{2}, x_2\right)$$

$$= 0.3438 + \tfrac{1}{4}(0.1562)$$

$$= 0.3828,$$

$$x_4 = x_3 + \tfrac{1}{4} f\left(\tfrac{3}{4}, x_3\right)$$

$$= 0.3828 + \tfrac{1}{4}(0.3672)$$

$$= 0.4746.$$

Thus Euler's method produces the estimate $x(1) \approx 0.4746$ when the computations are rounded to four decimal places.                                                                  ∎

Euler's method is easy to program. Figure 1.30 shows a complete Fortran program that implements Euler's method for the initial value problem Eq. (1.33). The equivalent C or Pascal program is similarly brief. A different initial condition can be specified by changing the initial values given to $x$ and $t$ in the program. A different equation can be integrated by changing the second statement in the function routine.

```
x=0.5                        function f(t,x)
t=0.0                        f=t-x
h=0.25                       return
n=4                          end
do j=1,n
    x=x+h*f(t,x)
    t=t+h
    print *, t,x
enddo
stop
end
```

**FIGURE 1.30**    A Fortran program to integrate Eq. (1.33) from $t = 0$ to $t = 1$ with a step size of $h = 1/4$.

## How Accurate Is Euler's Method?

Euler's method generates the sequence $x_1, x_2, \ldots$, where each value in the sequence is computed from the previous one. Only $x_0$ is known exactly (because it is the initial condition). The value of $x_1$ produced by Euler's method is only an approximation to the true value of the solution at $t_1$. Moreover, we obtain $x_2$ directly from $x_1$. Therefore, if there is some error in $x_1$, then there will be some error in $x_2$. At each step, there is necessarily some error in the approximation that propagates to subsequent steps. One important question, therefore, concerns the cumulative effects of the errors.

It is impossible to know exactly how much error is introduced in each step of any numerical method unless an analytic formula is known for the solution. An analytical solution of Eq. (1.32) can be found, and Table 1.1 shows how the error decreases with the step size. The absolute error in the numerical solution at $t = 1$ is defined as

$$|\hat{x}(1) - x(1)|,$$

where $\hat{x}(1)$ is the numerical approximation of the true solution at $t = 1$. In the case of Eq. (1.32), the absolute error is $|\hat{x}(1) - x(1)| = 0.0862$ when $h = 1/2$. When $h = 1/4$, the absolute error is reduced to 0.0469—about half as large. Table 1.1 suggests that the absolute error in the solution decreases linearly with the step size. In fact, the following statement is true whenever $h$ is sufficiently small:

> *The absolute error in the numerical solution generated by Euler's method is reduced approximately by a factor of 2 whenever $h$ is cut in half.*

This statement holds only if $h$ is not too large to begin with. In general, however, you can apply this rule of thumb when using Euler's method: If you want to double the accuracy of your solution, then you must take twice as many steps. Chapter 9 discusses the notion of errors in numerical methods in more detail.

## Exercises

**1.** Let $x' = t^2$.

(a) Apply Euler's method to approximate $x(2)$, assuming that $x(t)$ satisfies the initial condition $x(0) = 1$. Use step sizes of $1/2$, $1/4$, and $1/8$.

(b) Find the exact solution and generate a table similar to Table 1.1. Does the absolute error in the solution decrease linearly with the step size?

**2.** Find a class of differential equations for which Euler's method generates a *completely accurate* numerical solution, that is, where $x_n$ exactly equals the true solution $x(t_n)$ for every $n$. (Hint: For which differential equations do all solution curves coincide with the slope lines?)

**3.** Euler's method approximates the solution of a

given differential equation by a sequence of tangent lines.

(a) Suppose the graph of the solution curve is always concave upward. Is the approximation generated by Euler's method larger or smaller than the "true" solution?

(b) Repeat (a) under the assumption that the graph always is concave downward.

**4.** Let $x' = x^3$.

(a) Find an expression for $x''$ in terms of $x$.

(b) Suppose $x(0) = 1$. Is the solution curve concave up or concave down? Use the result in (a) to justify your answer.

(c) Does Euler's method overestimate or underestimate the true value of the solution at, say, $t = 0.1$? Explain. (Hint: See Exercise 3.)

**5.** Suppose you have a simple four-function calculator, that is, a calculator that can add, subtract, multiply, and divide but that cannot calculate transcendental functions such as $e^x$ with the touch of a key. Describe how you can use Euler's method to approximate $e^{0.3}$ to three significant digits.

**6.** Let $x' = -\sqrt{x}$ with initial condition $x(0) = 1$.

(a) Let $h = 1/2$. Show that you can take three Euler steps but not four. Explain the difficulty encountered in the fourth Euler step.

(b) Differential equations similar to the one in this problem model the draining of a cylindrical tank, where $x(t)$ is the height of the liquid in the tank at time $t$. Show how to extrapolate from the third Euler step to estimate the time at which the tank empties.

(c) Let $h = 1/4$. How many Euler steps can you take now? Use your results to estimate the time at which the tank empties.

(d) Using a suitable modification of the program in Fig. 1.30, a programmable calculator, or other software package, generate a numerical approximation of the solution using step sizes 1/8, 1/16, and 1/32. In each case, estimate the time at which the tank empties. Do your numerical solutions and your estimates of the emptying time of the tank appear to reach a limit? Explain.

**7.** Let $x' = -x$, where $x(0) = 5$.

(a) Use Euler's method to determine $x(1)$ using step sizes of $h = 0.1, 0.01, 0.001$, and $0.0001$.

(b) Find the exact value of $x(1)$.

(c) Determine the absolute error in the numerical solution that you found in (a) at $t = 1$ and plot your results. Does the absolute error decrease at a rate that is linearly proportional to the step size? Explain.

**8.** Apply Euler's method to the equation

$$\frac{dx}{dt} = \sin \frac{1}{x} \qquad (1.34)$$

with the initial condition $x(0) = 4/\pi$. Approximate $x(1)$ using $h = 1/2$, $h = 1/4$, $h = 1/8$, and $h = 1/16$. Use a graphics program or graph paper to plot all the computed solutions on the same axes. How might you be able to assess the accuracy of the computed solutions as a function of $h$?

**9.** Numerical methods like Euler's method may not work if the solution is not well behaved. Let $x' = x^2$ with the initial condition $x(0) = 1$. Use Euler's method to approximate $x(1)$ as follows.

(a) Let $h = 1/4$ and take four Euler steps.

(b) Let $h = 1/8$ and take eight Euler steps.

(c) Let $h = 1/16$ and take 16 Euler steps.

(d) Let $h = 1/32$ and take 32 Euler steps.

(e) Show by substitution that

$$x(t) = \frac{1}{1-t}$$

is a solution of the initial value problem. What is the difficulty with Euler's method here?

## 1.5   THE SEPARATION OF VARIABLES

So far, we have discussed graphical methods, which permit a qualitative analysis of the solutions of first-order differential equations, and Euler's method, which permits a quantitative approximation of solutions. In this section, we illustrate a procedure, called the *separation of variables*, that in some cases can be used to find an analytical formula for the solution of a first-order differential equation.

Loosely speaking, the separation of variables is an application of the chain rule in reverse. More precisely, suppose that $H$ is a function of $x$. If $x$ in turn is a function of $t$, then $H$ also depends on $t$. The chain rule implies that

$$\frac{d}{dt}H(x) = \frac{dH}{dx}\frac{dx}{dt}. \tag{1.35}$$

A first-order differential equation is *separable* if it can be written in the form

$$\frac{dx}{dt} = f(x)g(t). \tag{1.36}$$

The chain rule can be used to express $x$ as a function of $t$ in the following way.

If $f(x) \neq 0$, then we can rewrite Eq. (1.36) as

$$\frac{1}{f(x)}\frac{dx}{dt} = g(t). \tag{1.37}$$

Now suppose that $H$ is an antiderivative of $1/f(x)$, that is, $dH/dx = 1/f(x)$. Then Eq. (1.35) implies that Eq. (1.37) is equivalent to

$$\frac{d}{dt}H(x) = g(t). \tag{1.38}$$

The fundamental theorem of calculus yields

$$H(x) = \int g(t)\,dt + C, \tag{1.39}$$

where $C$ is an arbitray constant of integration. Equation (1.39) states $x$ as a function of $t$; in principle, Eq. (1.39) is a solution of Eq. (1.36).

On the other hand, because $H(x)$ is an antiderivative of $1/f(x)$, we have

$$H(x) = \int \frac{1}{f(x)}\,dx;$$

therefore, Eq. (1.39) can be expressed as

$$\int \frac{1}{f(x)}\,dx = \int g(t)\,dt + C.$$

The above expression segregates the factors involving $x$ on the left-hand side and the factors involving $t$ on the right. Formally, it appears that we have simply divided each side of Eq. (1.36) by $f(x)$ and multiplied each side by $dt$ before we integrate. The chain rule justifies this procedure, as explained above.

Equation (1.39) is an *implicit* solution of Eq. (1.36) insofar as it expresses a function of $x$ as a function of $t$. The constant $C$ is arbitrary, so the separation of variables yields a general solution. Each choice of $C$ leads to a different functional relationship between $x$ and $t$; the corresponding graphs are called *integral curves*. The value of $C$ usually can be determined uniquely if an initial condition of the form $x(t_0) = x_0$ is specified. The following example illustrates a typical application of the solution method.

✦ *Example 1*    Find an explicit solution of the initial value problem

$$\frac{dx}{dt} = xt, \qquad x(0) = -1.$$

**Solution**    The equation is separable, with $f(x) = x$ and $g(t) = t$. If $x \neq 0$, then we can write

$$\frac{1}{x}\frac{dx}{dt} = t. \tag{1.40}$$

We have

$$H(x) = \int \frac{1}{x}\, dx = \ln|x|.$$

We may integrate both sides of Eq. (1.40) with respect to $t$ to obtain

$$\ln|x| = \tfrac{1}{2}t^2 + C, \tag{1.41}$$

where $C$ is a constant of integration.

Equation (1.41) is a general solution of Eq. (1.40). It also expresses $x$ as an *implicit* function of $t$. That is, if a value of $t$ and a value of $C$ are specified, then the value of $\ln|x|$ is determined. The value of the logarithm implicitly determines the value of $|x|$, which is why we call Eq. (1.41) an implicit solution.

Of course, it is usually more convenient to have an *explicit* solution, that is, one in which we can read off the value of $x$ directly. It is not always possible to find an explicit solution from an implicit one; in this example, however, an explicit solution can be found by exponentiating both sides of Eq. (1.41) to obtain

$$|x| = C_1 e^{t^2/2},$$

where $C_1 = e^C$. An explicit solution is

$$x(t) = C_2 e^{t^2/2}, \tag{1.42}$$

where $C_2 = C_1$ if $x_0 > 0$ and $C_2 = -C_1$ if $x_0 < 0$. (Why?) We can check that Eq. (1.42) is a general solution by differentiating it with respect to $t$:

$$x' = tC_2 e^{t^2/2}$$

$$= tx.$$

The initial condition $x(0) = -1$ implies that

$$x(t) = -e^{t^2/2},$$

where $C_2 = -C_1 = -1$.                                                  ∎

The procedure outlined above is called the *separation of variables* because the antiderivatives of each side of the differential equation $x' = f(x)g(t)$ are determined separately: the left-hand side with respect to $x$ and the right-hand side with respect to $t$. Three important comments about the method are the following:

- The method can be used only if $f(x) \neq 0$, because $1/f(x)$ is undefined otherwise.
- If $f(x_0) = 0$, then $x_0$ is an equilibrium solution.
- The antiderivative of $1/f(x)$ may be difficult or impossible to find explicitly.

Despite these caveats, the separation of variables is useful in many situations.

✦ *Example 2*   Let

$$\frac{dx}{dt} = x^2. \tag{1.43}$$

Find a solution that satisfies the initial condition $x(0) = 1$ and a solution that satisfies the initial condition $x(0) = 0$.

**Solution**   Equation (1.43) is separable with $f(x) = x^2$ and $g(t) = 1$. If $x \neq 0$, then we can express Eq. (1.43) as

$$\frac{1}{x^2}\frac{dx}{dt} = 1. \tag{1.44}$$

We have

$$H(x) = \int \frac{1}{x^2}\, dx = -\frac{1}{x}.$$

We may integrate both sides of Eq. (1.44) to obtain the implicit solution

$$-\frac{1}{x} = t + C.$$

We can solve for $x$ to obtain the explicit solution

$$x = -\frac{1}{t + C}. \tag{1.45}$$

The initial condition $x(0) = 1$ implies that $C = -1$. Thus the corresponding solution is

$$x = -\frac{1}{t - 1} = \frac{1}{1 - t}.$$

The initial condition $x(0) = 0$ cannot be satisfied by any choice of $C$ in Eq. (1.45). The reason for the difficulty is the following: The function $1/x^2$ used to derive Eq. (1.45) is not defined when $x = 0$. Therefore, the separation of variables cannot be applied when the initial condition is $x(0) = 0$. Another method must be used to derive the solution.

In this case, the solution is apparent by inspection. If $x = 0$, then Eq. (1.43) implies that $x' = 0$. Thus the zero function is an equilibrium solution of Eq. (1.43). Therefore, the solution that satisfies the initial condition $x(0) = 0$ is $x(t) = 0$. ∎

✦ *Example 3*    The separation of variables cannot be applied to the differential equation

$$\frac{dx}{dt} = x^2 + t^2,$$

because the right-hand side cannot be written as a product of the form $f(x)g(t)$ for any functions $f$ and $g$. ∎

## Exercises

**1.** Use the method of separation of variables to derive the general solution of the exponential growth equation

$$\frac{dx}{dt} = kx.$$

For which values of $x_0$ does the method yield an explicit solution that satisfies the initial condition $x(0) = x_0$?

**2.** Find a solution of $x' = x^3$ for each of the following initial conditions.

   (a) $x(0) = 2$

   (b) $x(0) = -3$

   (c) $x(0) = 0$

**3.** Consider Example 1 with the initial condition $x(0) = 0$. What is the difficulty with the method of separation of variables? What is the solution of this initial value problem?

**4.** Can the method of separation of variables be used to find a solution of

$$x' = x^2 - t^2,$$

since the equation can be rewritten as

$$x' = (x + t)(x - t)?$$

**5.** Use the method of separation of variables to find explicit solutions of each of the following equations:

(a) $x' = x^{1/3}$         (b) $x' = x^{1/4}$

with $x(0) = x_0$. For each equation, answer the following questions:

- For which values of $x_0$ is the method of separation of variables valid?

- Let $x_0 = 1$. Is $x(t)$ defined for every positive $t$? If so, does $x(t)$ approach a limit as $t \to \infty$?

**6.** Let

$$\frac{dp}{dt} = p(1 - p).$$

(a) Find the general solution using the separation of variables. (Hint: Integrate by partial fractions.)

(b) Let $p(0) = p_0$. Suppose $0 < p_0 < 1$. What happens to $p(t)$ as $t \to \infty$?

(c) Let $p(0) = p_0$. Suppose $p_0 > 1$. What happens to $p(t)$ as $t \to \infty$?

(d) Draw a direction field and verify your answers in (b) and (c) graphically. Notice that the graphical analysis yields qualitative information more readily than the analytical solution.

**7.** For each equation in the following list,

- determine all equilibrium solutions;

- determine whether the method of separation of variables is applicable;

- If so, then find a solution satisfying $x(0) = 1$ and a solution satisfying $x(0) = -1$, where possible.

(a) $x' = x + 1$         (b) $x' = x^2 + 1$

(c) $x' = -\sqrt{x}$         (d) $x' = t + x$

**8.** Repeat Exercise 7 for each equation in the following list.

(a) $x' = x^t$         (b) $x' = t/x$

(c) $x' = e^{t-x}$         (d) $x' = te^{x^2}$

(e) $x' = \ln x$

**9.** Let

$$x' = \cos x. \qquad (1.46)$$

(a) Find an implicit general solution of Eq. (1.46).

(b) Find an explicit general solution of Eq. (1.46).

(c) Find a solution that satisfies the initial condition $x(0) = 0$.

(d) Find a solution that satisfies the initial condition $x(0) = \pi/2$.

(e) Determine the limit as $t \to \infty$ of the solutions that you found in (c) and (d).

**10.** Repeat Exercise 9 for the equation $x' = \sin x$.

**11.** Let

$$x' = x - x^2. \qquad (1.47)$$

(a) Find an implicit general solution of Eq. (1.47).

(b) Find an explicit general solution of Eq. (1.47).

(c) Find a solution that satisfies the initial condition $x(0) = 1/2$.

(d) Find a solution that satisfies the initial condition $x(0) = 2$.

(e) Determine the limit as $t \to \infty$ of the solutions that you found in (c) and (d).

**12.** Let

$$p' = p(p - 1)(2 - p). \qquad (1.48)$$

(See Example 7 in Section 1.3.)

(a) Find an implicit general solution of Eq. (1.48).

(b) Find an explicit solution that satisfies the initial condition $p(0) = 1/2$.

(c) Find an explicit solution that satisfies the initial condition $p(0) = 3/2$.

(d) Determine the limit as $t \to \infty$ of the solutions that you found in (b) and (c).

# Chapter 2

---

# EXPONENTIAL GROWTH

---

The goal of this chapter is to acquaint you with some of the many applications of exponential growth and decay. One very practical application is the time value of money, which refers to the value of an investment as interest accrues. Other physical applications include the dilution of chemical species and simple models of air resistance.

Section 2.1 reviews the fundamental properties of the exponential function and considers some simple applications of exponential growth and decay processes. Section 2.2 considers other applications of the exponential growth equation, including the time value of money, Newton's law of cooling, and a simple model of air resistance. Section 2.3 discusses a generalization of the basic exponential growth equation that can be solved analytically. Section 2.4 discusses some of the ways in which errors in measurement can affect predictions about the future state of a given system.

## 2.1  BASIC PROPERTIES AND APPLICATIONS

### The Equal Ratio Property

We use the term *exponential process* to refer to a process whose growth can be described by the exponential function. A useful example of an exponential process is unrestrained population growth, in which the rate of growth of the population is proportional to its present size. We now consider some fundamental properties of such processes.

Observe first that the most general form of an exponential process (function) is given by

$$p(t) = Ce^{rt}, \tag{2.1}$$

where $C$ and $r$ are constants. If $p$ represents a population, we can think of $p(t)$ as the size of the population at time $t$. Now consider the population at a later time, say $t + \delta$. The ratio of the two populations is

$$\frac{p(t+\delta)}{p(t)} = \frac{Ce^{r(t+\delta)}}{Ce^{rt}}$$

$$= \frac{Ce^{rt}e^{r\delta}}{Ce^{rt}}$$

$$= e^{r\delta}. \qquad (2.2)$$

A similar calculation shows that

$$\frac{p(t+2\delta)}{p(t+\delta)} = e^{r\delta}. \qquad (2.3)$$

The relations (2.2) and (2.3) show that *an exponential process grows by equal ratios in time intervals of equal length.* The equal ratio property makes it possible to ask about the *doubling time* of an exponential process, that is, the time $T_d$ for which

$$\frac{p(t+T_d)}{p(t)} = 2.$$

The next two examples illustrate this idea.

◆ *Example 1*    Figure 2.1 illustrates the doubling time of an exponential process graphically. In the figure, about 1.7 units of time are needed for the height of the curve to

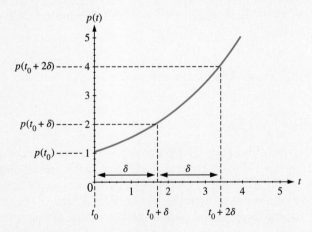

**FIGURE 2.1**    A graphical illustration of equal ratios in exponential growth. The interval $\delta$ corresponds to one doubling time.

double from 1 to 2. An additional 1.7 units of time are needed for the height of the curve to double from 2 to 4, and so on. The doubling time $T_d$ in this example is about 1.7. ∎

✦ *Example 2*   In some developing countries, the population is growing by 3 percent per year. Although future population trends are difficult to predict for a variety of reasons, we can ask what the population of such a country will be in, say, 30 years if the population growth is exponential and the current rate of increase continues unchanged. Under these assumptions, a 3 percent annual population increase implies that

$$p(t) = p_0(1.03)^t = p_0 e^{t \ln 1.03},$$

where $p(t)$ represents the population at time $t$. In other words, $p(t)$ is given by Eq. (2.1) with $r = \ln 1.03$ and $C = p_0$. It follows from Eq. (2.1) that

$$\frac{p(t + \delta)}{p(t)} = \frac{p_0 e^{(t+\delta) \ln 1.03}}{p_0 e^{t \ln 1.03}} = e^{\delta \ln 1.03} = (1.03)^\delta.$$

After 30 years, the population will be $(1.03)^{30} \approx 2.43$ times its original size. The doubling time of the population is the value $T_d$ for which $(1.03)^{T_d} = 2$, that is,

$$T_d = \frac{\ln 2}{\ln 1.03} \approx 23.4 \text{ yr.}$$
∎

The equal ratio property, given by Eq. (2.2), has three key features:

- The ratio holds for any time interval $\delta$. In particular, Eq. (2.2) is *not* a limit as $\delta \to 0$. The equal ratio property holds for long time intervals as well as short ones.
- The ratio (2.2) is independent of the initial value of $p(t)$, assuming of course that it is not zero.
- The doubling time $T_d$ is determined completely by the rate coefficient $r$. To see this, observe that $p(t + T_d)/p(t) = 2$ if and only if $e^{r T_d} = 2$, which implies that

$$r T_d = \ln 2,$$

that is,

$$T_d = \frac{\ln 2}{r}.$$

## The Half-Life of a Radioactive Isotope

The doubling time is frequently of interest in financial applications and population dynamics. In contrast, the decay of a radioactive isotope often is well approximated by an exponential function, for which the halving time—more commonly called the *half-life*—characterizes the rate of decay.

The precise instant at which a radioactive atom decays is essentially random, but the probability that a given atom decays within a fixed interval of time is constant. In many practical applications, a sample of some isotope may contain on the order of $10^{20}$ atoms or more. An accurate, continuous approximation of the rate of decay of such a large ensemble can be stated as follows:

*The rate at which the mass of the isotope decreases is proportional to the current size of the mass.*

We can express this relationship as the differential equation

$$\frac{dm}{dt} = rm,$$

where $m(t)$ is the mass at time $t$ of the isotope and $r$ is a *negative* constant of proportionality. The solution is $m(t) = m_0 e^{rt}$. Thus, the mass of the isotope decays exponentially, and the equal ratio property applies. The half-life of an isotope is the time needed for a given sample to decay to half its original mass.

✦ *Example 3*   Suppose that an 80-microgram ($\mu$g) sample of a radioactive isotope decays to 70 $\mu$g after 10 hours. What is the half-life of the isotope?

*Solution*   To determine the rate constant $r$, we use the equal ratio law for exponential processes,

$$\frac{m(t + \delta)}{m(t)} = e^{r\delta}. \tag{2.4}$$

Two measurements taken at a time $\delta$ apart determine the left-hand side of Eq. (2.4) and hence $r$. The stated information implies that

$$\frac{70}{80} = e^{10r}, \tag{2.5}$$

so $r \approx -0.01335$. The half-life is the time required for the mass to reach one-half of its initial size. Thus, the half-life is the time $\delta$ for which $e^{r\delta} = 1/2$. After taking logarithms, we find that

$$\delta = \frac{\ln \frac{1}{2}}{r} \approx 51.9 \text{ h.} \qquad \blacksquare$$

## Continuous Compounding of Interest

The exponential function arises in a natural way in considering the compounding of interest on a loan or a bank deposit. For concreteness, let us consider the mechanics of credit card debt. Suppose that you purchase something for $100 today and charge it to a credit card that carries an 18 percent annual interest rate.

The rate at which interest accrues is proportional to the size of the balance, and the constant of proportionality is the interest rate, by definition. If $B(t)$ is the balance owed in year $t$, then the growth of the balance is governed by the equation

$$\frac{dB}{dt} = rB, \tag{2.6}$$

where $r$ is the interest rate. (We refer to $r$ as the *nominal* interest rate.) Equation (2.6) is just another form of the exponential growth equation; the solution is

$$B(t) = B_0 e^{rt},$$

where $B_0 = B(0)$ is the initial balance.

Equation (2.6) describes a situation in which the interest is compounded continuously, as we explain below. Notice that a nominal annual interest rate $r$ means that the factor by which the balance has grown after one year is $e^r$. For example, if $r$ is 18 percent, then $e^r = e^{0.18} \approx 1.197217$. In other words, a *nominal* annual interest rate of 18 percent causes the balance to grow by approximately 19.7217 percent after one year. The quantity $e^r$, in this case 19.7217 percent, is called the *annual percentage rate*, or APR.

In practice, interest is compounded discretely, not continuously. (Discrete compounding is called *simple compounding*.) To illustrate the simplest case, suppose you borrow $100 and make no payments. Also suppose that the bank compounds the interest once a year. Let $B_1(t)$ denote the balance after $t$ years. (The subscript "1" denotes one compounding per year.) Then

$$B_1(0) = \$100,$$
$$B_1(1) = 1.18 \times \$100,$$
$$B_1(2) = 1.18 \times 1.18 \times \$100,$$

and in general,

$$B_1(t) = (1.18)^t \times \$100.$$

Under a regime of annual compounding, the annual percentage rate is the same as the nominal annual interest rate. Notice that the balance still grows exponentially, just as in the case of continuous compounding.

A more typical compounding scheme applies the interest charges monthly. After one month, the bank assesses an interest fee equal to $1/12$ of the nominal annual rate (in this case, $0.18/12$, or 1.5 percent). Thus, an initial $100 balance becomes $101.50 after one month. After 12 months, the balance owed is

$$(1.015)^{12} \times \$100 \approx \$119.56.$$

Let $B_{12}(t)$ denote the balance owed after $t$ years under a regime of monthly compounding. Then

$$B_{12}(t) = (1.015)^{12t} \times \$100.$$

Monthly compounding of interest at an 18 percent nominal annual rate leads to an APR of 19.56 percent. In effect, you pay interest on the interest that has already been assessed, and the monthly compounding adds an extra $\$1.56$ to the balance after one year compared to annual compounding.

Banks typically apply *daily compounding* if you use a credit card to borrow cash. The interest that accrues is $1/360$ of the nominal annual rate.[1] After $k$ days, an initial $\$100$ balance becomes

$$\left(1 + \frac{0.18}{360}\right)^{k} \times \$100,$$

so after $t$ years, the balance is

$$B_{360}(t) = \left(1 + \frac{0.18}{360}\right)^{360t} \times \$100.$$

(The subscript "360" reflects the daily compounding.) After one year, the balance is approximately $\$119.72$. The daily compounding increases the total interest charge compared to monthly compounding, but not by much: The APR for daily compounding is 19.72 percent, compared to 19.56 percent for monthly compounding. Notice also that the APR from daily compounding (19.72 percent) is nearly the same as the APR from continuous compounding (19.7217 percent).

Figure 2.2 shows the growth of a $\$100$ debt as a function of time in years. Each curve corresponds to a different compounding strategy: once per year (bottom curve), once per quarter (four times per year), once per month, and once per day. As the compounding becomes more frequent, the annual percentage rate rises. However,

*the annual percentage rate reaches a limit as the time interval between successive compoundings goes to zero.*

---

[1] For purposes of computing interest, banks usually assume that one year consists of twelve 30-day months.

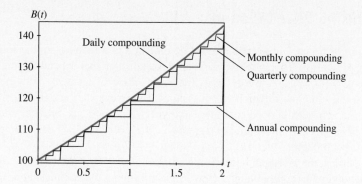

**FIGURE 2.2** The growth of principal under yearly compounding (bottom curve), quarterly compounding (second curve from bottom), monthly compounding (second curve from top), and daily compounding (top curve).

*Continuous compounding* refers to the limit as $n$ goes to infinity. To illustrate this limit, we observe that interest compounded $n$ times per year at a nominal annual rate $r$ leads to a balance $B_n(t)$ after one year that is given by

$$B_n(t) = \left(1 + \frac{r}{n}\right)^{nt} B_0,$$ (2.7)

where $B_0$ is the initial balance. Elementary calculus establishes the result that

$$\lim_{n \to \infty} \left(1 + \frac{r}{n}\right)^n = e^r.$$

Therefore,

$$\lim_{n \to \infty} B_n(t) = \lim_{n \to \infty} \left(1 + \frac{r}{n}\right)^{nt} B_0$$
$$= e^{rt} B_0.$$

Therefore, $B_n(t) \to e^{rt} B_0$ as $n \to \infty$, which is just the solution of Eq. (2.6). This result justifies the previous contention that Eq. (2.6) models the continuous compounding of interest.

Finally, we remark that Eq. (2.7) is simply Euler's method applied to Eq. (2.6) using a step size of $1/n$. The reasoning above proves that Euler's method converges to the true solution of Eq. (2.6) as $n \to \infty$. (See Exercise 17.)

## Remarks on Price Inflation

The growth of prices due to inflation can be modeled in a manner similar to the compounding of interest. That is, $p' = ip$, where $i$ is a constant that depends on the inflation rate. Thus, $p(t) = p_0 e^{it}$, where $p_0$ is the initial price of an item. However, when economists say that the inflation rate is 3 percent per year, they usually mean that the *net* increase in prices is 3 percent after one year. A 3 percent inflation rate implies that the price $p(t)$ of an item after $t$ years is $p(t) = (1.03)^t p_0$, which implies that $i = \ln 1.03 \approx 0.02956$.

In contrast, the principal in a savings account is $p(t) = e^{rt} p_0$ after $t$ years under a regime of continuous compounding. Here, $r$ by definition is the nominal interest rate. For instance, if the nominal interest rate is 3 percent annually, then $p(t) = e^{0.03t} p_0$. Inflation and continuous compounding of interest have the same mathematical description, but the constants of proportionality are interpreted differently.

## Basic Analytical Properties of the Exponential Function

The exponential function grows rapidly when its argument is positive and decays rapidly when its argument is negative. In the course of your scientific training, you may hear statements such as, "The exponential function grows faster than a polynomial function." The purpose of the discussion here is to explore the meaning of this informal statement from a graphical and analytical perspective.

Figure 2.3 shows a plot of the exponential function $e^t$ (solid curve) and a quadratic function, $q(t) = 3t^2 + 10$ (dashed curve). It is obvious from the graph that the exponential function is not always larger. In this example, $q(t) > e^t$ if $t$ is approximately between 0 and 4. On the other hand, if $t$ is larger than 4, then $e^t > q(t)$. In fact, $e^t$ eventually becomes very large in comparison to $q(t)$.

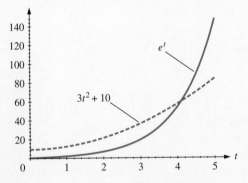

**FIGURE 2.3**    A plot of $3t^2 + 10$ (dashed curve) and $e^t$ (solid curve) for $0 \leq t \leq 5$.

One way to compare the relative size of two functions $f$ and $g$ is to examine their ratio $f(t)/g(t)$ (provided, of course, that $g(t) \neq 0$). Because $\lim_{t \to \infty} e^t$ and $\lim_{t \to \infty} q(t)$ are both infinite, l'Hôpital's rule applies,[2] and we have

$$\lim_{t \to \infty} \frac{e^t}{q(t)} = \lim_{t \to \infty} \frac{(e^t)'}{q'(t)}$$

$$= \lim_{t \to \infty} \frac{e^t}{6t}$$

$$= \lim_{t \to \infty} \frac{(e^t)'}{(6t)'}$$

$$= \lim_{t \to \infty} \frac{e^t}{6}$$

$$= \infty.$$

Thus, the value of $e^t$ becomes extremely large in comparison to the value of $q(t)$ as $t$ becomes large. The computation presented above is an example of an *asymptotic property*. Although it is not true that $e^t > q(t)$ for all values of $t$, $e^t$ becomes much larger than $q(t)$ when $t$ is sufficiently large.

In fact, the same conclusion can be drawn in comparing $e^t$ to any function of the form $q(t) = Ct^n + B$, where $B$ and $C$ are any positive constants and $n$ is any positive integer. For fixed values of $B$, $C$, and $n$, there is always a finite value $t_{max}$ such that $e^t > q(t)$ for $t > t_{max}$. Moreover, $\lim_{t \to \infty} e^t/q(t) = \infty$.

Similar reasoning can be applied to prove the following results.

1. Let $M$, $C$, $n$, and $k$ be positive constants. Then there exists a positive number $t_{max}$ such that $Me^{kt} > Ct^n$ for $t > t_{max}$.

2. Let $M$, $C$, $n$, and $k$ be positive constants. Then

$$\lim_{t \to \infty} \frac{Me^{kt}}{Ct^n} = \infty \tag{2.8}$$

and

$$\lim_{t \to \infty} \frac{Ct^n}{Me^{kt}} = \lim_{t \to \infty} \left(\frac{C}{M}\right) e^{-kt} t^n = 0. \tag{2.9}$$

The limit (2.9) is a precise restatement of the informal observation that "the exponential function goes to zero faster than any polynomial function as $t$ goes to infinity."

---

[2]If $\lim_{t \to \infty} f(t)$ and $\lim_{t \to \infty} g(t)$ are both infinite, and if $f$ and $g$ are continuously differentiable, then l'Hôpital's rule implies that $\lim_{t \to \infty} f(t)/g(t) = \lim_{t \to \infty} f'(t)/g'(t)$.

## Exercises

**1.** In (a)–(d), find a value of $C$ and a value of $k$ that make $Ce^{kt}$ identical to each of the following functions.

(a) $2^{t/2}$

(b) $\left(\sqrt{3}\right)^{3t}$

(c) $5 \times 10^t$

(d) $\left(\frac{7}{8}\right)^{t/4}$

(e) Show that Eq. (2.5) implies that $r = \frac{1}{10} \ln \frac{70}{80}$.

**2.** Let $f(t) = \left(\frac{7}{8}\right)^{t/9}$, $g(t) = \left(\frac{7}{8}\right)^{t/10}$, and $h(t) = \left(\frac{7}{8}\right)^{t/11}$.

(a) What are the domain and range of each function?

(b) Find the limit of each function as $t \to -\infty$; as $t \to +\infty$.

(c) Find the largest interval for which the value of all three functions is less than $1/2$.

(d) Can $f$, $g$, and $h$ be represented in the form $Ce^{kt}$ for appropriate values of $C$ and $k$? If so, find $C$ and $k$ for each of the three functions. (Derive exact expressions, not numerical approximations.) If no such expressions can be derived, then explain why not.

**3.** What is the doubling time of a credit card balance assuming an 18 percent annual interest rate and

(a) monthly compounding?

(b) daily compounding?

(c) continuous compounding?

**4.** Price inflation is not constant. Suppose that the inflation rate averages 3 percent per year in year 1; 4 percent in year 2; and 5 percent in year 3. If a basket of goods and services costs $100 initially, what will be the cost at the end of 3 years?

**5.** Assume that an investment carries an interest rate $r$, compounded continuously.

(a) Determine the doubling time of the investment as a function of $r$ and plot it for $0 \le r \le 0.25$.

(b) If you double the interest rate, do you halve the doubling time? Explain.

**6.** This exercise considers some basic problems in radioactive decay.

(a) One-quarter of a radioactive sample decays after one day. What is its half-life?

(b) Is the time required for a radioactive sample to decay to $3/4$ of its original mass equal to one-half of its half-life? Explain.

**7.** Following are the half-lives of some radioactive elements.

| | |
|---|---|
| Plutonium 239 | $2.411 \times 10^4$ yr |
| Americium 243 | $7.37 \times 10^3$ yr |
| Oxygen 14 | $70.6$ s |

(a) For each element, determine how much time is required for a 1000-microgram ($\mu$g) mass to decay by 1 $\mu$g.

(b) Determine how much time is required for the mass to decay to 125 $\mu$g.

**8.** This exercise asks you to investigate the asymptotic properties of the exponential function in more detail. For each of the functions $f(t)$ in the following list,

- Find an approximate interval for which $f(t) > e^t$.

- Show that $e^t > f(t)$ once $t$ is large enough.

(a) $f(t) = 100t$   (b) $f(t) = 1000t$

(c) $f(t) = 10^4t$   (d) $f(t) = t^3$

(e) $f(t) = t^4$   (f) $f(t) = t^5$

**9.** The asymptotic behavior of the exponential function often can be investigated by using l'Hôpital's rule.

(a) Use l'Hôpital's rule to find the limit

$$\lim_{t \to \infty} \frac{e^t}{at},$$

where $a$ is any positive constant. What does your result imply about the graph of $e^t$ compared to the graph of $at$ for large $t$?

(b) Use l'Hôpital's rule to find the limit

$$\lim_{t \to \infty} \frac{e^t}{at^2},$$

where $a$ is any positive constant.

(c) Use l'Hôpital's rule to find the limit

$$\lim_{t \to \infty} \frac{e^t}{at^n},$$

where $a$ is any positive constant and $n$ is any positive integer. (Use induction.) What does your result imply about the graph of $e^t$ compared to that of $at^n$ for large $t$?

(d) Let $p_n(t) = a_n t^n + a_{n-1} t^{n-1} + \cdots + a_1 t + a_0$. Find

$$\lim_{t \to \infty} \frac{e^t}{p_n(t)}.$$

What does your result imply about the graph of $e^t$ compared to that of any polynomial of the form $p_n(t)$?

(e) Discuss how your results in (a)–(d) change if you consider $e^{kt}$ instead, where $k$ is any positive constant.

**10.** In the same vein, the exponential function is often described informally as "going to zero faster than any polynomial."

(a) Consider the functions $ate^{-t}$, $at^2 e^{-t}$, and $at^3 e^{-t}$. Describe the behavior of each function as $t$ becomes large. (Suggestion: Plot these functions for various values of $a$.)

(b) Using calculus, find the maximum absolute value of each of the functions in (a) for $t \geq 0$.

(c) Use l'Hôpital's rule to confirm your observations in part (a).

(d) State and prove a theorem about the function $at^n e^{-t}$, where $n$ is a positive integer and $a$ is any constant.

(e) Discuss how your results in (a)–(d) change if you replace $e^{-t}$ with $e^{-kt}$, where $k$ is any positive constant.

**11.** Let $f(t) = Ke^{-t} \sin \omega t$, where $K$ and $\omega$ are any two positive constants.

(a) Explain why l'Hôpital's rule does not apply to $\lim_{t \to \infty} f(t)$.

(b) Even so, you can find the limit in (a). What is it?

**12.** An editorial in the *Washington Post* on highway accidents claimed that "the physical forces of impact double with every 10 miles per hour over 50."[3]

(a) Does the editorial imply that the severity of impact grows exponentially with speed? Why or why not?

(b) Although several factors determine the severity of a collision, the kinetic energy of the vehicle is the most important. (The kinetic energy is proportional to the square of the velocity of an object.) Is the newspaper account consistent with the physics? If not, what is a more accurate statement about the severity of impact as a function of speed?

**13.** Suppose a house cost $100,000 ten years ago and costs $150,000 today. Assuming that the price of the house reflects the real estate market as a whole and that prices continue to rise at the same rate, how much will the house cost ten years from now?

---

[3]"Lethal-speed trucks in Virginia," *Washington Post*, April 2, 1994, p. A16.

(a) Poll some of your friends, family members, and co-workers (ask at least five people). Tally their answers. What fraction of the people polled says $200,000?

(b) What is a better estimate of the sales price ten years from now? Explain.

**14.** Suppose that the value of an investment has increased from $36,000 five years ago to $48,000 today.

(a) Assume that the growth rate is exponential. Express the value of the investment in the form $p(t) = Cb^t$ for appropriate constants $C$ and $b$. Explain the units of time that you use.

(b) Assuming the same rate of growth, what will be the value of the investment five years from now?

(c) Suppose the value of the investment grows linearly instead. Determine a functional form for $p(t)$ and find the value of the investment five years from now.

**15.** An important and politically divisive issue concerns the disposal of radioactive waste from nuclear weapons production. A crucial ingredient in fission bombs is plutonium 239, a highly toxic and relatively long-lived radioactive isotope. Suppose that the remaining plutonium is relatively harmless after 99.9 percent of it has decayed. (This assumption is made to keep the problem simple. Most nuclear waste is a combination of different isotopes, and considerable controversy surrounds the question of what constitutes a "safe" level of radioactivity.)

(a) How long must the plutonium be stored, given that the half-life of plutonium 239 is 24,360 years?

(b) Is your answer independent of the initial mass of plutonium to be stored? Explain.

**16.** Simple mathematical ideas are often the ones of greatest practical importance. The method of

*carbon-14 dating,* for which the American chemist Willard Libby won the Nobel Prize in 1960, is an example. The method is based on measurements of the relative abundance of two carbon isotopes found in all living things. Most naturally occurring carbon is carbon 12, a stable isotope. However, the bombardment of the earth by cosmic rays continuously produces small amounts of carbon 14, a radioactive isotope with a half-life of approximately 5570 years. Carbon 14 is ingested by plants (as radioactive carbon dioxide) and works its way throughout the food chain. The ratio of carbon 14 to carbon 12 is roughly the same in all living things. After they die, the carbon 14 decays and the ratio decreases. Libby showed that the relative ratio of the isotopes in the organic remains can be used to estimate the time of death.

(a) Determine the decay constant assuming that the radioactive decay of carbon 14 can be modeled by Eq. (2.4).

(b) If the ratio of carbon 14 to carbon 12 in some archeological remains is 10 percent of its original value, how old are the remains?

(c) What is the maximum age of remains for which carbon 14 dating is practical, assuming that the ratio of the isotopes can be measured accurately down to $10^{-6}$ of its original value?

**17.** Let $B(t)$ denote the balance owed on a credit card as a function of $t$. Suppose $t$ is measured in years. If interest is compounded continuously, then $B' = rB$, where $r$ is the annual interest rate.

(a) Let $B_0$ be the initial balance. If interest is compounded four times per year, find an expression for the balance after one year.

(b) If interest is compounded once per month, find an expression for the balance after one year.

(c) If interest is compounded $n$ times per year, find an expression for the balance after one year.

(d) Show that your answer in (c) is equivalent to the approximation to $B(1)$ obtained by taking *n* steps of Euler's method with a step size of $1/n$.

**18.** This exercise considers some basic properties of inflation.

(a) Suppose that prices rise at an average rate of 5 percent per year. How long does it take for prices to double? (You can think of this time interval as the half-life of a dollar.)

(b) Suppose that prices rise by *i* percent per year (*i* is the annual inflation rate). Determine the doubling time of prices as a function of *i* and draw a graph. Is the curve increasing or decreasing? Is the second derivative positive or negative?

(c) If the inflation rate is twice as great, is the doubling time of prices cut in half?

(d) The costs of the war in Bosnia and international economic sanctions led to hyperinflation in the former Yugoslavia. The *New York Times* reported that inflation in November 1993 was running at an annual rate of 300 million percent. At that rate of inflation, what is the daily percentage rise in prices? (Remember that a doubling of prices corresponds to a 100 percent increase, not a 200 percent increase.) What is the doubling time of prices?

**19.** This exercise illustrates the basic computations that you make when planning for retirement.

(a) Suppose you invest $1000 today at an annual interest rate of 10 percent. What will be the value of your investment 30 years from now if you make no deposits or withdrawals? Assume continuous compounding.

(b) If prices rise at a net annual rate of 5 percent, then goods and services worth $1000 today will cost $1050 one year from now if the inflation rate stays constant. If prices continue to rise at a net annual rate of 5 percent, how much will today's $1000 worth of goods and services cost in 30 years?

(c) If the value of your investment is $v(t)$ and the general price level is given by $p(t)$, then the ratio $v(t)/p(t)$ gives the value of your investment relative to inflation. For instance, if you invest $1000 today, you can take $v(0) = 1000$ and $p(0) = 1000$. If the value of your investment after one year is $1050, and if the net increase in prices also is 5 percent, then $v(1)/p(1) = 1050/1050 = 1$. Hence, the value of your inflation-adjusted investment is unchanged.

Obviously, to build wealth, your rate of return must exceed the rate of inflation. How much does your investment in (a) grow relative to inflation after 30 years, assuming the rise in prices in (b)?

(d) Suppose that inflation averages 6 percent per year instead of 5 percent. What will be the value of the investment relative to inflation 30 years from now?

(e) Repeat your calculation in (d) assuming that inflation averages 3 percent per year. Is the inflation-adjusted value of your investment double that of the 6 percent case?

(f) Suppose the annual inflation rate is *i* and the interest rate on an investment is *r*. Derive an expression for the value of the investment 30 years from now relative to inflation.

**20.** Extremely large numbers often can be handled on pocket calculators with a careful application of the laws of exponents, as this exercise illustrates.

(a) A typical pocket calculator can compute $e^x$ only if *x* is not too large. (For instance, if the calculator cannot represent numbers larger than $10^{100}$, then *x* cannot be larger than about 230.) Show how such a calculator can be used to compute $e^{10,000}$ even though the value of the expression is too large to be represented on the calculator. (Suggestion: $e^{10,000} = (e^{100})^{100}$.)

(b) Show how to compute $e^{-10,000}$, even though the value of the expression is too small to be represented on the calculator.

(c) Generalize your approach to calculate exp $10^k$, where $k$ is any integer. Cast your result as a computer program or other explicit algorithm.

(d) Any real number $x$ can be written as

$$x = m \times 10^k$$

for an appropriate integer $k$ (called the *exponent*) and real number $m$ (called the *mantissa*). For any nonzero $x$, the values of $m$ and $k$ are unique if we require $1 \leq |m| < 10$. Show how to apply your results in (c) to compute

$$e^x = e^{m \times 10^k}$$

for any nonzero number $x$, regardless of size, using a pocket calculator.

## 2.2   APPLICATIONS OF EXPONENTIAL GROWTH

In this section, we introduce you to some applications of exponential growth and decay processes. Examples include the following:

- the growth of an annuity,
- Newton's law of cooling,
- simple models of air resistance, and
- simple population models.

One goal of this section is to discuss the mechanics of setting up and solving mathematical models of simple processes. An essential step in the solution procedure is to construct a differential equation that is an appropriate model. The next subsection discusses the issue of mathematical modeling in a general way. We conclude this section with some applications drawn from the preceding list.

### The Modeling Process

Most attempts to model the behavior of a given system mathematically, such as the spread of a disease, the growth of a population, or the flow of oil in a pipeline, involve some approximation. For instance, a model for fluid flow in a pipe might assume that the pipe has a perfectly circular cross section or is infinitely long.

A comprehensive discussion of the methods and limitations of mathematical models is beyond the scope of this text. Nevertheless, there are some basic steps in the construction of mathematical models that are applicable both to simple cases (our main interest in this text) and to more complicated problems in other disciplines. For the purpose of this discussion, we consider four basic steps.

### Step 1.  Make Appropriate Simplifications

Usually, it is necessary to idealize the problem in some way in order to get a tractable mathematical model. In some cases, a laboratory experiment can determine whether the idealizations lead to an accurate approximation of the actual process. (The interaction between laboratory experiment, mathematical theory, and numerical simulation is especially important in fields like fluid dynamics.) In other disciplines (such as the study of global climate change), laboratory experiments are not possible, and other means must be employed to determine whether a given model is realistic.

### Step 2.  Determine the Differential Equation and the Initial Conditions

Some useful rules of thumb are stated below.

### Step 3.  Solve the Differential Equation

In some cases (such as most of the problems considered in this chapter), an explicit analytical solution exists. More often, a simple formula for the solution cannot be found, and numerical methods must be used to estimate the solution. (Chapter 9 discusses numerical methods in more detail.)

### Step 4.  Compare the Predictions of the Model with Observations of the Phenomenon Being Modeled

Prediction is a basic goal of science in general and mathematical modeling in particular. The solutions of differential equations allow scientists to make inferences about the behavior of the process under study. In ideal cases, the predictions of the model can be compared with the results of laboratory experiments. If no experiments are possible, then the predictions can be compared with other observations or data of the underlying process.

Sometimes the predictions do not agree well with observations of the process being modeled. Even so, much useful information can be gained. For example, terms that initially were disregarded in formulating the model (Step 1 above) may need to be included. In other cases, the model may be accurate but uncertainties in measuring the initial conditions or the parameters of the model may lead to large prediction errors. Section 2.4 discusses the latter issues in more detail.

Mathematical modeling involves some art as well as science. If a model yields poor predictions, then one may need to return to Step 1 to derive an improved model. The following rules of thumb are intended as a guide to help you set up differential equation models.

**Check Simple Cases.**    For instance, if $T(t)$ represents the temperature of a kettle of water at time $t$, then $dT/dt$ must be negative if the water initially is hot and the kettle is placed on a table at room temperature. Similarly, $dT/dt > 0$ if the water initially is colder than room temperature.

**Determine Whether the Problem Has an Equilibrium Solution.**    Is there an initial condition for which the system does not change? For instance, the temperature of a kettle of water does not change if the water initially is at room temperature and the kettle is neither heated nor cooled. Thus, $dT/dt = 0$, and room temperature represents an equilibrium solution.

**Check the Units on Each Quantity.**    Each term must have consistent units. For instance, each term in a differential equation that models the rate of change of the value of an investment over time must have units such as dollars per year.

**Determine the Relevant Constants of Proportionality.**    If the rate of growth of a quantity is proportional to some other quantity, then a constant of proportionality is involved. You must then decide how the constant can be determined from the information presented in the problem.

**Sketch a Direction Field.**    A direction field can help you see inconsistencies in a given model. For instance, if your intuition suggests that the process has an equilibrium state, then the slope lines in the direction field cannot all point upward.

**Use Your Solutions to Make Predictions.**    Predictions are a useful way to assess whether a model is reasonable. The predictions should agree at least qualitatively with your intuition. If they don't, then you need to check all of your work. If you cannot find obvious algebraic or computational errors, then you need to inspect your model and its solution.

 The remainder of this section discusses some simple applications of differential equations. In each case, we show examples of how to apply the above analysis.

## The Time Value of Money

An annuity is a kind of savings account into which money is invested at regular intervals. The account earns a rate of return that is proportional to the size of the principal. Suppose you have an annuity paying an interest rate $r$ to which you add $A$ dollars per year. Let us consider the three steps that are needed to construct a model of this situation.

### Step 1. Find Appropriate Simplifications

The future value of a typical annuity is impossible to know precisely, because the value depends on unpredictable factors like interest rates. One appropriate simplification for the purpose of *estimating* the future value is to assume that the interest rate maintains a certain average value over the years. Similarly, although the rate at which you invest money is likely to vary over time, you might assume that, on the average, $A$ dollars per year are added to the account. Again, since we are interested in estimating the future value, we can assume that the interest is compounded continuously and that payments are made continuously.[4] We also assume that no withdrawals are made.

### Step 2. Determine the Differential Equation

A differential equation is used to model the annuity because only the *rate* of growth of the principal is known. Let $p(t)$ be the principal at time $t$. Then $dp/dt$ must always be positive, because money is always added to the account, interest is credited to the balance, and no withdrawals are made. For this reason, the differential equation can have no equilibrium points.

We have

$$\frac{dp}{dt} = \text{rate of return from interest} + \text{rate of contributions}.$$

Each term must have units of dollars per year. Because the annuity earns interest at a rate that is proportional to the amount of money that currently is in the account,

$$\text{rate of return from interest} = rp(t).$$

Here, $r$ is the constant of proportionality, which by definition is the interest rate and is positive. The units on $r$ are inverse time (e.g., 1/yr). The value of $r$ is determined by the assumptions that you make about the average value of the interest rate over the life of the annuity.

You add $A$ dollars per year to the annuity, so

$$\text{rate of contributions} = A.$$

The units of $A$ are dollars per unit time, for example, dollars per year. Thus, the growth rate of the principal is given by the differential equation

$$\frac{dp}{dt} = rp + A. \tag{2.10}$$

---

[4]In other words, the model that we derive approximates the situation in which you contribute small amounts of money at frequent intervals for a total of $A$ dollars per year rather than the case in which you make a single deposit of $A$ dollars once each year.

The initial condition is $p(0) = p_0$ and represents the initial amount of money that you invest in the annuity. (This initial investment is money added to the account above and beyond to the annual contributions.)

## Step 3. Solve the Differential Equation

Equation (2.10) is separable. Let $u = rp + A$. Then $du = r\,dp$, so

$$\frac{1}{r} \int \frac{du}{u} = \int dt,$$

which leads to

$$u(t) = Ce^{rt},$$

and back substitution yields

$$rp(t) + A = Ce^{rt}.$$

The initial condition implies that $C = rp_0 + A$, so

$$p(t) = \left( p_0 + \frac{A}{r} \right) e^{rt} - \frac{A}{r}. \tag{2.11}$$

## Step 4. Make Predictions

The solution (2.11) contains three constants: the interest rate $r$, the rate of contributions $A$, and the initial principal $p_0 = p(0)$. The solution lets you predict the effects of different investment scenarios, as the following examples show.

✦ *Example 1*  Suppose that you open a retirement annuity next January 1 and that you invest \$2000 per year. Let us assume that your contributions and the interest are added continuously, and suppose that the annual interest rate is 10 percent. Consider the outcomes of the following two possibilities:

1. You open the account on January 1 with an initial investment of \$2000.

2. You open the account on January 1 with no initial investment.

How much money will you have in the account after five years under these two possibilities?

*Solution*  The natural way to count time in this example is to let $t = 0$ correspond to next January 1; then $t = 5$ represents five years from next January 1. The difference between possibilities 1 and 2 is the difference between $p_0 = \$2000$ and $p_0 = \$0$, respectively. In both cases,

$$r = \frac{0.1}{\text{yr}} \quad \text{and} \quad A = \frac{\$2000}{\text{yr}},$$

so each term in Eq. (2.11) has units of dollars.

In scenario 1, the principal after five years is

$$p(5) = \left( \$2000 + \frac{\$2000}{0.10} \right) e^{0.5} - \frac{\$2000}{0.10}$$

$$= \$16{,}271.87.$$

In scenario 2, the principal after five years is

$$p(5) = \left( \$0 + \frac{\$2000}{0.10} \right) e^{0.5} - \frac{\$2000}{0.10}$$

$$= \$12{,}974.43.$$

If you subtract the initial $2000 advantage, scenario 1 still leaves you with $1297.44 more than scenario 2. The extra money results from the compounding of interest on the initial investment. Figure 2.4(a) shows a plot of $p(t)$ for the two cases. Figure 2.4(b) graphs the difference between the amount of money as a function of time in scenarios 1 and 2.

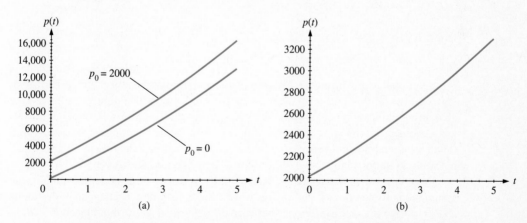

(a)                                                                    (b)

FIGURE 2.4   (a) A plot of the principal as a function of time under the two scenarios in Example 1. (b) The difference in the amount of money between the two scenarios as a function of time.

◆ *Example 2*    Suppose that you invest $150 per month in an annuity that pays an annual interest rate of 9 percent. How much interest will you earn over a 10-year period, assuming that interest and investments are paid continuously and that the account is opened initially with no money in it?

*Solution*    The relevant differential equation is

$$p' = rp + A,$$

where $A = \$150/\text{month}$. The interest rate is

$$r = \frac{0.09}{\text{yr}} \times \frac{1 \text{ yr}}{12 \text{ months}} = \frac{0.0075}{\text{month}},$$

which implies that

$$p' = 0.0075p + 150.$$

Notice that each term in the differential equation has units of dollars per month.

The initial condition $p(0) = \$0$ yields

$$p(t) = \$20{,}000\left(e^{0.0075t} - 1\right),$$

where $t$ measures time in months. After 10 years, the principal is $p(120) = \$29{,}192.06$. The total amount of money invested is $\$150 \times 120 = \$18{,}000$. The difference between the two figures represents the interest earned in the annuity, in this case $\$11{,}192.06$.    ■

## Newton's Law of Cooling

Newton's law of cooling states that the rate at which the temperature of an object changes is proportional to the difference between the temperature of the object and its surroundings. To model this problem, we must make some appropriate simplifications. One obvious simplification is to assume that the temperature is uniform throughout the object. Another simplification is that the temperature of the surroundings remains constant. (Such an assumption may be accurate for a cup of hot coffee in a large room, but it may not be accurate for a large bowl of hot soup placed into a small freezer.)

Next we need to determine the relevant differential equation under the simplifying assumptions listed above. The rate at which the temperature $T$ changes with time is $dT/dt$. The temperature difference between the object and its surroundings is $T - R$, so

$$\frac{dT}{dt} = k(T - R), \tag{2.12}$$

where $k$ is the constant of proportionality and $R$ is the constant temperature of the surrounding medium. The units of the right-hand side must be degrees per unit time, for example, degrees per minute; thus, $k$ has units of inverse time. The initial condition is $T(0) = T_0$, that is, the initial temperature of the object.

Let us check whether Eq. (2.12) is reasonable by examining some simple cases. If a hot object is placed in a cold room, then it cools off, so $dT/dt < 0$. Since $T - R > 0$, we must have $k < 0$. If a cold object is placed in a hot room, then it warms up, so $dT/dt > 0$, and $T - R < 0$, which again implies that $k < 0$. Hence, the constant of proportionality in Eq. (2.12) must be negative.

There must be an equilibrium solution, because if the temperature of the object is the same as the temperature of the surroundings, then the object neither warms nor cools. In this situation, $T - R = 0$, which implies that $dT/dt = 0$. So Eq. (2.12) has the equilibrium solution $T = R$, as expected.

Figure 2.5 shows a schematic representation of the direction field for Eq. (2.12). The equilibrium solution must be stable, because the temperature of the object eventually must approach the temperature of its surroundings. Notice also that Newton's law of cooling is an autonomous process; the rate of temperature change depends only on the current temperature difference.

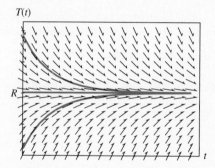

**FIGURE 2.5**   A plausible direction field describing Newton's law of cooling. The line $T = R$ represents room temperature.

Equation (2.12) is separable, and $T(t)$ can be found by a straightforward integration. Let

$$u = T - R. \tag{2.13}$$

Then

$$\int \frac{du}{u} = \int k\,dt,$$

which leads to the general solution, $u(t) = Ce^{kt}$. The substitution in Eq. (2.13) implies that $n(0) = T(0) - R = T_0 - R$, so

$$u(t) = (T_0 - R)e^{kt}.$$

This result implies that *the temperature difference between the object and its surroundings decays exponentially.* We use Eq. (2.13) to solve for $T$ and obtain

$$T(t) = (T_0 - R)e^{kt} + R. \tag{2.14}$$

The only quantity that we have not determined is the constant of proportionality $k$. It must be found by an independent measurement, as the next example shows.

◆ *Example 3*    Suppose that a boiling kettle of water, initially at 100°C, is placed in a room whose temperature is 20°C. When will the water reach 40°C?

*Solution*    This question cannot be answered unless we measure the temperature of the water some time after we take the kettle off the stove. Suppose that 10 minutes later, the temperature is 90 degrees. Then $T(10) = 90$, and we can solve for $k$ by substitution:

$$90 = (100 - 20)e^{10k} + 20,$$

which implies that $k = \frac{1}{10}\ln\frac{7}{8} \approx -0.01335$. Notice that $k < 0$, as required by our previous considerations. After substituting into Eq. (2.14) and simplifying, we find

$$T(t) \approx 80e^{-0.01335t} + 20. \tag{2.15}$$

A plot of the solution, Eq. (2.15), is shown in Fig. 2.6 with the equilibrium solution. The water reaches 40°C when $40 = 80e^{-0.01335t} + 20$, which implies $t \approx 104$. Hence, the water reaches 40 degrees Celsius after about 104 minutes.

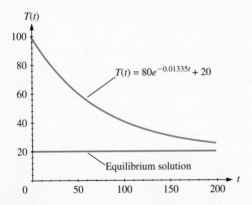

**FIGURE 2.6**    The solution curve for Example 3.

## Air Resistance

Acceleration is the rate of change of velocity. Newtonian mechanics says that the acceleration $a$ is proportional to the applied force $F$; that is, $F = ma$, where the mass is the proportionality constant. Acceleration, like velocity, is a vector-valued quantity that has a direction as well as a magnitude. We are interested here only in the vertical displacement of an object as it rises or falls. Therefore, we can treat the velocity and acceleration as scalars. They are signed quantities, where the sign indicates the direction (up or down).

Let the ground be position 0, and let $x > 0$ denote the distance of the object above the ground. In this coordinate system, a rising object has positive velocity and a falling object has negative velocity. Similarly, a force that is directed toward the earth implies a negative acceleration, and a force that is directed away from the earth implies a positive acceleration.

Suppose a ball is initially at a height $x(0) = x_0$ above the ground when it is dropped. If we ignore air resistance, then the only force acting on the ball is gravity. Gravity pulls the ball toward the earth, so according to our convention, we must have $F_{\text{gravity}} < 0$. Therefore,

$$F_{\text{gravity}} = m\frac{dv}{dt} = -mg,$$

which implies that

$$\frac{dv}{dt} = -g. \tag{2.16}$$

Equation (2.16) can be integrated directly to give

$$v(t) = -gt + v(0), \tag{2.17}$$

where $v(0)$ is the initial velocity. The solution (2.17) can be integrated again to give the position $x(t)$. Thus, it is easy to answer questions like how fast the ball is falling at a particular time (given its initial velocity) or when it hits the ground (given its initial position and velocity).

This simple calculus example ignores air resistance. What happens if we include air resistance? Air resistance is a relatively complicated phenomenon to model in practice. Many factors potentially are important, including air pressure, temperature, and the shape of the object. For the moment we consider the simplest case, in which we suppose that the object is a ball that encounters air resistance proportional to its velocity. As above, we consider only the vertical displacement of the ball.

Air resistance always opposes the motion. When the ball falls, the flow of air around the bottom gives it some lift; that is, the force of air resistance acts upward, as illustrated

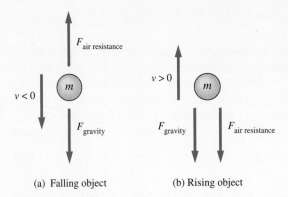

(a) Falling object          (b) Rising object

**FIGURE 2.7**    The forces of gravity and air resistance on (a) a falling ball; (b) a rising ball.

in Fig. 2.7(a). When the ball rises, the flow of air over the top of the object adds a drag; hence, air resistance acts in the same direction as gravity, as illustrated in Fig. 2.7(b).

In our coordinate frame, the force of air resistance must be positive when the ball falls and negative when the ball rises. We have assumed that the force of air resistance is proportional to the velocity. Suppose we take the constant of proportionality, $k$, to be positive. Then

$$F_{\text{air resistance}} = -kv.$$

(Why is $F_{\text{air resistance}} \neq +kv$ if we take $k$ as a positive number?)

The net force on the ball is the sum of the forces of gravity and air resistance, that is,

$$F_{\text{net}} = F_{\text{gravity}} + F_{\text{air resistance}}.$$

Our assumptions imply that the differential equation governing the velocity $v(t)$ of the ball is

$$mv' = -mg - kv. \tag{2.18}$$

Each term in Eq. (2.18) has units of force. In the meter-kilogram-second (MKS) system of measurement, the appropriate units are *newtons*; $1 \text{ N} = 1 \text{ kgm/s}^2$. (What are the units of $k$?)

Notice that the force of gravity in Eq. (2.18) is constant; this assumption is accurate if the object is not too far from the ground. There are four basic possibilities for the motion:

1. *The ball rises.* In this case, both gravity and air resistance are forces that point toward the ground. The net force is at least as great as that of gravity, so a rising

ball in the presence of air resistance eventually stops and begins to fall. See Fig. 2.7(b).

2. *The ball falls slowly.* In this case, $|F_{\text{gravity}}| > |F_{\text{air resistance}}|$, so the speed of the ball increases (it falls more rapidly). See Fig. 2.7(a).

3. *Air resistance balances gravity.* If the ball falls fast enough, then $|F_{\text{gravity}}| = |F_{\text{air resistance}}|$, so the net force on the ball is 0. In this case, the acceleration is 0, and the ball falls with constant velocity, called *terminal velocity*. Terminal velocity corresponds to an equilibrium solution of Eq. (2.18).

4. *The ball falls very fast.* If the ball initially is thrown downward at a sufficiently high speed, then $|F_{\text{air resistance}}| > |F_{\text{gravity}}|$. At a sufficiently high speed, the net force on the ball pushes up. The speed of the ball decreases until the ball approaches terminal velocity.

These considerations imply that the terminal velocity is a stable equilibrium point. Notice that

$$v_{\text{terminal}} = -\frac{mg}{k}.$$

Figure 2.8 shows what the direction field associated with Eq. (2.18) must look like.

Equation (2.18) is separable, and its solution is

$$v(t) = -\frac{mg}{k} + \left(v_0 + \frac{mg}{k}\right)e^{-kt/m}. \tag{2.19}$$

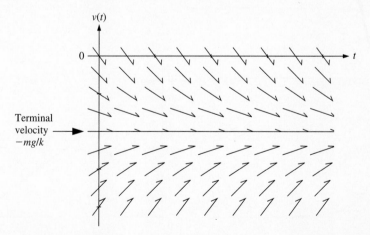

**FIGURE 2.8** A qualitative sketch of the direction field for a falling object when air resistance is proportional to velocity.

As $k$ and $m$ are positive constants, $\lim_{t\to\infty} v(t) = -mg/k$. Thus, $v(t)$ eventually approaches the terminal velocity, $-mg/k$, regardless of the initial velocity $v_0$. (This analysis confirms that the equilibrium point is stable.) Notice that the solution curves must "flatten out" in Fig. 2.8 as they approach the terminal velocity.

✦ *Example 4*    A beach ball of mass $1/10$ kg is dropped from a tall platform. Suppose that the force due to air resistance is proportional to the velocity of the ball with a proportionality constant $k = 1/2$. Find the velocity of the ball as a function of time.

*Solution*    Equation (2.18) yields

$$\tfrac{1}{10}v' = -\tfrac{1}{10}g - \tfrac{1}{2}v,$$

which is separable. The general solution is

$$v(t) = -\tfrac{1}{5}g + e^{-5t}\left(v_0 + \tfrac{1}{5}g\right), \tag{2.20}$$

where $v_0$ is the initial velocity. If the ball is dropped, then $v_0 = 0$.

Notice that the height of the platform is irrelevant in deriving $v(t)$. Of course, Eq. (2.20) is valid only until the ball hits the ground. The height of the platform affects the length of time for which the solution is valid, but it does not affect the formula for the solution $v(t)$.

Figure 2.9(a) shows a plot of the velocity Eq. (2.20), where $v_0 = 0$. Notice that the beach ball approaches terminal velocity after about 1 s. (As usual, we take $g = 9.8$ m/s$^2$.)

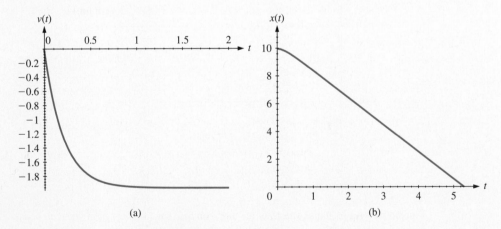

(a)                                              (b)

**FIGURE 2.9**    A plot of (a) Eq. (2.20); (b) Eq. (2.21).

We can determine the position of the ball at any instant if we know the height of the platform. The position $x(t)$ of the ball is obtained by solving the equation

$$\frac{dx}{dt} = v(t) = \tfrac{1}{5}g\left(-1 + e^{-5t}\right),$$

which is just Eq. (2.20) with $v_0 = 0$. If we suppose that the platform is 10 m above the ground, then $x_0 = 10$. The solution of this initial value problem is

$$x(t) = -\tfrac{1}{5}gt + \tfrac{1}{25}g\left(1 - e^{-5t}\right) + 10. \tag{2.21}$$

The ground corresponds to position 0. However, it is not possible to solve the relation $x(t) = 0$ explicitly for $t$. We can estimate the time of impact numerically, using bisection or Newton's method, to determine that the ball hits the ground after about 5.3 s. Figure 2.9(b) shows the position of the ball as a function of $t$.    ∎

## Chemical Mixing

The dilution of one chemical species in another often can be modeled as an exponential process. Suppose, for instance, that a natural gas leak is discovered in a building. The gas line is shut off, and fire department personnel set up an emergency ventilation system to pump fresh air into the building and flush out the natural gas. An important question is how long the ventilation system must be operated to reduce the concentration of the natural gas to an acceptably safe level. For a given rate at which fresh air is pumped in, the volume of air in the building, and the initial concentration of natural gas, one might want to know whether the predicted ventilation time is hours or days.

Two simplifying assumptions are needed to reduce the problem to one that can be modeled by a single ordinary differential equation. First, we assume that the chemical species do not react with each other. Second, we assume that the two species are well mixed, that is, that all mixing occurs completely and instantaneously. In the present example, of course, the first assumption implies that the natural gas does not react with the air (no explosion occurs). The second assumption is less realistic, but it still may provide an acceptably accurate estimate of the required ventilation time if the building has large open spaces. (In practice, one must worry about pockets of gas that may accumulate inside walls, machinery, and other enclosed spaces.)

The quantity of interest in the mathematical model is the concentration $c(t)$ of natural gas in the air in the building. Obviously, $c' \le 0$ because there is no source of natural gas after the gas line is shut off. The only equilibrium solution is $c = 0$. (Why?)

The concentration of natural gas is determined by the volume $v(t)$ of natural gas in the building as well as the volume $V$ enclosed by the building itself. Because we assume that the natural gas is uniformly mixed with the air, $c(t) = v(t)/V$. The total volume of air in the building remains constant; thus, the volume of air pumped into the building is the same as the volume of the air/natural gas mixture that is blown out. Therefore, the rate at which natural gas is vented from the building is

$$\frac{dv}{dt} = -c(t) \times \text{rate that mixture is vented}$$

$$= -\left(\frac{v(t)}{V}\right)B,$$                                    (2.22)

where $B$ is the rate at which the blower pumps air into the building, assumed to be constant. (Explain the rationale for the minus sign.)

✦ *Example 5*    A natural gas leak has filled a building enclosing 50,000 m³ with a 1 percent mixture of natural gas and air. The gas line is shut off, and an emergency ventilation system pumps in fresh air at the rate of 1000 m³/min. How long must the ventilation system be run to reduce the concentration of natural gas to 0.01 percent?

*Solution*    If we assume uniform, instantaneous mixing of air and gas throughout the building, then $c(t) = v(t)/50{,}000$, where $v$ is the volume and $c$ is the concentration of natural gas. The rate at which gas leaves the building is $c(t)B$, where $B = 1000$ m³/min. Thus,

$$\frac{dv}{dt} = -\left(\frac{1000}{50{,}000}\right)v,$$

which implies that the volume $v(t)$ of natural gas in the building is

$$v(t) = 500e^{-t/50}.$$

(Explain the factor of 500 in the solution.)
    We want to know the time $t_1$ at which the concentration of natural gas in the building reaches 0.01 percent. This condition implies that $v(t_1) = 5$ m³ or

$$t_1 = -50 \ln \tfrac{1}{100} \approx 230 \text{ min}.$$

Thus, the ventilation system must operate for about 4 hours.                                    ■

## *Exercises*

**1.** This exercise considers Example 1 in more detail.

(a) Suppose $2000 is added to the annuity each year and the interest rate is 10 percent. Explain why the differential equation

$$\frac{dp}{dt} = \frac{p}{10} + 2000t$$

is not a correct model for the rate of growth of the principal in the annuity. Explain what the expression $2000t$ means in this context.

(b) Suppose you deposit $2000 into a savings account that pays 10 percent interest per year. You make no additional deposits and no withdrawals. How much interest will the account earn in five years?

(c) Suppose that the interest rate is 6 percent per year and you contribute $150 per month to the annuity. Find an expression for $dp/dt$ assuming that $t$ is measured in months and that the interest and contributions are paid continuously.

**2.** You want to save $20,000 to buy a new car four years from now. You decide to open an account for this purpose, initially with no money in it, and you plan to deposit a fixed amount of money each month.

(a) Write a differential equation that describes the rate of growth, $dp/dt$, of the principal in the account, assuming that interest and deposits are added continuously. Determine the units on each constant and each variable in the equation.

(b) Suppose the account earns interest at the rate of 6 percent per year. Explain how this assumption determines the value of at least one of the constants in your equation in (a). What is the value of that constant?

(c) How much money do you need to deposit each month to save $20,000 in 4 years?

(d) If you could get 12 percent annual interest, could you cut your monthly contributions in half? Why or why not?

**3.** This exercise discusses the dynamics of borrowing money. Let $p(t)$ be the balance owed on a loan at time $t$, called the *principal*. Interest charges increase the principal at a rate proportional to its present size. The principal decreases at a rate that is equal to the rate at which payments are made.

(a) Argue that the direction field must look qualitatively like the one in Fig. 2.10. The five curves shown in the figure are some representative solutions. Interpret the meaning of the solutions in terms of the repayment of a loan.

(b) Find an expression for $dp/dt$ as a function of $p$, the interest rate $r$, and the rate of payments $m$.

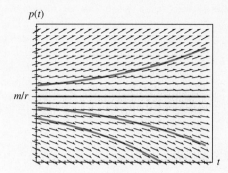

**Figure 2.10**   For Exercise 3.

———————

(c) Does the equation in (b) have an equilibrium point? If so, what is it? Is it stable or unstable? Explain.

(d) Suppose you borrow $10,000 at an interest rate of 9 percent per year. If you make continuous payments totaling $1000 per year, how long will it take you to repay the loan?

**4.** You have an outstanding balance of $2400 on your credit card and decide to pay it off at the rate of $100 per month.

(a) Let $b(t)$ be the balance owed at time $t$. Find an expression for $db/dt$.

(b) If the interest rate is 18 percent annually, compounded continuously, how long will it take you to eliminate the balance, assuming that you make no additional charges on the card?

(c) If you pay only $50 per month, does it take twice as long to pay back the balance? What if you pay $200 per month?

(d) Suppose you have an opportunity to transfer the balance to a card that charges 14 percent annual interest, compounded continuously. How long will it take you to pay off the balance at $100 per month?

**5.** As in Exercise 4, assume an outstanding balance of $2400.

(a) Assume an 18 percent annual interest rate, compounded continuously. Suppose you make payments continuously at the rate of $m$ dollars per month. We will call the *payoff time* the amount of time needed to reduce the outstanding balance to zero, assuming that you make no additional charges on the card. Find an expression for the payoff time as a function of $m$ and plot it.

(b) How much must you pay each month to pay off the balance in exactly two years?

(c) Is it possible for the payoff time to be infinite? Explain.

(d) Suppose you start with a $2400 balance and pay $100 per month. Find an expression for the payoff time as a function of the interest rate and plot it. If the interest rate is cut in half, is the payoff time also cut in half?

**6.** This exercise uses the same differential equation as Exercise 3 but with a different application. If you want to borrow money to buy a house, you can go to a bank or mortgage company to be qualified for a loan. The bank uses your financial information to find the maximum amount of money that you can afford to pay each month. This determines the maximum amount of money that you can borrow and limits the price of the houses that you can buy.

(a) Suppose you apply for a conventional, 30-year mortgage at a fixed interest rate. If you can afford $1000 per month in principal and interest, how large a mortgage can you take out, assuming 9 percent annual interest, compounded continuously?

(b) Repeat (a), but assume a 15-year mortgage.

(c) Assume a 30-year mortgage and a $1000 monthly payment. How large can the mortgage be if the interest rate is 6 percent annually? 12 percent?

(d) Assume 9 percent annual interest, compounded continuously, and a 30-year mortgage. If you can afford $2000 per month in mortgage payments, can you borrow twice as much money as in (a)?

(e) In general, which of the following possibilities allows you to borrow twice as much money, and why?

- Halve the interest rate.
- Double the monthly payment.
- Double the term of the loan (e.g., 30 years instead of 15 years).

**7.** Suppose you borrow $150,000 to buy a house at 9 percent annual interest, compounded continuously.

(a) What are your monthly payments, assuming a 30-year loan?

(b) What is the principal after one year? How much interest do you pay on the loan in the first year?

(c) How much interest do you pay on the loan in the 30th year?

(d) Find an expression for the interest paid in the $k$th year of the loan and plot it. Are the interest charges greatest at the beginning, the middle, or the end of the loan?

(e) What are your total interest costs over the life of the loan? Suppose the term of the loan is 15 years instead. Are the interest costs cut in half?

**8.** Find an expression for the total amount of interest charges paid over a 30-year mortgage, assuming a fixed annual interest rate $r$, monthly payment $m$, and initial principal $p_0$.

**9.** An *adjustable rate* mortgage is one whose interest rate fluctuates over the life of the loan. Typically the rate is fixed for one year at a time. On each anniversary of the loan, the interest rate is adjusted up or down according to prevailing market forces.

(a) Describe how to compute the monthly payments on a loan in this situation, even though no one knows what interest rates will be in the future.

(b) Suppose you borrow $100,000 under a 30-year adjustable rate mortgage. The interest rate is 6 percent annually in the first year of the loan. In the second year, it is 7 percent annually. What are the monthly payments in the first year of the loan? The second year?

**10.** Suppose you borrow $5000 to buy a car at a 4 percent annual interest rate for a period of 2 years.

(a) What are your monthly payments, assuming that the interest is compounded continuously?

(b) Until the 1960s, banks could advertise a 4 percent interest rate for new car loans, but figure the payments as follows: 4 percent of $5000 is $200. Therefore, the total interest is $400 for two years. Your total payments are $5400 over 24 months, or $225 per month. Explain why the practice is unethical. (It is now illegal.) (Hint: What are your interest costs if your payments are figured as in part (a)? What is the approximate annual percentage rate in (b)?)

**11.** Newton's law of cooling says that hot water cools at a faster rate than an equal amount of cold water when placed in a freezer (assuming that the freezer remains at constant temperature). If you want to make ice cubes as quickly as possible, should you start with hot water or cold water? Explain.

**12.** Suppose a pot of boiling water, initially at 100°C, cools to 80°C after 30 minutes. Is the proportionality constant $k$ in Newton's law of cooling therefore 40/h? Why or why not?

**13.** This exercise shows how the equal ratio property applies to Newton's law of cooling.

(a) The initial temperature difference between the kettle and the surroundings in Example 3 is 80°C. How long does it take for the temperature difference to decrease to 40°C? How long does it take for the temperature difference to decrease from 40°C to 20°C? From 20°C to 10°C?

(b) Show that the time needed for $T - R$ to reach half its present value is a useful way to characterize the rate of temperature change, regardless of the specific values of the initial temperature $T_0$, room temperature $R$, and constant of proportionality $k$.

**14.** Suppose alcohol is introduced into a 2-liter (L) beaker, which initially contains pure water, at the rate of 0.1 L/min. The well-stirred mixture is removed at the same rate.

(a) How long does it take for the concentration of alcohol to reach 50 percent? 75 percent? 87.5 percent?

(b) Suppose the current concentration of water in the beaker is $c$ and we ask how long it takes before the concentration is cut in half. Is this time interval the same, regardless of $c$?

**15.** Derive Eq. (2.19).

**16.** Consider an object with mass $m = 2$ kg that falls under the influence of gravity but which encounters air resistance of $k = 1/2$ kg/s.

(a) Suppose the object is dropped from a great height (zero initial velocity). Find the velocity of the object 5 s later.

(b) Does your answer in (a) depend on the height at which the object is dropped? Explain.

(c) Suppose the object initially is dropped from a height of 70 m. How fast is it going when it hits the ground?

(d) Suppose the object is shot upward from a height of 70 m with an initial velocity of 5 m/s. Determine the maximum height that the object attains and the time at which it hits the ground.

**17.** Consider Example 4.

(a) Suppose that, instead of a beach ball, we drop a bowling ball of mass 6 kg from a 10-m platform. Assume that the coefficient of air resistance is $k = 1/2$, as in Example 4. Find the velocity of the ball as a function of $t$ and plot it.

(b) Contrast your graph in (a) to Fig. 2.9(a). Does the bowling ball approach terminal velocity more or less quickly than the beach ball?

(c) When does the bowling ball hit the ground, and how fast is it falling at the time of impact? Does the bowling ball ever reach terminal velocity?

(d) If we conducted the experiment in a vacuum, what would be the differential equation that models the motion of the two balls? Does the equation predict that the two balls hit the ground at the same time if they are dropped at the same time? Explain.

**18.** Consider Example 4.

(a) Suppose the coefficient of air resistance is $k = 1$. Does this value of $k$ correspond to an increased or decreased effect of air resistance compared to the situation described in Example 4?

(b) Suppose $k = 1$. Does the beach ball approach terminal velocity more quickly or less quickly than in the case in which $k = 1/2$?

(c) In general, as $k$ increases, does the time it takes the beach ball to reach terminal velocity increase or decrease? Justify your answer.

(d) As $k$ increases, does the terminal velocity of the ball increase or decrease? Explain.

(e) Suppose we fix $k = 1/2$, as in Example 4. Does a heavier ball reach terminal velocity more quickly or less quickly than a lighter ball? Justify your answer. (Suggestion: Look at Exercise 17.)

(f) Compared to a lighter ball, does a heavier ball hit the ground at a greater or lesser speed than a lighter ball, assuming that the coefficient of

air resistance for both balls is the same and that both balls are dropped from the same height?

**19.** Experimental observations of falling objects in the presence of air resistance suggest that in some cases, the force due to air resistance is proportional to the *square* of the velocity. Thus, the velocity of a falling object in this situation can be modeled as

$$m\frac{dv}{dt} = -mg + kv^2, \qquad (2.23)$$

where $k$ is a positive constant of proportionality.

For purposes of this exercise, assume that $m = 1$ kg, $g = 10$ m/s$^2$, and $k = 1/10$ kg/m. (Let the ground be position 0, and let $x > 0$ denote a position $x$ meters above the ground.)

(a) Suppose that the object is dropped from a great height (initial velocity 0). Find $v(t)$. Does $v(t)$ depend on the height from which the object is dropped?

(b) What is the maximum speed attained by the object?

(c) Is Eq. (2.23) an accurate model if the initial velocity is positive, that is, if the ball initially is moving upward? If not, modify Eq. (2.23) so that it describes the forces on a rising object in a manner consistent with the coordinate system.

(d) Find the maximum height attained by an object that is shot upward from an altitude of 50 m above the ground at an initial speed of 10 m/s.

(e) When does the object in (d) hit the ground?

**20.** The objective of this exercise is to calculate some of the relevant quantities for the fall of a skydiver out of an airplane. We use the simplest model of air resistance, in which we suppose that the drag force through the air is proportional to the velocity. We also suppose that the skydiver simply topples out of the airplane. The relevant initial value problem is

$$v' = -g - \frac{k}{m}v, \qquad v(0) = 0. \qquad (2.24)$$

We let $x = 0$ represent the ground and let $x > 0$ correspond to positions above the ground. As $v = x'$, a falling skydiver has a negative velocity in this coordinate system. The gravitational constant is $g$, and the coefficient of air resistance is $k$. For algebraic simplicity, assume that $g = 10 \text{ m/s}^2$.

(a) Suppose that $k/m = 0.2$ before the skydiver opens the parachute. What is the terminal velocity for the free-falling skydiver?

(b) When does the skydiver reach 95 percent of the terminal velocity in (a)?

(c) Suppose the airplane is 2000 m above the ground when the skydiver jumps out. What is the skydiver's altitude at the time indicated in part (b)?

(d) To avoid serious injury, suppose that the skydiver must hit the ground at a velocity that is no greater than 6 m/s. What must the value of $k/m$ be to satisfy this requirement, assuming that the parachute is opened in sufficient time?

**21.** This exercise discusses some of the issues that arise in modeling the opening of a parachute. Assume that the differential equation that governs the velocity of the skydiver is given by

$$v' = -g - \frac{k}{m}v, \qquad (2.25)$$

where $g$ is the acceleration due to gravity. For simplicity, assume that $g = 10 \text{ m/s}^2$. Also suppose that $k/m = 0.2$ when the skydiver is in free fall and that $k/m = 1.5$ when the parachute has fully opened. Assume that the skydiver topples out of the airplane at an altitude of 2000 m and that the parachute opens instantaneously at an altitude of 500 m.

(a) Find the velocity, position, and acceleration of the skydiver during the free fall. How long does

it take for the skydiver to reach an altitude of 500 m?

(b) Find the velocity, position, and acceleration of the skydiver after the parachute opens. Explicitly identify the initial conditions that you use when computing the position and velocity. When does the skydiver hit the ground?

(c) Plot the acceleration for the entire time that the skydiver is in the air. Explicitly determine the acceleration that the skydiver experiences when the parachute opens, and discuss whether the acceleration is continuous at the instant that the parachute opens.

(d) Forces that accelerate the human body by more than $2g$ can cause injury or death. How realistic are the preceding assumptions about the motion of a real skydiver?

(e) Discuss more realistic assumptions about the opening of the parachute. Cast your assumptions in terms of a differential equation and solve it, either analytically or numerically. Discuss whether your improved model circumvents the difficulties outlined in (c) and (d).

**22.** Craig wants to lose 20 pounds before his high school reunion 100 days from now. His wife, Jenny, consults a diet book that says that he must consume 15 calories each day for every pound that he weighs to maintain his current weight. Jenny also notes that he will lose one pound of fat whenever he runs a cumulative deficit of 3500 calories. (In other words, if he consumes a total of 3500 calories less than what his body needs over any period of time, then he loses one pound.) Craig wants to go on a diet in which he consumes the same number of calories each day.

(a) Suppose Craig consumes $C$ calories per day and his current weight is $W$. Write an expression for his caloric deficit. What are the units on each term?

(b) Use your answer to (a) to write the differential equation that expresses the rate at which Craig loses weight as a function of his current weight.

(c) Craig supposes that since 20 pounds = 70,000 calories, he should consume 700 fewer calories each day for 100 days. Since Craig weighs 200 pounds today, he decides to consume 2300 calories per day (2300 = 15 × 200 − 700). Will his plan work? How much weight will Craig really lose in 100 days at 2300 cal/day?

(d) Determine the maximum number of calories that Craig should consume each day if he wants to weigh exactly 180 pounds 100 days from now.

(e) Is there an equilibrium solution? If so, what does it mean in the context of Craig's diet?

**23.** Suppose you add cold milk to a cup of hot coffee. Which scenario gives you cooler coffee, assuming that the volume and initial temperature of each liquid are the same in both cases?

- You add milk to the hot coffee and wait five minutes.

- You wait five minutes, then add milk to the coffee.

Answer the question under each of the following additional hypotheses about the problem.

(a) Suppose that the milk is at room temperature. Show that your answer is independent of

- the value of the proportionality constant $k$, assuming that $k$ is the same for the black coffee and the coffee-milk mixture;
- the temperature at which the coffee is brewed;
- the amount of milk that is added, provided that it is the same in each scenario;
- the amount of coffee in each cup, provided that it is the same in each scenario.

(b) Suppose that the milk is stored in a pitcher that rests on a bed of ice cubes, keeping the temperature of the milk constant and colder than the

room temperature. Show that your answer is independent of the factors listed in (a).

You may wish to experiment with specific values of $k$ and the initial conditions to build intuition before deriving the general answers. If you conduct some numerical experiments, try to find reasonable values of $k$ and explain how you derive them.

**24.** The exponential growth equation

$$\frac{dp}{dt} = kp \qquad (2.26)$$

can be regarded as a model of a population in which there is no mortality (for example, the rate of growth of bacteria in a petri dish with lots of nutrients). In other words, the rate of creation of new individuals is proportional to the current size of the population and the constant of proportionality is $k$.

(a) Suppose that the death rate is proportional to the current population. Call the constant of proportionality $m$ (for "mortality") and modify Eq. (2.26) so that it contains separate terms for births and deaths. (You don't need to solve this equation.) What are the units of $m$?

(b) Argue that your model in (a) is consistent with the idea that if the death rate exceeds the birth rate, then the population declines with time; and the population rises if the birth rate exceeds the death rate.

(c) Suppose the birth rate in a country is $b$, the death (mortality) rate is $m$, and the immigration laws permit 1,000,000 people per year to enter the country (independent of the current population). Modify your equation in (a) to include terms for births, deaths, and immigration.

(d) Explain the meaning of the expression 1,000,000$t$ in the context of (c).

(e) Solve the equation you got in (c). What will be the population of the country 3 years from now if the birth rate $b = 0.02$, the mortality rate $m = 0.01$, immigration is 1 million per year, and the population today is 100 million?

## 2.3   THE NONAUTONOMOUS CASE

In the previous sections, we considered various applications of the autonomous exponential growth equation

$$x' = cx + m, \tag{2.27}$$

where $m$ and $c$ are constants. Equation (2.27) can be solved by the separation of variables.

In many cases, $c$ and $m$ vary with time. For instance, the rate of return of an annuity generally is not constant, nor does one expect to contribute exactly the same amount of money each year. In the following discussion, we consider the solutions of differential equations of the form

$$x' + kx = f(t), \tag{2.28}$$

where $f$ is a continuous function of $t$ and $k$ is a constant. Equation (2.28) is not autonomous because the derivative depends explicitly upon time as well as on the current value of $x$. (It also is not separable. Why?)

Nonautonomous exponential growth equations appear frequently in applications. The most interesting mathematical aspect is the basic structure of solutions of Eq. (2.28), which is the focus of this section. A procedure for finding explicit solutions of Eq. (2.28) is outlined at the end of this section.

## Qualitative Analysis of Solutions

In many situations, we are interested in the question "What happens to the system after a while?" In mathematical terms, we want to know how the solution behaves when $t$ is relatively large. The term *qualitative behavior* refers to the gross features of the solution, such as whether it is a straight line, polynomial, or periodic function, as well as the corresponding slope, period, and the like. Let us consider this question in the following two examples.

✦ *Example 1*   Consider the differential equation

$$x' + x = t, \tag{2.29}$$

whose direction field is plotted in Fig. 2.11. The solution curves corresponding to the initial conditions $x_0 = -4$, $x_0 = -1$, and $x_0 = 2$ are displayed as the bottom, middle, and top curves, respectively. As $t$ increases, it appears that all three solutions converge to the middle curve, which is a straight line of unit slope. The middle curve is the

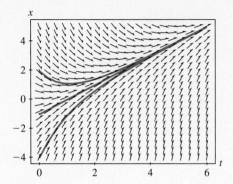

**FIGURE 2.11**   The direction field and three solutions of Eq. (2.29).

graph of $x(t) = t - 1$, which is a solution of Eq. (2.29) as the following substitution verifies:

$$(t-1)' + (t-1) = 1 + (t-1)$$
$$= t.$$

In fact, the analytical solution of Eq. (2.29) is

$$x(t) = (t-1) + e^{-t}(x_0 + 1). \tag{2.30}$$

(We show how to derive Eq. (2.30) in Example 4.)

Equation (2.30) confirms that all solutions eventually "converge" to the line $t - 1$, because $\lim_{t \to \infty} e^{-t}(x_0 + 1) = 0$. Qualitatively, we can say that all solutions eventually grow linearly with unit slope.                                                                    ∎

◆ *Example 2*   The general solution of

$$x' + x = \cos 3t \tag{2.31}$$

has a similar structure. Figure 2.12 shows the corresponding direction field and three solution curves. Each solution curve "decays" to a single oscillatory curve. In Example 5, we show that the analytical solution of Eq. (2.31) is given by

$$x(t) = e^{-t}\left(x_0 - \tfrac{1}{10}\right) + \tfrac{3}{10}\sin 3t + \tfrac{1}{10}\cos 3t. \tag{2.32}$$

Because $e^{-t}$ becomes very small as $t$ becomes large, the solution (2.32) confirms that $x(t)$ approaches a periodic function, regardless of the initial condition. Qualitatively, every solution eventually becomes periodic with period $2\pi/3$.

## A Procedure for Finding Explicit Solutions

The solution procedure described in this section is a special case of a more general one, called the *variation of parameters*, which is discussed in more detail in the appendix. We are interested in solutions of the nonautonomous exponential growth equation given by

$$x'(t) + kx(t) = f(t),  \tag{2.35}$$

where $k$ is a constant and $f(t)$ is a continuous function.

Equation (2.35) is not separable and cannot be solved analytically by using the methods we have described so far. However, it is possible to express $x(t)$ in terms of an integral involving $f(t)$. For this purpose, we introduce an *integrating factor*: a function that, when multiplied by both sides of the differential equation, yields an expression that can be integrated explicitly for $t$.

The method relies on a clever application of the chain and product rules to the exponential function. If $x$ is a differentiable function of $t$, then

$$\begin{aligned}
\left(x(t)e^{kt}\right)' &= x'(t)e^{kt} + x(t)\left(e^{kt}\right)' \\
&= x'(t)e^{kt} + x(t)ke^{kt} \\
&= e^{kt}\left(x'(t) + kx(t)\right),
\end{aligned}  \tag{2.36}$$

where we use the prime to denote differentiation with respect to $t$. Notice that Eq. (2.36) is the product of the left-hand side of Eq. (2.35) and $e^{kt}$.

If we multiply each side of Eq. (2.35) by $e^{kt}$, then we have

$$e^{kt}\left(x'(t) + kx(t)\right) = e^{kt}f(t).  \tag{2.37}$$

By (2.36), Eq. (2.37) is equivalent to

$$\frac{d}{dt}\left(e^{kt}x(t)\right) = e^{kt}f(t).  \tag{2.38}$$

An expression for $x$ can be determined by integrating both sides of Eq. (2.38). An antiderivative of the left-hand side of Eq. (2.38) is simply $e^{kt}x(t)$. The fundamental theorem of calculus implies that we can integrate both sides to obtain

$$\int_0^t \frac{d}{ds}\left(e^{ks}x(s)\right)ds = \int_0^t e^{ks}f(s)\,ds,$$

so

$$e^{kt}x(t) - x(0) = \int_0^t e^{ks}f(s)\,ds.  \tag{2.39}$$

Therefore,

$$x(t) = e^{-kt}\left(x(0) + \int_0^t e^{ks} f(s)\,ds\right) \qquad (2.40)$$

is a general solution of Eq. (2.35). The integral on the right-hand side of Eq. (2.40) can be evaluated if $f$ is a polynomial, simple exponential, sine, or cosine function; otherwise, the integral may be difficult or impossible to evaluate.

✦ *Example 4*    Find an explicit solution of the initial value problem

$$x' + x = t, \qquad x(0) = x_0. \qquad (2.41)$$

*Solution*    Here, $k = 1$, so we multiply both sides of Eq. (2.41) by $e^t$ to obtain

$$e^t\left(x' + x\right) = e^t t, \qquad (2.42)$$

where we observe that

$$e^t\left(x'(t) + x(t)\right) = \frac{d}{dt}\left(x(t)e^t\right).$$

Therefore, the integral of the left-hand side of Eq. (2.42) is

$$\int_0^t \frac{d}{ds}\left(e^s x(s)\right) ds = e^s x(s)\big|_0^t$$

$$= x(t)e^t - x(0),$$

by the fundamental theorem of calculus. The right-hand side of Eq. (2.42) can be integrated by parts to obtain

$$\int_0^t e^s s\,ds = e^t(t - 1) + 1.$$

Thus,

$$x(t)e^t - x_0 = e^t(t - 1) + 1,$$

so

$$x(t) = t - 1 + (x_0 + 1)e^{-t}. \qquad (2.43)$$

It is straightforward to check that Eq. (2.43) is a solution by back substitution into Eq. (2.41). ∎

✦ *Example 5*  Find an explicit solution of the initial value problem

$$x' + x = \cos 3t, \qquad x(0) = x_0. \tag{2.44}$$

*Solution*  We proceed as above and obtain

$$e^t \left( x' + x \right) = e^t \cos 3t. \tag{2.45}$$

The integral of the left-hand side of Eq. (2.45) is $x(t)e^t - x_0$, as in the previous example. The right-hand side of Eq. (2.45) can be integrated by parts twice to yield

$$\int_0^t e^s \cos 3s \, ds = \tfrac{1}{10} e^t \cos 3t + \tfrac{3}{10} e^t \sin 3t - \tfrac{1}{10}.$$

A straightforward simplification yields the solution

$$x(t) = \tfrac{1}{10} \cos 3t + \tfrac{3}{10} \sin 3t + e^{-t} \left( x_0 - \tfrac{1}{10} \right). \qquad \blacksquare$$

## Exercises

**1.** Show that $g(t) = t$ is not a solution of the differential equation in Example 1.

**2.** Suppose the initial condition in Example 4 is specified at $t = -1$ instead of at $t = 0$. Discuss what changes are needed to the solution procedure above. Solve the initial value problem $x' + x = t$, $x(-1) = 1$.

**3.** Let $x' + x = \sin t$.

(a) Show that the steady-state cannot be a function of the form $x_{ss}(t) = A \sin t$ for any constant $A$.

(b) Show that it is possible to find constants $A$ and $B$ such that

$$x_{ss}(t) = A \sin t + B \cos t$$

is a steady-state solution. (Suggestion: Substitute $x_{ss}$ into the differential equation and collect like terms. Choose $A$ and $B$ so that $x'_{ss} + x_{ss} = \sin t$.)

**4.** Let $x' + 2x = f(t)$.

(a) Let $f(t) = t^2$. Find a quadratic polynomial that is a steady-state solution $x_{ss}$.

(b) Let $f(t) = t^3 + t$. Find a cubic polynomial that is a steady-state solution $x_{ss}$.

(c) Let $f(t)$ be a polynomial of the form $f(t) = at^n + b$. Argue that $x_{ss}$ contains a term involving $t^n$.

**5.** Let $x' + x = f(t)$. For each function in the following list,

- Find the solution when $x(0) = 1$.

- Identify the transient and steady-state parts of the solution.

- On the same axes, plot the solution, the transient, and the steady-state.

- Describe the steady-state qualitatively. Consider questions such as: Is the steady-state constant, linear, or polynomial? Is it positive, negative, or zero? Is it periodic? If so, what are the period and amplitude?

(a) $f(t) = \cos 2t$       (b) $f(t) = e^{-t}$

(c) $f(t) = -e^t$          (d) $f(t) = 2$

(e) $f(t) = t^2$

**6.** The discussion in the text states that the solutions of the equation

$$x' + kx = f(t)$$

can be described as the sum of a transient and a steady-state part if $k > 0$. This problem shows what can happen if $k < 0$.

(a) Let $x' - x = t$. (Here, $k = -1$, and $f(t) = t$.) Find the solution $x(t)$ that satisfies the initial condition $x(0) = x_0$.

(b) Compare your answer to (a) with the solution of the initial value problem $x' + x = t$, $x(0) = x_0$.

(c) Does it make sense to characterize the solution in (a) as a sum of a transient and a steady-state? If not, how would you describe the solution?

**7.** Let $x' + 2x = mt$, where $m$ is any constant.

(a) Find the solution that satisfies the initial condition $x(0) = x_0$.

(b) Identify the transient and steady-state terms in the solution.

(c) Describe the qualitative properties of the solution $x(t)$ as $t$ becomes large. (For instance, does $x(t)$ approach a straight line? If so, what is the slope? Does $x(t)$ approach a periodic function? If so, what is the period? Does $x(t)$ decay exponentially to 0?)

**8.** Let us generalize the conclusions of the previous exercise. Let $x' + kx = mt$.

(a) Repeat Exercise 7 under the assumption that $k$ is any positive constant.

(b) Repeat Exercise 7 under the assumption that $k$ is any negative constant.

(c) State your conclusions in the form of a theorem. What happens in the case $k = 0$?

**9.** Let $x' + 2x = f(t)$. Without explicitly computing the general solution, find a function $f(t)$ for which the solution of the equation eventually

(a) oscillates with period $\pi$;

(b) oscillates with period $\pi/10$;

(c) oscillates with period 1;

(d) grows linearly with $t$;

(e) grows quadratically with $t$;

(f) decays exponentially to 0;

(g) approaches the constant value 10.

**10.** Suppose that a radioactive isotope, $P$, decays into another radioactive isotope, $Q$, which subsequently decays into $R$. Assume that $P$ has a half-life of one day and $Q$ has a half-life of two days.

(a) Find an expression for the rate of change of the mass of $Q$. (The rate of change is the rate at which $Q$ is formed minus the rate at which $Q$ decays.)

(b) If we start with 1 $\mu$g of $P$ and no $Q$, how much of $R$ will there be after three days?

**11.** Let $x' + kx = e^{rt}$, where $k$ and $r$ are any constants.

(a) Assume that $k \neq -r$. Find the solution that satisfies the initial condition $x(0) = x_0$.

(b) For which values of $k$ and $r$ does your solution in (a) tend to 0 as $t \to \infty$?

(c) Assume that $k = -r$. Find the solution that satisfies the initial condition $x(0) = x_0$.

(d) For which values of $k$, if any, does your solution in (c) tend to 0 as $t \to \infty$?

## 2.4   UNCERTAINTY AND PREDICTABILITY

Section 2.2 considers some of the applications of exponential growth and explains the basic steps in the formulation of mathematical models. One of the best ways to validate a given model is to compare its predictions with observations from the phenomenon under study. For instance, mathematical models of the weather are used to generate forecasts. Meteorologists refine the models in part by comparing the forecasts with actual observations. Mathematical models are used to make predictions about simpler processes as well, such as the motion of a simple pendulum or the decay of a radioactive isotope.

However, every laboratory or field measurement is subject to some uncertainty, and most mathematical models involve some approximations and idealizations. Therefore, no mathematical model is likely to yield completely accurate predictions. The question of how error affects one's ability to predict the future behavior of some process is a recurring theme in the study of differential equations. This section gives a brief introduction to the subject of errors and how they affect the predictability of various processes. We concentrate on the case of exponential growth, as it is the easiest to analyze.

There are at least three sources of error in any mathematical model. They include

- errors in measuring the initial conditions;
- errors in the parameters (constants) in the model;
- an incorrect model of the underlying process.

In many cases, all three types of errors are present. A comprehensive discussion of each type of error is beyond the scope of this text. However, in the case of exponential growth and decay, a rather complete description of the first two types of error is possible. We will consider this topic at the end of the section. First, let us consider two ways in which error can be quantified.

### Absolute Error and Relative Error

Suppose that the weight recorded by a scale is too high by 1 pound. In other words, if the true weight of the object is $x$ pounds, then the scale reads $x + 1$ pounds. We say that the *absolute error* in the weight measured by the scale is 1 pound. More generally,

$$\text{absolute error} = \epsilon_a = |\text{estimated} - \text{actual}|. \qquad (2.46)$$

Notice that the absolute error of the scale is the same if the recorded weights are too low by 1 pound.

How serious is a 1-pound error? The answer depends on the context. For instance, if the scale is a bathroom scale, and you weigh 150 pounds, then a 1-pound error is probably acceptable. However, a 1-pound error is intolerable for a countertop scale in a butcher shop. (You don't want to be charged for 3 pounds of steak when you buy only 2 pounds.)

The heuristic examples above can be quantified more precisely in terms of the *relative error*, defined as

$$\text{relative error} = \epsilon_r = \frac{\text{absolute error}}{|\text{actual}|}. \tag{2.47}$$

The relative error can be expressed as a percentage:

$$\text{percentage relative error} = \frac{\text{absolute error}}{|\text{actual}|} \times 100\%.$$

If your true weight is 150 pounds and the bathroom scale registers 151 pounds, then the relative error in the measurement is $1/150$, or approximately 0.67 percent. Put another way, the measurement is 99.33 percent accurate, which is very good for a bathroom scale. In contrast, a 1-pound absolute error in the purchase of a 2-pound steak corresponds to a relative error of 50 percent. Notice that the relative error of a measurement is undefined if the actual value in question is 0.

## Errors and the Exponential Growth Equation

Errors in a mathematical model can arise from incorrect measurements of the initial condition as well as errors in estimating the parameters in the model. (We omit a discussion of the effect of errors in the model itself for now.)

Consider the exponential growth equation

$$\frac{dp}{dt} = kp, \tag{2.48}$$

whose solution is

$$p(t) = p_0 e^{kt} \tag{2.49}$$

when the initial condition is specified at $t = 0$.

### Errors in Initial Conditions

First consider the effect of errors in measuring the initial condition. Imagine for the moment that Eq. (2.48) is an extremely accurate model for the growth of some bacteria in a petri dish but that the number of bacteria at $t = 0$ is not known precisely. Suppose that a technician estimates the initial population at 900,000 bacteria when the true initial population is 1 million. If $p(t)$ is measured in units of millions of bacteria, then $p_0 = 1$, but the measured value, denoted $\hat{p}_0$, is 0.9. Let $p(t)$ denote the "true" solution,

$$p(t) = p_0 e^{kt} = e^{kt}, \tag{2.50}$$

and let $\hat{p}(t)$ denote the "predicted" solution,

$$\hat{p}(t) = \hat{p}_0 e^{kt} = 0.9 e^{kt}. \tag{2.51}$$

Equation (2.51) predicts the population of bacteria based on the initial census. Table 2.1 shows the predicted and actual populations every half hour, assuming that $k = 1$ and $t$ measures time in hours. Notice that the absolute error between the predicted population and the actual observed population increases with time. Figure 2.13

| $t$ | $p(t)$ | $\hat{p}(t)$ | $\epsilon_a(t)$ | $\epsilon_r(t)$ |
|-----|--------|--------------|-----------------|-----------------|
| 0.0 | 1.000  | 0.900        | 0.100           | 0.100           |
| 0.5 | 1.649  | 1.484        | 0.165           | 0.100           |
| 1.0 | 2.718  | 2.446        | 0.272           | 0.100           |
| 1.5 | 4.482  | 4.034        | 0.448           | 0.100           |
| 2.0 | 7.389  | 6.650        | 0.739           | 0.100           |

TABLE 2.1   The true and predicted populations, Eqs. (2.50) and (2.51); respectively, their absolute error $\epsilon_a$; and their relative error $\epsilon_r$ as a function of $t$ when $k = 1$

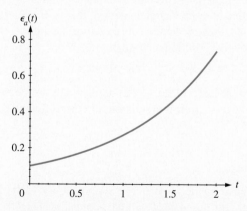

FIGURE 2.13   A plot of the absolute error, $\epsilon_a$, as a function of $t$, based on the values in Table 2.1.

shows a plot of $\epsilon_a$ as a function of $t$. The absolute error appears to grow exponentially. However, the relative error in the predicted population remains constant at 10 percent.

The following general analysis explains the observations in Table 2.1. Suppose that the relative error in the initial condition is $\epsilon_r(0)$, that is, let

$$\epsilon_r(0) = \left| \frac{p_0 - \hat{p}_0}{p_0} \right|,$$

where $\hat{p}_0$ denotes the measured (estimated) initial condition. Then the relative error in the solution at time $t$ is

$$\epsilon_r(t) = \left| \frac{p(t) - \hat{p}(t)}{p(t)} \right|$$

$$= \left| \frac{p_0 e^{kt} - \hat{p}_0 e^{kt}}{p_0 e^{kt}} \right|$$

$$= \left| \frac{p_0 - \hat{p}_0}{p_0} \right|$$

$$= \epsilon_r(0), \tag{2.52}$$

assuming, of course, that $p_0 \neq 0$. Equation (2.52) implies that

*if the only error is in the measurement of the initial condition, then the relative error in the solution of the exponential growth equation remains constant.*

The following are important points to keep in mind:

- The derivation of Eq. (2.52) implies that in the case of exponential growth, the solution is as accurate as the measurement of the initial condition; that is, if the relative error in the initial condition is small, then predictions of the future state of the system also are accurate insofar as the relative error remains small.

- The analysis is independent of the value of the rate constant $k$.

We emphasize that the analysis leading to Eq. (2.52) applies only to the exponential growth equation, Eq. (2.48). The relative error in solutions of other differential equations behaves differently.

## Errors in the Parameter

Next we consider the effect of errors in the parameter $k$. Suppose as above that Eq. (2.48) is an accurate model of bacterial growth and that the initial population is exactly 1 million. However, assume that the value of $k$ is estimated to be 0.9 when the true value is 1, corresponding to a relative error of 10 percent. (We write $k = 1$ and $\hat{k} = 0.9$, and we let

| $t$ | $p(t)$ | $\hat{p}(t)$ | $\epsilon_a(t)$ | $\epsilon_r(t)$ |
|-----|--------|--------------|-----------------|-----------------|
| 0.0 | 1.000  | 1.000        | 0.000           | 0.000           |
| 0.5 | 1.649  | 1.568        | 0.081           | 0.049           |
| 1.0 | 2.718  | 2.460        | 0.258           | 0.095           |
| 1.5 | 4.482  | 3.857        | 0.625           | 0.139           |
| 2.0 | 7.389  | 6.050        | 1.339           | 0.181           |

**TABLE 2.2**   The errors in the solutions $p(t)$ and $\hat{p}(t)$ when $k = 1$ and $\hat{k} = 0.9$, respectively

$p(t)$ and $\hat{p}(t)$ denote the corresponding solutions.) Table 2.2 shows the corresponding absolute and relative errors in the solution for selected values of $t$. In contrast to the case of an error in the initial condition, an error in the parameter results in growing absolute and relative errors as $t$ increases.

Table 2.3 shows corresponding values of the errors when $\hat{k} = 1.1$. As in the previous example, the absolute and relative errors grow as $t$ increases.

| $t$ | $p(t)$ | $\hat{p}(t)$ | $\epsilon_a(t)$ | $\epsilon_r(t)$ |
|-----|--------|--------------|-----------------|-----------------|
| 0.0 | 1.000  | 1.000        | 0.000           | 0.000           |
| 0.5 | 1.649  | 1.733        | 0.084           | 0.051           |
| 1.0 | 2.718  | 3.004        | 0.286           | 0.105           |
| 1.5 | 4.482  | 5.207        | 0.725           | 0.162           |
| 2.0 | 7.389  | 9.025        | 1.636           | 0.221           |

**TABLE 2.3**   The errors in the solutions $p(t)$ and $\hat{p}(t)$ when $k = 1$ and $\hat{k} = 1.1$, respectively

A general analysis of the error in $k$ proceeds as follows, where $\hat{k}$ denotes the estimated value. The relative error in the solution at time $t$ is

$$\epsilon_r(t) = \left| \frac{p(t) - \hat{p}(t)}{p(t)} \right|$$

$$= \left| \frac{p_0 e^{kt} - p_0 e^{\hat{k}t}}{p_0 e^{kt}} \right|$$

$$= \left| 1 - \exp\left(\hat{k} - k\right)t \right|. \tag{2.53}$$

There are two cases to consider:

**Case 1:** $\hat{k} - k < 0$. As $t$ becomes large, $\epsilon_r(t)$ approaches 1. In other words, the absolute error becomes as large as the solution itself. Figure 2.14(a) shows a plot of Eq. (2.53) as a function of time for the case $\hat{k} = 0.9$ and $k = 1$. Notice that the relative error in the solution is less than 20 percent for $0 < t < 2.2$ or so. The graph of the relative error is concave down, reflecting the fact that the relative error approaches an asymptotic value.

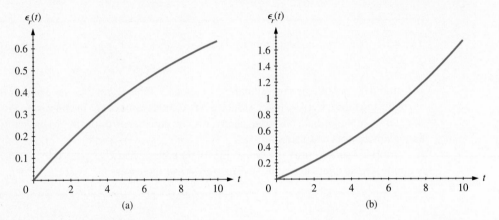

**FIGURE 2.14**    (a) A plot of the relative error as a function of time, when $\hat{k} = 0.9$, $k = 1$. (b) The relative error when $\hat{k} = 1.1$, $k = 1$.

**Case 2:** $\hat{k} - k > 0$. In this case, we can write

$$\epsilon_r(t) = \left[\exp\left(\hat{k} - k\right)t\right] - 1.$$

Notice that $\epsilon_r(t)$ grows exponentially because $\hat{k} > k$. In fact, once $t$ becomes large enough, the error dwarfs the true solution. This fact is reflected in Fig. 2.14(b), where the graph of $\epsilon_r$ is concave up.

In contrast to the case of errors in the initial condition, errors in the parameter in the exponential growth equation are serious because they lead eventually to large relative errors between the solutions. In particular, the relative error in the solution of the exponential growth equation is independent of the initial condition, provided that the initial condition is nonzero. (See Exercise 4 for another perspective on this situation.)

Again, we emphasize that this analysis applies only to the exponential growth equation, Eq. (2.48). The analysis of errors in other differential equations is more complicated. We will return to this topic from time to time in the remainder of the text.

Errors in parameters and in initial conditions can have profound consequences in many situations. In Chapter 8, we show examples of simple physical systems in which even tiny errors in measuring the initial condition make it impossible to predict the state of the system beyond the near future. Such systems are said to exhibit sensitivity to initial conditions. For instance, it is believed that the earth's weather depends sensitively on initial conditions. Because it is impossible to measure all variables of meteorological relevance precisely, no weather model, no matter how detailed, can make completely accurate predictions. In fact, the inevitable errors in measuring meteorological data make detailed weather forecasts impossible beyond a few days. To appreciate the implications, consider the problem of forecasting where a hurricane will reach land.

## When Does a Glass of Water Reach Room Temperature?

Suppose a glass of water, initially at $1°C$, is placed in a room whose temperature is $20°C$. Assume that the warming of the water obeys Newton's law, $T' = k(T - 20)$, where $k = -1/10$ and $t$ measures time in minutes. Then

$$T(t) = 20 - 19e^{-t/10}. \tag{2.54}$$

When does the water reach room temperature? If we set the left-hand side of Eq. (2.54) equal to 20 and solve for $t$, we find

$$e^{-t/10} = 0,$$

which is impossible. Thus, Eq. (2.54) implies that the water in the glass *never* reaches room temperature. This implication is contrary to everyday experience.

The apparent contradiction can be resolved by recognizing that Newton's law of cooling is a simplified mathematical model of a complex physical process. Among other issues,[5] the mathematical model assumes that we can measure temperature to any arbitrary accuracy. However, no thermometer is completely accurate.

---

[5] Newton's law of cooling ignores some physical processes that become important when we are interested in the finer details of the experiment. The temperature of the water near the surface and the sides of the vessel does not change at the same rate as the temperature of the water near the center of the vessel. In addition, there is evaporation where the water comes in contact with the air, which cools the water near the surface. The resulting temperature gradients stir up the water in the vessel. In other words, if we look closely enough, the water is not an inert mass whose temperature is uniform throughout. Therefore, this discussion concerns the limitations of the model as much as the limitations in measurement.

Suppose that the readings of the thermometer that is used to check the temperature of the water are subject to an absolute error of 0.5°. Figure 2.15 shows the graph of Eq. (2.54). The absolute error in the temperature readings is indicated by two horizontal lines at 19.5° and 20.5°. The area enclosed by these lines forms a "band of uncertainty" in the following sense. Once the true temperature of the water reaches 19.5°, the uncertainty in the thermometer readings implies that the true temperature of the water cannot be distinguished from the room temperature. In other words, *within our ability to tell the difference with an imperfect thermometer*, the water reaches room temperature when the graph falls within the band of uncertainty.

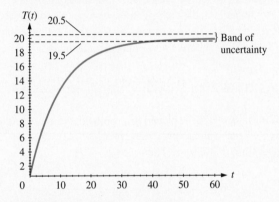

**FIGURE 2.15**    The graph of Eq. (2.54).

A straightforward calculation shows that the "true" temperature, as given by Eq. (2.54), reaches 19.5° when $t \approx 36.4$ min. If we ask an observer, who has a thermometer with an uncertainty of 0.5°, whether the water has reached room temperature, then the answer is a confident "no" only for the first 36.4 min of the experiment.

Table 2.4 shows a list of the times $t_b$ at which the true temperature of the water, as given by Eq. (2.54), reaches some selected values. Suppose we repeat our experiment but give the observer a thermometer with a 1° uncertainty. The observer can distinguish that the temperature of the water in the glass is different from the room temperature only

| °C    | 19.0 | 19.5 | 19.9 | 19.99 | 19.999 |
|-------|------|------|------|-------|--------|
| $t_b$ | 29.4 | 36.4 | 52.5 | 75.5  | 95.5   |

**TABLE 2.4**    The time $t_b$ at which the temperature $T$ reaches the indicated value, based on Eq. (2.54)

for the first 29.4 min of the experiment. If the thermometer is more accurate—say, if the uncertainty is reduced to $0.1°$—then the true temperature of the water can be distinguished from room temperature for a longer time, 52.5 min. Because the thermometer does not have absolute accuracy, the band of uncertainty always has finite width, and the solution curve enters it after a finite amount of time.

## Exercises

**1.** A rule of thumb, often suggested in books on finance, is that the doubling time for an investment is approximately 72 divided by the annual interest rate in percent. For instance, the "rule of 72" implies that money invested at a 6 percent annual interest rate will double in 12 years ($72/6 = 12$).

(a) What is the actual doubling time, assuming that interest is compounded continuously at an annual rate of 6 percent?

(b) What is the relative error in the doubling time as estimated by the "rule of 72" with respect to your answer in (a)?

(c) Is the relative error in the doubling time, as given by the "rule of 72," independent of the interest rate? Explain.

**2.** Discuss how the absolute error in the initial condition grows with time in Eq. (2.48). (Your solution is a special case of a more general result called *Gronwall's inequality*, which is discussed in more detail in the next chapter.)

**3.** The doubling time of an exponential process $p(t)$ is the time $T_d$ for which $p(t + T_d) = 2p(t)$. By virtue of the equal ratio property discussed in Section 2.1, $T_d$ does not depend on $t$.

(a) Consider the exponential process described by Eq. (2.48). What is the doubling time as a function of the parameter $k$?

(b) Find an expression for the relative error in the doubling time as a function of $k$ and the measured value $\hat{k}$.

(c) Suppose the relative error in $k$ is 10 percent. What is the relative error in the doubling time?

**4.** A large relative error is not necessarily important in some circumstances, such as when a process decays exponentially to 0. In this situation, the absolute value of the solution may be so small that a large relative error is irrelevant.

This exercise illustrates an example where a large relative error in a solution may not be considered serious. Let $x' = kx$.

(a) Let $k = -1$. Find the solution that satisfies the initial condition $x(0) = 1$.

(b) Let $\hat{k} = -1.1$. Find the solution that satisfies the initial condition $\hat{x}(0) = 1$.

(c) Complete the following table, where $\epsilon_a$ and $\epsilon_r$ denote the absolute and relative errors in the solution, respectively.

| $t$ | $x(t)$ | $\hat{x}(t)$ | $\epsilon_a(t)$ | $\epsilon_r(t)$ |
|-----|--------|--------------|-----------------|-----------------|
| 0.0 | | | | |
| 1.0 | | | | |
| 2.0 | | | | |
| 3.0 | | | | |
| 4.0 | | | | |
| 5.0 | | | | |

(d) Briefly describe what happens to the absolute error as $t$ gets larger.

(e) Find an expression for the relative error in the solution $x(t)$ as a function of $t$ and plot it. Determine the time, $t_{50}$, when the relative error exceeds 50 percent. Let the $t$ axis in your plot extend at least from $t = 0$ to $t = t_{50}$.

(f) Plot the absolute error in the solution $x(t)$ from $t = 0$ to $t = t_{50}$.

(g) Explain in what sense the large relative error might not be considered significant.

**5.** Newton's law of cooling has a potential forensic application, as follows.

(a) Suppose the temperature of the body of a homicide victim is 23°C when it is discovered in a room whose ambient temperature is 20°C. If nominal body temperature is 37°C and the body cools by 2° one hour after death, how long ago did the murder occur?

(b) In a celebrated criminal case, former football star O. J. Simpson was accused of murdering his ex-wife Nicole Brown Simpson and her friend Ronald Goldman between 9:45 and 10:30 one evening in June 1994. One of the critical points in the case was to establish the time of the murders more precisely. Newton's law of cooling was *not* used for this purpose, and the remainder of this exercise explores one reason why.

     The bodies were found at 6:30 the morning after the murders. If we assume the data in (a), what does your result in (a) imply about the time of the murders?

(c) The rate at which a body cools depends in part on its size and how it is clothed. In any particular case, there is some uncertainty about how much the body might cool in an hour. Suppose that the conditions were as described in (a) but that the bodies cooled by 3°C after one hour. Repeat your calculation of the time of death.

(d) Nominal body temperature varies among individuals. Assume the same set of circumstances as in (a), but suppose that the body temperature of the victims was 36°C at the time of the murders. How long ago did the murders occur?

(e) On the basis of your experience in this question, how good a forensic tool might Newton's law of cooling be? Consider the range of estimates that you obtain for the time of death as you formulate your answer.

**6.** A kettle of water obeys Newton's law of cooling. The initial temperature of the kettle is 95°C. The kettle is placed in a room whose temperature is constant at 20°C. After 10 minutes, the kettle has cooled to 80°C. We want to predict when the kettle will reach 40°C.

(a) Suppose each measurement of temperature is subject to an absolute error of 0.5°C. In other words, the true temperature of the water initially is between 94.5°C and 95.5°C. After 10 minutes, the true temperature of the water is between 79.5°C and 80.5°C. Given these uncertainties, what is the range of values that you can obtain for the rate constant $k$?

(b) What is the range of values that you get for the time at which the kettle reaches 40°C?

**7.** Consider the effect of errors in the initial condition in the case of Newton's law of cooling, given by Eq. (2.12). Let $R$ denote the ambient temperature, and let $T(t)$ be the temperature of the object at time $t$.

(a) Suppose the initial temperature is $T_0$ and the measured value is $\hat{T}_0$. What is the absolute error in $T(t)$ as a function of time?

(b) What is the relative error in $T(t)$ as a function of time?

(c) Is it ever possible for the relative error to go to infinity while the absolute error goes to 0? Explain.

**8.** Consider the effect of errors in the parameter $k$ in the case of Newton's law of cooling, given by

Eq. (2.12). Let $R$ denote the ambient temperature, and let $T(t)$ be the temperature of the object at time $t$.

(a) Suppose the true value of the parameter is $k$ and the measured value is $\hat{k}$. Determine the absolute error in the solution at time $t$. What happens to the absolute error as $t$ becomes large?

(b) Determine the relative error in the solution at time $t$. What happens to the relative error as $t$ becomes large?

(c) Is it ever possible for the relative error to go to infinity while the absolute error goes to 0? Explain.

**9.** Let $x' + x = t$.

(a) Let $x(0) = x_0$. Find the general solution.

(b) Suppose that the actual initial condition is $x_0 = 1$ and that the measured value is $\hat{x}_0 = 1.1$. Find an expression for the absolute error in the solution and plot it. Does the absolute error decay exponentially?

(c) Find an expression for the relative error in the solutions in (b) and plot it. Does the relative error decay exponentially?

(d) When, if ever, do the absolute and relative errors become negligible? Interpret "negligible" in terms of the resolution of your plots in (b) and (c).

**10.** Let $x' + kx = t$ with $x(0) = 1$.

(a) Find the solution.

(b) Suppose $x(t)$ is the solution corresponding to $k = 1$ and $\hat{x}(t)$ is the solution corresponding to $\hat{k} = 1.1$. Find an expression for the absolute error in the solution and plot it.

(c) Find an expression for the relative error in the solution and plot it.

(d) Do the absolute and relative errors decay exponentially? When, if ever, do they become negligible? (Interpret "negligible" in terms of the resolution of your plots.)

**11.** Let $x' + x = mt$ with $x(0) = 1$.

(a) Find the solution.

(b) Suppose $x(t)$ is the solution corresponding to $m = 1$ and $\hat{x}(t)$ is the solution corresponding to $\hat{m} = 1.1$. Find an expression for the absolute error in the solution and plot it.

(c) Find an expression for the relative error in the solution and plot it.

(d) Describe the qualitative behavior of the absolute and relative errors as $t$ becomes large.

**12.** Let $x' + x = \sin t$.

(a) Find the solution that satisfies the initial condition $x(0) = x_0$.

(b) Let $x_0 = 1$ and $\hat{x}_0 = 0.9$. Let $x(t)$ and $\hat{x}(t)$ denote the corresponding solutions. Find an expression for the absolute error in the solution and plot it.

(c) Find an expression for the relative error in the solution. Are there any problems with the definition of the relative error? If so, how might you circumvent them?

(d) Is the general solution sensitive to errors in the initial condition? Explain.

**13.** Let $x' + x = \sin \omega t$, where $\omega$ is a constant.

(a) Find the solution that satisfies the initial condition $x_0 = 1$. (You may find a table of integrals or a computer algebra system helpful.)

(b) Let $x(t)$ be the solution in (a) corresponding to $\omega = 1$ and let $\hat{x}(t)$ be the solution corresponding to $\omega = 1.1$. Plot the two solutions on the same axes for $0 \leq t \leq 20$.

(c) Describe how the solutions differ qualitatively for sufficiently large values of $t$.

(d) Is the relative error a good way to characterize the difference in the solutions? Explain.

**14.** Consider the nonautonomous exponential growth equation

$$x'(t) + kx = f(t),$$

where $k > 0$.

(a) Show that the initial condition always appears in the transient part of the solution, never in the steady state.

(b) Is the solution sensitive to errors in the initial condition? Explain.

(c) Is the solution sensitive to errors in the parameter $k$? Explain.

(d) Which terms in the equation govern the behavior of the steady-state term?

**15.** To determine the half-life of a radioactive substance, you must measure the mass of a sample of the substance over time. Every laboratory measurement necessarily involves some error, so there is always uncertainty in the estimate of the half-life. Let us consider this question in a simple example.

(a) Suppose a 100-microgram ($\mu$g) sample of some radioactive substance decays to 90 $\mu$g after one hour. Assuming that radioactive decay is exponential and follows Eq. (2.48), determine the rate constant $k$ and the half life $t_{1/2}$.

(b) Now assume that each measurement of the mass is subject to an absolute error of 1 $\mu$g.

(In other words, the initial measured mass is 100 $\mu$g, but the true value can be anywhere between 99 $\mu$g and 101 $\mu$g, and similarly for the second measurement.) What is the maximum absolute error in the rate constant $k$ and the half-life $t_{1/2}$ that can occur as a result of this uncertainty?

**16.** Suppose that you have a sample of a radioactive isotope whose true half-life is 10.0 h. You record the initial mass of the sample and the mass one hour later. These two measurements suffice in principle to determine the half life of the isotope. How accurate must each of your measurements be if you want to determine the half-life with a relative error of no more than 1 percent? (In other words, how much error can you tolerate in each measurement so that your estimate of the half-life is between 9.9 h and 10.1 h?). Assume a constant absolute error in each of the mass measurements.

(a) Answer the question under the assumption that the measured mass is 100 $\mu$g initially.

(b) Find a formula or algorithm that allows you to determine the maximum absolute error in the mass measurements that satisfies the requested accuracy of the half-life, given that the true value of the initial mass is $m_0$. Assume that each measurement of the mass is subject to the same absolute error $\epsilon_a$.

# Chapter 3

## FUNDAMENTAL THEORY

Chapter 2 was concerned primarily with applications of the exponential growth equation, whose solutions are simple exponential functions. However, most of the differential equations that arise in practice do not have solutions that can be written as simple functions of time. In fact, it is not always clear whether a given differential equation has a solution at all.

This chapter is devoted to four basic questions:

1. The solution of the exponential growth equation $x' = kx$ is $x(t) = x_0 e^{kt}$. The function $x(t)$ is defined, continuous, and differentiable for every number $t$. Are the solutions of other differential equations functions defined, continuous, and differentiable for every $t$?

2. Although $x(t) = x_0 e^{kt}$ is a solution of the exponential growth equation $x' = kx$, is there another, different function that also satisfies the same differential equation with the same initial conditions? If so, then predictions of the future state of an exponential process are impossible, because more than one outcome is consistent with the available information about the rate of change of the dependent variable.

3. Given an initial value problem such as

$$\frac{dx}{dt} = x^2 + t^2, \qquad x(0) = x_0, \tag{3.1}$$

whose analytical solution is not obvious, how do we know whether a solution exists at all? Equation (3.1) is not separable, and the method of integrating factors discussed in Section 2.3 does not work. (Try it.) Nevertheless, we can apply Euler's method with a suitably small step size to find a curve that appears to be consistent with the direction field, at least for values of $t$ near 0. How can we understand this situation, wherein we can find a numerical approximation of the solution but not an explicit formula for it? (See, for instance, Example 4 in Section 3.1.) From a mathematical point of view, then, we want to know which differential equations have solutions, at least in principle.

4. Errors in measurement are an unavoidable part of every scientific investigation. Even if a given differential equation is a highly accurate model of some physical process, the initial state of the process (and hence the initial condition of the differential equation) cannot be known with absolute precision. How quickly can errors in measuring the initial condition affect one's ability to predict the future state of the process based on the solution of the underlying differential equation? For example, if small errors in initial conditions grow rapidly, then the solution cannot predict the future state of the process except for a short period of time. More precisely, we need to know how solutions of a given differential equation change as a function of the initial condition.

We call this chapter *fundamental theory* because these questions can be answered satisfactorily *even if it is not possible to find an explicit formula for the solution of the differential equation.* As a result, most differential equations that arise in practice have solutions that behave in a reasonable way. Such assurance is essential to put the art of mathematical modeling with differential equations on a solid theoretical footing.

Much of the basic theory relies on ideas that are geometrical in origin. Throughout the chapter, we state why each result is important and what its implications are from the point of view of a scientist or engineer. As the chapter proceeds, we will explore the questions listed above in more detail.

## 3.1   INTERVALS OF EXISTENCE

The exponential growth equation

$$\frac{dx}{dt} = kx$$

arises whenever the rate of growth of some quantity is proportional to its present size. The solution, given by

$$x(t) = x_0 e^{kt}, \tag{3.2}$$

is well defined for every number $t$. We can characterize the solutions of the exponential growth equation as follows:

- For any given value of $t$, $x(t)$ is a well-defined, continuous function; $x'(t)$ exists; and $x(t) \neq 0$ except in the trivial case in which $x_0 = 0$.

- If $k > 0$, then the solution (3.2) tends to $\infty$ only in the limit as $t \to \infty$. We say that $x(t)$ *goes to infinity in infinite time.* If $k < 0$, then $x(t)$ tends to 0 only in the limit as $t \to \infty$.

Let us explain the second observation in more detail. Suppose $t = 10$. Then $x(10) = x_0 e^{10k}$, which is a finite number. Similarly, $x(10^4)$ and $x(-300)$ also are finite values. The value of $x(t)$ for a given $t$ may be enormously large or extremely small, but it is finite and it is computable. (See Exercise 20 in Section 2.1.)

Now consider the initial value problem

$$\frac{dx}{dt} = x^2, \qquad x(0) = 1, \tag{3.3}$$

whose solution is

$$x(t) = \frac{1}{1 - t}. \tag{3.4}$$

The solution can be characterized mathematically as follows:

- The solution (3.4) is not defined for $t = 1$. Equation (3.3) is an example of a differential equation whose solutions are functions that are not defined for all real numbers.
- The solution (3.4) tends to $\infty$ as $t \to 1$. We say that $x(t)$ *goes to infinity in finite time.*

Figure 3.1 shows a plot of Eq. (3.4). The vertical asymptote illustrates the idea that

$$\lim_{t \to 1} |x(t)| = \infty.$$

We say that $x(t)$ has a *singularity* at $t = 1$.

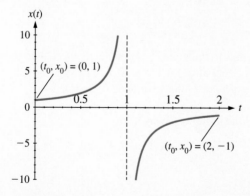

**FIGURE 3.1**   The graph of Eq. (3.4). The dashed line indicates the asymptote.

**Definition 3.1** *Let $x(t)$ be a solution of the differential equation $x' = f(t, x)$. We say that $x$ has a* singularity *at $s$ if either*

- $x(t)$ *is undefined or discontinuous at $t = s$ or*
- $\lim\limits_{t \to s} |x(t)| = \infty$.

The initial condition $x(0) = 1$ corresponds to the point $(0, 1)$ in Fig. 3.1. The curve containing $(0, 1)$ cannot be extended past $t = 1$ in a continuous way. Therefore, we say that the *interval of existence* of the solution $x(t)$ is the interval $(-\infty, 1)$.

**Definition 3.2** *Let $x(t)$ be a solution of the differential equation*

$$\frac{dx}{dt} = f(t, x)$$

*satisfying the initial condition $x(t_0) = x_0$. The interval of existence of $x(t)$ is the largest interval containing $t_0$ on which $x(t)$ is continuous and differentiable.*

◆ *Example 1*   The solution of the initial value problem

$$x' = 2x, \qquad x(0) = 1$$

is $x(t) = e^{2t}$, which is defined, continuous, and differentiable for every real number $t$. Hence, the interval of existence is the entire real line.                                    ∎

◆ *Example 2*   The solution of the initial value problem

$$x' = x^2, \qquad x(0) = 1 \tag{3.5}$$

is

$$x(t) = \frac{1}{1 - t}.$$

The initial condition corresponds to the point $(0, 1)$, which is located on the hyperbola to the left of the vertical asymptote at $t = 1$ in Fig. 3.1. The hyperbola is defined for $t < 1$. Hence the interval of existence of the solution is $(-\infty, 1)$.                   ∎

◆ *Example 3*   The solution of the initial value problem

$$x' = x^2, \qquad x(2) = -1$$

also is

$$x(t) = \frac{1}{1 - t}.$$

In this case, however, the initial value is specified at $t = 2$, corresponding to the point $(2, -1)$ in Fig. 3.1. The hyperbola containing $(2, -1)$ is defined only for $t > 1$. Thus, the interval of existence is $(1, \infty)$.                                                                  ∎

In Examples 2 and 3, neither interval of existence can be extended to include $t = 1$, because $x(t)$ has a singularity there. By definition, the set of all numbers with the number 1 excluded does not constitute an interval, because this set contains a gap. We commonly speak of the *domain* of a function as the set of all numbers over which the function is defined. However, *the domain of a function $x(t)$ is not the same set as the interval of existence*, as we now explain. For example, although the function

$$x(t) = \frac{1}{1 - t}$$

is defined for $t \neq 1$, its graph consists of two disjoint curves, only one of which contains a given initial condition. The function $x(t)$ is a solution of the initial value problem (3.5), where the initial condition is specified as $x(0) = 1$. The solution is the left-hand curve in Fig. 3.1, because only the left-hand curve contains the initial point $(0, 1)$. The right-hand curve is irrelevant to the solution satisfying the initial condition $x(0) = 1$. Therefore, the interval of existence is $(-\infty, 1)$.

The interval of existence depends on the initial condition as well as the differential equation. The interval of existence often can be estimated even when no explicit formula for the solution can be found, as the next example shows.

✦ *Example 4*   The comparison theorem (Theorem 1.2 in Section 1.2) can be used to show that the solution $x(t)$ of the initial value problem

$$\frac{dx}{dt} = x^2 + t^2, \qquad x(0) = 1 \tag{3.6}$$

has a singularity between $t = \pi/4$ and $t = 1$, as follows.

Let $x_a$ denote the solution of the initial value problem

$$x_a' = x_a^2, \qquad x_a(0) = 1, \tag{3.7}$$

and let $x_b$ denote the solution of the initial value problem

$$x_b' = x_b^2 + 1, \qquad x_b(0) = 1. \tag{3.8}$$

The functions $x$, $x_a$, and $x_b$ all start from the same initial value at $t = 0$, and $x_a'(t) < x'(t) < x_b'(t)$ when $0 < t < 1$ because $x^2 < x^2 + t^2 < x^2 + 1$. Therefore, the comparison theorem implies that

$$x_a(t) < x(t)$$

on the interval of existence for $x_a(t)$ and

$$x(t) < x_b(t)$$

on the interval of existence for $x_b(t)$. In other words, $x_a$ is a lower bound on the solution of the initial value problem (3.6), and $x_b$ is an upper bound.

Using the separation of variables, we find that the solution of (3.7) is

$$x_a(t) = \frac{1}{1-t},$$

and the solution of (3.8) is

$$x_b(t) = \tan\left(t + \frac{\pi}{4}\right).$$

The solution $x_a$ has a singularity at $t = 1$, and $x_b$ has a singularity at $t = \pi/4$. Since $x_a$ is a lower bound for the solution $x$ of Eq. (3.6) and $x_b$ is an upper bound, the solution $x$ has a singularity between $t = \pi/4$ and $t = 1$.

Figure 3.2 shows the graphs of $x_a$ and $x_b$ together with a numerical approximation of the solution $x$ of Eq. (3.6). Euler's method can be used to locate the singularity in the solution of Eq. (3.6) more precisely. See Exercise 3.

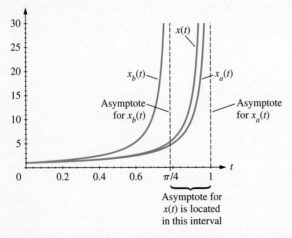

**FIGURE 3.2**   The solutions of Eqs. (3.6)–(3.8).

## Exercises

**1.** Let $x' = x^2$.

(a) Find the solution satisfying the initial condition $x(0) = 2$ and determine the interval of existence.

(b) Find the solution satisfying the initial condition $x(0) = -2$ and determine the interval of existence.

(c) Find the solution (if there is one) satisfying the initial condition $x(0) = 0$ and determine the interval of existence.

(d) Find an initial condition at $t = 0$ for which the solution has a singularity at $t = 1/10$.

(e) Let $x_0$ be a positive constant. Find the solution satisfying the initial condition $x(0) = x_0$ and determine the interval of existence.

(f) Let $x_0$ be a positive constant. Find the solution satisfying the initial condition $x(0) = -x_0$ and determine the interval of existence.

(g) Show that $x_0$ can be chosen so that the initial condition $x(0) = x_0$ leads to a solution has a singularity at $t = \delta$, where $\delta$ is an arbitrary positive number.

**2.** Let

$$x' = x + \cos x.$$

Show that the solution satisfying the initial condition $x(0) = 1$ does not have a singularity for $t > 0$. (Use the comparison theorem.)

**3.** Let

$$x' = x^2 + t^2, \qquad x(0) = 1.$$

(a) Use Euler's method to show that $x(0.9) \approx 14$. What is an appropriate step size?

(b) Argue that the numerical approximation produced by Euler's method underestimates the true value of $x(t)$.

(c) Find the solution of the initial value problem

$$x' = x^2, \qquad x(0.9) = 14.$$

(d) Use your results in (b) and (c) and Example 4 to show that the true solution has a singularity between $\pi/4$ and $34/35$. This is a sharper estimate of the location of the singularity than the one derived in Example 4.

**4.** Show that the solution of the initial value problem

$$x' = x^2 + t^2, \qquad x(0) = 2$$

has a singularity and estimate its location as accurately as you can. Use a strategy similar to that outlined in Example 4.

**5.** Let $x' = kx^2$.

(a) Let $k$ be a positive constant. Find the solution satisfying the initial condition $x(0) = 2$ and determine the interval of existence as a function of $k$.

(b) Find the solution satisfying the initial condition $x(0) = -2$ and determine the interval of existence as a function of the constant $k$.

(c) Repeat (a)–(b) assuming that $k$ is a negative constant.

**6.** Let $x' = x^{1/2}$.

(a) Find the solution satisfying the initial condition $x(0) = 1$ and determine the interval of existence.

(b) Let $x_0$ be any constant. For which values of $x_0$ is there a solution satisfying the initial condition $x(0) = x_0$?

(c) Can $x_0$ be chosen so that the initial condition $x(0) = x_0$ leads to a solution that has a singularity for a positive value of $t$?

(d) Is there a nonzero value of $x_0$ for which $x(t) = 0$ for some positive $t$?

**7.** Let $x' = x^3$.

(a) Find the solution satisfying the initial condition $x(0) = 1$ and determine the interval of existence.

(b) Find the solution satisfying the initial condition $x(0) = -1$ and determine the interval of existence.

(c) Let $x_0$ be a positive constant. Find the solution satisfying the initial condition $x(0) = x_0$ and determine the interval of existence.

(d) Let $x_0$ be a positive constant. Find the solution satisfying the initial condition $x(0) = -x_0$ and determine the interval of existence.

(e) For which constants $x_0$ does the initial condition $x(0) = x_0$ lead to a solution that has a singularity?

**8.** Repeat Exercise 7 for $x' = x^{1/3}$.

**9.** Let $x' = xt$. Show that the interval of existence of the solution satisfying the initial condition $x(0) = x_0$ is the entire real line, regardless of $x_0$.

**10.** Let

$$x' = \frac{xt}{t^2 - 1}. \qquad (3.9)$$

Find the solution, where possible, that satisfies the specified initial condition and determine its interval of existence.

(a) $x(0) = 1$        (b) $x(0) = -1$

(c) $x(0) = 0$        (d) $x(2) = 1$

(e) $x(2) = -1$

**11.** Suppose $p$ is an integer greater than 1 and let $x' = x^p$.

(a) Find the solution satisfying the initial condition $x(0) = 1$ and determine how the interval of existence depends on $p$.

(b) Let $x(0) = x_0$ be positive. Find the solution satisfying the initial condition $x(0) = x_0$ and determine the interval of existence.

**12.** Repeat Exercise 11 for the case $x' = x^{1/p}$, where $p$ is an integer greater than 1.

**13.** Repeat Exercise 11 for the case $x' = x^{-p}$, where $p$ is an integer greater than 1.

**14.** Repeat Exercise 11 for the case $x' = x^{-1/p}$, where $p$ is an integer greater than 1.

## 3.2   EXISTENCE AND UNIQUENESS OF SOLUTIONS

### The Uniqueness Problem

When it is possible to find a formula for or a numerical approximation to the solution of an initial value problem, a fundamental question is the uniqueness of the solution. In other words, given one solution, does there exist a second solution that also solves the initial value problem?

The practical implications of such a question are important. One of the main uses of differential equations is to make predictions about the future behavior of a system. If there is more than one solution, then it is impossible to make predictions, because more than one outcome is consistent with the initial state.

We will motivate the question with a specific physical example. Torricelli's law describes the rate at which fluid drains from a tank. Imagine a cylindrical bucket with a hole in the bottom. Torricelli's law says that the rate at which water drains from the bucket is proportional to the square root of the height of the water remaining inside. Let $h(t)$ denote the height of the water in the bucket at time $t$, and suppose for the sake of algebraic simplicity that all constants of proportionality are unity. Then Torricelli's law can be stated as

$$\frac{dh}{dt} = -h^{1/2}. \tag{3.10}$$

Notice that the bucket is empty when $h = 0$. (Explain the rationale for the minus sign.)

Suppose that the height of the water initially is 1. Equation (3.10) can be used to predict how long it will take the bucket to drain. Equation (3.10) is separable, and an implicit solution is

$$2\sqrt{h} = -t + C.$$

The initial condition $h(0) = 1$ implies that

$$2\sqrt{h} = -t + 2, \tag{3.11}$$

which for $0 \le t \le 2$ is equivalent to

$$h(t) = \left(1 - \tfrac{1}{2}t\right)^2. \tag{3.12}$$

Equations (3.11)–(3.12) imply that the bucket empties at $t = 2$. However, Eq. (3.11) is not a valid solution for $t > 2$, because the right-hand side becomes negative. Physically, of course, the height of the water remains 0 after the bucket empties. The expression for $h$ must be amended accordingly; we have

$$h(t) = \begin{cases} \left(1 - \tfrac{1}{2}t\right)^2, & 0 \le t \le 2, \\ 0, & t > 2. \end{cases} \tag{3.13}$$

Figure 3.3 shows a plot of Eq. (3.13). Let us check whether $h(t)$ solves Eq. (3.10) for all positive $t$. By the chain rule, the first expression in Eq. (3.13) gives

$$h'(t) = 2\left(1 - \tfrac{1}{2}t\right) \times \tfrac{1}{2} = 1 - \tfrac{1}{2}t = \sqrt{h(t)}.$$

The derivative of each expression in Eq. (3.13) is 0 when $t = 2$. Each expression satisfies Eq. (3.10), and the first expression satisfies the initial condition. Thus, $h(t)$ solves Eq. (3.10). The important point here is that we can predict when the bucket will empty, given a measurement of how much water it contains now.

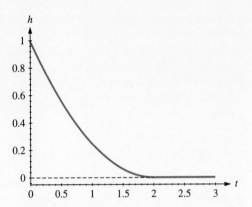

**FIGURE 3.3**    A plot of the solution (3.13).

The situation is not so simple if we want to make inferences about the past. Suppose that the bucket is initially empty. Clearly, the bucket could have drained at any point in the past. Indeed, the bucket may never have contained any water at all. In other words, the initial observation of an empty bucket provides no information about the past. Let us see how this difficulty is reflected mathematically.

Notice that the zero function $h(t) = 0$ is an equilibrium solution of Eq. (3.10). The zero function is consistent with the idea that the bucket has always been empty. On the other hand, the bucket could have emptied at $t = -1$; then we have a nonzero function whose graph is similar to that in Fig. 3.3, but with the time axis shifted. When $t < -1$, that is, before the bucket emptied, the height of the water inside would have been given by $x(t) = \left(\frac{1}{2}t + \frac{1}{2}\right)^2$, as a substitution into Eq. (3.11) shows. Thus, the function

$$h(t) = \begin{cases} \left(\frac{1}{2} + \frac{1}{2}t\right)^2, & t \le -1, \\ 0, & t > -1 \end{cases} \tag{3.14}$$

is another solution of Eq. (3.10) that is consistent with the initial condition $x(0) = 0$.

In fact, since the bucket could have emptied at any time in the past, say at $t = -A$ for some positive number $A$, there are infinitely many possible solutions of Eq. (3.10), each given by

$$h(t) = \begin{cases} \left[\frac{1}{2}(A + t)\right]^2, & t \le -A, \\ 0, & t > -A. \end{cases} \tag{3.15}$$

Figure 3.4 shows the direction field corresponding to Eq. (3.10) together with three representative solutions, each of which satisfies the initial condition $h(0) = 0$.

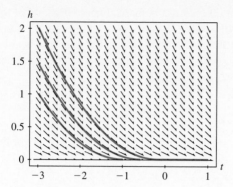

**FIGURE 3.4**   The direction field and three solutions corresponding to Torricelli's law, Eq. (3.10).

In this example, it is impossible to make an inference about the past state of an empty bucket; one scenario is as valid as another. Theorem 3.3, discussed later in this section, gives conditions under which an initial value problem is guaranteed to have a unique solution.

## The Existence Problem

### Two Motivating Examples

The first important mathematical question concerning an initial value problem is whether a solution exists at all. Consider Torricelli's law again,

$$\frac{dx}{dt} = -\sqrt{x}. \tag{3.16}$$

The preceding discussion shows the derivation of solutions that satisfy the initial conditions $x(0) = 1$ and $x(0) = 0$. However, the initial condition $x(0) = -1$ poses a difficulty: Equation (3.16) implies that $x'(0)$ is an imaginary number if $x(0) = -1$. Therefore, there is no solution that satisfies the initial condition $x(0) = -1$. (The solution, if it exists, is a complex-valued function.) This example shows that a given differential equation need not have a solution for an arbitrary initial condition.

Problems can arise when solutions of a given differential equation are approximated numerically. In practice, most of the differential equations that one encounters cannot be solved analytically. Frequently, a computer program is used to generate a numerical approximation of the solution of a given initial value problem. However, numerical methods are not always reliable, as we now discuss.

Let $x' = \tan x$. The curve in Fig. 3.5 is a numerical approximation of the solution satisfying the initial condition $x(0) = 0.1$. The numerical solution appears to be consistent with the direction field for $0 < t < 2.5$ or so. However, the curve is not consistent with the direction field for $2.5 < t < 3.5$. The plot suggests that the numerical solution does not reflect the behavior of the true solution.

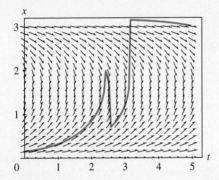

**FIGURE 3.5**   The direction field and a numerically generated approximation to the solution of $x' = \tan x$, $x(0) = 0.1$.

A moment's thought reveals one problem with the numerical solution. Notice that $\pi/2 \approx 1.5$. The derivative $x' = \tan x$ is not defined at $x = \pi/2$, because $\tan(\pi/2)$ is not defined. The slope lines become vertical as $x \to \pi/2$, and any numerical scheme (such as Euler's method) that attempts to approximate the solution by tangent lines is likely to lead to problems when the slopes of the tangent lines attain very large values. This example suggests that difficulties may arise when one tries to solve the differential equation

$$\frac{dx}{dt} = f(t, x),$$

either analytically or numerically, if $f(t, x)$ is not defined at some point.

## The Existence and Uniqueness Theorem

We now state a criterion that can be used to determine when solutions of a differential equation exist and are unique.

**Theorem 3.3**   *The Existence and Uniqueness Theorem*   *Let*

$$\frac{dx}{dt} = f(t, x),$$

and let S be the region $\{(t, x) : t \in [a, b], x \in [c, d]\}$. *Suppose that* $f$ *and* $\partial f / \partial x$ *are continuous at every point in S and that* $(t_0, x_0) \in S$. *Then a solution of the initial value problem* $x(t_0) = x_0$ *exists and is unique, at least as long as the graph of* $x(t)$ *remains in S.*

The next examples illustrate how the theorem can be applied.

✦ *Example 1*   The exponential growth equation is

$$\frac{dx}{dt} = f(t, x) = kx.$$

Then $\partial f / \partial x = k$. Suppose $x' = x$ and $x(0) = 1$. The hypotheses of Theorem 3.3 are satisfied, so any solution satisfying the initial condition $x(0) = 1$ is unique. Therefore, $x(t) = e^{kt}$ is the only solution.

An intuitive understanding of the mathematical theory can be gained by studying Fig. 3.6, which shows the direction field for $k = 1$ in a small box around $(0, 1)$. The slope lines suggest that it is possible to draw only one curve through $(0, 1)$ that is consistent with the direction field. The existence and uniqueness theorem confirms our intuition.

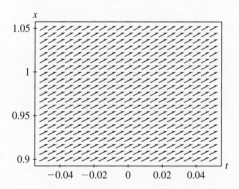

**FIGURE 3.6**   The direction field for the equation $x' = x$ in a region containing the point $(0, 1)$. ■

✦ *Example 2*   Theorem 3.3 does not apply to the initial value problem

$$\frac{dx}{dt} = f(t, x) = \sqrt{x}, \qquad x(0) = 0.$$

Observe that

$$\frac{\partial f}{\partial x} = \frac{1}{2\sqrt{x}},$$

which is not defined if $x = 0$. Therefore, the hypotheses of the theorem are not satisfied, and the theorem cannot be used to say anything about the solutions, if any.

Figure 3.7 shows the corresponding direction field in a small box containing $(0, 0)$. Obviously, the zero function is a solution satisfying the initial condition $x(0) = 0$, and its graph coincides with the horizontal axis. But the function $x(t)$ defined by

$$x(t) = \begin{cases} t^2/4, & t \geq 0, \\ 0, & t < 0, \end{cases}$$

also solves the initial value problem, as illustrated in Fig. 3.7. In this example, there are two curves through the origin that are both consistent with the direction field. This finding is not inconsistent with Theorem 3.3, because the theorem says nothing about the solutions of differential equations that do not satisfy its hypotheses.

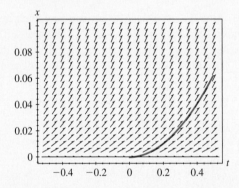

**FIGURE 3.7**    The direction field in a small box containing $(0, 0)$ for $x' = \sqrt{x}$. The curve represents a solution satisfying the initial condition $x(0) = 0$. The constant function $x(t) = 0$ is another solution.                                                                                    ■

◆ *Example 3*    The initial value problem

$$\frac{dx}{dt} = \sqrt{x}, \qquad x(0) = 1$$

*does* have a unique solution, in contrast to the previous example. Let $S$ be the box of width $1/2$ around $(0, 1)$. Then

$$\frac{\partial f}{\partial x} = \frac{1}{2\sqrt{x}}$$

is continuous at every point of $S$, because $1/2 \leq x \leq 3/2$. The hypotheses of Theorem 3.3 are satisfied, so the solution satisfying the initial condition $x(0) = 1$ is unique, at least for as long as the solution curve lies in $S$. Figure 3.8 shows the direction field in $S$ together with the solution curve. It appears that there is only one way to draw a solution curve through the initial point $(0, 1)$ in a manner that is consistent with the slope lines.

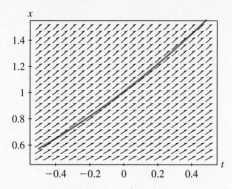

**FIGURE 3.8**   The direction field in a box of width $1/2$ about $(0, 1)$ for $x' = \sqrt{x}$. The curve represents a solution satisfying $x(0) = 1$.  ■

Examples 2 and 3 show that the existence and uniqueness of solutions depend on both the differential equation and the initial condition. A given differential equation may be guaranteed to have unique solutions for some initial conditions but not for others.

✦ *Example 4*   If we restrict attention to the set of all real-valued functions, then no solution of the initial value problem

$$x' = \sqrt{x}, \qquad x(0) = -1$$

exists, because $\sqrt{-1}$ is not a real number.  ■

✦ *Example 5*   Theorem 3.3 does not apply to the initial value problem

$$tx' = x, \qquad x(0) = 0, \tag{3.17}$$

for the following reason. The differential equation can be rewritten as

$$x' = f(t, x) = \frac{x}{t}.$$

We have

$$\frac{\partial f}{\partial x} = \frac{1}{t},$$

which tends to $\infty$ as $t \to 0$ and is undefined at $t = 0$. The initial condition is the origin.

Therefore, no conclusion about the existence and uniqueness of solutions can be drawn from Theorem 3.3.

Figure 3.9 shows the direction field corresponding to Eq. (3.17). Although Theorem 3.3 does not apply, $x(t) = 0$ is an equilibrium solution, so $x(t) = 0$ solves the initial value problem in Eq. (3.17).

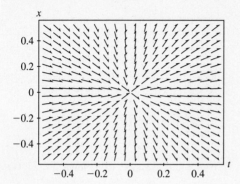

**FIGURE 3.9**   The direction field for $tx' = x$.

In fact, there are infinitely many solutions that satisfy the initial condition $x(0) = 0$. The isoclines in Fig. 3.9 are straight lines through the origin; they are lines of the form $x = mt$, where $m$ is a constant. On the other hand,

$$\frac{d}{dt}(mt) = m = \frac{mt}{t},$$

which shows that $x(t) = mt$ solves Eq. (3.17). Notice also that the graph of $x(t) = mt$ goes through the origin for any constant $m$. Hence, there are infinitely many solutions that satisfy the initial condition $x(0) = 0$.    ∎

If the hypotheses of Theorem 3.3 are not satisfied, then no conclusion about the existence and uniqueness of the solutions of a given initial value problem can be drawn. There may be no solution (as in Example 4), infinitely many solutions (as in Example 5), or a unique solution (see Exercise 7).

✦ *Example 6*   This example applies Theorem 3.3 to the qualitative analysis of differential equations. Let $S$ be a region for which the equation

$$x' = f(t, x)$$

satisfies the hypotheses of Theorem 3.3. Suppose $x_1$ and $x_2$ are two solutions. Then the graphs of $x_1$ and $x_2$ cannot intersect in $S$, for the following reason. Suppose the graphs did intersect at some point in $S$, say $(t_0, x_0)$. If $x_1$ and $x_2$ are distinct functions, then they are two different solutions satisfying the initial condition $x(t_0) = x_0$. But this conclusion contradicts Theorem 3.3. ∎

## Exercises

**1.** For each equation in the following list, determine the set of initial conditions for which Theorem 3.3 guarantees a unique solution.

(a) $x' = kx$

(b) $x' = \sin x$

(c) $x' = t/x$

(d) $x' = (\cos t) + x^3 - x$

**2.** Let $x' + kx = g(t)$, where $g$ is any continuous function and $k$ is any constant. Show that the solution satisfying any initial condition is unique.

**3.** Let $x' = x^{1/3}$.

(a) Show that the hypotheses of Theorem 3.3 do not hold in any box around the initial point $(0, 0)$.

(b) Using an argument similar to that for Torricelli's law, construct a set of distinct solution curves that all satisfy the initial condition $x(0) = 0$.

(c) Use Theorem 3.3 to show that the solution satisfying the initial condition $x(0) = 1$ is unique.

(d) Let $x(t_0) = x_0 \neq 0$. Show that the hypotheses of Theorem 3.3 are satisfied if $S$ is a sufficiently small box around $(t_0, x_0)$. Conclude that the initial value problem has a unique solution as long as the solution curve lies in $S$.

(e) Find an explicit formula for the function $x(t)$ in (d).

**4.** Let $x' = x^p$, where $0 < p < 1$.

(a) Show that the hypotheses of Theorem 3.3 do not hold for the initial condition $x(0) = 0$. (You may restrict attention to points where $x \geq 0$.)

(b) Show that the hypotheses of Theorem 3.3 are satisfied if $S$ is a sufficiently small box containing the initial point $(t_0, x_0) = (0, 1)$. Conclude that the solution satisfying the initial condition $x(0) = 1$ is unique for as long as the graph of the solution remains in $S$.

(c) Show that the hypotheses of Theorem 3.3 are satisfied if $S$ is a sufficiently small box containing the initial point $(t_0, x_0) = (0, x_0)$, provided that $x_0 > 0$. Conclude that the solution satisfying the initial condition $x(0) = x_0$ is unique as long as the graph of the solution remains in $S$, provided that $x_0 > 0$.

**5.** Let $x' = x^2$. Show that the hypotheses of Theorem 3.3 are satisfied when the initial condition is $x(0) = 1$ and that the solution is unique, even though the solution has a singularity.

**6.** Let $x' = -\sqrt{x}$ with $x(0) = 2$.

(a) Show that a unique solution exists for this initial value problem.

(b) Find an explicit solution of the initial value problem.

(c) Notice that

$$x_1(t) = \begin{cases} \left(\sqrt{2} - \frac{1}{2}t\right)^2, & 0 \leq t \leq \sqrt{8}, \\ 0, & t > \sqrt{8}. \end{cases} \tag{3.18}$$

and

$$x_2(t) = \left(\sqrt{2} - \frac{1}{2}t\right)^2 \tag{3.19}$$

are two different functions for which $x_1(0) = x_2(0) = 2$. Does the existence of these two functions imply that the initial value problem has more than one solution? Explain.

**7.** Let $tx' = -x$.

(a) Show that the hypotheses of Theorem 3.3 do not hold for the initial condition $x(0) = 0$.

(b) Plot the direction field around the origin. Argue that a unique solution exists satisfying the initial condition $x(0) = 0$. What is the solution?

**8.** Figure 3.10 shows numerical approximations of the solutions of the differential equation

$$x' = -x \tan(t - 1.3)$$

for the initial conditions $x(0) = 1$ and $x(0) = 0.9$. Do the numerical solutions have any features that suggest that the numerical solutions are *not* in reasonable agreement with the true solutions? Explain.

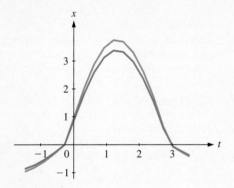

**FIGURE 3.10**    For Exercise 8.

**9.** Let $tx' = 4x$.

(a) Generate a direction field on the rectangle $-4 \le t \le 4, -4 \le x \le 4$.

(b) Use Euler's method with a step size of 0.1 to generate a numerical approximation of the solution $x(t)$ satisfying the initial condition $x(-1) = 1$. Print out the results at each time step. What difficulties, if any, do you encounter?

(c) The zero function is an equilibrium solution. Euler's method for the initial value problem

in (b) suggests that $x(-0.1) = 0$. Is this conclusion wrong?

(d) Let $x(0) = x_0$. Can you use Theorem 3.3 to determine whether a solution of this initial value problem exists and is unique?

(e) Find an explicit solution satisfying the initial condition $x(-1) = 1$. What is the interval of existence of the solution? Does your answer contradict your result in (d)?

(f) Show that there are infinitely many solutions that satisfy the initial condition $x(0) = 0$. (Hint: Show that the solution satisfying every initial condition of the form $x(t_0) = x_0, t_0 \ne 0$, goes through the origin.)

**10.** Let $tx' = -4x$.

(a) Generate a direction field on the rectangle $-4 \le t \le 4, -4 \le x \le 4$.

(b) Use Euler's method with a step size of 0.1 to generate a numerical approximation of the solution $x(t)$ satisfying the initial condition $x(-1) = 1$. Print out the results at each time step. What difficulties, if any, do you encounter?

(c) Let $x(t)$ be the solution of the initial value problem in (b). What does Euler's method imply about the value of $x(0)$? Is Euler's method wrong?

(d) Find an explicit solution of the initial value problem in (b). What is the interval of existence of the solution? What, if anything, can you say about the true value of $x(0)$?

(e) Show that your solution in (d) is unique.

(f) Argue that there is a unique solution satisfying the initial condition $x(0) = 0$. What is it? (Hint: Look carefully at the direction field.)

(g) Does Theorem 3.3 apply to initial conditions of the form $x(0) = x_0$? Explain. Does the fact that a unique solution exists for $x(0) = 0$ contradict the theorem?

## 3.3   DEPENDENCE ON INITIAL CONDITIONS

Section 2.4 discusses how errors in the initial condition affect the solutions of the exponential growth equation. We can regard $x(t)$ as the solution corresponding to the "true" state of the system and $\hat{x}(t)$ as the solution corresponding to a "predicted" state of the system based on an initial measurement. The absolute difference $\epsilon_a = |x(t) - \hat{x}(t)|$ can be regarded as the absolute error in the predicted state of the system at time $t$. Obviously, smaller values of $\epsilon_a$ correspond to more accurate predictions.

This section discusses a general theory, called Gronwall's inequality, that bounds the rate at which $\epsilon_a$ can grow with time. It is possible to regard Gronwall's inequality as a worst-case estimate of the size of the absolute error at some future time, given an error in the initial condition. An important aspect of the theory is that it can be applied without finding an explicit solution of the differential equation.

Figure 3.11(a) shows two solutions of the differential equation

$$x' = x - t. \tag{3.20}$$

One solution satisfies the initial condition $x_0 = 1$, and the other satisfies $\hat{x}_0 = 1.1$. The absolute difference between the two curves grows with time, although relatively slowly, over the indicated time interval.

Figure 3.11(b) shows two solutions of the differential equation

$$x' = x^2 \tag{3.21}$$

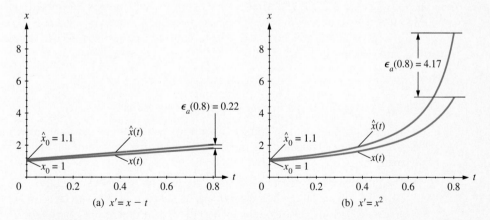

(a) $x' = x - t$                                        (b) $x' = x^2$

**FIGURE 3.11**   (a) Two solutions of Eq. (3.20). (b) Two solutions of Eq. (3.21) for the same initial conditions.

for the same initial conditions. In contrast to Eq. (3.20), these solution curves separate much more quickly. Put another way, the solutions of Eq. (3.21) appear to be more sensitive to errors in initial conditions than do those of Eq. (3.20). If we think of $x(t)$ as the "true" solution in each case and $\hat{x}(t)$ as a "predicted" solution based on a slightly erroneous measurement of the initial condition, then the absolute error in the solutions of Eq. (3.21) at $t = 0.8$ is about 19 times greater than that in the solutions of Eq. (3.20).

Figure 3.11 illustrates a basic problem:

> *Given two differential equations, is it possible to determine which of them is more likely to be sensitive to errors in the initial condition, and can this determination be made without deriving explicit formulas for the solutions?*

The goal of this section is to show that the question can be answered with a qualified "yes." In the following discussion, we introduce a result, called *Gronwall's inequality*, that applies to a very general class of differential equations. Gronwall's inequality gives an upper bound on how quickly the absolute error in a solution can grow when the true initial condition is not the same as the measured one.

## How Quickly Can Two Solutions Separate?

Suppose that the initial value problem

$$x' = f(x), \qquad x(0) = x_0,$$

models some physical process for which measurements of the initial state $x_0$ are subject to an absolute error of 0.1. (For simplicity, we consider autonomous differential equations.) In other words, if $\hat{x}_0$ is the measured initial condition, then $|\hat{x}_0 - x_0|$ may be as large as 0.1. We want to know how large the absolute error $\epsilon_a$ between the two solutions can become, where

$$\epsilon_a(t) = |x(t) - \hat{x}(t)|.$$

At least two factors determine the answer to this question:

- The size of $|\hat{x}_0 - x_0|$ is important. Intuitively, two solutions that start closer together are more likely to remain close for a longer time than are two solutions that start farther apart.

- The difference between two solutions depends on how long we follow them. For example, although the solution curves depicted in Fig. 3.11(b) are far apart at $t = 0.8$, they stay relatively close together for $0 \leq t \leq 0.2$. In other words, the longer we follow a given pair of solution curves, the larger the separation between them can become.

Gronwall's inequality incorporates these two factors, as we shall see.

Obviously, the differential equation itself also determines the rate of separation of two solutions. Consider one step of Euler's method applied to the equation $x' = f(x)$.

The initial condition $x(t_0) = x_0$ leads to

$$x_1 = x_0 + hf(x_0).$$

The initial condition $x(t_0) = \hat{x}_0$ leads to

$$\hat{x}_1 = \hat{x}_0 + hf(\hat{x}_0),$$

so

$$|\hat{x}_1 - x_1| = |\hat{x}_0 - x_0 + h[f(\hat{x}_0) - f(x_0)]|.$$

Suppose $x_0 = 1.4$ and $\hat{x}_0 = 1.6$. If $f(x) = 3x$, then

$$|\hat{x}_1 - x_1| = 0.2 + 0.6h.$$

However, if $f(x) = x^6$, then

$$|\hat{x}_1 - x_1| \approx 0.2 + 9.24h.$$

Thus differences in nearby solutions of $x' = x^6$ grow faster after one iteration of Euler's method than those in nearby solutions of $x' = 3x$. This observation is independent of the time step $h$.

Can we estimate the difference $|\hat{x}_1 - x_1|$ given an arbitrary function $f(x)$? The answer is yes, and the answer follows from an application of the mean value theorem.

**Theorem 3.4   *The Mean Value Theorem*** *Let $f(x)$ be a differentiable function on an interval containing $a$ and $b$. Then there exists a point $p$ between $a$ and $b$ such that*

$$f(a) - f(b) = \frac{df}{dx}(p)(a - b).$$

Figure 3.12 is a schematic depiction of the mean value theorem. The slope of the indicated tangent line is $f'(p)$, and the tangent line is parallel to the secant line connecting the points $(a, f(a))$ and $(b, f(b))$. (There is a second point where the tangent line to the function $f$ depicted in Fig. 3.12 is parallel to the secant line; where is it?)

We apply the mean value theorem to estimate $|\hat{x}_1 - x_1|$ for an arbitrary differentiable function $f(x)$ as follows. Suppose $f$ is differentiable on an interval $I$ containing $x_0$ and $\hat{x}_0$. Then

$$|\hat{x}_1 - x_1| = |\hat{x}_0 - x_0 + h\,[f(\hat{x}_0) - f(x_0)]\,|$$

$$= |\hat{x}_0 - x_0|\left|1 + h\left(\frac{df}{dx}(p)\right)\right| \tag{3.22}$$

for some point $p$ between $x_0$ and $\hat{x}_0$. Theorem 3.4 does not tell you how to find $p$, because $p$ depends on $x_1$, $x_2$, and $f$. However, it is often possible to determine an upper

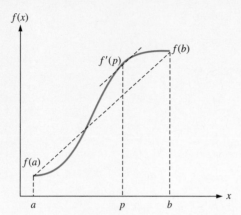

**FIGURE 3.12**    A schematic illustration of the mean value theorem, Theorem 3.4.

bound on the size of $|df(p)/dx|$ on $I$. If

$$K = \max_{x \in I} \left| \frac{df}{dx} \right|$$

exists, then

$$|\hat{x}_1 - x_1| \le |\hat{x}_0 - x_0||1 + hK|. \tag{3.23}$$

The upper bound $K$ is called a *Lipschitz constant* for $f$ on the interval $I$.

Notice that Eq. (3.22) is an equality and Eq. (3.23) is an inequality. The value of $df/dx$ depends on the value of $x$. In Eq. (3.23), we have replaced $df/dx$ with its maximum value over the interval of interest. This substitution turns the equality (3.22) into an inequality.

The right-hand side of Eq. (3.23) depends on the size of $K$ as well as the step size $h$. For a given $h$, the Lipschitz constant determines bounds for the absolute difference between the Euler approximations of two solutions that satisfy the initial conditions $x(t_0) = x_0$ and $\hat{x}(t_0) = \hat{x}_0$.

## Gronwall's Inequality

Gronwall's inequality is a precise statement of the intuitive idea presented above. We outline a heuristic argument at the end of this section that is based on Euler approximations in the limit as the step size shrinks to $0$.[1]

---

[1] Gronwall's inequality holds for nonautonomous differential equations as well as autonomous equations. However, for simplicity, the following examples are concerned with autonomous equations.

**Theorem 3.5   *Gronwall's Inequality***   *Let $x(t)$ and $\hat{x}(t)$ be two solutions of $x' = f(t, x)$ satisfying the initial conditions $x(t_0) = x_0$ and $\hat{x}(t_0) = \hat{x}_0$, respectively. Let $S$ be the rectangular region $S = \{(t, x) : t \in [a, b], x \in [c, d]\}$, and suppose $(t_0, x_0)$ and $(t_0, \hat{x}_0)$ are in $S$. Let*

$$K = \max_{(t,x) \in S} \left| \frac{\partial f}{\partial x}(t, x) \right|,$$

*provided that the maximum exists. Then*

$$\epsilon_a(t) = |\hat{x}(t) - x(t)| \le |\hat{x}_0 - x_0| \, e^{K(t - t_0)} \tag{3.24}$$

*as long as the graphs of $x(t)$ and $\hat{x}(t)$ remain in $S$.*

Figure 3.13 is a schematic illustration of Gronwall's inequality. The region $S$ is depicted by the rectangle. Equation (3.24) gives an upper bound on the size of the absolute error for as long as the graphs of both solutions lie in $S$. The inequality (3.24) may not be valid once one of the graphs leaves $S$.

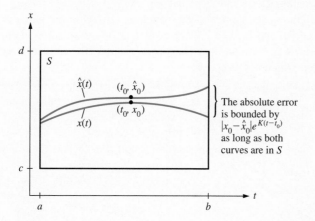

**FIGURE 3.13**   A schematic illustration of the region $S$ and initial conditions in Gronwall's inequality.

Gronwall's inequality has two important consequences:

1. The absolute error between two solutions can be estimated, without explicitly solving the differential equation; and

2. the solutions of a given initial value problem vary continuously with the initial condition.

The latter statement is very important in the context of mathematical modeling. Gronwall's inequality implies that in the limit as $|x_0 - \hat{x}_0| \to 0$, we have $\hat{x}(t) \to x(t)$ as long as $t$

lies in the interval of existence of both solutions. In other words, as a measurement of an initial condition becomes increasingly accurate, the predicted solution "converges" to the true solution.

The following examples illustrate several applications of Gronwall's inequality.

✦ *Example 1*    Let $S$ be the rectangular region defined by $1 \le x \le 2$ and $0 \le t \le 1$, and let

$$x' = \sqrt{x}.$$

Here $f(t, x) = \sqrt{x}$, so

$$K = \max_{(t,x)\in S} \left| \frac{\partial f}{\partial x} \right| = \max_{(t,x)\in S} \left| \frac{1}{2\sqrt{x}} \right| = \frac{1}{2}.$$

If $(t_0, x_0)$ and $(t_0, \hat{x}_0)$ are any two initial conditions in $S$, then Gronwall's inequality (3.24) implies that

$$\epsilon_a(t) = |\hat{x}(t) - x(t)| \le |\hat{x}_0 - x_0| \, e^{(t-t_0)/2} \tag{3.25}$$

for as long as the graphs of $x(t)$ and $\hat{x}(t)$ remain in $S$.

Figure 3.14(a) shows the graphs of the solutions satisfying the initial conditions $x(0) = 1$ (bottom curve) and $\hat{x}(0) = 1.1$ (top curve). The bottom curve in Fig. 3.14(b) shows the absolute error in the solutions. The top curve shows the graph of the Gronwall estimate of the error, which in this case is

$$\epsilon_a(t) \le \tfrac{1}{10} e^{t/2}. \tag{3.26}$$

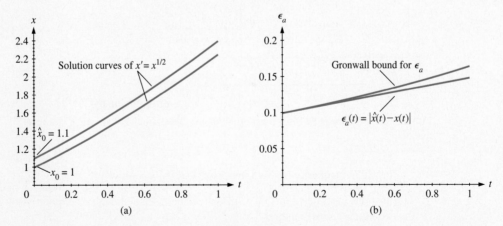

**FIGURE 3.14**    (a) The solutions of $x' = \sqrt{x}$ satisfying $x_0 = 1$ and $\hat{x}_0 = 1.1$. (b) Bottom curve: $|\hat{x}(t) - x(t)|$; top curve: the Gronwall estimate of the error. See Example 1.

The Gronwall estimate of the error is conservative insofar as it overestimates the actual error. However, the bound in Eq. (3.26) is not too far off, and it can be derived without finding explicit formulas for $x(t)$ and $\hat{x}(t)$.  ∎

◆ *Example 2*   Consider the situation in Example 1, but suppose instead that the absolute error in the initial condition is ten times smaller; that is, suppose $\hat{x}_0 = 1.01$. Then Gronwall's inequality implies that

$$\epsilon_a(t) = |\hat{x}(t) - x(t)| \leq \tfrac{1}{100} e^{t/2}.$$

This bound implies that the maximum possible distance between the solution curves is 10 times less than the bound given in Example 1.

Similarly, if $\hat{x}_0 = 1.001$, then the maximum possible distance between the solution curves is 100 times less than the bound given in Example 1. More generally, the bound (3.25) implies that $\hat{x}(t) \to x(t)$ as $\hat{x}_0 \to x_0$ at any fixed value of $t$ in the interval of existence of $x$ and $\hat{x}$.  ∎

◆ *Example 3*   Consider the initial value problem

$$x' = \sqrt{x}, \qquad x(0) = 0.$$

To apply Gronwall's inequality, the rectangular region $S$ must include a portion of the line $x = 0$. However,

$$\frac{\partial f}{\partial x} = \frac{1}{2\sqrt{x}}$$

tends to $\infty$ as $x$ decreases to 0 and is undefined at $x = 0$. Therefore, $f(x) = \sqrt{x}$ does not have a Lipschitz constant on $S$, and Gronwall's inequality cannot be used to estimate the absolute error between a solution that satisfies $x(0) = 0$ and a solution that satisfies $\hat{x}(0) = \hat{x}_0$, where $\hat{x}_0$ is some positive number.  ∎

◆ *Example 4*   The upper bound on the absolute error given by Gronwall's inequality (3.24) is reached by solutions of the exponential growth equation,

$$x' = kx. \tag{3.27}$$

Here, $f(t, x) = kx$, so $\partial f / \partial x = k$, and the Lipschitz constant $K$ equals $k$ regardless of the choice of the rectangular region $S$. Gronwall's inequality (3.24) yields the bound

$$\epsilon_a(t) = |x(t) - \hat{x}(t)| \leq |\hat{x}_0 - x_0| \, e^{kt},$$

where $x(t)$ and $\hat{x}(t)$ are solutions satisfying the initial conditions $x(0) = x_0$ and $\hat{x}(0) = \hat{x}_0$, respectively.

In this case, $x(t) = x_0 e^{kt}$ and $\hat{x}(t) = \hat{x}_0 e^{kt}$, so

$$\epsilon_a(t) = |x_0 - \hat{x}_0| \, e^{kt}.$$

Gronwall's inequality provides an *exact* bound on the growth of errors between solutions of the exponential growth equation. (See also Section 2.4.) ∎

✦ *Example 5*   Let

$$\frac{dx}{dt} = f(t, x) = tx \tag{3.28}$$

and let $S$ be the region $\{(t, x) : t \in [0, 2], \ x \in [0, 8]\}$. Here,

$$\max_{(t,x)\in S} \left| \frac{\partial f}{\partial x} \right| = \max_{(t,x)\in S} |t| = 2.$$

Gronwall's inequality implies that

$$|x(t) - \hat{x}(t)| \le |x_0 - \hat{x}_0| \, e^{2t} \tag{3.29}$$

as long as both solution curves lie in $S$. Explicit solutions of Eq. (3.28) can be computed by using the separation of variables; we have

$$\epsilon_a(t) = |x(t) - \hat{x}(t)| = |x_0 - \hat{x}_0| \, e^{t^2/2}. \tag{3.30}$$

Figure 3.15(a) shows the graphs of the solutions corresponding to the "true" initial condition $x_0 = 1$ and the "estimated" initial condition $\hat{x}_0 = 0.9$. The Gronwall

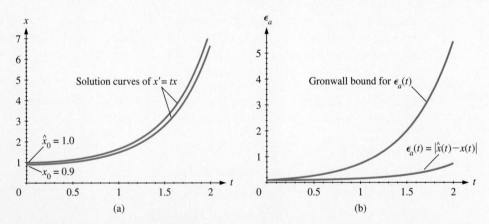

(a)    (b)

**FIGURE 3.15**    (a) Two solutions of Eq. (3.28). (b) The actual difference $\epsilon_a$, given by Eq. (3.30), between the solution curves in (a) and the Gronwall bound, Eq. (3.29).

bound (3.29) is valid for $0 \le t \le 2$, because both solution curves remain in $S$ for that interval of time. Figure 3.15(b) shows a plot of Eq. (3.30), which shows the actual difference between the solutions, together with the Gronwall bound (3.29).

Notice that Eqs. (3.29)–(3.30) imply that

$$\epsilon_a(t) \le |x_0 - \hat{x}_0| \, e^{2t}. \tag{3.31}$$

Gronwall's inequality guarantees that this relation is valid for $0 \le t \le 2$. However, Eq. (3.31) does not hold if $t^2/2 > 2t > 0$ (i.e., when $t > 4$). *This does not contradict Theorem 3.5*, because the portions of the graphs where $t > 4$ fall outside the region $S$. (In fact, the graphs leave $S$ before $t = 4$.) The Gronwall bound (3.29), and hence the relation (3.31), is guaranteed only as long as the solution curves lie in $S$.  ■

✦ *Example 6*   This example applies Gronwall's inequality to a differential equation whose solutions cannot be written explicitly. Let

$$\frac{dx}{dt} = f(t, x) = \sin\left(x^2\right). \tag{3.32}$$

Although Eq. (3.32) is separable, no simple formula for $x(t)$ can be found.

The equilibrium solutions of Eq. (3.32) are $x = \pm\sqrt{n\pi}$, where $n$ is any positive integer. (Figure 3.16 shows part of the phase diagram.) Suppose $0 < x(0) < \sqrt{\pi}$. Because $x = \sqrt{\pi}$ is a stable equilibrium solution (see Exercise 7), $x(t) \to \sqrt{\pi}$ as $t \to \infty$. Let $S$ be the rectangular region such that $0 \le x \le \sqrt{\pi}$ and $0 \le t \le 10$. Then

$$\max_{(t,x)\in S} \left| \frac{\partial f}{\partial x} \right| = \max_{(t,x)\in S} \left| 2x \cos\left(x^2\right) \right| = 2\sqrt{\pi}.$$

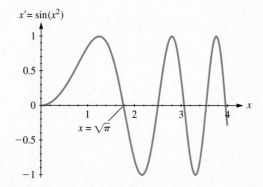

**FIGURE 3.16**   The phase diagram for $x' = \sin(x^2)$.

Hence, if $x_0$ and $\hat{x}_0$ are two initial values between 0 and $\sqrt{\pi}$, then a Gronwall bound for the difference between the respective solutions is

$$|x(t) - \hat{x}(t)| \le |x_0 - \hat{x}_0| \, e^{2\sqrt{\pi}\, t}. \qquad\qquad \blacksquare$$

## An Argument for Gronwall's Inequality

Euler's method provides a heuristic argument for Gronwall's inequality, as we now show. For simplicity, we assume that the differential equation is autonomous, that is,

$$x' = f(x),$$

and that the initial condition is specified at $t = 0$. (It is straightforward to generalize the argument below to nonautonomous differential equations and to other values of $t_0$.) Let $S$ be a rectangular region for which

$$\max_{(t,x)\in S} \left| \frac{\partial f}{\partial x} \right| = K,$$

and let $x(t_0) = x_0$, $\hat{x}(t_0) = \hat{x}_0$. As was discussed above, the difference between the two solutions $x$ and $\hat{x}$ after one Euler time step is

$$|\hat{x}_1 - x_1| = |\hat{x}_0 - x_0 + h\,[f\,(\hat{x}_0) - f(x_0)]\,|$$
$$= \left|\hat{x}_0 - x_0\right|\left|1 + hf'(p_1)\right|,$$

where $p_1$ is a point between $x_0$ and $\hat{x}_0$. If the true solutions stay inside $S$, then $x_1$ and $\hat{x}_1$ lie in $S$, provided that $h$ is a sufficiently small time step. Although we do not know $p_1$ precisely, we do know that $|f'(p_1)| \le K$ if the approximate solution stays in the region $S$. Hence,

$$|\hat{x}_1 - x_1| \le |\hat{x}_0 - x_0|\,|1 + hK|.$$

Similarly, we find that after two time steps,

$$|\hat{x}_2 - x_2| = |\hat{x}_1 - x_1 + h\,[f\,(\hat{x}_1) - f(x_1)]\,|$$
$$= \left|\hat{x}_1 - x_1\right|\left|1 + hf'(p_2)\right|,$$

where $p_2$ is a point between $x_1$ and $\hat{x}_1$. Hence,

$$|\hat{x}_2 - x_2| \le |\hat{x}_1 - x_1|\,|1 + hK|$$
$$\le |\hat{x}_0 - x_0|\,|1 + hK|^2,$$

provided that the true solutions stay inside $S$ and $h$ is small enough.

Continuing in this way for $n$ time steps, we find that

$$|\hat{x}_n - x_n| \le |\hat{x}_0 - x_0|\,|1 + hK|^n. \tag{3.33}$$

After $n$ time steps, $t = nh$. Therefore,

$$(1 + hK)^n = \left(1 + \frac{Kt}{n}\right)^n.$$

As $n \to \infty$ (and holding $t$ fixed so $h \to 0$), the right-hand side of Eq. (3.33) tends to $|\hat{x}_0 - x_0|\,e^{Kt}$, and the left-hand side tends to $|\hat{x}(t) - x(t)|$. This concludes the argument.

## Exercises

**1.** For each function $f$ in the following list, find $K = \max |df/dx|$ on the indicated interval.

(a) $f(x) = x^2$ on $[5, 6]$

(b) $f(x) = \sin x$ on $[0, 2\pi]$

(c) $f(x) = \sin x$ on $[0, \pi]$

(d) $f(x) = \sin 2x$ on $[0, \pi]$

(e) $f(x) = \tan x$ on $[-\pi/4, \pi/4]$

(f) $f(x) = e^x$ on $[0, 1]$

(g) $f(x) = e^{-x}$ on $[0, 1]$

(h) $f(x) = 1/x$ on $[1/(n + 1), 1/n]$, where $n$ is a positive integer

**2.** For each function in the following list,

- give an example of an interval for which $K = \max |df/dx|$ is finite and calculate it, and

- determine whether there is an interval on which $K$ is undefined or infinite.

(Suggestion: Graph each function and examine $df/dx$.)

(a) $f(x) = 1/(1 - x)$     (b) $f(x) = x^{1/3}$

(c) $f(x) = xe^{-x}$        (d) $f(x) = \sin(1/x)$

**3.** For each equation in the following list, determine a Gronwall bound on the rectangular region $0 \le x \le 2, 0 \le t \le 2$.

(a) $x' = tx^2$            (b) $x' = (x + t)^2$

(c) $x' = x^2 + t$

**4.** Show that the Gronwall bound in Example 1 is valid for every rectangular region $S$ that extends down as far as the line $x = 1$. Argue that the Gronwall bound (3.26) is valid for all $t \ge 0$.

**5.** Let $x' = t^2 + 1$.

(a) Determine a Gronwall bound on an appropriate region. What does your bound predict about the growth of absolute errors between two solutions?

(b) Find the solutions satisfying the initial conditions $x(0) = x_0$ and $\hat{x}(0) = \hat{x}_0$. What is $|x(t) - \hat{x}(t)|$? Is this difference consistent with the Gronwall bound in (a)? Explain.

**6.** Let $x' = x^2$.

(a) Let $S$ be the rectangular region where $0 \le x \le 2$ and $0 \le t \le 2$, and let $x(0) = x_0$, $\hat{x}(0) = \hat{x}_0$ be two initial conditions in $S$. Derive a Gronwall bound for $|x(t) - \hat{x}(t)|$. Show that the Gronwall bound does not depend on the width of $S$. (That is, the Gronwall bound is valid for $0 \le t \le a$, where $a$ is any positive number.)

(b) Let $x_0 = 0.5$ and $\hat{x}_0 = 0.6$. Find an exact expression for $|x(t) - \hat{x}(t)|$ and compare it to the Gronwall bound in (a) for $0 \leq t \leq 1$.

(c) What goes wrong with the Gronwall bound in (b) on the interval $0 \leq t \leq 1.66$? Explain.

(d) Give a precise upper bound for the length of time that the Gronwall bound in (a) is valid. (Hint: When does the graph of $x$ or the graph of $\hat{x}$ leave $S$?)

**7.** Consider Example 6.

(a) Show that $x = \sqrt{\pi}$ is a stable equilibrium solution. (Hint: See Theorem 1.4 in Section 1.3.)

(b) Draw a direction field for Eq. (3.32) that includes several equilibrium solutions. Describe the qualitative behavior of the nonequilibrium solutions of Eq. (3.32).

(c) Let $x(t)$ and $\hat{x}(t)$ denote the solutions satisfying the initial conditions $x(0) = x_0$ and $\hat{x}(0) = \hat{x}_0$, respectively. Is Gronwall's inequality more useful for estimating $|x(t) - \hat{x}(t)|$ in the short term or the long term? Explain.

**8.** Let $x' = x^2 + t^2$.

(a) Let $S$ be the region where $0 \leq x \leq 2$ and $0 \leq t \leq 2$, and let $x(0) = x_0$, $\hat{x}(0) = \hat{x}_0$ be two initial conditions in $S$. Derive a Gronwall bound for $|x(t) - \hat{x}(t)|$.

(b) Does the Gronwall bound in (a) change if $S$ is the region for which $0 \leq x \leq 2$ and $0 \leq t \leq 3$? Explain.

(c) Let $x_0 = 0.5$ and $\hat{x}_0 = 0.6$. Approximate $x$ and $\hat{x}$ for $0 \leq t \leq 0.8$ using an appropriate step size. What is $|\hat{x}(0.8) - x(0.8)|$, based on your numerical solutions? Is the difference consistent with the Gronwall bound in (a)?

(d) Let $x_0 = 0.5$. Using the Gronwall bound in (a), what is the maximum value of $|x_0 - \hat{x}_0|$ that ensures that $|x(0.8) - \hat{x}(0.8)| \leq 0.1$?

**9.** This problem considers an example in which Gronwall's inequality can be applied only in certain circumstances.

(a) Let $f(x) = x^{1/3}$. Plot $f$ over the interval $[-1, 1]$.

(b) Determine $df/dx$ on the same interval. Is $df/dx$ continuous everywhere?

(c) Let $x' = x^{1/3}$ and consider the solutions satisfying the initial conditions $x(0) = 0.01$ and $\hat{x}(0) = -0.01$. What, if anything, can you say about $|x(t) - \hat{x}(t)|$ without finding an explicit formula for $x$ and $\hat{x}$?

(d) Let $x' = x^{1/3}$ and consider the solutions that satisfy the initial conditions $x(0) = 0.01$ and $\hat{x}(0) = 0.02$. What, if anything, can you say about $|x(t) - \hat{x}(t)|$ without finding an explicit formula for $x$ and $\hat{x}$?

(e) Determine $|x(t) - \hat{x}(t)|$ explicitly in (c) and (d). Discuss how they are consistent or inconsistent with Gronwall's inequality.

**10.** Suppose the population of a protozoan is determined by the logistic equation

$$p' = f(t, p) = p(1 - p). \qquad (3.34)$$

Let $p$ be in units of millions of cells and $t$ be in hours.

(a) Let $p$ and $\hat{p}$ satisfy the initial values $p(0) = p_0$, $\hat{p}(0) = \hat{p}_0$. Find a Gronwall bound for $|p(t) - \hat{p}(t)|$ that is valid in the region $0 \leq p \leq 1$.

(b) There is always some uncertainty in the exact number of cells that are present at any instant. Suppose you want 1 million cells, plus or minus 50,000, in the culture 4 hours from now. Use Gronwall's inequality to estimate how accurate your initial census must be.

(c) Suppose the culture grows for a long time. Is the number of cells in the culture after a long time about the same regardless of how many are put in the culture initially? Explain. Is a "yes" answer inconsistent with the Gronwall bound?

**11.** Let

$$x' = 2x + t. \qquad (3.35)$$

(a) Find a Lipschitz constant $K$ for the right-hand side of Eq. (3.35) in a suitable region $S$ containing $t = 0$. Then use Gronwall's inequality to determine a bound on the rate of separation of two solutions satisfying the initial conditions $x(0) = x_0$ and $\hat{x}(0) = \hat{x}_0$.

(b) Determine an explicit solution of Eq. (3.35) that satisfies the initial condition $x(0) = x_0$.

(c) Find an exact expression for $|x(t) - \hat{x}(t)|$, where $\hat{x}(0) = \hat{x}_0$ and compare it to the Gronwall's bound in (a).

(d) Suppose $x_0 = 1$. What are the maximum and minimum values of $\hat{x}_0$ that can guarantee that $|x(1) - \hat{x}(1)| \leq 0.1$?

**12.** Without attempting to find explicit solutions, and using only Gronwall's inequality, determine which of the differential equations in the following list might exhibit the greatest sensitivity to initial conditions for solutions in the region $S = \{(t, x) : x \in [0, 2]\}$.

(a) $x' = x + 3t$            (b) $x' = 3x + t$

(c) $x' = 3t + \sin x$       (d) $x' = \frac{1}{2}x^2 + x$

**13.** Let

$$\frac{dx}{dt} = x^3 - x. \qquad (3.36)$$

(a) Find a Lipschitz constant $K$ for the right-hand side of Eq. (3.36) that is valid over the region $S = \{(t, x) : x \in [-2, 2]\}$.

(b) Use Gronwall's inequality to find a bound on the growth of errors in the initial conditions.

(c) Draw a direction field for Eq. (3.36).

(d) Using the direction field, identify subregions within $S$ where you expect that solutions are likely to be relatively insensitive to errors in the initial conditions.

(e) Identify subregions of $S$ where solutions are sensitive to initial errors.

**14.** Suppose you are trying to debug a program to find numerical approximations of the solutions of

$$x' = \sqrt[3]{x} + t^2.$$

Given the initial condition $x(0) = 0.5$, your program estimates $x(1/2) = 0.864$. Similarly, given the initial condition $x(0) = 0.4$, your program estimates $x(1/2) = 0.645$. Is there an error in the program, assuming that the true solutions of both initial value problems take on values between 0.4 and 1 for $0 \leq t \leq 1/2$? Explain.

**15.** The conclusion of Gronwall's inequality is valid only in the region $S$ over which the Lipschitz constant for $f$ is valid. Explain the step in the argument above that leads to this requirement.

# Chapter 4

---

# LINEARITY

---

The idea of a linear operation is one of the most important abstractions in higher mathematics. The word *linear* suggests a straight line; conversely, the word *nonlinear* suggests something that is crooked or bent. Linearity also suggests a notion of simple proportionality. Linear equations of various kinds are often relatively easy to solve because they possess a special structure. This structure is the subject of the rest of the chapter.

We begin with a discussion of linear functions. In Section 4.2, we define the notion of an operator and show that it is a natural extension of the idea of a function. We will see how a differential equation can be considered an operator in Section 4.3, where we discuss what is meant by the term *linear* differential equations. Section 4.4 discusses an important theoretical result, called the *superposition principle*, that applies to linear equations. Section 4.5 describes the structure of solutions of linear differential equations. Throughout this chapter, we stress that the concept of linearity extends beyond the notion of a simple straight line; in fact, it has applications in all of mathematics.

## 4.1 LINEAR FUNCTIONS

### Functions of a Single Variable

Every line through the origin can be written as $f(x) = mx$, where $m$ is a constant called the *slope*. The distributive law of multiplication yields

$$
\begin{aligned}
f(x + y) &= m(x + y) \\
&= mx + my \\
&= f(x) + f(y).
\end{aligned}
$$

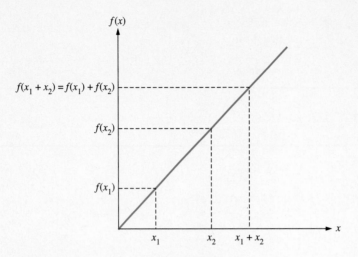

**FIGURE 4.1**   The graph of a linear function.

Figure 4.1 provides a graphical illustration of this additive property. In the same way, it is easy to verify the following proportionality property. Let $c$ be any constant. By the commutative law of multiplication,

$$f(cx) = m(cx)$$
$$= c(mx)$$
$$= cf(x).$$

These considerations motivate the following definition.

**Definition 4.1**   *The function $f$ is* linear *if for all numbers $x$ and $y$ and for every constant $c$,*

- $f(x + y) = f(x) + f(y)$,
- $f(cx) = cf(x)$.

The criteria in Definition 4.1 imply that a linear function is one whose graph is a straight line through the origin (see Exercise 1). Functions whose graphs are not straight lines through the origin are not linear.

◆ *Example 1*    Although the graph of the function $f(x) = mx + b$ with $b \neq 0$ is a straight line, it does not go through the origin. Hence, $f$ does not satisfy the additivity property in Definition 4.1.    ■

✦ *Example 2* Although the graph of $\sin x$ goes through the origin, $\sin x$ is not linear because $\sin(x + y) \neq \sin x + \sin y$ in general.  ∎

✦ *Example 3* The quadratic function $f(x) = x^2$ is not linear because $(x + y)^2 \neq x^2 + y^2$ in general.  ∎

✦ *Example 4* The law of diminishing returns is another example of a nonlinear function. Suppose, for instance, that the number of grade points, $g$, that you earn in a mathematics course as a function of the number of hours $t$ that you spend per week studying the course material is given by the curve in Fig. 4.2. In this case, you earn 30 points with little or no effort, 75 points with moderate effort, 85 points with somewhat more effort, and 90 points with great effort. The function $g(t)$ is not linear, because if you double the amount of effort, say from 5 to 10 hours per week, then the number of grade points earned does not double.

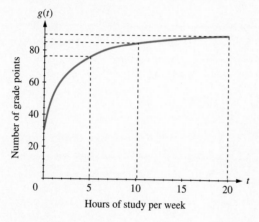

**FIGURE 4.2** An illustration of the law of diminishing returns.  ∎

## Domain and Range

A function $f$ assigns elements in one set, called the *domain*, to elements in another set, called the *range*. (The terms *map*, *transformation*, and *operator* are synonyms for *function*.) For every $x$ in the domain, the function $f$ assigns a *unique* element in the range, denoted $f(x)$. It is often useful to think of $f$ as a kind of "black box" or input-output system, as illustrated in Fig. 4.3. The domain consists of the set of allowed inputs. For each input $x$, the function $f$ outputs a $y$, called the *image* of $x$ under $f$. We say that $x$ is the *preimage* of $y$. The range of $f$ is the image of the domain. We say that $f$ is *one-to-one* if every element in the range has a unique preimage in the domain.

**FIGURE 4.3** A schematic illustration of a function $f$ as an input-output system.

The domain and range of a function can be any sets. In calculus and differential equations, we are usually interested in functions whose domain is the set of real numbers or the set of complex numbers. Other possibilities for the domain and range are illustrated in the following examples.

✦ *Example 5* Table 4.1 shows some familiar functions, their domains, and their ranges.

| $f(x)$ | Domain | Range | One-to-one |
|---|---|---|---|
| $3x$ | $(-\infty, \infty)$ | $(-\infty, \infty)$ | yes |
| $x^2$ | $(-\infty, \infty)$ | $[0, \infty)$ | no |
| $\ln x$ | $(0, \infty)$ | $(-\infty, \infty)$ | yes |
| $\sin x$ | $(-\infty, \infty)$ | $[-1, 1]$ | no |

**TABLE 4.1** The domain and range for a few functions of one variable. ■

✦ *Example 6* The domain of a function need not be an interval. Let $f$ be the function that associates with every day in 1990 the maximum temperature in degrees Fahrenheit recorded by the National Weather Service in Phoenix, Arizona. The domain of $f$ is the set of integers from 1 to 365. The range of $f$ is the set of integers from 45 to 122. The function $f$ is not continuous, because it is defined only on a discrete set of integers. Moreover, there is no convenient formula for $f$; it is defined entirely by the 365 pairs of the form (day, maximum temperature). Although (fortunately) there was only one day in 1990 in which the high temperature was 122°F, $f$ is not a one-to-one function because there were many days in 1990 in which the high temperature was 100°F. ■

✦ *Example 7* Functions can be defined on any number of variables. A function of geophysical interest is the map **B** that associates to each point **p** in North America the magnetic field emanating from the earth's surface at **p**. The magnetic field assumes only one (vector) value at any given point, so **B** is a function. The domain of **B** is a set of ordered pairs, because we can treat North America as a surface coordinatized by longitude and latitude. The range of **B** is a set of ordered triples; that is, **B(p)** gives the three components of the earth's magnetic field at **p**. As in the previous example, **B** is well defined but cannot be represented by a simple formula. ∎

## Linear Functions in the Plane

Linear functions in one variable have the form $f(x) = mx$. The situation is more interesting in the case of vector-valued functions.

Let

$$\mathbf{f(x)} = \mathbf{Ax},\tag{4.1}$$

where $\mathbf{x} = (x_1, x_2)$ is a vector in the plane and **A** is a $2 \times 2$ matrix,

$$\mathbf{A} = \begin{pmatrix} a_{11} & a_{12} \\ a_{21} & a_{22} \end{pmatrix}.$$

The matrix-vector product **Ax**, and hence **f(x)**, is defined as

$$\mathbf{Ax} = \begin{pmatrix} a_{11} & a_{12} \\ a_{21} & a_{22} \end{pmatrix} \begin{pmatrix} x_1 \\ x_2 \end{pmatrix} = \begin{pmatrix} a_{11}x_1 + a_{12}x_2 \\ a_{21}x_1 + a_{22}x_2 \end{pmatrix}.\tag{4.2}$$

The domain of **f** is the set of all vectors in the plane, and the range is a subset of all the vectors in the plane. Matrix-vector multiplication as defined in Eq. (4.2) has two important properties, whose proofs are left as exercises.

- $\mathbf{A}(c\mathbf{x}) = c(\mathbf{Ax})$, where $c$ is any constant (scalar).
- $\mathbf{A}(\mathbf{x} + \mathbf{y}) = \mathbf{Ax} + \mathbf{Ay}$, where **x** and **y** are any 2-vectors. (This property is called the distributive law of matrix multiplication over vector addition.)

These properties imply that **f** is linear in the sense of Definition 4.1. In fact, it can be shown that every linear function that is defined on the plane can be written in form of Eq. (4.1) for an appropriate matrix **A**.

✦ *Example 8*   Equation (4.2) implies that

$$\begin{pmatrix} 1 & 3 \\ 2 & 1 \end{pmatrix} \begin{pmatrix} 1 \\ -2 \end{pmatrix} = \begin{pmatrix} 1 - 6 \\ 2 - 2 \end{pmatrix} = \begin{pmatrix} -5 \\ 0 \end{pmatrix}.$$
∎

✦ *Example 9*   Let $\text{Rot}(x_1, x_2)$ be the function that rotates the vector $\mathbf{x} = (x_1, x_2)$ by $\pi/2$. The domain and range of Rot are all vectors in the plane. If $(x_1, 0)$ is any vector in the horizontal axis, then $\text{Rot}(x_1, 0) = (0, x_1)$. The rotation maps the horizontal axis onto the vertical axis.

Figure 4.4 illustrates two vectors **x**, **y** and their sum, **x** + **y**, in the first quadrant. Their respective images under Rot are in the second quadrant. The sum **x** + **y** is represented by the diagonal of a parallelogram two of whose sides are shown as dashed lines. The picture suggests that Rot preserves sums insofar as Rot(**x** + **y**) = Rot(**x**) + Rot(**y**). In other words, the image of the parallelogram is another parallelogram, suggesting that Rot satisfies the additive property of Definition 4.1.

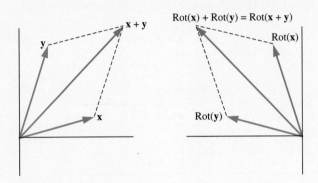

**FIGURE 4.4**    A rotation by $\pi/2$.

To show that Rot also satisfies the proportionality property, let **x** be an arbitrary nonzero vector, and consider the set of all scalar multiples of **x**, that is, the set of all vectors of the form $c\mathbf{x}$, where $c$ is any scalar. This set is a line, $L$, through the origin in the direction of **x**, as illustrated in Fig. 4.5. Suppose we rotate **x** by $\pi/2$; call the

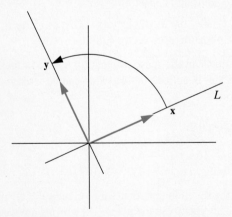

**FIGURE 4.5**    Rotation of a line through the origin by $\pi/2$.

resulting vector **y**. The set of all vectors of the form $c$**y** is a line that is a rotated copy of $L$. This observation suggests that $\text{Rot}(c\mathbf{x}) = c\,\text{Rot}(\mathbf{x})$.

   These geometrical considerations imply that Rot is a linear function, in the sense of Definition 4.1. It is not difficult to show that

$$\text{Rot}(x_1, x_2) = (-x_2, x_1) = \mathbf{A}\mathbf{x},$$

where

$$\mathbf{A} = \begin{pmatrix} 0 & -1 \\ 1 & 0 \end{pmatrix}.$$

The linearity of Rot follows from the linearity of matrix-vector multiplication.   ■

## *Exercises*

**1.** Explain why the criteria in Definition 4.1 imply that the graph of a linear function of a single real variable is a line that goes through the origin.

**2.** For each function in the following list, determine

- its domain;
- its range; and
- whether it is one-to-one.

(a) $f(x) = \cos(x)$

(b) $f(x) = (\sin t)x$, $t$ any constant

(c) $f(t) = \dfrac{3t - 4}{2t + 5}$

(d) $g(y) = a|y|$, $a$ any constant

(e) $g(y) = ay + b$, $a$ and $b$ any constants

(f) $f(x, y) = y \cos x$

(g) $f(x, t) = (\cos \pi/10)(x + t)$

(h) $f(t, s) = a(s + t)$, $a$ any constant

(i) $g(x, y) = axy$, $a$ any constant

(j) $g(x, y, z) = \sqrt{x^2 + y^2 + z^2}$

**3.** For each of the functions (a)–(e) in Exercise 2, determine whether it is linear in the sense of Definition 4.1. If the function is not linear, give an example in which one of the conditions in Definition 4.1 fails to hold.

**4.** Give an example of a function of the real numbers for which every negative number has two preimages.

**5.** Which of the curves in Fig. 4.6 depicts a graph of a function of $x$? Of $y$? Of both? Of neither?

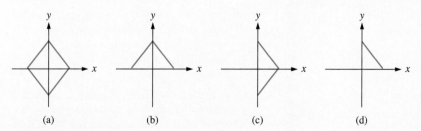

(a)          (b)          (c)          (d)

**FIGURE 4.6**   For Exercise 5.

**6.** Find the preimages, if any, of 1 for each of the functions in the following list.

(a) $f(x) = \sqrt{x}$

(b) $f(x) = \exp x$

(c) $f(x) = \ln x$

(d) $f(x) = \tan x$

(e) $f(x) = x(x-1)(x+1)$

**7.** Give an example of a function of the real numbers for which every number is its own preimage.

**8.** Let $f(x) = 3x - 2$.

(a) Draw the graph of $f$.

(b) Show that $f$ is not linear in the sense of Definition 4.1.

(c) Explain how to make $f$ into a linear function by a simple change of the $y$ coordinate. (Hint: Find a function $g(y)$ such that the composition $h(x) = g(f(x))$ is a linear function of $x$.)

(d) We often say that $y$ varies *linearly with x* if $y = ax + b$ for some constants $a$ and $b$. Explain why this terminology is justified in view of your results in (c).

**9.** Compute each matrix-vector product in the following list.

(a) $\begin{pmatrix} 1 & 2 \\ 3 & 4 \end{pmatrix} \begin{pmatrix} 5 \\ 6 \end{pmatrix}$    (b) $\begin{pmatrix} 10 & 0 \\ 0 & 2 \end{pmatrix} \begin{pmatrix} -2 \\ 1 \end{pmatrix}$

(c) $\begin{pmatrix} 0 & 10 \\ 2 & 0 \end{pmatrix} \begin{pmatrix} -2 \\ 1 \end{pmatrix}$    (d) $\begin{pmatrix} 2 & 4 \\ 1 & 2 \end{pmatrix} \begin{pmatrix} -2 \\ 1 \end{pmatrix}$

**10.** Let Rot be the rotation in the plane by $\pi/2$ as discussed in the text. Is Rot a one-to-one function? Explain.

**11.** Every line in the plane can be regarded as the set of all vectors of the form

$$\mathbf{a} + t\mathbf{x}, \qquad (4.3)$$

where $\mathbf{a}$ and $\mathbf{x}$ are fixed vectors and $t$ is any real number. We call Eq. (4.3) a *parametric representation* of the line. (In other words, as $t$ varies from $-\infty$ to $\infty$, the collection of points given by Eq. (4.3) forms a straight line.)

(a) Find a parametric representation of the line $y = 3x + 2$.

(b) Find a parametric representation of the line containing the points $(1, 1)$ and $(4, 3)$.

(c) Does a given line have a unique parametric representation? Explain.

**12.** Show that, for any 2-vectors $\mathbf{x}$ and $\mathbf{y}$ and any $2 \times 2$ matrix $\mathbf{A}$:

(a) $\mathbf{A}(c\mathbf{x}) = c(\mathbf{A}\mathbf{x})$, where $c$ is any constant (scalar).

(b) $\mathbf{A}(\mathbf{x} + \mathbf{y}) = \mathbf{A}\mathbf{x} + \mathbf{A}\mathbf{y}$.

(c) Let $\mathbf{f}(\mathbf{x}) = \mathbf{A}\mathbf{x}$, and let $L$ be a line in the plane. Show that $\mathbf{f}(L)$ is either a line or a single point. (Suggestion: Consider a parametric representation of $L$.)

**13.** Let $\mathbf{f}$ be any linear function from the plane to itself.

(a) Must $\mathbf{f}(\mathbf{0}) = \mathbf{0}$? Explain. (Use Definition 4.1.)

(b) Suppose $\mathbf{f}(\mathbf{x}) = \mathbf{0}$. Must it be the case that $\mathbf{x} = \mathbf{0}$? (Suggestion: See Exercise 9.)

**14.** Take a closer look at the law of diminishing returns.

(a) The curve in Fig. 4.2 is the graph of a nonlinear function $g$. Although the curve itself is not a straight line, can one or more pieces of the curve be approximated by straight lines? Explain.

(b) Is it reasonable to say that Fig. 4.2 models a situation in which the number of grade points increases approximately linearly with effort, at least in the case in which the amount of effort is small? Explain.

(c) Does the number of grade points increase approximately linearly with effort when the amount of effort is large? Explain.

(d) Explain how the law of diminishing returns might apply in a situation such as the crop yield per hectare as a function of the amount of fertilizer applied to the soil. Show what the graph of such a function might look like, and explain whether crop yield might be described as increasing approximately linearly with fertilizer application under certain circumstances.

**15.** Consider the vector-valued function **B** that represents the magnetic field on the surface of the earth, as discussed in Example 7. Choose an arbitrary point on the surface of the earth as the origin of your coordinate system.

(a) If **B** were linear, what would be the value of **B(0)**?

(b) Is it possible that **B** is a linear function? Why or why not? (Hint: If **B** were linear, what would happen to its magnitude as you move away from the origin?)

**16.** Let **f** be the map that takes any point $(x_1, x_2) \neq (0, 0)$ to its polar coordinate representation $(r, \theta)$. Is **f** a function? If so, determine the domain and range of **f** and determine whether **f** is linear.

**17.** Let $\mathbf{p}(t) = (\cos t, \sin t)$ be defined over the interval $[0, 2\pi)$. Is **p** a function? If so, what are the domain and range of **p**? Is **p** one-to-one? Is **p** linear?

**18.** Let $h(x, y) = xy$.

(a) We say that $f$ is *linear in x* if $f(x, c)$ is a linear function of $x$ when $c$ is a fixed constant. Determine whether $h$ is linear in $x$.

(b) Determine whether $h$, regarded as a function defined on the plane, is linear.

(c) For each of the functions $f$ in Exercise 2(f)–(h), determine whether $f$ is a linear function of its first argument. Is $f$ linear when you regard it as a function defined on the plane?

**19.** One consideration in the design of a stereo amplifier is that the amplitude of the output signal be a linear function of the input power.

(a) Discuss what this means and why it is a desirable feature.

(b) At a certain power level, the amplifier reaches a state at which the amplitude of its output no longer increases, regardless of the amount of input power. (This situation is sometimes called *clipping*.) Explain why clipping corresponds to a nonlinear function of the input power and what the implications are for the sound that is produced.

**20.** A useful technique in calculus is to approximate a differentiable function by its tangent line in a neighborhood of some point $x_0$ of interest. Suppose $f'$ is defined in an interval around $x_0$. The *linearization* of $f$ at $x_0$ is $\ell(h) = f(x_0) + hf'(x_0)$. (See also Section 8.5.)

(a) Show that $\ell(h) - f(x_0)$ is a linear function of $h$.

(b) Find the linearization of $f(x) = x^2$ at $x_0 = 3$.

(c) Use linearization to compute $(2.99)^2$, $(3.02)^2$, and $(2.98)^2$, accurate to two decimal places, without using a calculator, computer, or pencil and paper.

(d) Use linearization to compute the values of each of the following quantities to three significant digits without using a calculator, computer, or pencil and paper:

- $\ln 0.99$
- $(1.01)^3$
- $(1.98)^2$
- $\sqrt{15.96}$

**21.** A rotation of the plane by the angle $\phi$ maps the point $\mathbf{x} = (x_1, x_2)$ to the point $\text{Rot}_\phi(\mathbf{x})$ as follows:

$$\text{Rot}_\phi(x_1, x_2) = (x_1 \cos \phi - x_2 \sin \phi, \ x_1 \sin \phi + x_2 \cos \phi). \qquad (4.4)$$

(Here, $\phi$ is a fixed constant.)

(a) Derive Eq. (4.4) and show that, for each fixed constant $\phi$, $\mathrm{Rot}_\phi$ is a one-to-one function from the plane to itself.

(b) Show that, for each fixed constant $\phi$, $\mathrm{Rot}_\phi$ is a linear function of $\mathbf{x}$. (Hint: Show that $\mathrm{Rot}_\phi$ can be written in the form of Eq. 4.1.)

**22.** Let $\mathbf{f}(\mathbf{x}) = \mathbf{A}\mathbf{x}$, where

$$\mathbf{A} = \begin{pmatrix} 3 & 0 \\ 0 & 0.2 \end{pmatrix}.$$

(a) What are the range and the domain of $\mathbf{f}$? Is $\mathbf{f}$ one-to-one?

(b) Let $\mathbf{x} = (x_1, x_2)$ be any vector in the plane. What is the image of $\mathbf{x}$ under $\mathbf{f}$?

(c) Is $\mathbf{f}$ linear or nonlinear?

(d) Let $S$ be the square whose corners are the points $(0, 0)$, $(1, 0)$, $(0, 1)$, and $(1, 1)$. Explain why it is possible to determine the image of $S$ under $\mathbf{f}$ simply by computing $\mathbf{f}(0, 1)$ and $\mathbf{f}(1, 0)$.

(e) Describe the geometric relationship between $S$ and $\mathbf{f}(S)$. (In other words, how does $\mathbf{f}$ deform $S$?)

## 4.2  Linear Operators

The domain and range of most functions encountered in calculus are subsets of the real line or subsets of Euclidean 2- and 3-space. However, in much of mathematics, it is useful to consider functions whose domain and range are sets of functions rather than sets of numbers. This section is devoted to an explanation of the concept of "functions of functions," generally called *operators*.[1]

An operator is a function that takes a function $f$ as its input. The output of an operator can be another function, or it can be a number.

✦ *Example 1*    One example of an operator that is familiar from calculus is the differentiation operator, denoted $D$. For instance, $D(\sin x) = \cos x$. That is, $D$ maps one function (in this case, $\sin x$) to another function ($\cos x$). We can think of $D$ as a kind of input-output system, as illustrated schematically in Fig. 4.7. The domain of $D$ is the set of all differentiable functions, and the range is a subset of the set of continuous functions.

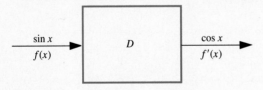

**Figure 4.7**    The differentiation operator regarded as an input-output system.                ∎

---

[1]The terms *function, operator, transformation,* and *map* are essentially synonymous. *Operator* typically is used to refer to a function that operates on functions, and *transformation* often refers to a function of points in Euclidean space.

✦ *Example 2*    Integration is another example of an operator. Let

$$\text{Int}(f) = \int_0^x f(s)\,ds.$$

For example, $\text{Int}(\cos x) = \int_0^x \cos s \, ds = \sin x$. ∎

✦ *Example 3*    Let $U$ be the operator defined by

$$U(f) = \int_0^1 f(x)\,dx. \tag{4.5}$$

The domain of $U$ consists of the set of all functions whose definite integral from 0 to 1 is defined and finite. The range of $U$ is the set of real numbers. For instance, $U(x^2) = \int_0^1 x^2\,dx = 1/3$. ∎

The concepts of domain, range, image, preimage, and one-to-one mapping apply to operators as well as to functions. In particular, the concept of linearity carries over to operators as well as functions.

**Definition 4.2**    *The operator* Op *is linear if for every function $f$ and $g$ in its domain,*

- $\text{Op}(f + g) = \text{Op}(f) + \text{Op}(g)$ *and*
- $\text{Op}(cf) = c\text{Op}(f)$,

*where $c$ is an arbitrary constant.*

Linear operators are often denoted by the letter $L$, much as functions are often denoted by the letter $f$. In this text, we use the notation Op to denote a generic operator; we use the notation $L$ exclusively to denote a linear operator.

If $L$ is linear, then Definition 4.2 implies that

$$L(af + bg) = aL(f) + bL(g), \tag{4.6}$$

where $a$ and $b$ are any constants. Equation (4.6) is called the *superposition property* of linear operators.

✦ *Example 4*    The differentiation operator $D$ is linear. Let $f$ and $g$ be differentiable functions and $c$ be any constant. The rules of differentiation from calculus imply that

$$D(f + g) = D(f) + D(g)$$

and

$$D(cf) = cD(f).$$

(We assume that $f$ and $g$ are functions of $x$ and that $D$ refers to differentiation with respect to $x$. Unless otherwise stated, this assumption is implicit whenever we discuss the differentiation operator.)                                                                ∎

✦ *Example 5*   The integration operator is linear.  Let $f$ and $g$ be integrable functions of $x$, and let $c$ be any constant. The rules of calculus imply that

$$\text{Int}(f+g) = \int_0^x \big(f(s)+g(s)\big)\,ds$$

$$= \int_0^x f(s)\,ds + \int_0^x g(s)\,ds$$

$$= \text{Int}(f) + \text{Int}(g),$$

and

$$\text{Int}(cf) = \int_0^x cf(s)\,ds$$

$$= c\int_0^x f(s)\,ds$$

$$= c\,\text{Int}(f).$$                                                                ∎

✦ *Example 6*   Let $\text{Sq}(f)$ map the function $f$ to its square; that is, $\text{Sq}(f(x)) = [f(x)]^2$. The square operator is not linear, because

$$\text{Sq}(f+g) = [f+g]^2$$

$$= f^2 + 2fg + g^2$$

$$\neq f^2 + g^2.$$

In fact, both conditions in Definition 4.2 fail. Let $c$ be a constant. Then $\text{Sq}(cf) = (cf)^2 = c^2f^2 \neq c\,\text{Sq}(f)$ in general.                                    ∎

Before we consider more complicated operators, let us consider two important nuances in Definition 4.2:

• To apply Definition 4.2, the quantity $f(x)+g(x)$ must be defined.  For instance, let $f(x) = \ln x$ and let $g(x) = \ln(-x)$. The domain of $f$ is the set of all positive real numbers, and the domain of $g$ is the set of all negative real numbers. However, $f(x)+g(x)$ is not defined for any real number $x$, so $D(f+g)$ is not defined.

• When we check the linearity of the differentiation operator $D$, we assume that $f$ and $g$ are functions of the same variable, say $x$. In contrast, we can ask what is meant by $D(f(x) + g(t))$. The answer depends on whether $D$ refers to differentiation with respect to $x$ or to $t$. In the same way, quantities such as $a$ and $b$ in Eq. (4.6) must be constant relative to the action of the operator in question. For example, if $D$ refers to differentiation with respect to $x$, then we can treat the function $a(t)$ as a constant; for example, $D(a(t)f(x)) = a(t)D(f(x))$.

## The Laplace Transform

There are many other kinds of linear operators that transform functions in ways that are rather different in nature from the operators that we have discussed up to now. An example is given by the *Laplace transform*, which arises in applications of particular interest in electrical engineering and control theory. (The Laplace transform leads to a useful algorithm for finding explicit solutions of certain kinds of common linear differential equations, as discussed in Chapter 10.) Here we will only define the Laplace transform operator and show that it is linear.

**Definition 4.3**   *Let $f(t)$ be a function. The Laplace transform of $f$, denoted by $\mathcal{L}(f)$ or $\mathcal{L}(f(t))$, is the function $F(s)$ defined by*

$$F(s) = \mathcal{L}(f(t)) = \int_0^\infty e^{-st} f(t)\, dt, \qquad (4.7)$$

*whenever the improper integral exists.*

Equation (4.7) has two important features. First, the Laplace transform takes a function $f$ of the variable $t$ and replaces it with a new function, $F$, of the variable $s$.[2] The following examples illustrate this idea in more detail.

Second, Eq. (4.7) involves an improper integral. It can be shown (see Chapter 10) that the Laplace transform is defined for all functions $f$ of *exponential order*. The function $f$ is of exponential order if there exist positive constants $M$ and $k$ such that

$$|f(t)| \le Me^{kt}$$

for every $t \ge 0$. Many common functions, such as polynomials, sines, cosines, and exponentials, are of exponential order. The following examples show how to compute the Laplace transform of some simple functions.

---

[2]Strictly speaking, $s$ should be regarded as a complex number in Definition 4.3. In other words, $\mathcal{L}$ maps the real function $f$ to the complex function $F$. For simplicity, we consider $s$ to be a real number.

✦ *Example 7*    Let $f$ be the constant function 1, that is, $f(t) = 1$ for all $t$. Clearly, $f$ is
of exponential order. (Let $M = 1$ and $k = 0$.) The Laplace transform of $f$ is

$$\mathcal{L}(f(t)) = \mathcal{L}(1)$$

$$= \int_0^\infty e^{-st}\, dt$$

$$= \lim_{t \to \infty} -\frac{e^{-st}}{s} + \frac{1}{s}$$

$$= \frac{1}{s} \tag{4.8}$$

whenever $s > 0$. In this example, the Laplace operator maps the constant function 1
(defined for all values of $t$) to the function $F(s) = 1/s$ in a manner that is defined for
all positive numbers $s$.                                                                                   ∎

✦ *Example 8*    Let $f(t) = e^{kt}$, where $k$ is any fixed constant. By definition, $f$ is of
exponential order. The Laplace transform of $f$ is

$$\mathcal{L}(f(t)) = \mathcal{L}\left(e^{kt}\right)$$

$$= \int_0^\infty e^{-st} e^{kt}\, dt$$

$$= \int_0^\infty e^{(k-s)t}\, dt$$

$$= \frac{1}{k-s} e^{(k-s)t} \Big|_{t=0}^{t=\infty}$$

$$= \frac{1}{s-k} \tag{4.9}$$

whenever $s > k$. (The improper integral diverges if $s \le k$.) Thus, the Laplace operator
maps the exponential function $f(t) = e^{kt}$, which is defined for all values of $t$, to the
function $F(s) = 1/(s-k)$ in a manner that is defined for $s > k$.                          ∎

**Theorem 4.4**    *The Laplace transform is a linear operator.*

**Proof**    Suppose that $f$ and $g$ are two functions whose Laplace transforms $\mathcal{L}(f)$ and
$\mathcal{L}(g)$ are defined. We must show that both conditions of Definition 4.2 are satisfied. The
definition of the Laplace transform leads to

$$\mathcal{L}(f+g) = \int_0^\infty e^{-st}(f(t) + g(t))\, dt$$

$$= \int_0^\infty e^{-st} f(t)\, dt + \int_0^\infty e^{-st} g(t)\, dt$$

$$= \mathcal{L}(f) + \mathcal{L}(g)$$

by the linearity of integration. Next let $c$ be an arbitrary constant. Then

$$\mathcal{L}(cf) = \int_0^\infty e^{-st}[cf(t)]\,dt$$

$$= c\int_0^\infty e^{-st}f(t)\,dt$$

$$= c\mathcal{L}(f),$$

again by the linearity of integration. Hence, $\mathcal{L}$ is a linear operator.   ∎

## Exercises

**1.** Notice that we do not require a linear operator $L$ to satisfy $L(fg) = L(f)L(g)$ wherever multiplication is defined, such as when the domain and range of $L$ consist of functions of real numbers. Does a linear function $f$ of one variable satisfy $f(xy) = f(x)f(y)$ in general?

**2.** Let $D$ denote the differentiation operator; that is, if $f$ is a function of $x$, then $D(f) = df/dx$. Compute $D(f)$ for each function in the following list.

  (a) $f(x) = \sin 3x$

  (b) $f(x) = e^{2x}$

  (c) $f(x) = x^2 + C$, $C$ any constant

**3.** Let $D$ denote the differentiation operator. Find the image of each function $f$ in the list below under $D$. Assume that all functions are differentiable and that $D$ means differentiation with respect to $t$.

  (a) $f(t) = g(t)h(t)$      (b) $f(t) = \sin\pi/10$

  (c) $f(t) = g'(t)$          (d) $f(t) = g(h(t))$

**4.** Is the differentiation operator $D$ one-to-one? If not, find two distinct functions $f$ and $g$ such that $D(f) = D(g)$.

**5.** Let $f$ be a function of $x$, and let

$$U(f) = \int_0^1 f(x)\,dx.$$

Compute $U(f)$ for each function $f$ in the following list (see Example 4.5).

  (a) $f(x) = x^2$          (b) $f(x) = x^2 + 1$

  (c) $f(x) = e^x$          (d) $f(x) = e^x + 1$

**6.** Let $x$ be a function of $t$, and let

$$\mathrm{Op}(x) = \frac{dx}{dt} + 3x.$$

Compute $\mathrm{Op}(x)$ for each function $x$ in the following list.

  (a) $x(t) = \sin 3t$        (b) $x(t) = 3\sin t$

  (c) $x(t) = e^{-3t}$         (d) $x(t) = -3e^t$

**7.** Consider the definite integration operator $U$ as defined in Eq. (4.5).

  (a) Show that $U$ is not one-to-one.

  (b) Show that $U$ is linear.

**8.** In the following list, $f(t)$ refers to a scalar function of the single variable $t$. Each operator replaces $f$ with a new function, $\mathrm{Op}(f)$, that also is a function of $t$. Determine whether each operator is linear by verifying each condition in Definition 4.2.

  (a) $\mathrm{Op}(f(t)) = \sin f(t)$

  (b) $\mathrm{Op}(f(t)) = 2\int_0^t f(s)\,ds$

  (c) $\mathrm{Op}(f(t)) = tf(t)$

  (d) $\mathrm{Op}(f(t)) = d^2f(t)/dt^2$

**9.** The function $f(t)$ is *square integrable* if $\int_{-\infty}^{\infty} [f(t)]^2 \, dt$ exists and is finite. The norm operator, $\| \cdot \|$, is an operator from the set of all square integrable functions to the set of nonnegative numbers defined by

$$\| f \| = \left( \int_{-\infty}^{\infty} [f(t)]^2 \, dt \right)^{1/2}.$$

Determine whether $\| \cdot \|$ is linear.

**10.** Floor$(x)$ is the function whose value is the largest integer $n$ such that $n \le x$. Let Op be the operator defined by $\text{Op}(f(x)) = \text{Floor}(f(x))$, where $f$ is any function of $x$. Is Op linear?

**11.** Can a linear operator map the zero function to a nonzero function? Explain.

**12.** Let $\mathbf{p}(t) = (f(t), g(t))$, where $f$ and $g$ are two scalar functions of $t$. The function $\mathbf{p}$ defines a parametric curve in the plane. Let Op be the operator that takes $\mathbf{p}$ and maps it to a new parametric curve given by

$$\text{Op}(\mathbf{p}(t)) = (3f(t) + 5g(t), -4f(t) + 5g(t)).$$

Is Op linear? Explain.

**13.** A *transformation* of the plane is a function that maps the plane to itself. Let $\mathbf{T}$ be the transformation that reflects the point $(x, y)$ across the horizontal axis; that is, $\mathbf{T}$ interchanges the top and the bottom half-plane.

(a) Find an explicit expression for $\mathbf{T}(x, y)$.

(b) Determine whether $\mathbf{T}$ is linear.

**14.** Let $f(n)$ be a function that is defined only for integers $n$. (Such functions arise in electrical engineering when one considers signals that are sent only at certain discrete times $t_n$.) Let $z$ be any number. The *Z-transform* of $f$ is the operator defined

by

$$Z(f(n)) = \sum_{n=0}^{\infty} \frac{f(n)}{z^n}$$

whenever the sum converges. Show that the $Z$ transform is a linear operator.

**15.** Let $\mathbf{T}$ be the transformation from the $x_1 x_2$ plane to itself that maps $\mathbf{x} = (x_1, x_2)$ to $\mathbf{T}(\mathbf{x}) = (4x_1 - 2x_2, 3x_1 + 4x_2)$. Is $\mathbf{T}$ linear?

**16.** Show how to use the linearity property of the Laplace operator to determine $\mathcal{L}(10)$ and $\mathcal{L}(5e^t + 7)$ directly from the results of Examples 7 and 8, that is, without computing the transform from an integral.

**17.** The hyperbolic cosine and the hyperbolic sine are defined as

$$\cosh x = \tfrac{1}{2}(e^x + e^{-x})$$

and

$$\sinh x = \tfrac{1}{2}(e^x - e^{-x}),$$

respectively. Use Example 8 and the linearity of the Laplace transform to compute $\mathcal{L}(\cosh kx)$ and $\mathcal{L}(\sinh kx)$, where $k$ is any constant.

**18.** The Fourier transform is important in the theory of partial differential equations. It is an integral transform, just as the Laplace transform is an integral transform. The Fourier transform of a function $f(x)$ is defined as

$$F_k(f) = \frac{1}{2\pi} \int_{-\infty}^{\infty} f(x) e^{-ikx} \, dx \qquad (4.10)$$

whenever the improper integral exists. Here, $i = \sqrt{-1}$ and $k$ is a constant.

(a) Use Eq. (4.10) to find $F_k(f)$, where

$$f(x) = \begin{cases} 0 & \text{if } x < -1, \\ 1 & \text{if } -1 \le x \le 1, \\ 0 & \text{if } x > 1. \end{cases}$$

(b) Show that $F_k$ is a linear operator.

## 4.3   DIFFERENTIAL EQUATIONS AS OPERATORS

Chapter 2 contains a detailed discussion of the exponential growth equation

$$\frac{dx}{dt} = kx, \tag{4.11}$$

whose solution is $x(t) = x_0 e^{kt}$ for any constant $k$. The goal of this section is to consider differential equations like Eq. (4.11) in terms of operators.

For this purpose, we rewrite Eq. (4.11) in the form

$$\frac{dx}{dt} - kx = 0. \tag{4.12}$$

The left-hand side of Eq. (4.12) defines an operator that is called a *differential operator* because it involves a derivative. The resulting operator, Op, is a well-defined rule for producing a new function from an old one: Op replaces $f$ with $df/dt - kf$. For example,

$$\text{Op}(t^2) = 2t - kt^2,$$

$$\text{Op}(\sin t) = \cos t - k \sin t,$$

$$\text{Op}(Ce^{kt}) = 0.$$

The last example in the list above is special: Op maps every function of the form $Ce^{kt}$ to the zero function. We can think of the function $Ce^{kt}$ as a "root" of Op in much the same way that the number 1 is a root of the function $f(t) = t^2 - 1$.

In algebra and calculus, we often want to find a number $t$ such that $f(t) = a$ for some function $f$ and a constant $a$. In the study of differential equations, we are interested in finding functions $x(t)$ for which

$$\text{Op}(x) = f. \tag{4.13}$$

Here, Op is an operator that involves the derivatives of $x$ as well as $x$ itself, and $f$ is a function of the independent variable $t$. A solution of Eq. (4.13) is analogous to a solution of $f(t) = a$, except that we are looking for a function instead of a single number. Here are two examples.

✦ *Example 1*   The equation $p' = rp + A$ describes the growth of an annuity in the case of continuously compounded interest. (Here, $p(t)$ is the principal at time $t$, $r$ is the interest rate, and $A$ is the rate of investment. We assume that $A$ and $r$ are constants.) We can rewrite this equation as

$$\frac{dp}{dt} - rp = A, \tag{4.14}$$

where the associated differential operator Op is the same as that produced by the exponential growth equation (4.12) with $r$ instead of $k$. A solution of Eq. (4.14) is a function $p(t)$ for which $\text{Op}(p) = A$. It is straightforward to verify that $p(t) = e^{rt}(p_0 + A/r) - (A/r)$ is the solution. (See Section 2.2.)                                                                     ∎

✦ *Example 2*   The logistic equation, $p' = p(1 - p)$, can be rewritten in the form

$$\frac{dp}{dt} - p + p^2 = 0. \tag{4.15}$$

The left-hand side of Eq. (4.15) defines a differential operator. The solutions of the logistic equation are precisely the functions $p(t)$ for which $\text{Op}(p(t)) = 0$, where Op is defined as the left-hand side of Eq. (4.15).                                                     ∎

A fundamental question, which we will address repeatedly in the rest of the text, is whether the operator associated with a given differential equation is linear. As we will see, linear differential operators impart a special structure to the solutions of the associated differential equation. Before we discuss the notion of the linearity of differential operators, we need the following definition.

**Definition 4.5**   *The differential operator* Op *is* homogeneous *if* $\text{Op}(0) = 0$. *(In other words,* Op *is homogeneous if* Op *maps the zero function to itself.) The differential equation* $x' = f(t, x)$ *is* homogeneous *if the zero function is a solution. A differential equation is* nonhomogeneous *if it is not homogeneous.*

✦ *Example 3*   Each of the differential equations in the following list is homogeneous, because the zero function is a solution:

1. $x' + kx = 0$
2. $x' + tx = \sin x$
3. $x' + t^2 x = x^2$
4. $x'' + x = 0$

Each of the equations in the following list is nonhomogeneous, because the zero function is not a solution:

5. $x' + kx = 1$
6. $x' + tx = \sin t$
7. $x' - x^2 = t^2$
8. $x'' + x = \sin t$                                                                     ∎

Most differential equations can be split into a homogeneous part and a nonhomogeneous part.[3] The homogeneous part of the equation consists of precisely those terms that are zero when the dependent variable is the zero function.

✦ *Example 4*  Let

$$x' + tx = \sin x - \sin t. \tag{4.16}$$

Equation (4.16) is not homogeneous, because the zero function is not a solution. The dependent variable is $x$. If $x = 0$ is the zero function, then Eq. (4.16) becomes

$$0' + t \cdot 0 = \sin 0 - \sin t,$$

all of whose terms (except $\sin t$) are zero. Therefore, the homogeneous part of Eq. (4.16) is

$$x' + tx = \sin x.$$

We can rewrite Eq. (4.16) with the homogeneous part on the left and the nonhomogeneous part on the right:

$$x' + tx - \sin x = \sin t. \qquad\blacksquare$$

✦ *Example 5*  The homogeneous part of each of (5)–(8) in Example 3 consists of those terms to the left of the equals sign. The nonhomogeneous part of each of (5)–(8) in Example 3 consists of those terms to the right of the equals sign.  $\blacksquare$

The examples above suggest that the homogeneous part of every differential equation defines a homogeneous differential operator of the dependent variable.

✦ *Example 6*  The homogeneous differential operators associated with Eqs. (5)–(8) in Example 3 are the following:

5'. $\mathrm{Op}(x) = x' + kx$

6'. $\mathrm{Op}(x) = x' + tx$

7'. $\mathrm{Op}(x) = x' - x^2$

8'. $\mathrm{Op}(x) = x'' + x.$  $\blacksquare$

**Definition 4.6**  *A differential equation is* linear *if it can be written in the form*

$$\mathrm{Op}(x) = f(t),$$

*where* $\mathrm{Op}(x)$ *is linear and $f$ is either a constant function or a function of the independent*

---

[3]We exclude equations like $x' + 1/x = f(t)$, which are not defined if $x$ is the zero function.

*variable t. Otherwise, the differential equation is* nonlinear. *If* Op *is linear, we often write L(x) instead of* Op(x).

The following examples illustrate how to check the linearity of a differential equation.

◆ *Example 7*   The exponential growth equation, $x' = kx$, is linear, because we write it in the form

$$Op(x) = x' - kx = 0.$$

We verify that Op is linear by checking the additivity and proportionality properties in Definition 4.2, as follows. Let $x$ and $y$ be any differentiable functions. Then

$$\begin{aligned}
Op(x + y) &= (x + y)' - k(x + y) \\
&= x' + y' - kx - ky \\
&= (x' - kx) + (y' - ky) \\
&= Op(x) + Op(y),
\end{aligned}$$

Let $c$ be any constant. Then

$$\begin{aligned}
Op(cx) &= (cx)' - k(cx) \\
&= c(x' - kx) \\
&= c\,Op(x).
\end{aligned}$$

We often write $L(x) = x' - kx$ to emphasize the linearity of the operator.                ∎

◆ *Example 8*   The equation $p' = rp + A$ is linear and nonhomogeneous. To check linearity, we rewrite it as

$$Op(p) = p' - rp = A. \tag{4.17}$$

Here, Op is a linear, homogeneous operator. In fact, it is the same operator as in Example 7 with $k$ replaced by $r$. Hence, Eq. (4.17) is linear.                ∎

◆ *Example 9*   The logistic equation, $p' = p(1 - p)$, defines the operator $Op(p) = p' - p + p^2$, which is homogeneous. We proceed as above to check the linearity of Op:

$$\begin{aligned}
Op(p + q) &= (p + q)' - (p + q) + (p + q)^2 \\
&= p' + q' - p - q + p^2 + 2pq + q^2 \\
&= (p' - p + p^2) + (q' - q + q^2) + 2pq \\
&= Op(p) + Op(q) + 2pq \\
&\neq Op(p) + Op(q).
\end{aligned}$$

Hence, the logistic equation is not linear.                ∎

✦ *Example 10*   The simplest, and one of the most important, second-order differential equations is

$$\frac{d^2x}{dt^2} + \omega^2 x = 0, \tag{4.18}$$

where $\omega^2$ is a positive constant. Equation (4.18) is second order because the highest-order derivative term in the equation is second order. Equation (4.18) defines the differential operator

$$\mathrm{Op}(x) = x'' + \omega^2 x,$$

which is homogeneous. The operator also is linear, because if $x$ and $y$ are any twice-differentiable functions of $t$, then

$$\begin{aligned}
\mathrm{Op}(x + y) &= (x + y)'' + \omega^2(x + y) \\
&= x'' + y'' + \omega^2 x + \omega^2 y \\
&= \left(x'' + \omega^2 x\right) + \left(y'' + \omega^2 y\right) \\
&= \mathrm{Op}(x) + \mathrm{Op}(y)
\end{aligned}$$

and

$$\begin{aligned}
\mathrm{Op}(cx) &= (cx)'' + c\omega^2 x \\
&= c\left(x'' + \omega^2 x\right) \\
&= c\,\mathrm{Op}(x).
\end{aligned}$$   ∎

It is important to note that a given differential equation may be homogeneous, linear, both, or neither. Homogeneity refers to the question of whether the zero function is a solution. Linearity refers to the properties of additivity and proportionality that are stated in Definition 4.2.

Linear differential equations have a special form that is easy to recognize with practice. The following theorem demonstrates the general form of linear first- and second-order differential equations. The proof is left as an exercise.

**Theorem 4.7**   *Let $a(t)$, $b(t)$, and $f(t)$ be continuous functions of $t$. The relation*

$$\frac{dx}{dt} + a(t)x = f(t) \tag{4.19}$$

*defines a linear first-order differential equation for $x$. The relation*

$$\frac{d^2x}{dt^2} + a(t)\frac{dx}{dt} + b(t)x = f(t) \tag{4.20}$$

*defines a linear second-order differential equation for $x$. Equations (4.19) and (4.20) are homogeneous if $f(t)$ is the zero function. Otherwise, they are nonhomogeneous.*

## *Exercises*

**1.** Rewrite each differential equation in the following list in the form

$$\text{Op}(x) = f(t).$$

In each case, determine whether Op is homogeneous.

(a) $x' = -2x + t$

(b) $x' = x - 10$

(c) $x' + e^x = 5$

(d) $x' + t^2x = x$

(e) $x' = tx^2 - \cos t$

(f) $x'' = 4x' + 3x + \sin t$

(g) $x'' = -4x' - 3\sin x + 2\sin t$

(h) $x'' = -\cos x + \cos t$

**2.** Determine which of the equations in Exercise 1 is linear.

**3.** Repeat Exercise 1 for each differential equation in the following list. The names $a$ and $b$ refer to continuous functions of their respective arguments.

(a) $x' = a(t)x + b(t)$

(b) $x' = a(x)x + b(t)$

(c) $x'' = 4x' + 3x + \sin t$

(d) $(\cos t)x'' + x^2 = \sin t$

(e) $x' = xe^{kx}$

(f) $x' - (\sin t)x - e^t = 0$

**4.** Give an example of a first-order differential equation that:

(a) is linear but not homogeneous.

(b) is homogeneous but nonlinear.

(c) is homogeneous and linear.

(d) is nonhomogeneous and nonlinear.

**5.** (a) Show that if $\text{Op}(x) = f(t)$ is a linear differential equation, then $\text{Op}(x)$ is a homogeneous operator.

(b) Find a counterexample to the following claim: If $\text{Op}(x)$ is homogeneous, then $\text{Op}(x) = f(t)$ is a linear differential equation.

**6.** Let $m$ and $n$ be positive integers. Show that every differential equation containing a term of the form $(d^n x/dt^n)x^m$ is nonlinear.

**7.** Consider Eqs. (4.19)–(4.20).

(a) Show that they are homogeneous if $f$ is the zero function.

(b) Show that they are linear.

**8.** State and prove a theorem regarding the linearity of the differential equation

$$a(t)x'(t) + b(t)x(t) = f(t),$$

where $a$, $b$, and $f$ are continuous functions of $t$.

## 4.4   THE SUPERPOSITION PRINCIPLE

The notion of linearity gives a straightforward way to describe the solutions of a wide variety of ordinary differential equations. Linearity is an abstract idea, but it exposes patterns that arise irrespective of the details of the individual differential equation. In this section, we describe the basic theory behind the structure of solutions of linear homogeneous ordinary differential equations. We begin with the case of the exponential growth equation,

$$\frac{dx}{dt} = kx. \tag{4.21}$$

The corresponding linear, homogeneous operator is

$$L(x) = x' - kx.$$

We know that $x_1(t) = e^{kt}$ is a solution of Eq. (4.21), that is, $L(x_1) = 0$. The function $x_2(t) = 2e^{kt} = x_1(t) + x_1(t)$ also is a solution, because the linearity of $L$ implies that

$$L(x_2) = L(x_1 + x_1) = L(x_1) + L(x_1) = 0 + 0.$$

A similar argument shows that the function $x_3(t) = x_1(t) + x_2(t)$ also solves Eq. (4.21).

Figure 4.8 shows the direction field corresponding to $x' = x$, together with the solutions $x_1(t) = e^t$, $x_2(t) = 2e^t$, and $x_3(t) = 3e^t$. From bottom to top, the solution curves correspond to the initial value problems $x_1(0) = 1$, $x_2(0) = 2$, and $x_3(0) = 3$. Notice that the top curve can be obtained by adding the lower two curves together. (When $t > 1$, the top curve is cropped from the figure. However, if we extended the top part of the figure farther up, we would still see that the top curve is the sum of the bottom two.)

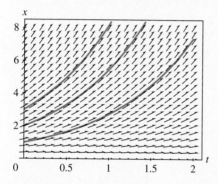

**FIGURE 4.8**  The direction field of $x' = x$ and three solutions satisfying the initial conditions $x_0 = 1$, $x_0 = 2$, and $x_0 = 3$.

There is nothing special about the choice of the initial values 1, 2, and 3. Let $c$ be any constant, and let $x$ be a solution of Eq. (4.21) satisfying the initial value $x(0) = c$. It is easy to see that $x(t) = cx_1(t)$ is also a solution, because the linearity of $L$ implies that

$$L(x) = L(cx_1) = cL(x_1) = 0.$$

These considerations prove the following statement. Suppose $x_1$ and $x_2$ are any two solutions of $x' = kx$. Then the functions $x_1 + x_2$ and $cx_1$ also are solutions, where $c$ is any constant.

A similar statement can be made about any homogeneous linear differential equation, as the following theorem shows.

**Theorem 4.8**   *The Superposition Principle*    *Let $L(x)$ be a linear, homogeneous differential operator, and suppose that $x_1$ and $x_2$ are functions for which*

$$L(x_1) = L(x_2) = 0.$$

*If $c_1$ and $c_2$ are any constants, then*

$$L(c_1 x_1 + c_2 x_2) = 0.$$

**Proof**   Because $L$ is linear,

$$\begin{aligned} L(c_1 x_1 + c_2 x_2) &= L(c_1 x_1) + L(c_2 x_2) \\ &= c_1 L(x_1) + c_2 L(x_2) \\ &= 0. \end{aligned}$$                                                                  ∎

◆ *Example 1*    The direction fields of first-order, linear homogeneous differential equations often can be distinguished visually from the direction fields of nonlinear homogeneous equations. Figure 4.9(a) shows the direction field and three solutions of a linear homogeneous differential equation. The displayed solutions are constant multiples of

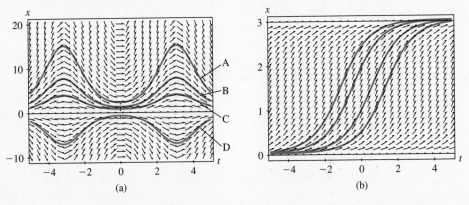

(a)                                                          (b)

**FIGURE 4.9**   The direction field corresponding to homogeneous differential equations that are (a) linear; (b) nonlinear.

each other. From top to bottom, the curves are labeled $A$, $B$, $C$, and $D$. Curve $B$ is the mirror image of curve $D$.

In addition, the slope lines suggest that a solution satisfying the initial condition $x(0) = c$ is a constant multiple of the solution satisfying the initial condition $x_1(0) = 1$. For instance, when $t = 0$, the height of curve $A$ is five times greater than the height of curve $C$. When $t = 3$, the height of curve $A$ remains 5 times greater than the height of curve $C$. In fact, the height of curve $A$ is 5 times greater that the height of curve $C$ for every value of $t$. Similar statements can be made about the relationship between each of the curves in Fig. 4.9(a).

Figure 4.9(b) shows a contrasting situation. Here the solution curves approach a constant value (slightly above 3) once $t$ is larger than about 2. Unlike Fig. 4.9(a), the curves in Fig. 4.9(b) cannot be constant multiples of each other. For instance, all the curves are at about the same height when $t > 2$, but they are not at the same height when $t = 0$. In other words, the ratio of the height of one curve to the height of another does not remain constant for every value of $t$. ∎

◆ *Example 2* Consider

$$x' = x \sin t \tag{4.22}$$

with $x(0) = M$. Equation (4.22) is separable, and it is easy to check that

$$x(t) = e^{1 - \cos t}$$

satisfies the initial condition $x(0) = 1$. The associated differential operator is

$$L(x) = x' - x \sin t,$$

which is linear and homogeneous. Theorem 4.8 implies that $x(t) = M e^{1 - \cos t}$ is a solution that satisfies the initial condition $x(0) = M$. ∎

◆ *Example 3* Higher-order linear differential equations have distinct solutions that are not constant multiples of each other. Consider

$$\frac{d^2 x}{dt^2} + x = 0. \tag{4.23}$$

Two solutions are $x_1(t) = \sin t$ and $x_2(t) = \cos t$, as is easily verified by substitution.

The associated differential operator is $L(x) = x'' + x$, which is linear and homogeneous. The superposition principle implies that any function of the form

$$x(t) = a \cos t + b \sin t \tag{4.24}$$

solves Eq. (4.23), where $a$ and $b$ are any constants. ∎

Formulas like Eq. (4.24) are so common in practice that we give them a special name.

**Definition 4.9**   *A function of the form*

$$h(t) = af(t) + bg(t),$$

*where $a$ and $b$ are constants, is a* linear combination *of the functions $f$ and $g$.*

Linear second-order differential equations, such as Eq. (4.23), generally have two homogeneous solutions that are not constant multiples of each other. (More details are given in Chapter 5.) We call Eq. (4.24) a *two-parameter family* of solutions of Eq. (4.23). In other words, for every choice of the two constants $a$ and $b$, there is a solution of Eq. (4.23).

✦ *Example 4*   Two solutions of the differential equation

$$t^2 x'' + tx' + x = 0 \qquad\qquad (4.25)$$

are $x_1(t) = \cos(\ln|t|)$ and $x_2(t) = \sin(\ln|t|)$, as you can readily verify by substitution. Equation (4.25) is homogeneous, because $x = 0$ is a solution. The associated differential operator is

$$L(x) = t^2 x'' + tx' + x,$$

which is linear. Therefore, Theorem 4.8 guarantees that if $a$ and $b$ are any constants, the function

$$x(t) = a\cos(\ln|t|) + b\sin(\ln|t|)$$

also is a solution of Eq. (4.25).                                                                  ■

✦ *Example 5*   It is straightforward to apply the ideas of linearity to third- and higher-order differential equations. Let

$$x''' - 6x'' + 11x' - 6x = 0. \qquad\qquad (4.26)$$

Equation (4.26) is a third-order differential equation because the highest-order derivative of $x$ is $x'''$. Equation (4.26) defines the differential operator

$$L(x) = x''' - 6x'' + 11x' - 6x,$$

which is linear. (The verification proceeds as in the previous examples.)

It is straightforward to verify that the functions $x_1(t) = e^t$, $x_2(t) = e^{2t}$, and $x_3(t) = e^{3t}$ are all solutions of Eq. (4.26). Since $L$ is linear, the linear combination

$$x(t) = ax_1(t) + bx_2(t) + cx_3(t) = ae^t + be^{2t} + ce^{3t}$$

is a solution of Eq. (4.26), where $a$, $b$, and $c$ are any constants.    ■

## Physical Applications of the Superposition Principle

The superposition principle, which is a direct consequence of linearity, has important technological applications. We present two examples from physics.

✦ *Example 6*   Electromagnetic radiation obeys the superposition principle to a good approximation as it travels through the earth's atmosphere. (We say that the atmosphere provides a linear medium.) As a result, many complex signals, such as those from radio broadcasts, can be treated as linear combinations of sines and cosines.

Figure 4.10 illustrates the phenomenon. The bold curve is a periodic, but reasonably complicated, curve. However, it is a linear combination of three sine curves, which are represented by the thin curves in the plot. Each of the three sine curves is easily characterized by its amplitude and period. The bold curve can be analyzed in terms of these simpler functions. This idea is the basis of a subject called *Fourier analysis*, which is quite important in signal-processing applications.

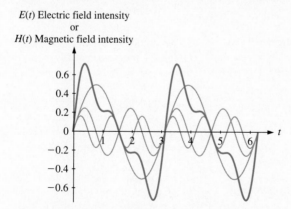

FIGURE 4.10   The superposition principle allows a complex wave form (bold curve) to be treated as a sum of simpler wave forms (thin curves).    ■

✦ *Example 7*   The surface of a pool behaves like a linear medium for the transmission of water waves in certain circumstances. For example, suppose that you and a friend stand on opposite sides of a swimming pool whose water is still. Each of you throws in a pebble. The resulting waves radiate in circles from the points where the rocks enter the water. The waves meet somewhere in the middle of the pool, but they do not interfere with each other. When two waves cross, they momentarily add to give a bigger wave, but then they appear to pass through one another. This description of water waves is very accurate as long as the waves are relatively small, that is, when the amplitude of the disturbance (wave) is sufficiently small that the restoring force of the elastic media (the water) is proportional to its displacement above or below the surface of the water. The situation is more complicated if you and your friend each tip a large boulder into the pool at the same time.                                                                        ∎

The mathematical description of the linear transmission of waves requires the use of partial differential equations, because the motion of waves depends on space as well as time. Nevertheless, an analysis similar to what we have described in this chapter applies to the resulting partial differential equations. Moreover, the superposition principle holds for the solutions of linear partial differential equations in much the same way as for linear ordinary differential equations.

## Exercises

**1.** For each differential equation in the following list:

- Plot a direction field.

- Find explicit solutions satisfying the initial conditions $x_1(0) = 1$, $x_2(0) = 2$, and $x_3(0) = 3$ and add their graphs to the direction field.

- Can you add the curves graphically? Explain.

(a) $x' = tx$

(b) $x' = x \sin t$

(c) $x' = x^2$

**2.** This exercise considers Example 2 in more detail.

(a) Show that Eq. (4.22) is homogeneous.

(b) Show that Eq. (4.22) is linear.

(c) Let $M$ be any constant. Apply the superposition principle to show that $Mx(t)$ is a solution

of Eq. (4.22) whenever $x(t)$ is a solution of Eq. (4.22).

**3.** Which direction fields in Fig. 4.11 correspond to linear, homogeneous differential equations? Explain.

**4.** Figure 4.12 shows two functions that are mirror images of each other (that is, one curve is $-1$ times the other). Does the direction field correspond to a linear differential equation? Explain.

**5.** This problem illustrates the special structure of the solutions of linear differential equations, at least for one tractable example. Let

$$x' + tx = 0. \qquad (4.27)$$

(a) Show that Eq. (4.27) is homogeneous and linear.

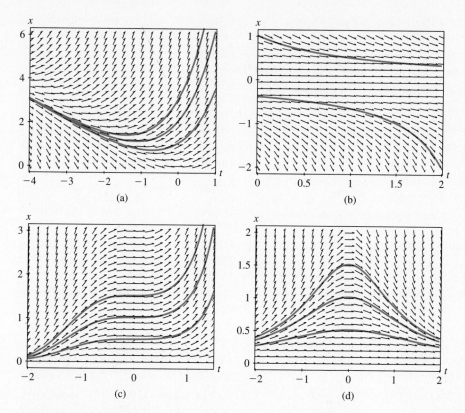

**FIGURE 4.11**   For Exercise 3.

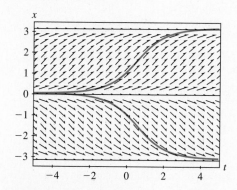

**FIGURE 4.12**   For Exercise 4.

(b) Suppose that $x_1(t)$ satisfies the initial condition $x_1(0) = 1$ and that $x_2(t)$ satisfies the initial condition $x_2(0) = 2$. Without attempting to find $x_1$ and $x_2$ explicitly, adapt your verification in (a) to show that the solution satisfying the initial condition $x(0) = 3$ must be the function $x(t) = x_1(t) + x_2(t)$.

(c) Suppose you want to find the solution satisfying the initial condition $x(0) = C$, where $C$ is any constant. Must it be the case that $x(t) = C x_1(t)$, where $x_1(t)$ satisfies the initial condition $x_1(0) = 1$? Answer the question without deriving an explicit solution of Eq. (4.27).

(d) Now find an explicit solution of Eq. (4.27) and use it to verify your answers to (b) and (c).

**6.** Let $x' = x^2$.

(a) Show that the differential equation is nonlinear.

(b) Find the solution $x_1(t)$ satisfying the initial condition $x_1(0) = 1$.

(c) Find the solution $x_2(t)$ satisfying the initial condition $x_2(0) = 2$.

(d) Find the solution $x_3(t)$ satisfying the initial condition $x_3(0) = 3$.

(e) Suppose you want to find the solution satisfying the initial condition $x(0) = C$, where $C$ is any constant. Is it true that $x(t) = Cx_1(t)$? Explain.

**7.** Let $x' = f(t, x)$ be a homogeneous, linear differential equation, and suppose that $f$ is not the zero function.

(a) Explain in what sense the direction field must be symmetric about the axis corresponding to the independent variable.

(b) Explain why $x = 0$ must be an equilibrium solution, and that $x = 0$ is the only equilibrium solution.

**8.** Let

$$x'' + \omega^2 x = 0. \qquad (4.28)$$

(a) Identify the associated differential operator and show that it is linear.

(b) Show by substitution that $\sin \omega t$ and $\cos \omega t$ are solutions of Eq. (4.28).

(c) Use the superposition principle to find a two-parameter family of solutions of Eq. (4.28).

**9.** Consider Example 5.

(a) Show that $L$ is a linear operator.

(b) Show that $L$ is homogeneous.

(c) Verify that the linear combination $ae^t + be^{2t} + ce^{3t}$ is a solution of Eq. (4.26).

**10.** Let

$$x'' - \omega^2 x = 0. \qquad (4.29)$$

(a) Identify the associated differential operator $L$ and show that it is linear.

(b) Find the corresponding image of each of the functions in the following list under $L$.

   (i) $x(t) = \sin \omega t$       (ii) $x(t) = \cos \omega t$

   (iii) $x(t) = \omega^2 t^2$      (iv) $x(t) = e^{\omega t}$

   (v) $x(t) = e^{-\omega t}$

(c) Find a two-parameter family of solutions of Eq. (4.29) and explain why the constant in Eq. (4.29) is written as $-\omega^2$.

(d) Find a solution satisfying the initial values $x(0) = 1$ and $x'(0) = -1$.

**11.** Let $x'' + 4x = 0$.

(a) Using the result of Exercise 8, or by trial and error, find a two-parameter family of solutions.

(b) Find the solution satisfying the initial values $x(0) = 2$ and $x'(0) = 1$.

**12.** Let

$$t^2 \frac{d^2 x}{dt^2} + t \frac{dx}{dt} + 4x = 0. \qquad (4.30)$$

(a) Identify the associated differential operator and verify that it is linear and homogeneous.

(b) Show that $x_1(t) = \cos(2 \ln |t|)$ and $x_2(t) = \sin(2 \ln |t|)$ are homogeneous solutions of Eq. (4.30).

(c) Find the solution satisfying the initial values $x(1) = 1$ and $x'(1) = 1$.

(d) Describe the solution qualitatively. Does it oscillate? If so, is it periodic? Does the solution have a constant amplitude? Does the solution approach a limit as $t \to \infty$? as $t \to 0$?

**13.** Let

$$x'' + 3x' + 2x = 0. \qquad (4.31)$$

(a) Identify the associated differential operator, $L$, defined by Eq. (4.31) and show that $L$ is linear.

(b) Let $k$ be a constant and find $L(e^{kt})$.

(c) Determine the values of $k$ for which $L(e^{kt}) = 0$ and use them to find a two-parameter family of solutions of Eq. (4.31).

(d) Use the result in (c) to find the solution of Eq. (4.31) satisfying the initial conditions $x(0) = 1$ and $x'(0) = 2$.

**14.** Let

$$x'' + ax' + bx = 0. \qquad (4.32)$$

(a) Identify the differential operator, $L$, defined by Eq. (4.32) above and show that $L$ is linear.

(b) Let $k$ be a constant and find $L(e^{kt})$. Under what conditions is $e^{kt}$ a solution of Eq. (4.32)?

(c) Find a two-parameter family of solutions of Eq. (4.32). Can you always use the superposition principle to find such a two-parameter family?

## 4.5   THE SOLUTIONS OF LINEAR EQUATIONS

This section discusses the structure of solutions of first- and second-order linear differential equations. As we will see, linearity allows us to add together various functions to construct solutions that can satisfy an arbitrary initial condition.

### The Solutions of First-order Linear Equations

We begin with a discussion of the nonhomogeneous first-order differential equation

$$\frac{dx}{dt} = rx + A, \qquad (4.33)$$

which describes the growth of an annuity (see Chapter 2). For simplicity, assume that $r$ and $A$ are constants. Equation (4.33) can be recast in the form $L(x) = f(t)$, where $L$ is a linear operator:

$$L(x) = \frac{dx}{dt} - rx = A. \qquad (4.34)$$

Equation (4.33) is nonhomogeneous if $A \neq 0$. One solution, which can be determined by inspection, is the constant solution $x_p = -A/r$. A solution of the nonhomogeneous equation is called a *particular* solution, often denoted by the subscript $p$.

Equation (4.33) is homogeneous if $A = 0$; in this case, the function $x_h(t) = e^{rt}$ is a solution. The subscript $h$ denotes a homogeneous solution. (A homogeneous solution is sometimes called a *complementary* solution.)

Because $L$ is linear, we have

$$L(x_p + Cx_h) = L(x_p) + L(Cx_h)$$
$$= L(x_p) + CL(x_h)$$
$$= L(-A/r) + CL(e^{rt})$$
$$= A + 0$$
$$= A,$$

where $C$ is any constant. Moreover, if an initial condition of the form $x(t_0) = x_0$ is specified, then $C$ can be determined uniquely. Hence, the function

$$x(t) = x_p(t) + Cx_h(t) = -\frac{A}{r} + Ce^{rt} \tag{4.35}$$

defines a family of solutions of Eqs. (4.33)–(4.34).

The following definition summarizes the basic terminology.

**Definition 4.10**  *Let*

$$L(x) = x' + a(t)x = f(t). \tag{4.36}$$

*A homogeneous solution of Eq. (4.36) is a function $x_h$, different from the zero function, for which $L(x_h) = 0$. (In other words, $x_h' + a(t)x_h = 0$.) A particular solution of Eq. (4.36) is a function $x_p$ for which $L(x_p) = f(t)$.*

The following theorem shows that the sum of a homogeneous solution and a non-homogeneous solution of every linear first-order differential equation is also a solution.

**Theorem 4.11**  *Let $a(t)$ and $f(t)$ be continuous functions of $t$, and let*

$$\frac{dx}{dt} + a(t)x = f(t). \tag{4.37}$$

*Suppose that $x_p(t)$ is a particular solution of Eq. (4.37) and that $x_h(t)$ solves the homogeneous equation*

$$\frac{dx}{dt} + a(t)x = 0.$$

*Then the function*

$$x(t) = x_p(t) + Cx_h(t) \tag{4.38}$$

*solves Eq. (4.37), where $C$ is any constant.*

The proof of Theorem 4.11 is outlined in Exercise 12.

✦ *Example 1*   Let

$$x' + x = \cos 3t. \tag{4.39}$$

The initial condition $x(0) = x_0$ implies that

$$x(t) = \tfrac{1}{10} \cos 3t + \tfrac{3}{10} \sin 3t + e^{-t}\left(x_0 - \tfrac{1}{10}\right). \tag{4.40}$$

(See Example 2 in Section 2.3.)

Equation (4.39) defines the linear, homogeneous differential operator $L(x) = x' + x$, so we can rewrite it as

$$L(x) = \cos 3t.$$

Notice that $L\left(e^{-t}\left(x_0 - \tfrac{1}{10}\right)\right) = 0$. Hence,

$$x_h(t) = e^{-t}\left(x_0 - \tfrac{1}{10}\right)$$

is a homogeneous solution of Eq. (4.39). The remaining terms in Eq. (4.40) constitute the particular solution,

$$x_p(t) = \tfrac{1}{10} \cos 3t + \tfrac{3}{10} \sin 3t.$$

It is straightforward to verify that $L(x_p) = \cos 3t$. Equation (4.40) is a general solution of Eq. (4.39), because it incorporates the initial condition and because

$$L(x) = L(x_p + x_h) = L(x_p) + L(x_h) = \cos 3t + 0.$$

In this example, the homogeneous solution $x_h$ corresponds to a transient solution, because $x_h(t) \to 0$ as $t \to \infty$. The particular solution $x_p$ does not decay to 0 as $t \to \infty$, so $x_p$ corresponds to a steady-state solution.                                                                      ■

✦ *Example 2*   Let

$$x' = -2x + 5\sin t. \tag{4.41}$$

The corresponding differential operator is

$$L(x) = x' + 2x,$$

which is linear and homogeneous. Thus, Eq. (4.41) can be written as

$$L(x) = 5\sin t.$$

A homogeneous solution is $x_h(t) = e^{-2t}$. A particular solution is $x_p(t) = 2\sin t - \cos t$, because

$$L(2\sin t - \cos t) = (2\sin t - \cos t)' + 2(2\sin t - \cos t)$$
$$= (2\cos t + \sin t) + (4\sin t - 2\cos t)$$
$$= 5\sin t.$$

(The solution procedure described in Section 2.3 can be used to derive $x_p$.)

Because $L$ is linear and homogeneous, the superposition principle (Theorem 4.8) implies that

$$x(t) = Cx_h(t) + x_p(t)$$

also solves Eq. (4.41), where $C$ is any constant. Notice also that $x$ can be characterized as the sum of a transient and a steady-state solution. In this case, $x_h$ is the transient and $x_p$ is the steady state.                                                                        ∎

✦ *Example 3*     Theorem 4.11 states that a family of solutions of first-order linear equations can be written in the form

$$x(t) = Cx_h(t) + x_p(t),$$

where $C$ is any constant. This example shows that it is generally *not* possible to multiply the particular solution by an arbitrary constant and still have a solution of the differential equation; that is, functions of the form

$$x(t) = x_h(t) + Cx_p(t)$$

are *not* solutions for typical linear differential equations, where $C$ is an arbitrary constant. Let

$$x' = -2x + 5\sin t. \tag{4.42}$$

In Example 2, the solution is written as

$$x(t) = Cx_h + x_p = Ce^{-2t} + (2\sin t - \cos t),$$

where the terms in parentheses correspond to the particular solution $x_p$.

Now consider the function

$$\hat{x}(t) = x_h + Cx_p = e^{-2t} + C(2\sin t - \cos t),$$

where we have multiplied the particular solution (instead of the homogeneous solution) by an arbitrary constant. The resulting function, $\hat{x}$, is *not* a solution. To see this, observe that

$$L(\hat{x}) = L(x_h + Cx_p)$$
$$= L(x_h) + L(Cx_p)$$
$$= 0 + CL(x_p)$$
$$= C(5 \sin t).$$

This calculation shows that $\hat{x}$ does not satisfy Eq. (4.42) if $C$ is an arbitrary constant. The function $\hat{x}$ solves Eq. (4.42) only if $C = 1$, which is not an arbitrary constant.   ∎

## The Solutions of General First-order Linear Equations

This subsection illustrates an important theoretical result:

*Every first-order linear differential equation with continuous coefficient functions has a unique solution that can be expressed in terms of an integral.*

We are interested in finding analytical representations of the solutions of initial value problems of the form

$$\frac{dx}{dt} + a(t)x = f(t), \qquad x(t_0) = x_0, \tag{4.43}$$

where $a$ and $f$ are continuous functions of $t$ in some interval containing the initial time $t_0$. The solution procedure here is analogous to that explained in Section 2.3. The difference is that the coefficient of $x$ can be a function of $t$ instead of just a constant.

Let

$$A(t) = \int_{t_0}^{t} a(s)\,ds.$$

The fundamental theorem of calculus implies that $dA/dt = a(t)$. The chain and product rules imply that

$$\left(x(t)e^{A(t)}\right)' = x'(t)e^{A(t)} + x(t)\left(e^{A(t)}\right)'$$
$$= x'(t)e^{A(t)} + x(t)\left[A'(t)e^{A(t)}\right]$$
$$= x'(t)e^{A(t)} + x(t)a(t)e^{A(t)}$$
$$= \left(x'(t) + a(t)x(t)\right)e^{A(t)}, \tag{4.44}$$

where the prime denotes differentiation with respect to $t$. Notice that Eq. (4.44) is the product of the left-hand side of Eq. (4.43) with $e^{A(t)}$.

Suppose we multiply each side of Eq. (4.43) by $e^{A(t)}$ to get

$$e^{A(t)}\left(x' + a(t)x\right) = e^{A(t)}f(t). \tag{4.45}$$

By (4.44), this is equivalent to

$$\frac{d}{dt}\left(e^{A(t)}x(t)\right) = e^{A(t)}f(t). \tag{4.46}$$

An antiderivative of the left-hand side of Eq. (4.46) is simply $e^{A(t)}x(t)$. By the fundamental theorem of calculus,

$$\int_{t_0}^{t}\frac{d}{ds}\left(e^{A(s)}x(s)\right)ds = \int_{t_0}^{t}e^{A(s)}f(s)\,ds = e^{A(t)}x(t) - x_0.$$

Therefore, the solution of the initial value problem (4.43) may be expressed as

$$x(t) = x_0 e^{-A(t)} + e^{-A(t)}\int_{t_0}^{t}e^{A(s)}f(s)\,ds. \tag{4.47}$$

The integral on the right-hand side of Eq. (4.47) cannot be evaluated explicitly except in special cases. However, it often can be approximated numerically to high accuracy.

The solution (4.47) can be regarded as the sum of a homogeneous and a particular solution. The homogeneous equation

$$x' + a(t)x = 0$$

is separable, and the general solution is

$$x_h(t) = Ce^{-A(t)}, \tag{4.48}$$

where $C$ is an arbitrary constant. Notice that $x_h(t)$ is precisely the first term on the right-hand side of Eq. (4.47) with $x_0$ replaced by $C$. It is straightforward to verify that the remaining term on the right-hand side of Eq. (4.47) is a particular solution. (See Exercise 13.)

It is possible to show that Eq. (4.47) is the *only* solution of Eq. (4.43) in an interval containing $t_0$ by verifying that the hypotheses of the existence and uniqueness theorem (Theorem 3.3) are satisfied. We can rewrite Eq. (4.43) as

$$x' = g(t, x) = f(t) - a(t)x.$$

Let $S$ be the box

$$S = \{(t, x) : t_0 - \alpha \le t \le t_0 + \alpha, \ x_0 - \beta \le x \le x_0 + \beta\},$$

where $\alpha$ and $\beta$ are positive constants and $\alpha$ is chosen so that $a$ is continuous on the interval $[t_0 - \alpha, t_0 + \alpha]$. Then

$$\frac{\partial g}{\partial x} = a,$$

which is continuous if $S$ is suitably chosen. The hypotheses of Theorem 3.3 are satisfied, a result that implies that the solution of Eq. (4.43) exists and is unique.

✦ *Example 4*   Let

$$x' + 2tx = t, \qquad x(0) = 1. \tag{4.49}$$

The solution is derived as follows.

We have $a(t) = 2t$, so $A(t) = \int_0^t 2s\, ds = t^2$. We multiply both sides of Eq. (4.49) by $e^{t^2}$ to obtain

$$e^{t^2}\left(x' + 2tx\right) = e^{t^2} t,$$

which by the chain and product rules is equivalent to

$$\frac{d}{dt}\left(e^{t^2} x(t)\right) = e^{t^2} t.$$

Therefore,

$$\int_0^t \frac{d}{ds}\left(e^{s^2} x(s)\right) ds = \int_0^t e^{s^2} s\, ds. \tag{4.50}$$

An antiderivative of the left-hand side is simply $e^{t^2} x(t)$. We use the fundamental theorem of calculus to evaluate the left-hand side:

$$\int_0^t \frac{d}{ds}\left(e^{s^2} x(s)\right) ds = e^{t^2} x(t) - x(0).$$

The right-hand side of Eq. (4.50) yields

$$\int_0^t e^{s^2} s\, ds = \frac{1}{2} \int_0^{t^2} e^u\, du$$

$$= \frac{1}{2} e^{t^2} - \frac{1}{2},$$

where we use the substitution $u = t^2$. Therefore,

$$e^{t^2} x(t) - x(0) = \frac{1}{2} e^{t^2} - \frac{1}{2},$$

which implies that

$$x(t) = \tfrac{1}{2} + \tfrac{1}{2}e^{-t^2}$$

after we substitute the initial condition $x(0) = 1$ and simplify.                     ∎

## The Solutions of Second-order Linear Equations

The linearity properties of differential operators can be used to characterize the solutions of second-order linear equations as the sum of a homogeneous and a particular solution, just as in the case of first-order equations. The following example illustrates the idea.

✦ *Example 5*   Let

$$x'' + x = t. \tag{4.51}$$

The left-hand side of Eq. (4.51) defines the homogeneous linear operator $L(x) = x'' + x$. The function $x_p(t) = t$ is a particular solution of Eq. (4.51), because

$$L(x_p) = L(t) = t'' + t = t.$$

The corresponding homogeneous equation is

$$L(x) = x'' + x = 0,$$

for which $\cos t$ and $\sin t$ are solutions. Because $L$ is linear, the function

$$x(t) = x_h(t) + x_p(t) = (a \cos t + b \sin t) + t$$

is also a solution:

$$
\begin{aligned}
L(x) &= L(a \cos t + b \sin t + t) \\
&= aL(\cos t) + bL(\sin t) + L(t) \\
&= 0 + 0 + t.
\end{aligned}
$$

The initial condition $x(0) = x_0$ implies that $a = x_0$, and the initial condition $x'(0) = v_0$ implies that $b = v_0 - 1$. Thus, any initial condition can be satisfied with an appropriate choice of the constants $a$ and $b$. Therefore, $x(t)$ is the general solution of Eq. (4.51).                     ∎

✦ *Example 6*   The solution of the homogeneous equation

$$x'' + \omega^2 x = 0$$

is

$$x(t) = c_1 \sin \omega t + c_2 \cos \omega t.$$

It will be shown in Chapter 5 that the constants $c_1$ and $c_2$ can be chosen to satisfy any initial conditions of the form $x(t_0) = x_0$, $x'(t_0) = v_0$.
For instance, the solution of $x'' + 4x = 0$ is

$$x(t) = c_1 \sin 2t + c_2 \cos 2t.$$

The initial condition $x(0) = 1$ implies that $c_2 = 1$, and the initial condition $x'(0) = 4$ implies that $c_1 = 2$. Here, $x$ is a homogeneous solution, because $L(x) = x'' + 4x = 0$. ∎

The following theorem describes the structure of general solutions for second-order linear differential equations. (Also see Exercise 14.)

**Theorem 4.12**   *Let $a(t)$, $b(t)$, and $f(t)$ be continuous functions of $t$, and let*

$$\frac{d^2x}{dt^2} + a(t)\frac{dx}{dt} + b(t)x = f(t). \tag{4.52}$$

*Let $x_p(t)$ be a particular solution of Eq. (4.52) and let $x_h(t)$ be a homogeneous solution, that is,*

$$\frac{d^2x_h}{dt^2} + a(t)\frac{dx_h}{dt} + b(t)x_h = 0.$$

*Then the function*

$$x(t) = x_p(t) + Cx_h(t)$$

*is a solution of Eq. (4.52) for any constant $C$.*

## Exercises

**1.** This problem is intended to help clarify the meaning of Theorem 4.11. Let

$$x' + x = t. \tag{4.53}$$

(a) Identify the linear homogeneous differential operator $L$ associated with Eq. (4.53).

(b) Show by substitution that $x_p(t) = t - 1$ is a particular solution; that is, show that $L(x_p) = t$.

(c) Find a homogeneous solution $x_h(t)$.

(d) Show by substitution that $L(x_h + x_p) = t$. Conclude that $x_h(t) + x_p(t)$ is a solution of Eq. (4.53).

(e) Prove that $Cx_h(t) + x_p(t)$ is also a solution of Eq. (4.53), where $C$ is any constant.

(f) Suppose the initial condition is specified as $x(t_0) = x_0$. Determine $C$.

**2.** This problem provides another illustration of Theorem 4.11. Let

$$x' + 3x = 5\cos t. \qquad (4.54)$$

(a) Identify the linear, homogeneous differential operator defined by Eq. (4.54).

(b) Find a homogeneous solution $x_h$ of Eq. (4.54).

(c) Show by substitution that $x_p(t) = \frac{1}{2}\sin t + \frac{3}{2}\cos t$ is a particular solution of Eq. (4.54).

(d) Use the linearity of the associated differential operator to show that $Cx_h + x_p$ is a solution of Eq. (4.54), where $C$ is any constant.

(e) Explain why $x_h + Cx_p$ is not a solution of Eq. (4.54), where $C$ is any constant.

(f) Suppose the right-hand side of Eq. (4.54) is replaced by $10\cos t$. How does the particular solution $x_p$ change? How does the homogeneous solution $x_h$ change?

**3.** Let

$$x' = (x/t) + 2t. \qquad (4.55)$$

(a) Identify the linear, homogeneous differential operator defined by Eq. (4.55).

(b) Find a homogeneous solution $x_h$ of Eq. (4.55).

(c) Show by substitution that $x_p(t) = 2t^2$ is a particular solution of Eq. (4.55).

(d) Use the linearity of the associated differential operator to show that $Cx_h + x_p$ is a solution of Eq. (4.55), where $C$ is any constant.

(e) Explain why $x_h + Cx_p$ is not a solution of Eq. (4.55), where $C$ is any constant.

(f) Use the result of part (e) to find a family of solutions of $x' = (x/t) + 10t$.

(g) Use the result of part (e) to find a family of solutions of $x' = (x/t) + Kt$, where $K$ is any constant.

(h) Can each solution of the family identified in (f) and (g) be characterized as the sum of a transient and a steady state? Explain.

**4.** Let

$$x'' + 4x = 2. \qquad (4.56)$$

(a) Identify the linear, homogeneous differential operator defined by Eq. (4.56).

(b) Find a homogeneous solution $x_h$ of Eq. (4.56) that is a linear combination of two functions.

(c) Show by substitution that $x_p(t) = 1/2$ is a particular solution of Eq. (4.56).

(d) Let $x_h$ be the solution that you found in (b). Show that $x_h + x_p$ is a two-parameter family of solutions of Eq. (4.56).

(e) Explain why $x_h + Cx_p$ is not a solution of Eq. (4.56), where $C$ is an arbitrary constant.

**5.** Let

$$x'' + x = e^{-t}. \qquad (4.57)$$

(a) Show that Eq. (4.57) is a linear second-order differential equation.

(b) Find a homogeneous solution $x_h$ of Eq. (4.57) that is a linear combination of two solutions.

(c) Show by substitution that $x_p(t) = Ae^{-t}$ is a particular solution of Eq. (4.57) for an appropriate value of $A$. What is it?

(d) Show that $Cx_h + x_p$ is a solution of Eq. (4.57), where $C$ is an arbitrary constant.

(e) Explain why $x_h + Cx_p$ is not a solution of Eq. (4.57), where $C$ is an arbitrary constant.

**6.** Suppose that a falling object encounters air resistance that is proportional to its velocity. The resulting first-order differential equation for the velocity is

$$v' = -g - kv, \qquad (4.58)$$

where $g$ and $k$ are constants.

(a) Find a homogeneous solution, a particular solution, and a family of solutions of Eq. (4.58).

(b) Find an expression for $x(t)$, the position of the object at time $t$. Identify the particular and homogeneous solutions in your expression.

(c) Show that your solution in (b) solves the second-order differential equation $x'' + kx' + g = 0$.

(d) Explain the physical significance of the two constants in your solution in (b).

**7.** Set up and solve, if possible, the integrals needed to form the general solution of each of the following differential equations.

(a) $x' + kx = t$, $k$ a constant

(b) $x' + kx = \sin t$, $k$ a constant

(c) $x' + kx = e^t$, $k$ a constant

(d) $x' + tx = t$

(e) $x' + tx = \sin t$

(f) $x' + (\cos t)x = f(t)$, $f(t)$ arbitrary

**8.** Let $x' + x/t = 1$.

(a) Find the solution satisfying the initial condition $x(1) = 1$.

(b) What is the interval of existence of the solution in (a)?

**9.** Identify the particular and homogeneous parts of the solution in Example 4.

**10.** Let $x' + (\tan t)x = \cos t$.

(a) Find the solution satisfying the initial condition $x(0) = 1$.

(b) What is the interval of existence of the solution in (a)?

**11.** Some savings instruments, such as certificates of deposit, pay interest rates that increase with time. Suppose a bank offers an initial annual interest rate of 5 percent that increases linearly to 7 percent after four years. Suppose that, at the start, you contribute at the rate of $5000 annually and increase your contributions by $500 annually. For simplicity, assume that interest is compounded continuously and that funds are invested continuously.

(a) How much money would you have in the account after four years if you open the account with no money initially?

(b) How much money have you contributed, and how much interest have you earned after four years?

**12.** Prove Theorem 4.11. (Suggestion: Determine the linear, homogeneous differential operator defined by Eq. (4.37) and use the superposition principle.)

**13.** This exercise discusses some additional details involved in finding solutions of the initial value problem

$$x' + a(t)x = f(t), \qquad x(t_0) = x_0, \qquad (4.59)$$

where $a$ is a continuous function of $t$.

(a) Let $A(t) = \int_{t_0}^t a(s)\, ds$. Show that the general solution of the equation

$$x' + a(t)x = 0$$

is given by Eq. (4.48). (Hint: Use the separation of variables and the fundamental theorem of calculus.)

(b) Show that

$$x_p(t) = e^{-A(t)} \int_{t_0}^t e^{A(s)} f(s)\, ds \qquad (4.60)$$

is a particular solution of Eq. (4.59), where $A(t)$ is an antiderivative of $a(t)$.

(c) The general solution of Eq. (4.59) has the form $x(t) = Cx_h(t) + x_p(t)$, where $C$ is an arbitrary constant of integration. Determine the value of $C$ such that $x$ solves the initial value problem (4.59) and compare it to Eq. (4.47).

**14.** Prove Theorem 4.12. (Suggestion: Determine the homogeneous differential operator defined by Eq. (4.52), then use the superposition principle, Theorem 4.8.)

# Chapter 5

---

# OSCILLATORY MOTION

---

Many natural phenomena exhibit periodic behavior. Examples include the rotation of the earth around the sun, the motion of a pendulum or a mass-spring system, the buildup of charge in an electrical circuit, and periodic outbreaks of childhood diseases. This chapter discusses how periodic functions arise naturally as solutions of second-order linear differential equations. The focus of our discussion will be on three simple physical models: the mass-spring system, an LRC electrical circuit, and the linear pendulum. Although these systems represent very different physical situations, idealized models of their behavior lead to identical mathematical descriptions.

As we discussed in Chapter 2, mathematical modeling involves some art as well as science. We frequently ignore aspects of a given problem that complicate the derivation of a mathematical description. (For instance, in our derivations of the equations of motion of mass-spring systems, we imagine that all the mass in the system is concentrated at a single point.) Nevertheless, our crude mathematical models are often useful approximations to the behavior that is seen in laboratory experiments. There are many questions that one might ask about the predicted behavior of such systems. Examples of questions that are relevant to a mass-spring system include the following:

- Can a mass-spring system be expected to oscillate, or does it just relax to some equilibrium state without oscillating?

- What is the maximum displacement of the mass?

- What is the maximum velocity attained by the mass?

- Is it ever possible for the amplitude of the oscillations to become arbitrarily large?

- How do physical parameters such as the size of the mass, the stiffness of the spring, and the coefficient of friction affect the motion?

- How do the initial conditions affect the motion?
- Are predictions of the future position and velocity of the mass sensitive to errors in measuring the initial conditions or physical parameters?

The focus of our discussion is on the motion of mass-spring systems and the simple pendulum. We use these examples because it is comparatively easy to visualize how the details of the motion change as physical parameters are varied. However, we also consider oscillatory behavior that arises in certain electrical circuits.

## 5.1  SOME SIMPLE PHYSICAL MODELS

### Newton's Laws of Motion

In Chapter 2 we discussed a modeling process that we used to set up mathematical models that describe simple annuities, Newton's law of cooling, and other examples. The modeling process requires us to determine a rate equation—a differential equation that states the rate of change of some quantity of interest in terms of its current value. The goal of this subsection is to model the motion of an idealized mass under the influence of gravity or other force. Our considerations lead to second-order differential equations, whose solutions are the subject of the remainder of the chapter.

Isaac Newton made fundamental discoveries about the origin of motion. (The question of how objects move was an important philosophical and scientific problem for hundreds of years.) In 1686, Newton published three laws of motion that provide a highly accurate mathematical model for a variety of physical phenomena.

**Newton's first law:** A net force must act on an object to put the object into motion or to change the direction or speed of the motion.

**Newton's second law:** The acceleration produced by a net force on an object is proportional to the mass of the object. That is,

$$\text{net force} = \text{mass} \times \text{acceleration},$$

often expressed as $F = ma$. In particular, $a = dv/dt$, so the velocity $v$ of an object is constant unless the acceleration is nonzero.

**Newton's third law:** If an object $A$ exerts a force on another object $B$, then $B$ also exerts a force on $A$. The forces are equal in magnitude and act in opposite directions.

Although Newton's first law is a corollary of the second, it is useful to state them separately. Newton's laws of motion can be applied to derive simple models of the motion of a mass on a spring and a pendulum.

## Mass-Spring Systems

A spring exerts a force when it is stretched or compressed from its natural length. This force is called the *restoring force*, because it acts to restore the spring to its natural, unstretched length. In the 1680s, Robert Hooke found a useful approximation, now called *Hooke's law*, that models the restoring force of real springs in certain situations. Hooke's law says that the magnitude of the restoring force is proportional to the amount of stretching or compression. The constant of proportionality $k$, which by convention is positive, is called the *spring constant*.

Hooke's law is only an approximation of the behavior of a real spring. If the spring is stretched too far (for example, if we attach a sufficiently large mass to it), then the restoring force may no longer be linearly proportional to the displacement; similarly, every spring can be compressed by only a finite amount. A spring for which Hooke's law applies is called a *linear spring*.

Let us now describe an idealized physical model of a mass and a spring together with its mathematical formulation. Consider a linear spring that is fixed at one end and is free to move at the other end as shown in Fig. 5.1. We attach a mass $m$ to the free end of the spring. The weight of the mass is $mg$, where $g$ is the acceleration due to gravity. The weight lengthens the spring by $\ell_{eq}$. If we attach the mass carefully so that it does not move and remains at rest when released, then Newton's first law implies that the forces acting on the mass sum to zero. In particular, the restoring force of the spring balances the weight of the mass. Hence,

$$F_{gravity} + F_{restoring} = 0.$$

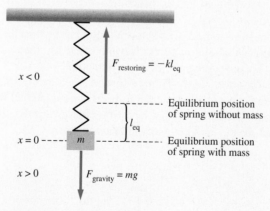

FIGURE 5.1   A mass-spring system.

To describe the forces that act on the mass, we must define a coordinate system. We let $x = 0$ be the equilibrium position of the mass after it is attached to the spring. (If the mass is placed at $x = 0$ and given no initial velocity, then the mass remains at $x = 0$ indefinitely.) This position corresponds to a total elongation of $\ell_{eq}$ with respect to the length of the spring without the mass (see Fig. 5.1). We let $x > 0$ denote that the mass is pulled down from the equilibrium by $x$ units (stretched spring). We let $x < 0$ denote a displacement of $|x|$ units up from the equilibrium (compressed spring).

Since gravity causes the mass to stretch the spring, we have

$$F_{\text{gravity}} = mg,$$

where $m$ and $g$ are positive constants. The restoring force acts to compress the spring, so

$$F_{\text{restoring}} = -k\ell_{eq}. \tag{5.1}$$

(The minus sign indicates that the restoring force acts in the direction opposite gravity.) Therefore, when the mass is at rest at the equilibrium position, Newton's first law implies that

$$mg - k\ell_{eq} = 0. \tag{5.2}$$

✦ *Example 1*    Suppose that a linear spring yields a restoring force of 10 N when it is stretched 10 cm. By Hooke's law,

$$10\,\text{N} = k(10\,\text{cm}),$$

which implies that the spring constant is

$$k = \frac{10\,\text{N}}{0.1\,\text{m}} = 100\,\text{N/m}. \qquad\blacksquare$$

If we pull down on the mass to a position below its equilibrium, then the magnitude of the restoring force exceeds that of gravity. By Newton's first law, the mass begins to move after it is released. We now derive the equation of motion.

When the mass is at position $x$, the restoring force exerted by the spring is $F = -k(\ell_{eq} + x)$. If the spring is linear and ideal, that is, if there is no internal friction, and if air resistance is negligible, then the restoring force of the spring and the weight of the mass are the only relevant forces. Newton's second law implies that

$$\begin{aligned}
m\frac{d^2x}{dt^2} &= mg - k\left(\ell_{eq} + x\right) \\
&= mg - k\left(\frac{mg}{k} + x\right) \qquad \text{by Eq. (5.2)} \\
&= -kx.
\end{aligned}$$

These considerations lead to a second-order differential equation for the idealized motion of a mass on a spring:

$$\frac{d^2x}{dt^2} = -\frac{k}{m}x. \tag{5.3}$$

Equation (5.3) is a linear, second-order, homogeneous differential equation.

To complete a description of the motion of the mass, we must specify its initial velocity as well as its position. A moment's thought shows why both the initial position and velocity are important. If the mass starts at $x = 0$ with no initial velocity, then the mass never moves. The outcome is different if we give the mass at $x = 0$ an initial push. The initial conditions can be specified as $x(0) = x_0$ and $x'(0) = v_0$.

## The Effects of Friction

Our considerations so far have assumed an ideal situation in which the motion occurs without friction. This assumption is not realistic in most physical settings. Various names are given to frictional forces. A *damped* system is one in which some kind of frictional or resistive force acts. An *undamped* system has no friction.

Friction refers to forces that resist the motion of an object. (Friction dissipates energy from physical systems in the form of heat.) An important question is how the effects of friction can be incorporated into our models. Unfortunately, a satisfactory answer is not easy to formulate in many cases.

Physicists distinguish between *static friction*, which refers to the frictional forces that act between two surfaces at rest, and *kinetic friction*, which refers to the frictional forces between two surfaces that are in motion with respect to one another. In any real system, such as a block on a table, a small force is necessary to set the block into motion. However, once the block starts moving, a smaller force usually is necessary to keep the block moving at a uniform velocity.

In general, the kinetic friction acting on an object is a function of the current velocity and the position, especially if the surface is not uniform. The simplest functional relationship from a mathematical perspective is *linear friction*, where we suppose that the frictional force is proportional to the current velocity of the object.

Let us consider how to incorporate a linear frictional term into this idealized model of a mass-spring system. Figure 5.2 shows that the relevant forces acting on the mass at any instant are gravity, the restoring force of the spring, and friction. In our coordinate system, forces that act downward are positive (they lengthen the spring) and forces that act upward are negative (they shorten the spring). Here,

$$F_{\text{net}} = F_{\text{gravity}} + F_{\text{restoring}} + F_{\text{friction}}.$$

Friction always opposes the motion. In particular, if $x' > 0$ (that is, the mass moves

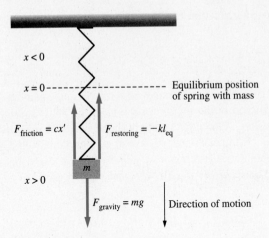

**FIGURE 5.2**    The forces acting on the mass when friction is considered.

down), then friction acts in an upward direction; conversely, friction acts in a downward direction if $x' < 0$. Hence, a linear frictional force is modeled as

$$F_{\text{friction}} = -cx',$$

where $c$ is a positive constant of proportionality. As was discussed above,

$$F_{\text{restoring}} = -k\left(\ell_{\text{eq}} + x\right).$$

Newton's second law implies that

$$F_{\text{net}} = mx''.$$

Therefore, the equation of motion is

$$mx'' = mg - cx' - k\left(\ell_{\text{eq}} + x\right)$$

$$= mg - cx' - k\left(\frac{mg}{k} + x\right)$$

$$= -cx' - kx,$$

so

$$mx'' + cx' + kx = 0. \tag{5.4}$$

Equation (5.4) models the motion of the mass-spring system, assuming a linear spring and linear friction. The constants $m$, $c$, and $k$ are positive.

## The Linear Pendulum

Consider the pendulum with a bob of mass $m$ attached to one end of a massless rod of length $\ell$ that is connected to a fixed pivot (Fig. 5.3). We imagine a point mass.[1]

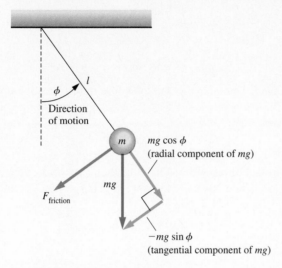

FIGURE 5.3 A simple pendulum.

Gravity induces a force of magnitude $mg$ on the bob, which can be decomposed into a radial component (acting along the length of the pendulum) and a tangential component (perpendicular to the radial component). If the rod is rigid, then the distance between the bob and the pivot remains constant (that is, there is no net force in the radial direction, by Newton's first law). Therefore, we consider only the tangential component of the weight of the bob in deriving the equation of motion.

As in the case of the mass-spring system, we need a coordinate system to derive a model for the motion. We define the equilibrium (straight-down) position of the bob as $\phi = 0$ and define $\phi > 0$ as a position $\phi$ radians in an anticlockwise direction from the vertical. With this convention, anticlockwise motion corresponds to positive angular velocity.

Frictional forces also are important in any real pendulum. However, they are difficult to model accurately for the reasons discussed above. In the simplest case, we suppose

---

[1] In fact, an actual body—even one as large as an automobile or rocket—may be regarded as a point particle for the purpose of analyzing its motion, provided that the effect of any rotation of the body about its center of mass is negligible.

that there is friction only in the pivot and that the magnitude of the friction is proportional to the angular velocity of the bob. If $\phi(t)$ denotes the angular position of the bob at time $t$, then the friction is proportional to $\phi'(t)$.

The net tangential forces on the bob are the tangential component of the weight and the force of friction. Therefore,

$$F_{\text{net}} = F_{\text{tangential}} + F_{\text{friction}}.$$

In addition,

$$F_{\text{tangential}} = -mg \sin \phi,$$

where the minus sign denotes the idea that the force acts in the direction opposite the current displacement. Similarly, because friction always opposes the motion, a linear frictional force at the pivot implies that

$$F_{\text{friction}} = -c\phi',$$

where $c$ is a positive constant of proportionality. Finally, $F_{\text{net}} = ma_\phi$, where $a_\phi$ is the tangential acceleration of the bob. Because the rod is rigid, $a_\phi = \ell \phi''$, so the equation of motion is

$$m\ell \phi'' = -mg \sin \phi - c\phi',$$

that is,

$$m\phi'' + \left(\frac{c}{\ell}\right)\phi' + \left(\frac{mg}{\ell}\right) \sin \phi = 0. \tag{5.5}$$

Unfortunately, Eq. (5.5) is not linear, and it is not possible to derive a convenient closed-form expression for $\phi(t)$. Often, however, we are interested in the motion of the pendulum as it oscillates near the equilibrium (downward) position. (For example, imagine the motion of the pendulum of a grandfather clock.) If $\phi(t)$ is always small, then we can *linearize* Eq. (5.5) by assuming that $\sin \phi \approx \phi$. In other words, we approximate the nonlinear equation (5.5) by the linear equation

$$m\phi'' + \left(\frac{c}{\ell}\right)\phi' + \left(\frac{mg}{\ell}\right)\phi = 0. \tag{5.6}$$

Notice that Eq. (5.6) is identical to the mass-spring equation (5.4) except that the interpretation of the constants is different.

## Electrical Circuits

The mass-spring system and the linear pendulum have identical mathematical descriptions, despite the differences in their physical details. In this section, we describe a simple electrical circuit, called an *LRC circuit*, which has the same mathematical description as the linear pendulum. A schematic diagram of an LRC circuit is shown in Fig. 5.4.

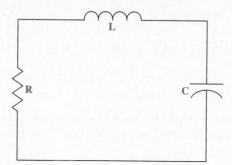

FIGURE 5.4   Diagram of an LRC circuit.

The total charge $Q$ in an electrical circuit is measured in *coulombs*, abbreviated C. The current $I$ is defined as the time rate of change of charge, $I = dQ/dt$. Current is measured in *amperes*, abbreviated A; 1 A = 1 C/s. The change in electrical potential across two components of a circuit is measured in *volts*, abbreviated V.

The physical principle that governs the flow of an electrical charge in a circuit is called *Kirchhoff's law* and states that the sum of changes in potential in a complete traversal of a circuit must be zero. (This statement, sometimes called the *loop theorem*, is a consequence of the conservation of energy in the circuit.) One simple application of Kirchhoff's law is to electrical circuits containing a resistor, a capacitor, and an inductor in series.

The resistance $R$ of a resistor is defined to be the ratio of the electrical potential $U_R$ across the resistor to the amount of current flowing through it:

$$R = U_R/I. \tag{5.7}$$

Resistance is measured in units of *ohms*, abbreviated $\Omega$; 1 $\Omega$ = 1 V/A = 1 V s/C.

The capacitance $C$ of a capacitor[2] is defined as the ratio of the charge on the internal plate to the electrical potential $U_C$:

$$C = Q/U_C. \tag{5.8}$$

Capacitance typically is measured in units called *microfarads*, abbreviated $\mu$F; 1 $\mu$F = $10^{-6}$ C/V.

---

[2]The use of the italic letter $C$ to denote capacitance is standard. The Roman letter C is the recognized abbreviation for coulombs. The notation is unfortunate because it is potentially confusing. The C in *LRC circuit* refers to the capacitor.

The inductance $L$ of an inductor is defined as the ratio of the electrical potential $U_L$ to the rate at which the current flow changes:

$$L = \frac{U_L}{dI/dt}. \tag{5.9}$$

Inductance is measured in units called *henrys*, abbreviated H; $1\ \text{H} = 1\ \text{V s}^2/\text{C}$.

The loop theorem implies that

$$U_R + U_C + U_L = U_B, \tag{5.10}$$

where $U_B$ is the electrical potential created by a battery, generator, or other external device. In this example, there is no external potential, so $U_B = 0$. Equation (5.10), together with the definitions (5.7)–(5.9), leads to

$$\begin{aligned} 0 &= U_L + U_R + U_C \\ &= LQ'' + RI + Q/C \\ &= LQ'' + RQ' + (1/C)Q, \end{aligned} \tag{5.11}$$

which is a second-order, linear differential equation with constant coefficients. Each term in Eq. (5.11) has units of volts, as each term is a measure of the change in electrical potential. Notice that Eq. (5.11) is identical in form to the mass-spring equation (5.4). If no resistor is included, then we call the circuit an *LC circuit*.

## Some Basic Theory

A basic question is whether the solutions of a second-order differential equation like Eq. (5.11) always have solutions. The same issues regarding the existence and uniqueness of solutions arise for second-order equations as for the first-order equations that we discuss in Chapter 3.

Consider a mass-spring system like the one depicted in Fig. 5.2. As we discuss above, the mass can be placed in any initial position and released with a specified initial velocity. Intuitively, it seems reasonable that if the initial position and velocity are given, then the motion of the mass is completely determined. The following theorem, which we state without proof, confirms this intuition.

**Theorem 5.1** *Let $a$, $b$, and $c$ be constants, and assume that $a \neq 0$. Then the linear, second-order initial value problem*

$$ax'' + bx' + cx = 0, \qquad x(0) = x_0, \qquad x'(0) = v_0, \tag{5.12}$$

*has a unique solution whose interval of existence is the entire real line.*

In the next four sections, we discuss the solutions of Eq. (5.12) in detail. Theorem 5.1 implies that if we can find a formula that solves the initial value problem, then we have found the only solution.

## Exercises

**1.** A mass of 500 g stretches a spring 5 cm. Find the spring constant.

**2.** A stiff spring exerts a large restoring force for a small displacement. Does a stiff spring have a large spring constant or a small spring constant? Explain.

**3.** In the derivation of the linear pendulum equation (5.6), we made the approximation $\phi \approx \sin\phi$. For what values of $\phi$ is the relative error in the approximation less than 1 percent? 5 percent? 10 percent? (Express each answer in radians and in degrees.)

**4.** Give two examples of factors that might be important in modeling a mass-spring system that are not taken into account in the derivation of Eq. (5.4).

**5.** One end of a linear spring is attached to a massive floor. A mass is attached to the free end and moves vertically up and down. (Think of the spring as a shock absorber.) Show that the motion is governed by Eq. (5.4) under a suitable idealization of the problem.

**6.** You decide to go camping and load the trunk of your car with 200 kg of equipment, which causes the beam of the headlights in the garage to move 10 cm up along the wall. Assume that the distance from the headlights to the garage wall is 2 m, that the length of the car is 4 m, and that the weight of the gear is borne entirely by the shock absorbers on the rear axle. Find the spring constant for the shock absorbers if you treat the equipment as a point mass on the back end of the vehicle.

**7.** In Chapter 2, we consider the motion of a falling body under the influence of gravity and air resistance.

The first-order equation governing the velocity $v$ of the body, assuming linear air resistance, is

$$m\frac{dv}{dt} = mg - kv. \tag{5.13}$$

(a) Write Eq. (5.13) as an equivalent second-order differential equation in terms of the position $x$.

(b) Equation (5.13) can be solved by the separation of variables to obtain

$$v(t) = \frac{mg}{k} + c_1 e^{-kt/m}, \tag{5.14}$$

where the constant $c_1$ depends on the initial velocity. Find $x(t)$ by integrating Eq. (5.14).

(c) Show by substitution that the function $x(t)$ that you derived in (b) solves the differential equation that you derived in (a).

**8.** A mass is placed on a track and attached to a linear spring as shown in Fig. 5.5. Here, $x = 0$ corresponds to the natural length of the spring (that is, the restoring force is 0). Assume that the mass is free to slide along the track in a straight line and that friction is a linear function of the velocity of the mass. Show that the motion of the mass is governed by Eq. (5.4).

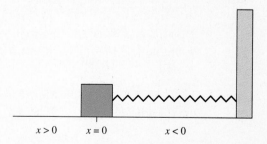

$x > 0$      $x = 0$         $x < 0$

**Figure 5.5**   A mass on a track, for Exercise 8.

## 5.2    SIMPLE HARMONIC MOTION

The term *simple harmonic motion* refers to the motion of systems in which friction or other resistive forces are negligible. An example of simple harmonic motion is the motion of a pendulum about its equilibrium position. If the force of friction at the pivot is very small, then the pendulum moves periodically. In this simple case the motion can be described in terms of sine and cosine functions.

✦ *Example 1*    Suppose a mass of 1 kg is attached to one end of a spring whose spring constant is 4 N/m. The spring is stretched 10 cm and released. Let us determine the position of the mass as a function of time, assuming that the frictional forces in the spring and the effects of air resistance are negligible.

The motion of the mass is governed by Eq. (5.3) with $m = 1$ kg and $k = 4$ N/m; we have

$$x'' + 4x = 0. \tag{5.15}$$

The solutions of Eq. (5.15) are functions $x(t)$ for which $x'' = -4x$. One such function is $x_1(t) = \sin 2t$, because $(\sin 2t)'' = -4 \sin 2t$. The function $x_2(t) = \cos 2t$ is another solution. Equation (5.15) is linear, so the superposition principle implies that

$$x(t) = c_1 \sin 2t + c_2 \cos 2t \tag{5.16}$$

is a solution for any constants $c_1$ and $c_2$.

In this example, we assume that the mass initially is stretched 10 cm and released. The initial conditions are $x(0) = 0.1$ m and $x'(0) = 0$. There is a unique choice of $c_1$ and $c_2$ in Eq. (5.16) that satisfies the initial conditions. We have $x(0) = 0.1 = c_2$ and $x'(0) = 0 = 2c_1$. Therefore, the position of the mass at time $t$ is given by $x(t) = \frac{1}{10} \cos 2t$.    ■

The existence and uniqueness theorem (Theorem 5.1) implies that the function $x(t) = \frac{1}{10} \cos 2t$ is the only solution of Eq. (5.15) that satisfies the given initial conditions. Therefore, we can make some basic predictions about the motion of the mass:

- The motion of the mass-spring system is periodic with period $\pi$.

- The mass crosses the equilibrium position at 0 infinitely many times, at least in principle.

Example 1 is one illustration of a general class of equations, called *harmonic oscillators*, which often are written as

$$x'' + \omega^2 x = 0. \tag{5.17}$$

It is straightforward to show by substitution that $\sin \omega t$ is a solution of Eq. (5.17). We have

$$(\sin \omega t)'' + \omega^2 \sin \omega t = -\omega^2 \sin \omega t + \omega^2 \sin \omega t = 0.$$

Similarly, it is easy to verify that $\cos \omega t$ is also a solution. Equation (5.17) is linear, so the superposition principle implies that

$$x(t) = c_1 \sin \omega t + c_2 \cos \omega t \tag{5.18}$$

is a solution. Notice that the solution (5.18) is a periodic function of period $2\pi/\omega$. Equation (5.17) and its solutions are so fundamental that we give them a special name.

**Definition 5.2** *The motion modeled by Eq. (5.17) is called simple harmonic motion. Systems whose motion can be described by a linear combination of sine and cosine functions are called harmonic oscillators.*

Equation (5.18) is called the *general solution* of Eq. (5.17), because there exists a unique choice of the constants $c_1$ and $c_2$ that satisfy any initial conditions of the form $x(0) = x_0$ and $x'(0) = v_0$. To see this, we substitute the initial values into Eq. (5.18) to obtain

$$x_0 = c_2,$$

$$v_0 = c_1 \omega.$$

Theorem 5.1 guarantees that the solution is unique. The following theorem summarizes our analysis.

**Theorem 5.3** *The solution of*

$$\frac{d^2x}{dt^2} + \omega^2 x = 0$$

*with the initial conditions $x(0) = x_0$ and $x'(0) = v_0$ is*

$$x(t) = x_0 \cos \omega t + \left(\frac{v_0}{\omega}\right) \sin \omega t.$$

*For a fixed $\omega$, the amplitude of the solution depends only on the initial conditions. The period of the solution does not depend on the initial conditions.*

Because the solutions of equations like (5.15) are unique, we can predict the future state of the system if the initial conditions are specified. In the case of a linear mass-spring system, for instance, we can predict the time at which the mass first crosses the equilibrium position. We can also predict the maximum displacement of the mass from the equilibrium position and when that displacement is reached, the maximum speed that the mass attains, and so on.

✦ *Example 2*   A mass of 1 kg stretches a spring 9.8 cm. Let $x = 0$ denote the equilibrium position of the mass after it is attached to the spring. (See Fig. 5.1.) Suppose that the spring acts linearly if it is not stretched or compressed more than 1 m from its length before the mass is attached. If the spring stretches more than this amount, then it no longer obeys Hooke's law (that is, the spring is deformed if it stretches too much). Ignore friction and air resistance. Consider the following questions:

1. Use Newton's laws to derive a differential equation that describes the motion. (See Fig. 5.1.)

2. What is the period of the motion?

3. The spring is stretched to an initial position $x(0) = x_0$ and released with zero initial velocity. For what values of $x_0$ will the spring not be stretched so much that it deforms?

4. The spring is pushed from the equilibrium position with initial velocity $v_0$. For what values of $v_0$ will the spring not be damaged?

### Solution

1. The gravitational force on the mass (that is, its weight) is $mg = 9.8$ N, which exactly balances the restoring force of the spring when the mass is at rest at the equilibrium position. The mass stretches the spring by 0.098 m, so the spring constant is $k = 9.8\,\text{N}/0.098\,\text{m} = 100\,\text{N/m}$. Friction is negligible, so the restoring force of the spring equals the weight of the mass. Therefore,

$$mx'' = -F_{\text{restoring}} = -kx,$$

so

$$x'' + 100x = 0. \tag{5.19}$$

2. The general solution of Eq. (5.19) is

$$x(t) = c_1 \cos 10t + c_2 \sin 10t. \tag{5.20}$$

Thus, the period of the motion is $2\pi/10$ s.

3. We substitute the initial values into the general solution (5.20) to obtain $c_1 = x_0$ and $c_2 = 0$. Hence, the maximum displacement of the mass is $x_0$ meters from the equilibrium. To avoid damaging the spring, we must not stretch it more than 1 m from its initial length before the mass is attached. The mass stretches the spring when it is attached; in this case, the equilibrium position $x = 0$ corresponds to an elongation of 0.098 m. Thus, the spring is not damaged if $|x_0| < 1 - 0.098$.

4. The initial conditions imply $c_1 = 0$, $c_2 = v_0/10$. The maximum amplitude of the spring is $v_0/10$ m. The spring is undamaged if $|v_0/10| < 1 - 0.098$, that is, $|v_0| < 9.02$ m/s.                                                                                          ∎

Mathematical models are useful for assessing the effect of errors in the assumed value of the constants in a given problem. For instance, the actual performance of many electrical components can vary appreciably from their rated values. The next example shows you what to expect in one simple situation.

✦ *Example 3*   A capacitor is rated at 1.2 $\mu$F and an inductor at 0.5 H. Suppose that the actual values may differ by as much as 5 percent in either direction from the rated values. Thus, the actual capacitance is between 1.14 and 1.26 $\mu$F, and the actual inductance is between 0.475 and 0.525 H. What is the period of the fluctuations in the charge in a circuit consisting of such an inductor and capacitor in series? Assume that the electrical resistance is negligible.

*Solution*   The potential in the circuit is modeled by Eq. (5.11) with $R = 0$; that is,

$$Q'' + \frac{Q}{LC} = 0, \tag{5.21}$$

which is equivalent to Eq. (5.17) with $\omega^2 = 1/(LC)$. The general solution of Eq. (5.21) is

$$Q(t) = c_1 \sin \frac{t}{\sqrt{LC}} + c_2 \cos \frac{t}{\sqrt{LC}},$$

which implies that the fluctuations of the charge in the circuit are periodic with period $T = 2\pi \sqrt{LC}$. The period is longest when $LC$ is largest and is shortest when $LC$ is smallest. We have

$$T_{\min} = 2\pi \sqrt{0.475 \times 1.14 \times 10^{-6}} \approx 4.62 \times 10^{-3} \text{ s}$$

and

$$T_{\max} = 2\pi \sqrt{0.525 \times 1.26 \times 10^{-6}} \approx 5.11 \times 10^{-3} \text{ s,}$$

so the actual period of the oscillations lies between $T_{\min}$ and $T_{\max}$.

If both components perform at their rated values, then the circuit oscillates with period $T = 2\pi \sqrt{0.5 \times 1.2 \times 10^{-6}} \approx 4.87 \times 10^{-3}$ s. We can think of $T$ as the predicted period, assuming that both components perform as rated, and $T_{\min}$ and $T_{\max}$ as the extreme cases that can occur. The extreme cases represent a variation of 5 percent from the period that would be expected if both components performed exactly as rated. ■

## The Phase-Amplitude Form of Solutions

Simple harmonic motion is described by sine and cosine functions. It is useful to represent sine and cosine functions in the *phase-amplitude* form

$$f(t) = A \cos (\omega t - \delta), \tag{5.22}$$

because the constants $A$, $\omega$, and $\delta$ are directly related to the graph of the function.

- The *amplitude A* determines the maximum value of $|f|$ because the cosine function is bounded between $-1$ and $+1$. By convention, $A > 0$.
- The period of the function $f$ in Eq. (5.22) is $T = 2\pi/\omega$, and the frequency is $\nu = \omega/(2\pi)$. Because $\omega$ is proportional to $\nu$, $\omega$ often is called the *circular frequency*. By convention, $\omega > 0$.
- The *phase shift* is $\delta/\omega$; it determines the horizontal displacement of the graph compared to an unshifted function. (The phase shift is expressed in radians. By convention, $-\pi \le \delta < \pi$.) If the phase shift is negative, then the curve is shifted to the left; if the phase shift is positive, then the curve is shifted to the right.

The horizontal displacement of the graph of Eq. (5.22) depends on $\omega$ as well as on $\delta$. To see this, observe that

$$A \cos (\omega t - \delta) = A \cos[\omega(t - \delta/\omega)],$$

so the change of variable $u = t - \delta/\omega$ leads to

$$A \cos (\omega t - \delta) = A \cos \omega u.$$

The $u$ axis is displaced by $\delta/\omega$ units from the $t$ axis.

✦ *Example 4*    Figure 5.6 shows the graph of $3 \cos 2t$ as the solid curve and the graph of $3 \cos (2t - 1)$ as the dashed curve. Here, $A = 3$, $\omega = 2$, and $\delta = 1$. The dashed curve is identical to the solid curve, except that it is displaced by $1/2$ unit to the right; the phase shift is $\delta/\omega = 1/2$. The period of both functions is $\pi$, and the amplitude of both functions is 3. (How would the graph of $3 \cos (2t + 1)$ compare to the solid curve?)

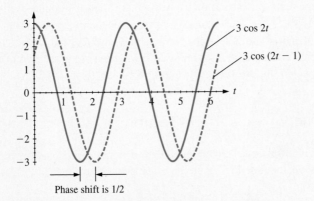

**FIGURE 5.6**    The solid curve shows the graph of $3 \cos 2t$; the dashed curve shows the graph of $3 \cos (2t - 1)$. The dashed curve is displaced to the right from the solid curve by $1/2$ unit. See Example 4.    ■

The solution of the initial value problem

$$x'' + \omega^2 x = 0, \qquad x(0) = x_0, \qquad x'(0) = v_0 \tag{5.23}$$

is

$$x(t) = x_0 \cos \omega t + \left(\frac{v_0}{\omega}\right) \sin \omega t.$$

We can rewrite the solution in the phase-amplitude form

$$x(t) = A \cos(\omega t - \delta) \tag{5.24}$$

as follows. The initial conditions imply that[3]

$$x_0 = A \cos(-\delta) = A \cos \delta, \tag{5.25}$$

$$v_0 = -A\omega \sin(-\delta) = A\omega \sin \delta. \tag{5.26}$$

Observe that

$$x_0^2 + \left(\frac{v_0}{\omega}\right)^2 = A^2 \cos^2 \delta + A^2 \sin^2 \delta = A^2.$$

Therefore,

$$A = \sqrt{x_0^2 + \left(\frac{v_0}{\omega}\right)^2}. \tag{5.27}$$

We take the positive square root because the amplitude is positive by convention.
Equations (5.25)–(5.26) imply that

$$\frac{v_0/\omega}{x_0} = \tan \delta,$$

so

$$\delta = \tan^{-1}\left(\frac{v_0}{x_0 \omega}\right). \tag{5.28}$$

There are two solutions of Eq. (5.28) in $[-\pi, \pi)$ that differ by $\pi$ radians. Only one of them satisfies Eqs. (5.25)–(5.26); the following example shows how to choose the correct one.

✦ *Example 5*  Let

$$x'' + x = 0, \qquad x(0) = -4, \qquad x'(0) = 3. \tag{5.29}$$

Since $\omega = 1$, the solution of Eq. (5.29) is $2\pi$-periodic. Therefore, the phase-amplitude form of the solution is

$$x(t) = A \cos(t - \delta).$$

---

[3]Recall that $\cos(-t) = \cos t$ and that $\sin(-t) = -\sin t$ for any $t$.

The initial conditions imply that

$$-4 = A\cos\delta \qquad \text{because } x(0) = -4,$$
$$3 = A\sin\delta \qquad \text{because } x'(0) = 3.$$

Therefore,

$$A = \sqrt{4^2 + 3^2} = 5,$$

and

$$\delta = \tan^{-1}\left(-\tfrac{3}{4}\right). \tag{5.30}$$

By definition, $\delta$ is the angle that the line of slope $-3/4$ makes with the horizontal axis. However, there are two such angles in $[-\pi, \pi)$. A calculator yields the numerical approximation

$$\tan^{-1}\left(-\tfrac{3}{4}\right) \approx -0.6435,$$

so one of the angles is $\delta_1 \approx -0.6435$. The other angle is $\delta_2 = \delta_1 + \pi \approx 2.4981$, as illustrated in Fig. 5.7. The angle $\delta_1$ lies in the fourth quadrant, where $\cos\delta_1 > 0$ and $\sin\delta_1 < 0$. This solution cannot satisfy the initial conditions, because we have $A\cos\delta < 0$ and $A\sin\delta > 0$. Therefore, we take the solution that lies in the second quadrant. In other words, $\delta \approx 2.4981$ is the solution of Eq. (5.30) that satisfies both initial conditions. Hence, the phase-amplitude form of the solution of the initial value problem (5.29) is approximately

$$x(t) = 5\cos(t - 2.4981). \tag{5.31}$$

Notice also that we may write the solution equivalently as

$$x(t) = -4\cos t + 3\sin t. \tag{5.32}$$

You can always check that you have the correct value of $\delta$ by plotting both forms of the solution.

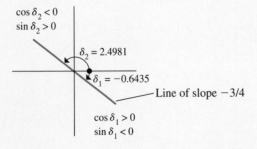

**FIGURE 5.7**    The correct solution of Eq. (5.30) depends on the initial conditions. ■

The main advantages of the phase-amplitude form

$$x(t) = A \cos(\omega t - \delta)$$

over the linear combination

$$x(t) = c_1 \cos \omega t + c_2 \sin \omega t$$

are the following:

- All the information needed to sketch a graph of the function can be read off from the constants $A$, $\omega$, and $\delta$.

- If we think of the initial value problem as a model of a mass-spring system, then the phase-amplitude form makes it easy to determine when the mass crosses the equilibrium ($x = 0$) position for the first time. The crossing time is the value of $t$ for which $\cos(\omega t - \delta) = 0$.

- In a similar way, the phase-amplitude formulation makes it easy to determine the maximum speed that is attained by the mass and when the maximum speed is attained.

✦ *Example 6*   Suppose the initial value problem (5.29), discussed in the previous example, models a frictionless mass-spring system in which distance is measured in meters and time in seconds. Let us address each of the aspects of the phase-amplitude formulation mentioned above.

- The graph of the solution (5.31) is the same as that of the function $y(t) = 5 \cos t$, except that the graph of $x$ is displaced by 2.4981 units to the right of the graph of $y$. (See Fig. 5.8.)

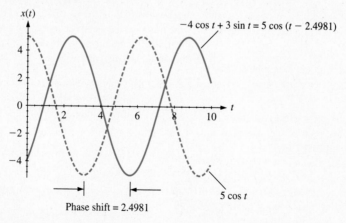

FIGURE 5.8   The graph of Eq. (5.31) and the graph of $5 \cos t$.

- The mass crosses the equilibrium position for the first time when $\cos(t - 2.4981) = 0$, or, in this case, when $t - 2.4981 = -\pi/2$. The first crossing time is at $t \approx 0.927$ s.
- The speed of the mass at any instant is

$$|x'(t)| = |5\sin(t - 2.4981)| \le 5.$$

- The maximum speed is 5 m/s and is reached when $t - 2.4981$ is an integer multiple of $\pi/2$. In this case, the maximum speed is reached for the first time when the mass crosses the equilibrium position for the first time.
- The maximum distance of the mass from the equilibrium position is 5 m, because $|x(t)| \le 5$.

On the other hand, the phase amplitude formulation shows that the initial position of the mass is $x(0) = 5\cos 2.4981$. It is easier to read off the initial position and velocity from Eq. (5.32) than from Eq. (5.31). ∎

The point of Example 6 is to show that the phase-amplitude form of a solution is not inherently "better" than another equivalent representation of the solution. However, the phase-amplitude form in some cases is a more convenient way to express periodic solutions, depending on the questions that you want to answer.

## Exercises

**1.** The metric system was developed in France in the late 1700s in response to economic pressures for a uniform standard of weights and measures. Originally, it was proposed that the meter be equal to the length of an ideal frictionless pendulum of period 1 s.

(a) How long would this unit of length be relative to the contemporary meter, assuming a linear pendulum and $g = 9.8$ m/s$^2$?

(b) The proposal was rejected in part because the value of $g$ varies by about 0.5 percent from a nominal value of approximately 9.8 m/s$^2$, depending on one's location on the earth. (Such variations were known to eighteenth century physicists.) What is the range of lengths that can be assumed by a linear pendulum with a period of 1 s, given these variations in $g$?

**2.** Find the position $x(t)$ of an 80-kg mass that is attached to a spring with a spring constant of 20 N/m.

Assume that friction is negligible and that the mass starts at the equilibrium with an initial velocity of 0.1 m/s. Plot the position, the velocity, and the acceleration of the mass as a function of $t$. Determine the times at which each quantity is at a relative maximum or minimum. How are the times of maximum position, velocity, and acceleration related?

**3.** Find the general solution of each of the differential equations in the following list. In each case, state the period of the solution.

(a) $x'' + 4x = 0$          (b) $x'' + 3x = 0$

(c) $x'' + 16x = 0$          (d) $x'' + 8x = 0$

(e) $x'' = -9x$              (f) $x'' = -25x$

(g) $x'' + \pi^2 x = 0$      (h) $x'' + 4\pi^2 x = 0$

(i) $x'' + \sqrt{2}x = 0$    (j) $x'' = -5x$

**4.** Suppose that each initial value problem in the following list is a model of an ideal, frictionless mass-spring system in which distance is measured in meters and time in seconds. For each problem,

- find the solution in phase-amplitude form;
- find the time at which the mass crosses the equilibrium position for the first time;
- find the maximum speed of the mass.

(a) $x'' + x = 0$, $x(0) = 2$, $x'(0) = 2$

(b) $x'' + 9x = 0$, $x(0) = 1$, $x'(0) = -3$

(c) $x'' + 4x = 0$, $x(0) = -2$, $x'(0) = 3$

(d) $x'' + 2x = 0$, $x(0) = 1$, $x'(0) = -\sqrt{2}$

**5.** If you want to double the period of a linear pendulum, how must you change its length?

**6.** Suppose you give a linear pendulum of length 1 m an initial push from its equilibrium position at $\phi = 0$. For what values of the initial angular velocity will the maximum displacement of the pendulum exceed 10 degrees from the vertical, assuming that friction is negligible?

**7.** Let $x'' + \omega^2 x = 0$. This exercise explores the relationship between the solution of the form

$$x(t) = c_1 \cos \omega t + c_2 \sin \omega t \qquad (5.33)$$

and the phase-amplitude form

$$x(t) = A \cos(\omega t - \delta). \qquad (5.34)$$

(a) Apply the trigonometric identity

$$\cos(\alpha - \beta) = \cos \alpha \cos \beta + \sin \alpha \sin \beta$$

to Eq. (5.34).

(b) Use the result in (a) to conclude that $c_1 = A \cos \delta$ and $c_2 = A \sin \delta$.

(c) Express $A$ in terms of $c_1$ and $c_2$.

(d) Express $\delta$ in terms of $c_1$ and $c_2$. Discuss how the initial conditions allow you to determine which of the two possible solutions is the correct one.

**8.** Suppose a given LC circuit oscillates with frequency $\nu$. If you want to double the frequency and

can change only one component in the circuit, what are your options?

**9.** Suppose that the true capacitance of a capacitor rated $1.0 \, \mu F$ can vary between $0.95 \, \mu F$ and $1.05 \, \mu F$ and that the true inductance of an inductor rated $1.0 \, H$ can vary between $1.1 \, H$ and $0.9 \, H$. What is the range of values of the period of the oscillations of an LC circuit with these components? (Assume that resistance is negligible.)

**10.** Suppose the measured inductance of an inductor is $0.950 \, H$. You want to add a capacitor to create an LC circuit that oscillates with period $T = 0.00100 \pm 0.00005$ s. What range of capacitances can be used to satisfy this tolerance?

**11.** In Example 2, we considered the maximum initial displacement and velocity that could be applied to a particular spring without damaging it. However, the maximum initial displacement was determined by assuming that $v_0 = 0$, and the maximum initial velocity was determined by assuming that $x_0 = 0$. Suppose that $x_0$ and $v_0$ are both nonzero. Derive an expression that determines the range of initial positions and velocities for which the spring is not damaged. (Hint: The set of initial conditions for which the spring is undamaged is enclosed by the ellipse in Fig. 5.9.)

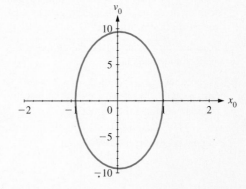

**Figure 5.9** Curve in the space of initial conditions that do not damage the spring (see Exercise 11).

**12.** Consider a swing of length 2 m. You start swinging at $\phi_0 = \pi/5$ with an initial angular velocity of $\omega_0 = 0.5$ rad/s. Suppose that friction at the pivot is negligible and that you do not try to make yourself swing higher or lower.

(a) What is the maximal amplitude of the swing?

(b) What is the period of the oscillation?

(c) What is the minimal horizontal space that you need so that the swing does not hit anything?

**13.** Suppose a linear spring with spring constant 10 N/m is hung from the ceiling and a 3-kg mass is attached to the free end. The mass is 1 m from the floor and 2 m from the ceiling when it is at rest at its equilibrium position. Now suppose the mass is given some initial velocity $v_0$ from the equilibrium. For what values of $v_0$ will the mass not hit the floor, assuming that frictional forces are negligible?

**14.** A 2-kg mass stretches a spring by 39.2 cm.

(a) Find the equation of motion, assuming that $g = 9.8$ m/s$^2$ and that friction is negligible.

(b) Suppose the mass starts from its equilibrium position in a downward direction with an initial speed of 2 cm/s. Find the position $x(t)$.

(c) When does the mass in (b) cross the equilibrium, moving upward, for the first time? When does the mass cross the equilibrium moving downward for the first time?

(d) What are the amplitude and the period of the motion of the mass in (b)?

(e) Suppose the mass is started from its equilibrium position. What must the initial velocity be for the amplitude of the motion to be 10 cm?

**15.** Is it possible to express the function

$$x(t) = 3 + 2\sin t + \cos t$$

in the form

$$x(t) = B + A\cos(\omega t - \delta)$$

for appropriate choices of $A$, $B$, $\omega$, and $\delta$? If not, explain why not. If so, determine the appropriate constants.

**16.** Let $x'' + \omega^2 x = 0$ with $x(0) = x_0$, $x'(0) = v_0$. Suppose we write the solution in the phase-amplitude form $x(t) = A\cos(\omega t - \delta)$. Explain how the signs of $x_0$ and $v_0$ determine the value of $\delta$.

**17.** Let $x'' + x = 0$, $x(0) = 0$, $x'(0) = 1$.

(a) Find the answer as a linear combination of the form $x(t) = c_1\cos t + c_2\sin t$ for appropriate values of $c_1$ and $c_2$.

(b) Show how to write the solution in the phase-amplitude form $x(t) = A\cos(t - \delta)$, even though you encounter a difficulty when you follow the procedure outlined in the text.

**18.** A phase-amplitude formulation of simple harmonic motion can be given equally well with the sine function instead of the cosine function.

(a) Let $x'' + \omega^2 x = 0$, $x(0) = x_0$, $x'(0) = v_0$. Show that the solution can be written in the alternative phase-amplitude form

$$x(t) = A\sin(\omega t + \phi) \qquad (5.35)$$

by finding $A$ and $\phi$ in terms of $x_0$ and $v_0$.

(b) If $\phi > 0$, is the graph of Eq. (5.35) shifted to the right or to the left of the graph of the function $A\sin\omega t$? What if $\phi < 0$?

(c) By how much is the graph of $A\sin(\omega t + \phi)$ shifted horizontally compared to the graph of $A\sin\omega t$, assuming that $-\pi < \phi < \pi$?

**19.** Suppose that each initial value problem in the following list is a model of an ideal, frictionless mass-spring system in which distance is measured in meters and time in seconds. For each problem,

• find the solution in the phase-amplitude form given by Eq. (5.35) in Exercise 18;

- find the time at which the mass crosses the equilibrium position for the first time;
- find the maximum speed of the mass.

(a) $x'' + x = 0$, $x(0) = 1$, $x'(0) = 1$

(b) $x'' + 9x = 0$, $x(0) = 2$, $x'(0) = -4$

(c) $x'' + 4x = 0$, $x(0) = -3$, $x'(0) = 2$

(d) $x'' + 2x = 0$, $x(0) = 2$, $x'(0) = -1$

**20.** Let

$$x(t) = A \cos(\omega t + \psi). \tag{5.36}$$

What is the relationship between the constant $\psi$ in Eq. (5.36) and the constant $\delta$ in Eq. (5.22)?

## 5.3 OVERDAMPED MOTION

In this section and the two that follow, we consider the behavior of linear second-order differential equations in which friction is present. Friction dissipates kinetic energy in the form of heat, and for this reason we say that damped systems are *dissipative*. Frictionless systems, such as those that exhibit simple harmonic motion, are examples of *conservative* processes, so called because kinetic energy is conserved.

### The Characteristic Equation

Simple harmonic motion is straightforward to analyze, because the solutions involve sine and cosine functions. In this section, we consider the solutions of the equation

$$mx'' + cx' + kx = 0, \tag{5.37}$$

where $c$ represents a positive coefficient of friction. Equation (5.37) is homogeneous, because the zero function is a solution. When $c = 0$, the solutions of Eq. (5.37) are sine and cosine functions. When $c \neq 0$, the solutions are less easy to determine by inspection.

In Chapters 1–2, we show that the solutions of the linear first-order differential equation $x' = kx$ are exponentials (that is, $x(t) = x_0 e^{kt}$), provided that $k$ is a constant. Equation (5.37) is second order, but it has constant coefficients. Let us determine whether the solutions of Eq. (5.37) also can be exponential functions.

Suppose

$$x(t) = e^{\lambda t}. \tag{5.38}$$

If we substitute (5.38) into Eq. (5.37), we find

$$\left(m\lambda^2 + c\lambda + k\right)e^{\lambda t} = 0. \tag{5.39}$$

As $e^{\lambda t}$ is never 0 for any real numbers $\lambda$ and $t$, Eq. (5.39) is satisfied if and only if

$$m\lambda^2 + c\lambda + k = 0. \tag{5.40}$$

Equation (5.40) is called the *characteristic equation* of Eq. (5.37). If $\lambda$ is a root (solution) of Eq. (5.40), then $x(t) = e^{\lambda t}$ is a solution of Eq. (5.37). The roots of the characteristic equation are given by the quadratic formula,

$$\lambda = \frac{-c \pm \sqrt{c^2 - 4mk}}{2m}.$$

Depending on the constants $m$, $c$, and $k$, Eq. (5.40) can have distinct real roots, a double real root, or complex conjugate roots. We consider the case of real, distinct roots in the remainder of this section.

## Real, Distinct Roots

If the characteristic equation (5.40) has real, distinct roots $\lambda_1$ and $\lambda_2$, then $x_1(t) = e^{\lambda_1 t}$ and $x_2(t) = e^{\lambda_2 t}$ are solutions of Eq. (5.37). As Eq. (5.37) is linear, the superposition principle implies that any function of the form

$$x(t) = c_1 e^{\lambda_1 t} + c_2 e^{\lambda_2 t} \tag{5.41}$$

is also a solution of Eq. (5.37). In fact, Eq. (5.41) is the general solution of Eq. (5.37), because there is a unique choice of the constants $c_1$ and $c_2$ that satisfies any initial condition, as the next example illustrates. (See also Exercise 9.)

◆ *Example 1*    Let

$$x'' + 10x' + 9x = 0. \tag{5.42}$$

The corresponding characteristic equation is

$$\lambda^2 + 10\lambda + 9 = 0,$$

whose roots are $\lambda_1 = -1$ and $\lambda_2 = -9$. Therefore, two solutions of Eq. (5.42) are $x_1(t) = e^{-t}$ and $x_2(t) = e^{-9t}$, as you can verify readily by substitution. The superposition principle implies that

$$x(t) = c_1 e^{-t} + c_2 e^{-9t} \tag{5.43}$$

also solves Eq. (5.42), where $c_1$ and $c_2$ are any constants.

Equation (5.43) is called the *general solution* of Eq. (5.42) because the values of $c_1$ and $c_2$ can be chosen to satisfy any choice of initial conditions. Suppose $x(0) = x_0$ and $x'(0) = v_0$. In this example, the initial conditions imply that

$$
\begin{aligned}
x_0 &= c_1 + c_2, \\
v_0 &= -c_1 - 9c_2.
\end{aligned}
\tag{5.44}
$$

The system (5.44) is a linear simultaneous system of equations whose solution is

$$c_1 = (9x_0 + v_0)/8,$$
$$c_2 = -(x_0 + v_0)/8.$$

Therefore, for any choice of the initial conditions $x(0)$ and $x'(0)$, we can determine two constants $c_1$ and $c_2$ such that Eq. (5.43) solves the initial value problem. ∎

The existence and uniqueness theorem (Theorem 5.1 in Section 5.1) guarantees that Eq. (5.43) is the only solution of Eq. (5.42) once we have found $c_1$ and $c_2$ from the initial conditions. As a result, if we think of Eq. (5.42) as an accurate model of a mass-spring system, then we can predict the position, velocity, and acceleration of the mass at any time in the future.

We summarize our discussion in the following theorem.

**Theorem 5.4** *Let*

$$m\frac{d^2x}{dt^2} + c\frac{dx}{dt} + kx = 0 \tag{5.45}$$

*where $m$, $c$, and $k$ are real constants. If the characteristic equation*

$$m\lambda^2 + c\lambda + k = 0$$

*has real, distinct roots $\lambda_1$ and $\lambda_2$, then the function*

$$x(t) = c_1 e^{\lambda_1 t} + c_2 e^{\lambda_2 t} \tag{5.46}$$

*is a general solution of Eq. (5.45). The initial conditions $x(0) = x_0$ and $x'(0) = v_0$ uniquely determine the values of $c_1$ and $c_2$.*

## Physical Interpretation of Overdamped Solutions

When the characteristic equation has real, distinct roots, the corresponding solutions do *not* oscillate, because the general solution (5.46) contains only exponential terms. In the physically realistic case in which $m$, $c$, and $k$ are positive constants, *all solutions eventually tend to 0 regardless of the initial conditions*. In such a case, we say that the motion is *overdamped*. (The notion of an overdamped solution applies only when $m$, $c$, and $k$ are all positive.) The solution (5.46) yields considerable information about the nature of overdamped linear systems, as the next example shows.

◆ *Example 2*    Consider a linear pendulum,

$$\phi'' + \left(\frac{c}{m\ell}\right)\phi' + \left(\frac{g}{\ell}\right)\phi = 0. \tag{5.47}$$

Suppose $g/\ell = 10$ and $c/(\ell m) = 7$. (What are the units of each of these ratios?) The associated characteristic equation is

$$\lambda^2 + 7\lambda + 10 = 0,$$

whose roots are $\lambda_1 = -2$ and $\lambda_2 = -5$. The general solution of Eq. (5.47) is

$$\phi(t) = c_1 e^{-2t} + c_2 e^{-5t}. \tag{5.48}$$

Each of the following statements about the motion of the pendulum is a direct consequence of Eq. (5.48):

- The pendulum does not oscillate.
- If the pendulum is placed at its equilibrium position ($\phi = 0$) with no initial velocity, then the pendulum remains at the equilibrium position forever.
- If at least one of $c_1$ and $c_2$ is nonzero, then the pendulum tends to the equilibrium position at an exponential rate. In the limit as $t \to \infty$, the pendulum approaches the equilibrium position, $x = 0$.
- The decay constants $-2$ and $-5$ that appear in the exponential terms in Eq. (5.48) are determined by $\ell$, $c$, and $m$. The decay constants do *not* depend on the initial conditions.
- For any set of initial conditions, the bob of the pendulum crosses the equilibrium at most once. To see this, suppose $\phi(t) = 0$. Then Eq. (5.48) implies that

$$-\frac{c_2}{c_1} = e^{3t},$$

which is satisfied for at most one value of $t$. We are interested only in the case $t \geq 0$, because the pendulum is set in motion at $t = 0$. Given the restriction $t \geq 0$, the bob crosses the equilibrium exactly once if $c_2/c_1 < -1$, and the bob never crosses the equilibrium if $c_2/c_1 > -1$. (See Exercise 2.) Because $c_1$ and $c_2$ are determined uniquely by the initial conditions, the issue of whether the bob ever crosses the equilibrium is determined completely by the initial position and velocity of the bob.

Suppose the pendulum initially is at the equilibrium position but is given an anti-clockwise angular velocity of 3 rad/s. The initial conditions $\phi(0) = 0$ and $\phi'(0) = 3$ imply that $\phi(t) = e^{-2t} - e^{-5t}$, whose graph is shown in Fig. 5.10. In this example, the bob never crosses the equilibrium position. Its maximum displacement is about 0.33 rad from the vertical.

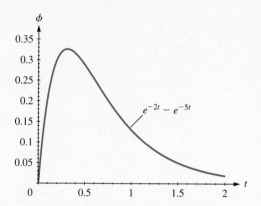

**FIGURE 5.10**    The graph of the solution (5.48) corresponding to the initial condition $\phi(0) = 0$, $\phi'(0) = 3$.    ∎

More generally, we can determine the values of $m$ and $\ell$ for which the linear pendulum does not oscillate. The characteristic equation corresponding to Eq. (5.47) is

$$\lambda^2 + \left(\frac{c}{m\ell}\right)\lambda + \frac{g}{\ell} = 0. \tag{5.49}$$

The roots are given by

$$\lambda = \frac{-c \pm \sqrt{c^2 - 4m^2 g\ell}}{2m\ell}.$$

If the coefficient of friction $c$ is large enough, then $c^2 > 4m^2 g\ell$, so the roots of Eq. (5.49) are real and distinct. In particular, they are both negative, because $c > \sqrt{c^2 - 4m^2 g\ell}$. Hence, both terms of the general solution (5.46) tend to 0 as $t$ becomes large.

## Exercises

**1.** For each of the initial value problems in the following list,

- find an explicit solution;

- if we interpret $x(t)$ as the position of a mass on a linear spring at time $t$, determine whether the mass ever crosses the equilibrium position at $x = 0$;

- find the maximum distance of the mass from the equilibrium position and the time at which this distance is reached.

(a) $x'' + 4x' + 3x = 0$, $x(0) = 1$, $x'(0) = 2$

(b) $\frac{1}{2}x'' + 3x' + 4x = 0$, $x(0) = 1$, $x'(0) = -3$

(c) $2x'' + 10x' + 12x = 0$, $x(0) = -1$, $x'(0) = 4$

**2.** Consider the pendulum described in Example 2.

(a) Suppose $\phi(0) = 0$. Is it possible to give the bob an initial velocity sufficient to make it cross the equilibrium position at least once? Explain.

(b) Suppose $c_2/c_1 = -1$. Can the pendulum ever be exactly at the equilibrium position? Explain.

**3.** Let the motion of a linear pendulum be governed by the equation

$$x'' + 4x' + 3x = 0. \qquad (5.50)$$

(a) Suppose the pendulum initially is at the equilibrium position, that is, $x(0) = 0$, and that $x'(0) = v_0 \neq 0$. Does the pendulum ever cross the equilibrium? Explain why or why not.

(b) Suppose $x(0) = 1$. Show that the pendulum crosses the equilibrium once if $v_0 < -3$.

(c) Are there any initial conditions for which the pendulum crosses the equilibrium position exactly twice? Explain why or why not.

(d) Find the solution satisfying the initial conditions $x(0) = x_0$, $x'(0) = v_0$.

**4.** Let $x'' + x' - 2x = 0$ with $x(0) = 1$, $x'(0) = v_0$.

(a) Find the solution.

(b) What happens to the solution in (a) as $t \to \infty$?

(c) For what values of $v_0$, if any, does the solution tend to 0 as $t \to \infty$?

(d) Is this differential equation a physically plausible model of a mass-spring system? Explain why or why not.

**5.** Let $x'' - 3x' + 2x = 0$.

(a) Determine the general solution.

(b) Find the solution satisfying the initial conditions $x(0) = 1$, $x'(0) = -2$. Discuss what happens to the function $x$ as $t \to \infty$.

(c) Is there an initial condition for which $x \to 0$ as $t \to \infty$? Explain.

**6.** Let

$$x'' + \tfrac{10}{3}x' + x = 0.$$

(a) Find the general solution.

(b) Describe the behavior of the general solution as $t \to \infty$.

(c) Find the solution satisfying the initial conditions $x(0) = 2$, $x'(0) = -3$.

(d) Find the solution satisfying the initial conditions $x(0) = 2$, $x'(0) = -6$.

(e) Plot the solutions in (c) and (d) on the same axes. How do the decay rates of the two solutions differ?

(f) Let $x(0) = x_0$, $x'(0) = -3x_0$. Show that the solution is a multiple of $e^{-3t}$.

(g) Show that, if the initial conditions are not related as in (f), then the solution has a term of the form $e^{-t/3}$.

**7.** Let $x'' - \tfrac{8}{3}x' - x = 0$.

(a) Find the general solution.

(b) Show that solutions grow exponentially for most initial conditions.

(c) Identify the set of initial conditions for which solutions go to zero as $t \to \infty$.

**8.** Kirchhoff's law leads to the differential equation

$$LQ'' + RQ' + (1/C)Q = 0$$

for the charge $Q$ in an LRC circuit with no external power source.

(a) Suppose $L = 1$ H, $R = 10$ $\Omega$, and $C = 1$ $\mu$F. Can the circuit be considered mathematically equivalent to an overdamped pendulum? Explain.

(b) Suppose $L = 1$ H and $C = 1$ $\mu$F. For what range of values of $R$ can the circuit be considered mathematically equivalent to an overdamped pendulum?

(c) Derive a general relationship between $L$, $R$, and $C$ that determines when the charge in the circuit does not oscillate as a function of time.

**9.** Let $mx'' + cx' + kx = 0$, and assume that the solutions of the characteristic equation are real and distinct; call them $\lambda_1$ and $\lambda_2$. The general solution is $x(t) = c_1 e^{\lambda_1 t} + c_2 e^{\lambda_2 t}$, where $c_1$ and $c_2$ depend on the initial conditions $x(0) = x_0$ and $x'(0) = v_0$.

(a) Identify the linear system of equations that must be solved to determine $c_1$ and $c_2$.

(b) Show that the linear system in (a) always has a unique solution, that is, that $c_1$ and $c_2$ can be written in terms of $x_0$ and $v_0$ in a unique way. Conclude that the general solution provides a unique solution of any initial value problem, at least in the case in which the characteristic equation has real distinct roots.

**10.** Suppose that a mass-spring system is governed by the equation

$$x'' + 4x' + 3x = 0,$$

where distance is measured in meters and time is measured in seconds.

(a) Suppose that you pull the mass 1 unit away from its equilibrium position (that is, $x_0 = 1$) and release it. (That is, the initial velocity $v_0$ is 0.) Find an expression for the position $x$ of the mass as a function of $t$.

(b) Suppose that, instead of simply releasing the mass, you accidentally impart a small initial velocity, say $x'(0) = 0.1$, to the mass. Find an expression for the position $\hat{x}$ of the mass as a function of $t$.

(c) Plot $x$ and $\hat{x}$ on the same axes. Does your accidental push in (b) have a significant impact on the outcome of the experiment?

(d) One way to quantify your answer in (c) is to think of $x(t)$ as the "desired" outcome and $\hat{x}$ as the "actual" outcome that results from some initial error. Plot the absolute error in $x$ as a function of $t$ and briefly describe its behavior.

(e) Plot the relative error in $x$ as a function of $t$ and briefly describe its behavior. How do the absolute and relative errors support your conclusions in (c)?

**11.** Suppose you have a spring whose spring constant is known to be $k = 2$ N/m. A mass of 1 kg is attached to one end of the spring, and the other end of the spring is attached to the ceiling. You wish to measure the damping coefficient, assuming that the frictional forces in the system are approximately linear. Suppose that you pull the mass to a position 10.0 cm away from its equilibrium position and let go. You then measure the position of the mass after 1 s and find that it has moved 4.0 cm closer to the equilibrium point. Can you estimate the damping coefficient $c$, at least in principle, from this experiment? If so, find $c$; if not, explain why not.

## 5.4   UNDERDAMPED MOTION

We say that the linear equation

$$mx'' + cx' + kx = 0 \tag{5.51}$$

is *underdamped* if the roots of the characteristic equation corresponding to Eq. (5.51) are complex conjugates. (The notion of an underdamped solution applies only to the case in which $m$, $c$, and $k$ are positive. The case $c = 0$ corresponds to simple harmonic motion, which is a special case of underdamped motion.) As we will see, underdamped systems oscillate. One question that we must answer is whether it is sensible to write a general solution in the form

$$x(t) = c_1 e^{\lambda_1 t} + c_2 e^{\lambda_2 t}$$

when $\lambda_1$ and $\lambda_2$ are complex numbers. The answer is "yes," as we now explain.

## Euler's Formula

Let

$$x'' + \omega^2 x = 0, \tag{5.52}$$

which is Eq. (5.51) with $c = 0$, $m = 1$, and $k = \omega^2$. Equation (5.52) describes simple harmonic motion, and its general solution is

$$x(t) = c_1 \cos \omega t + c_2 \sin \omega t \tag{5.53}$$

for appropriate constants $c_1$ and $c_2$. The characteristic equation corresponding to Eq. (5.52) is

$$\lambda^2 + \omega^2 = 0,$$

whose roots are $\lambda = \pm \omega i$, where $i = \sqrt{-1}$. Our considerations in Section 5.3 suggest that the general solution of Eq. (5.52) also can be written in the form

$$x(t) = b_1 e^{i\omega t} + b_2 e^{-i\omega t}. \tag{5.54}$$

Assume for the moment that Eq. (5.54) makes sense. Theorem 5.1 in Section 5.1 guarantees that the solutions of Eq. (5.52) are unique. Therefore, Eqs. (5.53)–(5.54) represent the same function. In fact, Eqs. (5.53)–(5.54) can be used to *define* the exponential of a complex number, as we now show.

Let $x(0) = 1$ and $x'(0) = 0$. The solution of Eq. (5.52) that satisfies these initial conditions is $x(t) = \cos \omega t$. Next let $x(0) = 0$ and $x'(0) = \omega$; then $x(t) = \sin \omega t$. We now apply these same sets of initial conditions to Eq. (5.54) to derive an expression for $e^{i\omega t}$.

Assume that the exponential function obeys the usual rules of calculus when its argument is a complex number; that is,

$$x'(t) = i\omega b_1 e^{i\omega t} - i\omega b_2 e^{-i\omega t}.$$

With this assumption, the initial conditions $x(0) = 1$ and $x'(0) = 0$ applied to the general solution (5.54) lead to the system

$$b_1 + b_2 = 1,$$

$$i\omega b_1 - i\omega b_2 = 0,$$

which yields $b_1 = b_2 = 1/2$. Since Eq. (5.53) implies that $x(t) = \cos \omega t$, we have

$$\cos \omega t = \tfrac{1}{2} e^{i\omega t} + \tfrac{1}{2} e^{-i\omega t}. \tag{5.55}$$

The initial conditions $x(0) = 0$ and $x'(0) = \omega$ lead to the system

$$b_1 + b_2 = 0,$$
$$i\omega b_1 - i\omega b_2 = \omega,$$

which yields $b_1 = 1/(2i)$ and $b_2 = -1/(2i)$. Since Eq. (5.53) implies that $x(t) = \sin \omega t$, we have

$$\sin \omega t = \tfrac{1}{2i} e^{i\omega t} - \tfrac{1}{2i} e^{-i\omega t}. \tag{5.56}$$

We can solve Eqs. (5.55)–(5.56) for $e^{i\omega t}$ to obtain

$$e^{i\omega t} = \cos \omega t + i \sin \omega t. \tag{5.57}$$

Equation (5.57) is called *Euler's formula.* Similarly,

$$e^{-i\omega t} = \cos \omega t - i \sin \omega t. \tag{5.58}$$

Equation (5.58) shows that $e^{-i\omega t}$ is the complex conjugate[4] of $e^{i\omega t}$.

The usual laws of exponents apply to complex numbers. Let $z = \alpha + i\omega$, where $\alpha$ and $\omega$ are real. Euler's formula implies that

$$e^{zt} = e^{(\alpha + i\omega)t} = e^{\alpha t} e^{i\omega t} = e^{\alpha t} (\cos \omega t + i \sin \omega t).$$

Euler's formula also can be used to represent the general solution of a given differential equation in two different ways.

## The Real and Complex Form of Oscillatory Solutions

Suppose that the roots of the characteristic equation corresponding to Eq. (5.51) are $\lambda = \alpha + i\omega$ and $\bar{\lambda} = \alpha - i\omega$. The general solution is

$$x(t) = b_1 e^{\lambda t} + b_2 e^{\bar{\lambda} t}, \tag{5.59}$$

where $b_1$ and $b_2$ are complex numbers. Euler's formula implies that

$$\begin{aligned}
x(t) &= b_1 e^{\lambda t} + b_2 e^{\bar{\lambda} t} \\
&= b_1 e^{\alpha t} (\cos \omega t + i \sin \omega t) + b_2 e^{\alpha t} (\cos \omega t - i \sin \omega t) \\
&= e^{\alpha t} [(b_1 + b_2) \cos \omega t + i(b_1 - b_2) \sin \omega t] \\
&= e^{\alpha t} (c_1 \cos \omega t + c_2 \sin \omega t), \tag{5.60}
\end{aligned}$$

---

[4]Let $z = x + iy$, where $x$ and $y$ are real. The complex conjugate of $z$, denoted $\bar{z}$, is $\bar{z} = x - iy$. The real and imaginary parts of $z$ are $\operatorname{Re} z = x$ and $\operatorname{Im} z = y$, respectively.

where $c_1 = b_1 + b_2$ and $c_2 = i(b_1 - b_2)$. We emphasize that both Eq. (5.59) and Eq. (5.60) represent the *same* function, although the analytical representations look different. The coefficients $b_1$ and $b_2$ in Eq. (5.59) are complex valued, and Eq. (5.59) is called the *complex form* of the solution. The coefficients $c_1$ and $c_2$ in Eq. (5.60) are real valued, so Eq. (5.60) is called the *real form* of the solution. (See Exercise 4.)

Equation (5.60) is the product of an exponential function and a sum of sine and cosine functions. As a result, the solutions of underdamped systems oscillate.

◆ *Example 1*  Let

$$x'' + 4x = 0 \tag{5.61}$$

with $x(0) = 2$ and $x'(0) = 2$. The characteristic equation is $\lambda^2 + 4 = 0$ with roots $\lambda = \pm 2i$. One form of the general solution is

$$x(t) = b_1 e^{2it} + b_2 e^{-2it}.$$

The initial conditions yield the system

$$b_1 + b_2 = 2,$$

$$2i b_1 - 2i b_2 = 2,$$

so

$$x(t) = \left(1 - \tfrac{1}{2}i\right) e^{2it} + \left(1 + \tfrac{1}{2}i\right) e^{-2it}. \tag{5.62}$$

Equation (5.62) is the complex form of the solution of Eq. (5.61).

On the other hand, Eq. (5.61) describes simple harmonic motion. Therefore, the general solution is

$$x(t) = c_1 \cos 2t + c_2 \sin 2t,$$

for which the initial conditions yield

$$x(t) = 2 \cos 2t + \sin 2t. \tag{5.63}$$

Equation (5.63) is the real form of the solution. Both Eq. (5.62) and Eq. (5.63) represent the same function, as you can easily check by applying Euler's formula to Eq. (5.62) and collecting terms.                                                                     ∎

◆ *Example 2*   Suppose a mass of 80 kg is attached to a linear spring whose spring constant is 25 N/m. If the force of friction is proportional to the current velocity of the mass with a proportionality constant of 40 kg/s, then the differential equation governing the motion is

$$80x'' + 40x' + 25x = 0. \tag{5.64}$$

The corresponding characteristic equation is

$$80\lambda^2 + 40\lambda + 25 = 0,$$

whose roots are $\lambda = -\frac{1}{4} \pm \frac{1}{2}i$. We can write the general solution of Eq. (5.64) in the form

$$x(t) = e^{-t/4}\left(c_1 \cos \tfrac{1}{2}t + c_2 \sin \tfrac{1}{2}t\right).$$

The initial conditions $x(0) = x_0$ and $x'(0) = v_0$ lead to the system

$$\begin{aligned} c_1 &= x_0, \\ -\tfrac{1}{4}c_1 + \tfrac{1}{2}c_2 &= v_0. \end{aligned} \tag{5.65}$$

If $x_0 = 1$ and $x'(0) = 3$, then $c_1 = 1$ and $c_2 = 13/2$, so the corresponding solution is

$$x(t) = e^{-t/4}\left(\cos \tfrac{1}{2}t + \tfrac{13}{2}\sin \tfrac{1}{2}t\right).$$

Notice that the system (5.65) can be solved uniquely for any set of initial conditions.   ∎

## The Phase-Amplitude Formulation of Underdamped Solutions

If the roots of the characteristic equation corresponding to $mx'' + cx' + kx = 0$ are complex conjugates, that is, if

$$\lambda_1 = \alpha + i\omega \qquad \text{and} \qquad \lambda_2 = \alpha - i\omega,$$

then the real form of the general solution is

$$x(t) = e^{\alpha t}(c_1 \cos \omega t + c_2 \sin \omega t). \tag{5.66}$$

Although Eq. (5.66) implies that the solution oscillates, more detailed information can be gleaned from the phase amplitude form of the solution, which is

$$x(t) = Ae^{\alpha t}\cos(\omega t - \delta). \tag{5.67}$$

The constants $A$ and $\delta$ can be computed from the initial conditions by using a procedure identical to that discussed in Section 5.2.

   The phase amplitude form (5.67) allows you to determine various properties of the motion easily by examining the constants $A$, $\alpha$, $\omega$, and $\delta$:

- If $\alpha \neq 0$, then Eq. (5.67) is *not* periodic. Although the cosine function is periodic with period $2\pi/\omega$, we have

$$\begin{aligned} x\left(t + \frac{2\pi}{\omega}\right) &= Ae^{\alpha(t+2\pi/\omega)}\cos\left(\omega\left(t + \frac{2\pi}{\omega}\right) - \delta\right) \\ &= Ae^{\alpha t}e^{2\alpha\pi/\omega}\cos(\omega t - \delta) \\ &= e^{2\alpha\pi/\omega}x(t), \end{aligned}$$

which is not equal to $x(t)$ if $\alpha \neq 0$. (See Example 3.)

- We say that Eq. (5.67) is *pseudoperiodic*, because it behaves like a periodic function except that its amplitude is not constant. The *pseudoperiod* is $2\pi/\omega$, the period of the cosine term.

- Equation (5.67) may be regarded as a cosine function with an exponentially decaying amplitude when $\alpha < 0$. The term $Ae^{\alpha t}$ is called the *envelope*. The graph of $Ae^{\alpha t}$ and the graph of $-Ae^{\alpha t}$ enclose the graph of $x(t)$, as the next example illustrates.

- Equation (5.67) allows you to determine by inspection when the mass crosses the equilibrium; you need only determine the times $t$ for which $\omega t - \delta$ is an odd multiple of $\pi/2$.

◆ *Example 3*    The oscillating curve in Fig. 5.11 is the graph of

$$x(t) = 10e^{-t/10}\cos 3t, \tag{5.68}$$

which is Eq. (5.67) with $A = 10$, $\alpha = -1/10$, $\omega = 3$, and $\delta = 0$. The envelope is $\text{env}(t) = 10e^{-t/10}$. Notice that the graphs of $\text{env}(t)$ and $-\text{env}(t)$ enclose the graph of $x(t)$. The function $x(t)$ is not periodic, because the amplitude of the oscillations decays exponentially. However, the pseudoperiod is $2\pi/3$, the period of the cosine term.

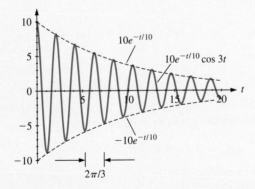

**FIGURE 5.11**    The graph of Eq. (5.68) and its envelope. The time interval between successive minima (and successive maxima) is the pseudoperiod, $2\pi/3$.    ■

---

◆ *Example 4*    Let us express the solution of the initial value problem discussed in Example 2 in phase-amplitude form. We have

$$80x'' + 40x' + 25x = 0, \qquad x(0) = 1, \qquad x'(0) = 3.$$

Since the roots of the characteristic equation are $\lambda = -\frac{1}{4} \pm \frac{1}{2}i$, the phase-amplitude form of the solution is

$$x(t) = Ae^{-t/4} \cos\left(\tfrac{1}{2}t - \delta\right),$$

where $A$ and $\delta$ must be determined. First observe that

$$x'(t) = -\tfrac{1}{4}x(t) - \tfrac{1}{2}Ae^{-t/4}\sin\left(\tfrac{1}{2}t - \delta\right);$$

therefore,

$$x'(0) = -\tfrac{1}{4}x(0) + \tfrac{1}{2}A\sin\delta.$$

The initial conditions $x(0) = 1$ and $x'(0) = 3$ imply that

$$1 = A\cos\delta,$$
$$\tfrac{13}{2} = A\sin\delta.$$

Consequently,

$$A = \sqrt{1 + \left(\tfrac{13}{2}\right)^2} = \tfrac{1}{2}\sqrt{173} \approx 6.576,$$

and

$$\delta = \tan^{-1}\tfrac{13}{2}. \tag{5.69}$$

The initial conditions imply that $\cos\delta$ and $\sin\delta$ are both positive; hence, we take the solution of Eq. (5.69) that lies in the interval $(0, \pi/2)$. Numerically, we find $\delta \approx 1.4181$. The phase-amplitude form of the solution is

$$x(t) = \tfrac{1}{2}\sqrt{173}\, e^{-t/4} \cos\left(\tfrac{1}{2}t - 1.4181\right). \tag{5.70}$$

Figure 5.12 shows a graph of Eq. (5.70), given by the solid curve. The dashed curves show the envelope; they are the graphs of the functions $Ae^{-t/4}$ and $-Ae^{-t/4}$.

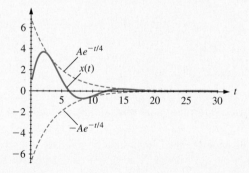

**Figure 5.12**   Solid line: the graph of Eq. (5.70); dashed lines: the envelope.

The phase-amplitude form (5.70) lets you calculate the answers to many different questions about the motion with relative ease. Here are three examples:

- The mass crosses the equilibrium for the first time when the cosine term in Eq. (5.70) first becomes 0, that is, when $\frac{1}{2}t - 1.4181 = \pi/2$ or $t \approx 6.0$ s.

- The mass crosses the equilibrium at $x = 0$ at intervals of $2\pi$ s. The pseudoperiod of the motion is $4\pi$, but the mass crosses twice during each pseudoperiod: once in one direction and once in the other.

- The velocity is

$$x'(t) = -\tfrac{1}{4}\sqrt{173}\, e^{-t/4}\left[\tfrac{1}{2}\cos\left(\tfrac{1}{2}t - 1.4181\right) + \sin\left(\tfrac{1}{2}t - 1.4181\right)\right].$$

It is possible to determine analytically the times at which the mass stops moving momentarily. In this case, we determine when the term in square brackets is 0. Therefore, the mass pauses at those values of $t$ for which

$$\tan\left(\tfrac{1}{2}t - 1.4181\right) = -\tfrac{1}{2}.$$

The mass stops momentarily at $t \approx 1.91$ s, and this instant corresponds to the maximum distance from the equilibrium, 3.65 m.                                      ∎

We summarize the main results in the following theorem.

**Theorem 5.5**   *Let $m$, $c$, and $k$ be real constants and let*

$$mx'' + cx' + kx = 0. \tag{5.71}$$

*Suppose that the corresponding characteristic equation has two complex conjugate roots, $\lambda_1 = \alpha + i\omega$ and $\lambda_2 = \bar{\lambda}_1 = \alpha - i\omega$. The general solution of Eq. (5.71) can be written in three equivalent forms:*

- *the* complex form $x(t) = b_1 \exp \lambda_1 t + b_2 \exp \lambda_2 t$,
- *the* real form $x(t) = e^{\alpha t}(c_1 \cos \omega t + c_2 \sin \omega t)$,
- *and the* phase-amplitude form $x(t) = A e^{\alpha t}\cos(\omega t - \delta)$.

*The constants $b_1$, $b_2$, $c_1$, $c_2$, $A$, and $\delta$ depend on the initial conditions. The constants $\alpha$ and $\omega$ do not depend on the initial conditions; their values are determined by $m$, $c$, and $k$.*

Theorem 5.5 implies that an underdamped linear system oscillates forever, at least in principle. Let us consider this idea in more detail. The general solution can be written in a phase-amplitude form, where the oscillations are enclosed by an exponentially decaying envelope. It is obvious that the amplitude of the oscillations becomes very small with time. Although *in principle* the mass oscillates forever if the motion is governed by

Eq. (5.70), eventually it is not possible to distinguish a mass that moves with extremely small oscillations from a mass that stands still if your powers of observation are limited.

Consider again the motion of the mass-spring system discussed in Example 4 and graphed in Fig. 5.12. The mass moves as much as 3.65 m away from the equilibrium position and oscillates forever, at least in principle. However, the amplitude of the oscillations decays exponentially. The oscillations of the mass after about 25 s are quite small in comparison to the oscillations that occur during the first 10 s. Once $t$ is large enough, the graph of the solution curve cannot be distinguished from the $t$ axis to within the resolution of the plot. These considerations suggest that if you watch such a mass-spring system in action, you might regard the oscillations as essentially negligible once their amplitude falls below some threshold value.

✦ *Example 5*   Consider the graph of Eq. (5.70), shown in Fig. 5.12. (See Example 4.) It is reasonable to ask when the amplitude of the oscillations becomes negligible. For the sake of this illustration, we regard oscillations as "negligible" once their amplitude becomes less than 1 cm.

The envelope equation provides an easy way to estimate the time past which the solutions are guaranteed to be negligible. From Eq. (5.70),

$$|x(t)| = \left| \tfrac{1}{2}\sqrt{173}\, e^{-t/4} \cos\left(\tfrac{1}{2}t - 1.4181\right) \right|$$
$$\leq \tfrac{1}{2}\sqrt{173}\, e^{-t/4}$$
$$\leq 0.01 \qquad \text{if } t > 4\ln\left(50\sqrt{173}\right) \approx 26.$$

Thus, the oscillations are less than 1 cm in amplitude after about 26 s.   ∎

## *Exercises*

**1.** For each differential equation in the following list, determine whether functions of the form $e^{\lambda t}$ can be solutions, where $\lambda$ is a suitable constant. If so, determine the permissible values of $\lambda$. If not, explain why the exponential function cannot be a solution.

(a) $3x'' = -2x - x'$   (b) $x'' + 5x + 3x' = 0$

(c) $x'' + x^2 = 0$   (d) $x'' + 2x' = 15x$

(e) $x'' + 9x = 6x$   (f) $x'' + \sin x = 0$

(g) $x'' + 6x' + 5x = 0$   (h) $x'' + x' = -\sin x$

(i) $x'' + (x')^2 + x = 0$   (j) $x'' + \ln x = 0$

**2.** Let $mx'' + cx' + kx = 0$. Suppose the charac-teristic equation has the following roots. For each pair of roots, determine all the possible values that $\lim_{t\to\infty} |x(t)|$ might assume if at least one of the initial conditions is nonzero.

(a) $\lambda_1 = 3, \lambda_2 = 5$

(b) $\lambda_1 = -3, \lambda_2 = -5$

(c) $\lambda_1 = -3, \lambda_2 = +5$

(d) $\lambda_1 = -3 + 5i, \lambda_2 = -3 - 5i$

(e) $\lambda_1 = 3 + 5i, \lambda_2 = 3 - 5i$

(f) $\lambda_1 = a, \lambda_2 = b$, where $a$ and $b$ are positive constants

**3.** Show that Eq. (5.62) is equivalent to Eq. (5.63).

**4.** Consider Eq. (5.59). Show that if $x(0)$ and $x'(0)$ are real, then $b_1 + b_2$ and $i(b_1 - b_2)$ must be real. This shows that the coefficients $c_1$ and $c_2$ in Eq. (5.60) must be real.

**5.** Let $x'' + 2x' + 9x = 0$ with initial conditions $x(0) = 1$, $x'(0) = 2$. Suppose $x(t)$ is the position at time $t$ of a mass attached to a linear spring with a linear coefficient of friction. The equilibrium position corresponds to $x = 0$; $x > 0$ denotes a stretched spring.

(a) Write the solution of the initial value problem in phase-amplitude form.

(b) Identify the envelope of the solution and graph it together with the solution.

(c) Identify the pseudoperiod of the oscillation.

(d) Find the time $t_0$ at which the mass first crosses the equilibrium position. What are the velocity and the acceleration at $t = t_0$?

(e) Find the time $t_1$ at which the mass reaches its maximum displacement in either direction from the equilibrium position. Identify the maximum displacement, the velocity at time $t_1$, and the acceleration at time $t_1$.

(f) What is the maximum compression of the spring and when does it occur?

(g) Suppose the motion of the mass is undetectable if the amplitude of the oscillations is less than 0.001. Approximately how long does it take before the motion becomes undetectable?

**6.** Let $x'' + 8x' + 20x = 0$ with $x(0) = 3$, $x'(0) = -2$.

(a) Find the complex form of the solution.

(b) Find the real form of the solution.

(c) Use Euler's formula to verify that your answers to (a) and (b) represent the same function.

**7.** Find the solution of each initial value problem in the following list. In each case,

• Express the solution in complex form.

• Express the solution in real form.

• Express the solution in phase-amplitude form.

• If we think of each equation as describing a linear mass-spring system, determine how often the mass crosses the equilibrium position.

• Find the time at which the mass first crosses the equilibrium position.

• Estimate the time $t$ for which $|x(t)| < 1/100$.

(a) $x'' + 4x' + 8x = 0$, $x(0) = 1$, $x'(0) = 2$

(b) $x'' + 2x' + 10x = 0$, $x(0) = 1$, $x'(0) = 1$

(c) $4x'' + 4x' + 13x = 0$, $x(0) = -1$, $x'(0) = -1$

**8.** Repeat Exercise 7 for each initial value problem in the following list:

(a) $16x'' + 8x' + 17x = 0$, $x(0) = 4$, $x'(0) = 0$

(b) $\frac{1}{2}x'' + \frac{1}{2}x' + \frac{1}{2}x = 0$, $x(0) = 2$, $x'(0) = -1$

(c) $x'' + x' + 2x = 0$, $x(0) = 1$, $x'(0) = 0$

**9.** Suppose you have a linear pendulum of length 1 m and damping coefficient $c = 0.1$ N s/m. Let $\phi(0) = 1/2$ rad and $\phi'(0) = 0$ rad/s. Assume that $g = 10$ m/s$^2$.

(a) Find the solution given that the mass of the bob is $m$.

(b) Find the envelope of the solution and use it to estimate the amplitude of the oscillations after 1 min if the mass is 1 kg.

(c) Use the envelope equation to estimate the size of the mass that is required so that the amplitude of the oscillations is at least 0.1 rad after 1 min.

(d) Suppose we attach a large mass to the pendulum. For the initial conditions above, does a larger mass move for a longer time before the oscillations become negligible, as compared to a smaller mass? Explain.

**10.** Suppose that the motion of a pendulum is governed by Eq. (5.6). Suppose that the mass of the bob and the coefficient of friction at the pivot are fixed and that the pendulum swings back and forth when its length is $\ell$.

(a) Can the oscillations of the pendulum be suppressed by altering the initial conditions? Explain.

(b) Can the oscillations of the pendulum be suppressed by altering its length? If so, should you lengthen the pendulum or shorten it?

**11.** A linear pendulum has a bob of mass 1 kg, a length of 1 m, and a coefficient of friction at the pivot equal to 1 N s/m. Assume that $g = 10$ m/s$^2$.

(a) Find the equation of motion.

(b) Show that the pendulum is underdamped and find the pseudoperiod.

(c) Suppose that the pendulum initially is at the equilibrium position and is given an initial velocity of 1 rad/s. Estimate the time $t_{max}$ past which the oscillations always remain within 0.01 rad of the equilibrium.

(d) Can you halve the pseudoperiod by changing the length of the pendulum? If so, explain how; if not, explain why not.

(e) If you alter the length of the pendulum, can you make it overdamped instead of underdamped? Explain.

**12.** Consider the model of the motion of a damped linear pendulum, Eq. (5.6).

(a) Determine the relation that the constants $m$, $c$, and $\ell$ must satisfy for the motion to be underdamped.

(b) Find the pseudoperiod of the motion in terms of $m$, $c$, and $\ell$ assuming that they satisfy the relation in (a).

(c) When the motion is underdamped, the phase amplitude form of the general solution is $\phi(t) = Ae^{\alpha t} \cos(\omega t - \delta)$. Does $\alpha$ depend on the initial conditions? If so, explain why. If not, find $\alpha$ in terms of $m$, $c$, and $\ell$.

(d) Suppose that a linear pendulum of length $\ell$ is underdamped and that the coefficient of friction at the pivot is $c$. How does the pseudoperiod of the motion change as $m$ is increased? Does the pseudoperiod approach a limit as $m$ becomes very large? If so, what is it?

(e) If the motion is underdamped, and $\ell$ and $c$ are fixed, does a heavier mass lead to a faster exponential decrease in the amplitude than does a lighter mass? Explain.

(f) Suppose that a linear pendulum of length $\ell$ and mass $m$ is underdamped and that the coefficient of friction at the pivot is $c$. How does the pseudoperiod of the motion change as $\ell$ is increased?

(g) If the motion is underdamped and $m$ and $c$ are fixed, does a longer pendulum lead to a faster exponential decrease in the amplitude compared to a shorter pendulum? Explain.

**13.** Suppose that for a given inductor, resistor, and capacitor, the voltage in an LRC circuit oscillates with time.

(a) Let the inductance and resistance be held fixed. How does the behavior of the circuit change as the capacitance is increased?

(b) Let the capacitance and resistance be held fixed. How does the behavior of the circuit change as the inductance is increased?

(c) Let the capacitance and inductance be held fixed. How does the behavior of the circuit change as the resistance is increased?

**14.** Let the mass-spring systems discussed below be governed by the differential equation

$$mx'' + cx' + kx = 0,$$

and assume that the systems are underdamped.

(a) Show that $\lim_{t \to \infty} x(t) = 0$ regardless of the initial conditions.

(b) Imagine two mass-spring systems that are identical except that one has a stiffer spring than the other. Discuss how the behavior of the two systems differs, assuming that they are set into motion with the same initial conditions.

(c) Imagine two mass-spring systems that are identical except that the coefficient of friction in one is larger than that in the other. Discuss how the behavior of the two systems differs, assuming that they are set into motion with the same initial conditions.

(d) Imagine two mass-spring systems that are identical except that the mass of one is larger than the mass of the other. Discuss how the behavior of the two systems differs, assuming that they are set into motion with the same initial conditions.

**15.** Let

$$x'' + cx' + kx = 0. \qquad (5.72)$$

Figure 5.13 shows a plot of the solution of Eq. (5.72) for a particular set of initial conditions.

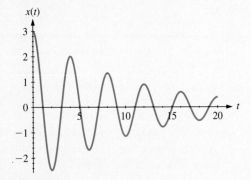

**FIGURE 5.13**   For Exercise 15.

(a) What is the phase-amplitude form of the general solution of Eq. (5.72)?

(b) Suppose the data used to generate Fig. 5.13

reveal the following maximum values:

| t | x(t) |
|------|------|
| 0.0 | 3.00 |
| 4.0 | 2.03 |
| 8.0 | 1.37 |
| 12.0 | 0.92 |

Show how your answer in (a) can be used to estimate the value of $c$ using the data in the table.

(c) Discuss how your answer in (b) allows you to estimate $k$.

**16.** Suppose that the equation $x'' + cx' + kx = 0$ defines an underdamped system.

(a) What is the time increment $T$ between successive maxima in the solution? (Suggestion: Look at the phase-amplitude form of the general solution.)

(b) The exponential decay of the amplitude and the equal time increments between relative maxima imply that the ratio between successive maxima is a constant, due to the equal ratio property of the exponential function. What is the ratio between each maximum and its successor?

(c) The logarithm of your answer in (b) is called the *logarithmic decrement*. Discuss how the logarithmic decrement can be used to estimate the values of $c$ and $k$ for an underdamped system.

## 5.5   CRITICALLY DAMPED MOTION

We say that the linear equation

$$mx'' + cx' + kx = 0 \qquad (5.73)$$

is *critically damped* if the roots of the characteristic equation corresponding to Eq. (5.73) are real and equal. Critically damped systems do not oscillate, as we will see.

For algebraic simplicity, we rewrite Eq. (5.73) as

$$x'' + ax' + bx = 0. \tag{5.74}$$

(Divide each term in Eq. (5.73) by $m$ and set $a = c/m$ and $b = k/m$.) The characteristic equation corresponding to Eq. (5.74) is

$$\lambda^2 + a\lambda + b = 0. \tag{5.75}$$

If the motion is critically damped, then $a^2 - 4b = 0$, and $\lambda = -a/2$ is a double real root. Our previous analysis suggests that $x_1(t) = e^{-at/2}$ is a solution of Eq. (5.74), as we can easily check:

$$\begin{aligned}
x_1'' + ax_1' + bx_1 &= \left(e^{-at/2}\right)'' + a\left(e^{-at/2}\right)' + be^{-at/2}\\
&= \left(a^2/4 - a^2/2 + b\right)e^{-at/2}\\
&= 0 \qquad \text{because } b = a^2/4.
\end{aligned}$$

Equation (5.74) is linear, so any constant multiple of $x_1(t)$ is also a solution. However, $x_1$ is not a general solution, as the next example illustrates.

✦ *Example 1*   Let

$$x'' + 2x' + x = 0. \tag{5.76}$$

The characteristic equation is

$$\lambda^2 + 2\lambda + 1 = 0,$$

which has the double root $\lambda_1 = \lambda_2 = -1$. Therefore, $x(t) = ce^{-t}$ is a solution of Eq. (5.76). However, it is not a general solution, for the following reason. Select two initial conditions, say $x(0) = 1$ and $x'(0) = 2$. The initial condition $x(0) = 1$ implies that $c = 1$. Having chosen $c$, we find that there is no way to satisfy the initial condition $x'(0) = 2$. Moreover, it is not possible to circumvent the difficulty by rewriting the solution as

$$x(t) = c_1 e^{-t} + c_2 e^{-t}.$$

The initial conditions imply that

$$\begin{aligned}
1 &= c_1 + c_2 \qquad \text{because } x(0) = 1,\\
2 &= -c_1 - c_2 \qquad \text{because } x'(0) = 2.
\end{aligned}$$

These two equations form an *inconsistent* system for $c_1$ and $c_2$: It is not possible to choose two constants that satisfy both relations simultaneously.   ∎

To solve equations like (5.76), we need to find a second solution $x_2$ that somehow is "different" from $x_1$. More precisely, given a pair of initial conditions $x(0) = x_0$

and $x'(0) = v_0$, the second solution must allow us to find a consistent system of linear equations for the constants $c_1$ and $c_2$. We say that the solutions $x_1$ and $x_2$ must be *linearly independent*. (Section 5.9 discusses this theoretical question in more detail.)

To find a second solution of Eq. (5.74), we treat the equation as a differential operator (see Chapter 4). Equation (5.74) defines the linear differential operator $L$,

$$L(x) = \left( \frac{d^2}{dt^2} + a\frac{d}{dt} + b \right)(x). \tag{5.77}$$

The function $x_1(t) = e^{-at/2}$ solves Eq. (5.74) because $L(e^{-at/2}) = 0$.

There is an apparent correspondence between the form of the operator $L$ given in Eq. (5.77) and the characteristic equation (5.75). In particular, we can rewrite Eq. (5.75) as

$$(\lambda + r)^2 = 0,$$

where $r = a/2$. This suggests that the operator $L$ can be written as the composition

$$L(x) = \mathrm{Op}(\mathrm{Op}(x)),$$

where

$$\mathrm{Op}(x) = \left( \frac{d}{dt} + r \right)(x) = 0,$$

as we now verify:

$$\begin{aligned}
\mathrm{Op}(\mathrm{Op}(x)) &= \mathrm{Op}\big(x' + rx\big) \\
&= \big(x' + rx\big)' + r\big(x' + rx\big) \\
&= x'' + 2rx' + r^2 x \\
&= x'' + ax' + bx
\end{aligned}$$

because $r = a/2 = \sqrt{b}$.

The operator Op maps the function $x_1(t) = e^{-rt}$ to the zero function, because $x_1$ is a solution of the corresponding differential equation $x' + rx = 0$. Now suppose that $x_2$ is a function for which $\mathrm{Op}(x_2) = x_1$. Then

$$\begin{aligned}
L(x_2) &= \mathrm{Op}(\mathrm{Op}(x_2)) \\
&= \mathrm{Op}(x_1) \\
&= 0.
\end{aligned}$$

Thus, if $\mathrm{Op}(x_2) = x_1$, then $x_2$ also solves Eq. (5.74). This observation allows us to determine a formula for $x_2$ as follows.

The relation $\mathrm{Op}(x_2) = x_1$ is equivalent to the first-order differential equation

$$x' + rx = e^{-rt}, \tag{5.78}$$

which can be solved by using the method of integrating factors discussed in Section 2.3. We multiply both sides of Eq. (5.78) by $e^{rt}$ to obtain

$$e^{rt}(x' + rx) = e^{rt}e^{-rt} = 1,$$

which implies that

$$\frac{d}{dt}(e^{rt}x) = 1.$$

Therefore, $e^{rt}x = t + C$, so a solution of Eq. (5.78) is

$$x(t) = te^{-rt} + Ce^{-rt}.$$

The derivation of $x$ ensures that $L(x) = 0$. Because $L$ is linear,

$$L\left(te^{-rt}\right) + L\left(Ce^{-rt}\right) = 0$$

for any constant $C$. In particular, if $C = 0$, then $L(te^{-rt}) = 0$. Hence, the function $x_2(t) = te^{-rt}$ is a second solution of Eq. (5.74).

◆ *Example 2*   Let

$$x'' + 2x' + x = 0.$$

One solution is $x_1(t) = c_1 e^{-t}$, as shown in Example 1. The preceding discussion implies that a second solution is

$$x_2(t) = te^{-t},$$

as we can verify by substitution:

$$\begin{aligned}
x_2'' + 2x_2' + x_2 &= \left(te^{-t}\right)'' + 2\left(te^{-t}\right)' + te^{-t} \\
&= \left(te^{-t} - 2e^{-t}\right) + 2\left(e^{-t} - te^{-t}\right) + te^{-t} \\
&= 0.
\end{aligned}$$

Moreover, the function

$$x(t) = c_1 e^{-t} + c_2 te^{-t}$$

is a general solution. Let $x(0) = x_0$ and $x'(0) = v_0$. Then

$$x_0 = c_1,$$
$$v_0 = -c_1 + c_2,$$

which has the unique solution $c_1 = x_0$ and $c_2 = x_0 + v_0$.                ■

We summarize our discussion in the following theorem.

**Theorem 5.6**  *Let*

$$\frac{d^2x}{dt^2} + a\frac{dx}{dt} + bx = 0, \tag{5.79}$$

*where $a^2 - 4b = 0$. Then the corresponding characteristic equation has the double real root $\lambda = -a/2$. The general solution of Eq. (5.79) is*

$$x(t) = e^{-at/2}(c_1 + tc_2) \tag{5.80}$$

*for appropriate constants $c_1$ and $c_2$ that depend on the initial conditions.*

✦ *Example 3*    Consider the initial value problem

$$x'' + x' + \tfrac{1}{4}x = 0, \qquad x(0) = 1, \qquad x'(0) = 1. \tag{5.81}$$

The corresponding characteristic equation has the double root $\lambda = -1/2$. Thus, one solution of Eq. (5.81) is $x_1(t) = e^{-t/2}$. A second solution is $x_2(t) = te^{-t/2}$. Because Eq. (5.81) is linear, the superposition principle implies that the linear combination

$$x(t) = c_1 x_1(t) + c_2 x_2(t) = e^{-t/2}(c_1 + c_2 t)$$

is also a solution of Eq. (5.81). The initial conditions imply $c_1 = 1$ and $c_2 = 3/2$. (The details are left as an exercise.) Therefore, the solution of Eq. (5.81) is

$$x(t) = e^{-t/2}\left(1 + \tfrac{3}{2}t\right). \tag{5.82}$$

Figure 5.14 shows a plot of Eq. (5.82). Notice that the solution increases initially but gradually decreases to 0 with time. In fact, l'Hôpital's rule implies that $\lim_{t\to\infty} x(t) = 0$. Notice also that the solution does *not* oscillate.

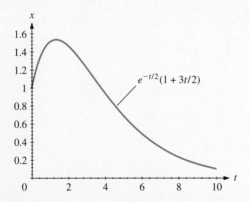

**FIGURE 5.14**   A plot of Eq. (5.82).

✦ *Example 4*  The motion of the damped linear pendulum is governed by the equation

$$m\phi'' + \left(\frac{c}{\ell}\right)\phi' + \left(\frac{mg}{\ell}\right)\phi = 0. \tag{5.83}$$

The roots of the corresponding characteristic equation are

$$\lambda = -\frac{c \pm \sqrt{c^2 - 4\ell m^2 g}}{2\ell m}.$$

Therefore, the pendulum is critically damped if $c^2 = 4\ell m^2 g$, in which case $\lambda = -c/(2\ell m)$. The general solution of Eq. (5.83) is

$$x(t) = e^{-tc/(2\ell m)}(c_1 + tc_2),$$

where the constants $c_1$ and $c_2$ are determined by the initial conditions.

The motion of a critically damped pendulum is similar to that of an overdamped pendulum. In particular,

• the bob does not oscillate around the equilibrium ($x = 0$) position;

• as $t \to \infty$, the bob tends to the equilibrium position; and

• the bob crosses the equilibrium position at most once.

It is difficult in practice to achieve critical damping, because the relation $c^2 = 4m^2 g\ell$ must be satisfied precisely. Even if we assume that the frictional forces are perfectly proportional to the current angular velocity of the pendulum, any error in measuring the mass or the length of the pendulum means that we are likely to be in the overdamped or underdamped case.                                                                        ∎

## Exercises

**1.** Suppose that each of the initial value problems in the following list models an idealized mass-spring system. For each initial value problem,

• find the solution;

• determine whether the mass ever crosses the equilibrium and, if so, determine the velocity of the mass at the instant it crosses the equilibrium;

• determine whether the mass ever crosses the equilibrium if the stated initial velocity is cut in half.

(a) $x'' + 2x' + x = 0$, $x(0) = 2$, $x'(0) = -3$

(b) $x'' + 4x' + 4x = 0$, $x(0) = 2$, $x'(0) = -4$

(c) $4x'' + 4x' + x = 0$, $x(0) = 2$, $x'(0) = -4$

**2.** Let $x'' + ax' + bx = 0$, where $a$ and $b$ are constants.

(a) What must be the relationship between $a$ and $b$ for the roots of the characteristic equation to be real and equal?

(b) If $a$ and $b$ are positive and satisfy the relationship that you found in (a), show that

$\lim_{t\to\infty} x(t) = 0$, regardless of the initial conditions.

(c) Let the initial conditions be $x(0) = x_0$ and $x'(0) = v_0$, and assume that at least one of $x_0$ and $v_0$ is nonzero. Assuming that $a$ and $b$ satisfy the relation in (a), for which values, if any, of $a$ and $b$ is $\lim_{t\to\infty} x(t)$ infinite? (Don't assume, as in (b), that $a$ and $b$ are both positive.)

**3.** Consider a mass-spring system that is modeled by Eq. (5.73).

(a) Determine the relation that $m$, $c$, and $k$ must satisfy if the motion is critically damped.

(b) Show by substitution that

$$x_1(t) = e^{\lambda_1 t} = e^{-ct/2m}$$

solves Eq. (5.73) if $m$, $c$, and $k$ satisfy the relation in (a).

**4.** Let a linear mass-spring system be governed by

$$x'' + 2x' + x = 0.$$

(a) Suppose $x(0) = 1$. For what values, if any, of the initial velocity $x'(0)$ will the mass cross the equilibrium at least once?

(b) Suppose $x(0) = 1$. For what values, if any, of the initial velocity $x'(0)$ will the mass cross the equilibrium at most once?

**5.** Suppose that a mass-spring system is governed by the equation $x'' + 2x' + x = 0$.

(a) Find the solution satisfying the initial conditions $x(0) = x_0$ and $x'(0) = v_0$.

(b) Prove or disprove: If $x_0$ and $v_0$ have the same sign, then the mass never crosses the equilibrium.

(c) Determine the set of initial conditions for which the mass crosses the equilibrium at least once.

**6.** Consider an ideal linear pendulum of length 1 m whose bob has mass $m$ and for which the coefficient of friction at the pivot is $c = 1$ N s/m.

(a) For what values of $m$, if any, does the pendulum cross the equilibrium more than once?

(b) For what values of $m$, if any, does the pendulum cross the equilibrium at most once?

**7.** Suppose that the hypotheses of Theorem 5.6 hold. Show that a unique pair of constants $c_1$ and $c_2$ can be determined so that the general solution (5.80) satisfies the initial conditions $x(0) = x_0$ and $x'(0) = v_0$. (Suggestion: Examine the linear system of equations that arises when $t = 0$ is substituted into $x(t)$ and $x'(t)$. Under what conditions is the linear system guaranteed to have a unique solution?)

**8.** Consider an LRC circuit consisting of an inductor of 1 H, a resistor of 2000 $\Omega$, and a capacitor of 1 $\mu$F. Suppose that initially there is no charge in the circuit and that the current initially is 1 A.

(a) Identify the corresponding initial value problem.

(b) Show that the circuit is critically damped, and find an expression for the charge in the circuit as a function of time.

(c) Electrical components frequently vary by several percent from their rated values. Suppose, for instance, that the actual resistance of the resistor is 2100 $\Omega$. Is such a circuit overdamped, underdamped, or critically damped? Find an expression for the charge in the circuit as a function of time.

(d) Suppose that the actual resistance of the resistor is 1900 $\Omega$ and repeat part (c).

(e) Plot the three solutions that you have found on the same axes. Argue that the charge remaining in the circuit after 10 milliseconds is negligible.

(f) In what sense is it reasonable to speak of a system as being "nearly critically damped"?

**9.** Let the position $x(t)$ of a mass on a linear spring be governed by the equation $x'' + cx' + x = 0$ with the initial conditions $x(0) = 2$ and $x'(0) = -3$.

(a) Find $x(t)$ assuming that $c = 2$. (See Exercise 1(a).)

(b) Find $x(t)$ assuming that $c = 1.9$. Is the equation overdamped or underdamped?

(c) Graph your solutions in parts (a) and (b) on the same axes for $0 \le t \le 10$. Compare and contrast the plots that you obtain.

(d) In principle, how many times does the mass cross the equilibrium position in (a)?

(e) In principle, how many times does the mass cross the equilibrium position in (b)?

(f) Despite the differences in the solutions that are quantified in (d)–(e), it is reasonable to regard the differences as "negligible" for most practical purposes. Explain why.

## 5.6 FORCING AND THE METHOD OF UNDETERMINED COEFFICIENTS

Section 5.2 shows that a mass-spring system with linear friction, the linear pendulum, and a simple LRC electrical circuit have identical mathematical descriptions, at least in the idealized case. This mathematical equivalence still holds if external forces are applied. For example, an external force might be a torque applied at the pivot of a linear pendulum or an electromotive force applied to an electrical circuit.

Various kinds of oscillating electrical circuits are of considerable technological importance. Electricity is distributed from generating stations through the power grid as alternating current; radio and television tuners contain circuits that oscillate at selected frequencies; digital computers are controlled by clocks that cycle power through the system in a periodic fashion millions of times per second.

Although many of these circuits have complex mathematical descriptions that are beyond the scope of this text, it is possible to describe the behavior of LRC circuits that are subjected to externally imposed potentials. These circuits are modeled by differential equations of the form

$$x'' + ax' + bx = f(t), \tag{5.84}$$

called a *forced* second-order equation.[5] The function $f(t)$ is called the *forcing*; in practice, $f$ corresponds to an applied external potential or other source of energy to the system. Equation (5.84) is linear, so the theory developed in Chapter 4 applies. We may regard the general solution of Eq. (5.84) as the sum of a homogeneous and a particular solution; that is, the general solution has the form

$$x(t) = x_h(t) + x_p(t),$$

---

[5] In the following discussion, we assume that $b \ne 0$ unless explicitly noted otherwise. The equation $x'' + ax' = f(t)$ is a second-order equation; but if we define $y(t) = x'(t)$, then it is also equivalent to the first-order equation $y' + ay = f(t)$.

where $x_h$ is a solution of the homogeneous equation

$$x'' + ax' + bx = 0$$

associated with Eq. (5.84).

The previous three sections have explained how to derive solutions of homogeneous second-order equations, at least in the case in which the coefficients $a$ and $b$ are constant. This section considers the question of how to derive solutions of the nonhomogeneous equation (5.84).

There are two types of forcing functions of primary technological interest:

- constant functions and

- linear combinations of sine and cosine functions.

Constant forcing functions arise in models of electrical circuits to which a constant electromagnetic potential is applied, such as when the circuit is connected to a battery. Forcing functions that are linear combinations of sine and cosine functions often arise when a periodically varying potential is applied. (Household electricity is an everyday example of such a potential.) Periodic forcing is particularly interesting; Sections 5.7 and 5.8 discuss this subject in detail.

Because of the special interest in constant and periodic forcing, we restrict attention to the solution of Eq. (5.84) when $f$ has the form $f(t) = $ constant or $f(t) = A \cos \omega t + B \sin \omega t$. The so-called *method of undetermined coefficients* can be used to solve Eq. (5.84) when $f$ is one of these two types. (In fact, the method is applicable to a broader class of forcing functions; the appendix contains the details.)

The method of undetermined coefficients is simply a matter of making educated guesses about the form of a particular solution. Equation (5.84) defines the linear operator $L(x) = x'' + ax' + bx$. If $x$ is a constant function, say $x(t) = A$, then

$$L(x) = L(A) = bA,$$

because the derivatives of a constant function are 0. In other words, $L$ maps constant functions to constant functions. Thus, if $E$ is a constant, then a particular solution of the equation

$$x'' + ax' + bx = E$$

must also contain a constant term.

✦ *Example 1*   Let

$$x'' + 10x' + 9x = 18. \qquad (5.85)$$

The corresponding differential operator is $L(x) = x'' + 10x' + 9x$. Let $x(t) = A$ be a constant function; then $L(x) = L(A) = 9A$. To solve Eq. (5.85), we must have

$$L(A) = 9A = 18.$$

A particular solution, therefore, is

$$x_p(t) = A = 2.$$

Let $x_h$ be a solution of the homogeneous equation associated with Eq. (5.85); then $L(x_h) = 0$, by definition. Therefore, any function of the form $x_p + x_h$ is again a solution of Eq. (5.85), because $L$ is linear. The characteristic equation associated with Eq. (5.85) is

$$\lambda^2 + 10\lambda + 9 = 0,$$

and the roots are $\lambda_1 = -1$ and $\lambda_2 = -9$. Hence, the homogeneous solution has the form

$$x_h(t) = c_1 e^{-t} + c_2 e^{-9t}.$$

The general solution of Eq. (5.85) is

$$x(t) = c_1 e^{-t} + c_2 e^{-9t} + 2.$$

The values of $c_1$ and $c_2$ can be adjusted to satisfy any initial conditions. Suppose that $x(0) = 16$ and $x'(0) = 2$. The initial condition $x(0) = 16$ implies that

$$16 = c_1 + c_2 + 2, \tag{5.86}$$

and the initial condition $x'(0) = 2$ implies that

$$2 = -c_1 - 9c_2. \tag{5.87}$$

Equations (5.86) and (5.87) imply that $c_1 = 16$ and $c_2 = -2$. Therefore, the solution of the initial value problem is

$$x(t) = 16e^{-t} - 2e^{-9t} + 2.$$

In most cases, a particular solution corresponding to periodic forcing can be determined in an analogous manner. Let

$$L(x) = x'' + ax' + bx. \tag{5.88}$$

Then

$$L(\cos \omega t) = \left(b - \omega^2\right) \cos \omega t - a\omega \sin \omega t,$$

and

$$L(\sin \omega t) = \left(b - \omega^2\right) \sin \omega t + a\omega \cos \omega t.$$

In other words, $L$ maps $\cos \omega t$ and $\sin \omega t$ to a linear combination of $\cos \omega t$ and $\sin \omega t$. This observation implies that a particular solution of the equation

$$x'' + ax' + bx = A \cos \omega t + B \sin \omega t$$

also is a linear combination of $\cos \omega t$ and $\sin \omega t$. (There is an exception to this general statement; see Example 5.)

◆ *Example 2*   Let

$$x'' + 10x' + 9x = 85 \sin 2t. \tag{5.89}$$

We have $L(x) = x'' + 10x' + 9x$. The previous discussion suggests that a particular solution has the form

$$x_p(t) = A \sin 2t + B \cos 2t.$$

We substitute $x_p$ into Eq. (5.89) and find that

$$L(x_p) = x_p'' + 10x_p' + 9x_p = (5A - 20B) \sin 2t + (20A + 5B) \cos 2t.$$

Therefore, $x_p$ is a solution of Eq. (5.89) if

$$5A - 20B = 85 \qquad \text{and} \qquad 20A + 5B = 0.$$

(Simply equate the coefficients of the sine and cosine terms.) The latter two conditions imply that $A = 1$ and $B = -4$. Therefore, a particular solution is

$$x_p(t) = \sin 2t - 4 \cos 2t.$$

The general solution is

$$x(t) = c_1 e^{-t} + c_2 e^{-9t} + \sin 2t - 4 \cos 2t,$$

because the homogeneous solution is the same as in Example 1.

We can choose $c_1$ and $c_2$ to satisfy any set of initial conditions. Suppose that $x(0) = 2$ and $x'(0) = 8$. The former condition implies that

$$2 = c_1 + c_2 - 4,$$

and the latter implies that

$$8 = -c_1 - 9c_2 + 2.$$

These two equations can be solved simultaneously to obtain $c_1 = 15/2$ and $c_2 = -3/2$. Therefore, the solution of the initial value problem is

$$x(t) = \tfrac{15}{2}e^{-t} - \tfrac{3}{2}e^{-9t} + \sin 2t - 4 \cos 2t. \qquad \blacksquare$$

The linearity of the differential operator can be exploited to find particular solutions of more complicated forcing functions, as the next example shows.

✦ *Example 3*   Let

$$x'' + 10x' + 9x = 18 + 85 \sin 2t, \tag{5.90}$$

where $L(x) = x'' + 10x' + 9x$. Here, the forcing function may be regarded as the sum

$$f(t) = f_1(t) + f_2(t),$$

where $f_1(t) = 18$ and $f_2(t) = 85 \sin 2t$. Let $x_p^{(1)}$ be a particular solution of the equation

$$L(x) = f_1 = 18,$$

and let $x_p^{(2)}$ be a particular solution of the equation

$$L(x) = f_2 = 85 \sin 2t.$$

Then a particular solution of Eq. (5.90) is

$$x_p = x_p^{(1)} + x_p^{(2)},$$

because the linearity of $L$ implies that

$$\begin{aligned} L\left(x_p^{(1)} + x_p^{(2)}\right) &= L\left(x_p^{(1)}\right) + L\left(x_p^{(2)}\right) \\ &= f_1 + f_2 \\ &= 18 + 85 \sin 2t. \end{aligned}$$

Examples 1 and 2 show that $x_p^{(1)} = 2$ and $x_p^{(2)} = \sin 2t - 4 \cos 2t$. Therefore, a particular solution of Eq. (5.90) is

$$x_p(t) = 2 + \sin 2t - 4 \cos 2t.$$

The general solution of Eq. (5.90) is

$$x(t) = c_1 e^{-t} + c_2 e^{-9t} + 2 + \sin 2t - 4 \cos 2t.$$

As in the previous examples, the constants $c_1$ and $c_2$ can be chosen to satisfy any initial conditions. ∎

✦ *Example 4*   Let

$$x'' + 2x' + x = 25 \cos 2t + 50 \sin 3t. \tag{5.91}$$

The right-hand side of Eq. (5.91) is a forcing function that is the sum of two sine and cosine functions of *different* periods. Equation (5.91) defines the linear operator

$L(x) = x'' + 2x' + x$. If we can find functions $x_p^{(1)}$ and $x_p^{(2)}$ such that

$$L\left(x_p^{(1)}\right) = 25\cos 2t \qquad \text{and} \qquad L\left(x_p^{(2)}\right) = 50\sin 3t,$$

then the linearity of $L$ implies that a particular solution of Eq. (5.91) is

$$x_p(t) = x_p^{(1)}(t) + x_p^{(2)}(t).$$

See Exercise 4.                                                                                    ∎

Examples 3 and 4 imply that the method of undetermined coefficients can be applied to each piece of a given forcing function. Roughly speaking, we break up $f(t)$ as needed into a sum of the form

$$f(t) = \text{constant} + f_1(t) + f_2(t) + \cdots + f_n(t),$$

where the terms $f_1, \ldots, f_n$ are linear combinations of sine and cosine functions, each with distinct periods.

◆ *Example 5*   The method of undetermined coefficients as described above can fail in certain situations, as this example shows. Let

$$x'' + 25x = 10\cos 5t, \tag{5.92}$$

which defines the linear operator $L(x) = x'' + 25x$.

The discussion above implies that a particular solution has the form

$$x_p(t) = A\cos 5t + B\sin 5t.$$

However,

$$L(x_p) = x_p'' + 25x_p$$
$$= (-25A\cos 5t - 25B\sin 5t) + 25(A\cos 5t + B\sin 5t)$$
$$= 0.$$

In other words, $x_p$ is a homogeneous solution. Therefore, the above guess for $x_p$ is not correct. The method of undetermined coefficients can be applied to Eq. (5.92), but no particular solution contains a linear combination of $\cos 5t$ and $\sin 5t$. Section 5.7 discusses this situation in more detail.                                                        ∎

The method of undetermined coefficients can be cast as an algorithm as follows:

1. Write the differential equation in the form

$$L(x) = x'' + ax' + bx = f(t). \tag{5.93}$$

(Note: The differential equation must be linear with constant coefficients. See Exercise 9.)

2. If $f(t)$ is a constant function, then guess a particular solution of the form $x_p(t) = A$. Substitute $A$ into Eq. (5.93) and determine the value of $A$ that leads to an equality.

3. If $f(t)$ is a function of the form

$$f(t) = \alpha \cos \omega t + \beta \sin \omega t,$$

where $\alpha$, $\beta$, and $\omega$ are constants, then guess a particular solution of the form

$$x_p(t) = A \cos \omega t + B \sin \omega t.$$

If $L(x_p) = 0$, that is, if $x_p$ is also a homogeneous solution of Eq. (5.93), then the guess for $x_p$ must be modified in a way that is described in Section 5.7. Otherwise, substitute $x_p$ into the left-hand side of Eq. (5.93). Equate the coefficients of each cosine term and each sine term, and determine the values of $A$ and $B$ accordingly.

4. If $f$ is a sum of the form

$$f(t) = \text{constant} + f_1(t) + \cdots + f_n(t),$$

where $f_i(t) = \alpha_i \cos \omega_i t + \beta_i \sin \omega_i t$, then apply Steps 2 and 3 to each term in the sum. That is, first find a particular solution $x_p^{(0)}$ that solves

$$L\left(x_p^{(0)}\right) = \text{constant},$$

then find a particular solution $x_p^{(1)}$ that solves

$$L\left(x_p^{(1)}\right) = f_1(t),$$

and so on. A particular solution that solves $L(x) = f(t)$ is

$$x_p(t) = x_p^{(0)} + x_p^{(1)} + \cdots + x_p^{(n)}.$$

5. Determine a homogeneous solution; that is, find a solution $x_h$ of the equation $L(x) = 0$. This solution contains two arbitrary constants when $L$ defines a second-order linear differential operator.

6. The general solution of Eq. (5.93) has the form

$$x(t) = x_h(t) + x_p(t).$$

If initial conditions are specified, then substitute them into the general solution and solve for the arbitrary constants.

## Exercises

**1.** Find the solution of the differential equation

$$x'' + 2x' + x = 2$$

that satisfies the initial conditions $x(0) = 1$ and $x'(0) = -1$.

**2.** Find the solutions of each of the following differential equations that satisfy the initial conditions $x(0) = 0$ and $x'(0) = 1$:

(a) $x'' + 4x = 2$

(b) $x'' + 3x = 1$

**3.** Use the method of undetermined coefficients where possible to find the solutions of each initial value problem in the following list. If the method fails or is not applicable, then indicate the reason for the difficulty.

(a) $x'' + 4x = 3\cos t$, $x(0) = 1$, $x'(0) = 1$

(b) $x'' + 4x = 3\cos 2t$, $x(0) = 1$, $x'(0) = 1$

(c) $x'' + 3x' + 2x = 10\sin t$, $x(0) = 1$, $x'(0) = -1$

(d) $x'' - 4x = 8 + 16\sin 2t$, $x(0) = 1$, $x'(0) = -2$

(e) $x'' + \sin x = 4$, $x(0) = 1$, $x'(0) = 0$

(f) $x'' + 5x' + 6x = 4 + 10\cos t$, $x(0) = 2$, $x'(0) = 3$

**4.** Find the solution of the equation

$$x'' + 2x' + x = 25\cos 2t + 50\sin 3t$$

that satisfies the initial conditions $x(0) = 1$ and $x'(0) = 2$.

**5.** The method of undetermined coefficients works just as well for first-order equations as for second-order equations. Illustrate this statement by applying the method to solve the initial value problem

$$x' + x = \cos t, \quad x(0) = 2.$$

**6.** This exercise illustrates a substitution that replaces a second-order equation with an equivalent first-order equation in special cases.

(a) Let

$$x'' + ax' = 3, \tag{5.94}$$

where $a$ is a constant. Show that the substitution $y(t) = x'(t)$ makes Eq. (5.94) equivalent to the first-order equation $y' + ay = 3$.

(b) Use the method of undetermined coefficients to find the general solution of $y' + ay = 3$, and use your result to find the general solution of Eq. (5.94).

(c) Use your results in (a) and (b) to find the general solution of the equation $x'' + ax' = 3 + \cos t$.

**7.** Let $x'' + ax' + bx = \cos t$, where $a$ and $b$ are nonzero constants. A common error in applying the method of undetermined coefficients is to guess a particular solution of the form $x_p(t) = A\cos t$. Explain why this guess does not work and why the correct guess must be of the form $x_p(t) = A\cos \omega t + B\sin \omega t$ instead.

**8.** Let

$$x'' + ax' + bx = f(t), \tag{5.95}$$

where $a$ and $b$ are constants. This exercise asks you to show that the method of undetermined coefficients can always be used to find a particular solution of Eq. (5.95) under certain circumstances.

(a) Suppose $f(t) = \alpha \cos \omega t + \beta \sin \omega t$, where $\alpha$, $\beta$, and $\omega$ are constants. Show that if $a \neq 0$, then a guess of the form $x_p(t) = A\cos \omega t + B\sin \omega t$ always yields a particular solution.

(b) Let $f$ be as in part (a), and suppose that $a = 0$ and $b \neq 0$. Under what circumstances does a guess of the form $x_p(t) = A\cos \omega t + B\sin \omega t$ fail to yield a particular solution?

(c) Let $f$ be as in part (a). Show that with the exception of the circumstances that you found in (b), the method of undetermined coefficients yields a particular solution when $a = 0$ and $b \neq 0$.

(d) Let $f$ be a constant function. Show that the method of undetermined coefficients always yields a particular solution if $b \neq 0$.

9. The linearity of the differential operator is es-sential to the method of undetermined coefficients. Show that a guess of the form $x_p = A \cos t + B \sin t$ does *not* yield a particular solution of the differential equation $x'' + x^2 = \cos t$.

## 5.7 FORCED, UNDAMPED, LINEAR OSCILLATORS

### Introduction

As is mentioned in Section 5.6, a common example of a forced oscillator is an electrical circuit to which an external potential is applied. Sometimes it is useful to regard the applied potential as the input signal to the circuit and the response of the circuit as the output signal. In the case of an LRC circuit, the situation can be modeled by a linear second-order differential equation of the form

$$Lx'' + Rx' + x/C = f(t),$$

where $L$, $R$, and $C$ are constants. The function $f(t)$ often is called the *forcing function* or simply the *forcing*. The solution $x(t)$ often is called the *response*. Figure 5.15 illustrates the situation schematically.

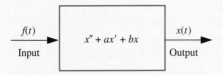

$f(t)$     $x'' + ax' + bx$     $x(t)$

Input                     Output

FIGURE 5.15    A representation of an electrical circuit as an input-output system.

In the rest of this section, we discuss the behavior of systems in which the damping is negligible and a periodic forcing is applied. That is, we investigate equations of the form

$$x'' + \omega^2 x = f(t), \tag{5.96}$$

where $f$ is a sine or cosine function. (Section 5.8 considers equations with damping.) Our interest in periodic forcing functions is motivated in part by the technological and scientific applications of oscillating systems. We will address the following questions:

1. How do the frequency and amplitude of the forcing affect the response?

2. How does the period of the response depend on the period of the forcing?

3. How can we characterize the oscillations of the response?

As we will see, the homogeneous solution $x_h$ of Eq. (5.96) is important to the analysis. Recall that $x_h$ solves the equation for simple harmonic motion, $x'' + \omega^2 x = 0$. As was discussed in Section 5.2, $x_h$ is periodic with period $2\pi/\omega$. There are two basic cases that we consider in turn:

- If $f$ is a sine or cosine function whose period is *not* equal to the period of $x_h$, then the solutions of Eq. (5.96) produce a phenomenon called *beating*.

- If $f$ is a sine or cosine function whose period is *equal* to the period of $x_h$, then the solutions of Eq. (5.96) produce a phenomenon called *resonance*.

## Remarks on Finding Solutions

The method of undetermined coefficients, discussed in Section 5.6, can be used to solve equations like Eq. (5.96) when $f$ is a constant function, sine, cosine, or sum of such terms. The method of Laplace transforms, discussed in Chapter 10, provides an alternative solution method.

One disadvantage of these symbolic methods is that they often involve lengthy algebraic calculations, particularly if the coefficients in the differential equation are not small integers. A computer algebra system can compute the answers quickly and reliably in such cases. Alternatively, numerical methods can be used to approximate the solutions.

In the remainder of this chapter, we are not concerned with the details of how the solutions are computed. Instead, our focus is on the interpretation of the solutions to understand what they imply about the behavior of the underlying system.

## The Phenomenon of Beats

The solid curve in Fig. 5.16(a) is the graph of the solution of the initial value problem

$$x'' + 25x = 11\cos 6t, \qquad x(0) = 0, \qquad x'(0) = 0. \tag{5.97}$$

Notice the following features of the solution:

- Although the initial conditions in Eq. (5.97) are both zero, the solution is not the zero function. In contrast, if there is no forcing, then zero initial conditions imply a zero solution.

- Although the oscillations are more complicated than those of a simple sine or cosine function, the solution is periodic with a period that is slightly larger than 6. (In fact, the period is $2\pi$, as we will see.)

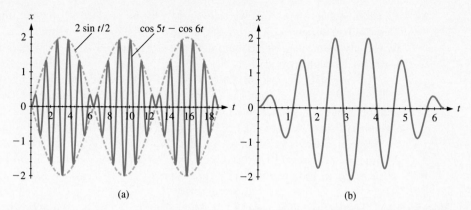

**FIGURE 5.16** (a) The solid curves indicate the solution of Eq. (5.97); the dashed curves indicate the envelope. (b) A magnified view of the solution.

- The successive peaks in the solution form clusters that appear to be contained within the graph of envelope functions, indicated in Fig. 5.16 by the dashed curves. In other words, the relative extrema of the oscillations vary in a periodic way.
- The oscillations of the solution appear to consist of two types: a "fast oscillation" that crosses the $t$ axis very frequently and a "slow oscillation" (indicated by the envelope) that crosses the $t$ axis once during each period of the solution.

The solution illustrated in Fig. 5.16(a) is an example of a phenomenon called *beating*. Beating phenomena arise when two oscillators of different frequencies interact.

We can analyze functions like the one depicted in Fig. 5.16(a) in considerable detail. The following two identities are useful for this purpose:

$$\cos\alpha - \cos\beta = -2\sin\left(\frac{\alpha+\beta}{2}\right)\sin\left(\frac{\alpha-\beta}{2}\right), \tag{5.98}$$

$$\sin\alpha - \sin\beta = 2\cos\left(\frac{\alpha+\beta}{2}\right)\sin\left(\frac{\alpha-\beta}{2}\right). \tag{5.99}$$

✦ *Example 1*   The solution of the initial value problem (5.97) is

$$x(t) = \cos 5t - \cos 6t, \tag{5.100}$$

which is $2\pi$-periodic. (See the appendix to this section for details on how to compute the period.) The identity (5.98) gives an equivalent expression for $x$:

$$x(t) = 2\sin\tfrac{11}{2}t \sin\tfrac{1}{2}t. \tag{5.101}$$

Equation (5.101) shows that we can regard $x$ as the product of two sine functions: a "fast oscillation," given by $\sin \frac{11}{2} t$, and a "slow oscillation," given by $2 \sin \frac{1}{2} t$. The latter function is the envelope of the solution (5.100).[6] The period of the envelope is $4\pi$, which is exactly twice the period of $x$.

The period of the fast oscillation is $4\pi/11$ and is reflected in the graph of $x$ in the following way. The function $\sin \frac{11}{2} t$ completes 11 full periods in the interval $[0, 4\pi]$, and the graph of the function crosses the $t$ axis twice during each period. Therefore, the value of $\sin \frac{11}{2} t$ is zero at 22 equally spaced values of $t$ in $[0, 4\pi]$. This implies that the graph of $x$ also is zero at those points. Therefore, the graph of $x$ crosses the $t$ axis 11 times in the interval $[0, 2\pi]$, which is one full period of $x$. Figure 5.16(b) shows the graph of Eq. (5.100) over $[0, 2\pi]$.                                                                  ∎

The identity (5.98) used in the above analysis can be applied only when the two cosine terms have equal amplitudes, as in Eq. (5.100). Nevertheless, beating phenomena are present even when the cosine terms do not have the same amplitude, although the relevant formulas are more complicated. See Exercise 12.

✦ *Example 2*     The analysis presented in the previous example can be generalized. Consider the initial value problem

$$x'' + 25x = A \cos \alpha t, \qquad x(0) = 0, \qquad x'(0) = 0. \tag{5.102}$$

An equation like (5.102) might arise as a model of an LC electrical circuit to which a periodic potential is applied. An interesting question in such cases is how the circuit responds as the amplitude and frequency of the potential are varied.

If $\alpha \neq 5$, then the solution of Eq. (5.102) is

$$x(t) = \left( \frac{A}{\alpha^2 - 25} \right) (\cos 5t - \cos \alpha t), \tag{5.103}$$

which we may express equivalently as

$$x(t) = \left( \frac{2A}{\alpha^2 - 25} \right) \sin \left( \frac{(\alpha + 5)t}{2} \right) \sin \left( \frac{(\alpha - 5)t}{2} \right). \tag{5.104}$$

We can regard $x$ as a product of a "fast oscillation" of period $4\pi/(\alpha + 5)$ and a "slow oscillation" of period $4\pi/|\alpha - 5|$. (We set $\alpha = 6$ in Example 1.) The envelope of the solution is

$$\mathrm{env}(t) = \left( \frac{2A}{\alpha^2 - 25} \right) \sin \left( \frac{(\alpha - 5)t}{2} \right).$$

---

[6]In fact, the graphs of both $2 \sin \frac{1}{2} t$ and $-2 \sin \frac{1}{2} t$ are shown in Fig. 5.16(a).

This analysis demonstrates the following facts about the forced oscillator described by Eq. (5.102):

- As the frequency $\alpha$ of the applied potential increases from 0 to 5, the amplitude of the response increases, and the period also increases.
- As the frequency $\alpha$ approaches 5, the amplitude of the response tends to infinity, and the period also tends to infinity. (The case in which $\alpha = 5$ leads to a situation called *resonance*, which is described in more detail below.)
- As the frequency $\alpha$ of the applied potential increases past 5, the amplitude of the response decreases, and the period also decreases.

Figure 5.17 shows the graphs of Eq. (5.103) for a fixed value of $A$. Figures 5.17(a) and (b) show what happens for two values of $\alpha$ that approach the critical value of 5.

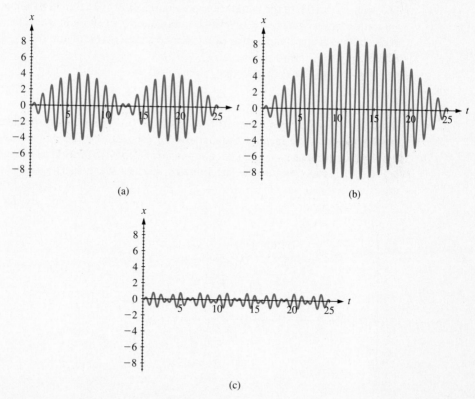

**FIGURE 5.17** The graph of Eq. (5.103) for $A = 11$ and (a) $\alpha = 5.5$; (b) $\alpha = 5.25$; (c) $\alpha = 7$.

Notice that the period of the solution is twice as great for $\alpha = 5.25$ as it is for $\alpha = 5.5$. The amplitude of the solution also is much larger for $\alpha = 5.25$ than for $\alpha = 5.5$. If we choose a larger value of $\alpha$, such as $\alpha = 7$, then the amplitude is relatively small and the period of the solution is relatively short, as shown in Fig. 5.17(c).                                                                      ■

Musicians use the beating phenomenon to tune string instruments. For instance, if a string is out of tune with respect to a tuning fork, then one hears a sequence of "beats" when the two are sounded together. The beats are simply the envelope that encloses the oscillations, and it is possible to quantify the degree to which the string is out of tune. See Exercise 14.

## Resonance

The solution (5.103) of the initial value problem (5.102) discussed in Example 2 is valid only in the case in which $\alpha \neq \omega$. We now consider the situation in which the frequency of the forcing is the same as that of the simple harmonic oscillator, that is, the case in which $\alpha = \omega$.

Figure 5.18 shows the solution of the equation

$$x'' + 25x = 11 \cos 5t \tag{5.105}$$

satisfying the initial condition $x(0) = x'(0) = 0$. The solution oscillates, but its amplitude appears to increase indefinitely. In contrast to the previous examples, *the envelope functions are straight lines that tend to infinity with time.* The situation depicted in Fig. 5.18 is called *resonance*. The following discussion explains how the phenomenon arises.

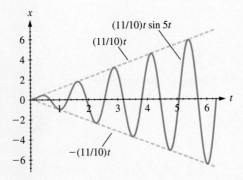

**FIGURE 5.18**  A graph of the solution of Eq. (5.105) satisfying the initial condition $x(0) = x'(0) = 0$.

Figures 5.19(a)–(c) show plots of the function

$$x(t) = \left( \frac{11}{\alpha^2 - 25} \right) (\cos 5t - \cos \alpha t) \tag{5.106}$$

for $\alpha = 5.1$, $\alpha = 5.05$, and $\alpha = 5.025$. Notice the following features of the graph as $\alpha$ approaches 5:

- the period increases;
- the maximum amplitude increases;
- the envelope function for the initial oscillations (say for $0 < t < 30$) increasingly resembles a straight line over the displayed interval.

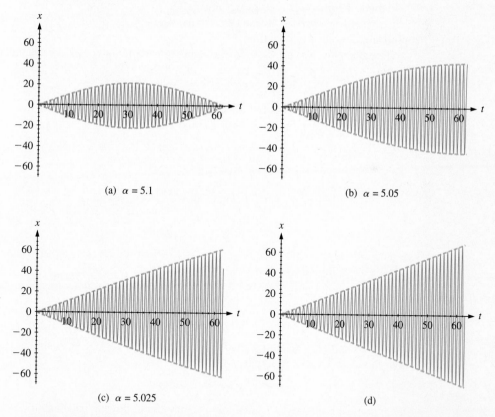

(a) $\alpha = 5.1$

(b) $\alpha = 5.05$

(c) $\alpha = 5.025$

(d)

**FIGURE 5.19** The graph of Eq. (5.106) for (a) $\alpha = 5.1$; (b) $\alpha = 5.05$; (c) $\alpha = 5.025$. Graph (d) shows a plot of $x(t) = \frac{11}{10} t \sin 5t$. The graphs of the corresponding envelope functions are shown as dashed lines.

What happens in the limit as $\alpha \to 5$? The answer is provided by l'Hôpital's rule, which can be used to find the limit as $\alpha \to 5$ as follows:

$$\lim_{\alpha \to 5} \frac{11(\cos 5t - \cos \alpha t)}{\alpha^2 - 25} = \lim_{\alpha \to 5} \frac{\dfrac{d}{d\alpha}[11(\cos 5t - \cos \alpha t)]}{\dfrac{d}{d\alpha}(\alpha^2 - 25)}$$

$$= \lim_{\alpha \to 5} \frac{11t \sin \alpha t}{2\alpha}$$

$$= \tfrac{11}{10} t \sin 5t.$$

Figure 5.19(d) shows a plot of the function $x(t) = \tfrac{11}{10} t \sin 5t$ together with its envelope.

Figure 5.19 provides a graphical illustration of this limiting behavior as $\alpha \to 5$, which is called *resonance*. Resonance is characterized by oscillatory motion whose amplitude grows linearly with time.

More generally, we can ask about the nature of the solutions of the forced undamped oscillator

$$x'' + \omega^2 x = \cos \alpha t \tag{5.107}$$

as $\alpha \to \omega$. The method of undetermined coefficients can be used to show that the solution that satisfies the initial conditions $x(0) = 0$ and $x'(0) = 0$ is

$$x(t) = \frac{\cos \omega t - \cos \alpha t}{\alpha^2 - \omega^2}, \tag{5.108}$$

provided that $\alpha \neq \omega$. We can use l'Hôpital's rule to show that

$$\lim_{\alpha \to \omega} x(t) = \frac{t \sin \omega t}{2\omega}. \tag{5.109}$$

This analysis suggests that the solutions of Eq. (5.107) oscillate with linearly increasing amplitude in the limit as $\alpha \to \omega$.

In fact, it is straightforward to verify by substitution that Eq. (5.109) is a solution of the differential equation

$$x'' + \omega^2 x = \cos \omega t$$

and that it also satisfies the initial conditions $x(0) = 0$ and $x'(0) = 0$. A similar analysis applies to the differential equation

$$x'' + \omega^2 x = \sin \alpha t$$

as $\alpha \to \omega$. (See Exercise 19.) Therefore, we have the following conclusion:

> *Resonance occurs when an undamped oscillator is forced by a function that is a homogeneous solution. The response is characterized by linearly increasing amplitudes that become arbitrarily large with time.*

## Summary

We summarize the theoretical results of this section in the following theorem.

**Theorem 5.7**  *Let*

$$x'' + \omega^2 x = f(t).$$

- *If f is a periodic function of the form $A \cos(\alpha t + \delta)$, where $A$, $\alpha$, and $\delta$ are constants, and if $\alpha \neq \omega$, then the solution $x(t)$ is a linear combination of functions of the form $\sin \alpha t$, $\cos \alpha t$, $\sin \omega t$, and $\cos \omega t$. The amplitude of the solution depends on A and the initial conditions. The solution can be described as a beating solution.*

- *If f is a linear combination of $\cos \omega t$ and $\sin \omega t$, then the solution exhibits resonance: The amplitude of $x(t)$ grows linearly with time and becomes arbitrarily large as $t \to \infty$.*


## Resonance and the Method of Undetermined Coefficients

The method of undetermined coefficients, as discussed in Section 5.6, fails if the guess for a particular solution is the same as the homogeneous solution. The following example illustrates the required modification to the method in the case of resonant solutions.

✦ *Example 3*  Let

$$x'' + 25x = 10 \cos 5t, \qquad x(0) = 0, \qquad x'(0) = 1. \tag{5.110}$$

Equation (5.110) defines the undamped, unforced oscillator $L(x) = x'' + 25x$. The homogeneous solution is

$$x_h(t) = c_1 \cos 5t + c_2 \sin 5t,$$

with period $2\pi/5$. The circular frequency is $\omega = 5$, which is the same as that of the forcing $f$. Therefore, the solution exhibits resonance. Because the amplitude of the solution grows linearly with time, a particular solution is of the form

$$x_p(t) = t(A \cos 5t + B \sin 5t).$$

(See Example 5 in Section 5.6.)

We substitute this guess for $x_p$ into Eq. (5.110) to obtain

$$\begin{aligned}
x_p'' + 25x_p &= [(-25At + 10B) \cos 5t - (10A + 25Bt) \sin 5t] \\
&\quad + 25[t(A \cos 5t + B \sin 5t)] \\
&= 10B \cos 5t - 10A \sin 5t,
\end{aligned}$$

which, when we equate to the right-hand side of Eq. (5.110) and collect like terms, implies that $A = 0$ and $B = 1$. Therefore, a particular solution is $x_p = t \sin 5t$.

The general solution is

$$x(t) = x_p(t) + x_h(t) = t \sin 5t + c_1 \cos 5t + c_2 \sin 5t.$$

The initial conditions imply that

$$x(t) = \left(t + \tfrac{1}{5}\right) \sin 5t. \tag{5.111}$$

The graph of Eq. (5.111) and its envelope are shown in Fig. 5.20.

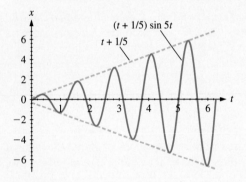

**FIGURE 5.20**    A plot of Eq. (5.111) (solid curve) and its envelope (dashed lines). ∎

Let

$$x'' + ax' + bx = f(t),$$

where $a$ and $b$ are constants, $b \neq 0$, and $f$ is a constant function, a sine function, a cosine function, or a linear combination of these. The method of undetermined coefficients outlined in Section 5.6 works for all such functions $f$ if we modify Step 3 as follows:

3. If $f(t)$ is a function of the form

$$f(t) = \alpha \cos \omega t + \beta \sin \omega t,$$

where $\alpha$, $\beta$, and $\omega$ are constants, then guess a particular solution of the form

$$x_p(t) = A \cos \omega t + B \sin \omega t.$$

If $L(x_p) = 0$, that is, if $x_p$ is a homogeneous solution, then multiply the guess by $t$;

that is, guess a particular solution of the form

$$x_p(t) = t(A \cos \omega t + B \sin \omega t).$$

Substitute $x_p$ into the differential equation, equate the coefficients of like sine and cosine terms, and determine the values of $A$ and $B$ accordingly.

## Appendix: A Review of Periodic Functions

The purpose of this appendix is to remind you how to determine whether the sum of two sine and cosine functions is a periodic function and, if it is, to find the period. The function $f$ is said to be *periodic* if there exists a positive constant $s$ such that $f(t_0 + s) = f(t_0)$ for every $t_0$ in the domain of $f$. The smallest $s$ for which this condition holds is called the *period* of $f$ and is denoted by $T$. The *frequency* of $f$ is $v = 1/T$. For brevity, we say that $f$ is $T$-*periodic* if $f$ is periodic with period $T$.

If $f(t) = c_1 \sin \omega t + c_2 \cos \omega t$, then its period is $2\pi/\omega$. We are interested in the case in which $f$ is the sum of two sine and cosine functions of *different* periods. Such functions occur commonly as solutions of forced, undamped, linear oscillators.

✦ *Example 4*   Let

$$f(t) = 4 \cos 2.8t + 4 \cos 3t, \qquad (5.112)$$

whose graph is shown in Fig. 5.21. The graph suggests that the period of $f$ is approximately 30. It is possible to compute the period precisely, as follows. The period of $\cos 2.8t$ is $T_1 = 2\pi/2.8 = 5\pi/7$ and the period of $\cos 3t$ is $T_2 = 2\pi/3$. The relation $f(t_0 + s) = f(t_0)$ is satisfied if $s$ is a positive integer multiple of both $T_1$ and $T_2$. The

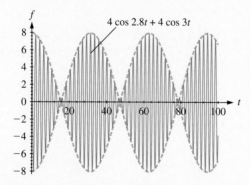

FIGURE 5.21   The graph of Eq. (5.112).

period of $f$ is the least common multiple of $T_1$ and $T_2$. Therefore, we seek the smallest positive integers $m$ and $n$ such that $mT_1 = nT_2$, that is, $m(5\pi/7) = n(2\pi/3)$. We have

$$\frac{5m}{7} = \frac{2n}{3},$$

which implies that $m/n = 14/15$. Thus, the period of $f$ is $T = 14T_1 = 15T_2 = 10\pi \approx 31.4$, which is reflected in Fig. 5.21.

Figure 5.21 also shows that the graph of $f$ is enclosed by the graphs of two periodic functions of period $20\pi$, called the *envelope*. The function $f$ and its envelope intersect on the $t$ axis at points that are $10\pi$ apart.                                                  ■

◆ *Example 5*   The period of $\cos 2t$ is $T_1 = \pi$, and the period of $\cos \sqrt{2}\, t$ is $T_2 = \pi \sqrt{2}$, but the linear combination

$$x(t) = \cos 2t + \cos \sqrt{2}\, t \qquad (5.113)$$

is not periodic.  To be periodic, there must be positive integers $m$ and $n$ such that $mT_1 = m\pi = nT_2 = n\pi \sqrt{2}$. This requires that

$$m = n\sqrt{2},$$

which cannot be satisfied for any integers $m$ and $n$ because $\sqrt{2}$ is irrational. (That is, $\sqrt{2}$ is not a ratio of integers.) The function $x$ defined by Eq. (5.113) is called a *quasiperiodic* function because its graph resembles that of a periodic function.  Figure 5.22 shows a plot of $x$.  Notice how the graph never quite repeats itself, although portions of the graph closely resemble each other over certain intervals.

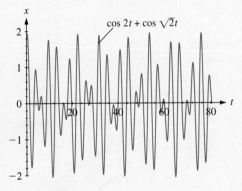

**FIGURE 5.22**   The graph of Eq. (5.113).                                                  ■

## *Exercises*

**1.** Consider the solution of the equation

$$x'' + 25x = 11 \sin 6t$$

satisfying the initial conditions $x(0) = 0$ and $x'(0) = -1$.

(a) Find $x(t)$ and plot it.

(b) Use the identity (5.99) to express $x(t)$ as a product of a long-period function and a short-period function. Identify the period of $x$, the long-period function and the short-period function.

(c) Determine the envelope from your answer in (b) and add it to your graph.

(d) Use your answer in (b) to prove that the graph of $x$ crosses the $t$ axis 11 times during each period of $x$.

**2.** Consider the solution (5.103) of the initial value problem (5.102).

(a) Determine the envelope and the period of the solution for $\alpha = 5\frac{1}{8}$, $\alpha = 5\frac{1}{16}$, and $\alpha = 5\frac{1}{32}$.

(b) Find a formula for the envelope and the period of the solution when $|\alpha - 5| = 2^{-n}$, where $n$ is a positive integer.

(c) Describe what happens to the envelope and the period of the solution as $\alpha \to 5$.

**3.** This exercise and the next explore the possibilities for oscillatory solutions in cases in which the forcing function does not oscillate. Let $x'' + 9x = E$, where $E$ is a nonzero constant. (This equation might model an LC circuit that is connected to a constant electrical potential, such as a battery.)

(a) Find the general solution and show that it oscillates periodically. What is the period?

(b) Show that there exists an initial condition for which the solution in (a) does not oscillate and identify it.

(c) It is reasonable to say that "almost all" initial conditions produce oscillating solutions. Explain why.

**4.** Let $x'' + 4x = e^{-t}$.

(a) Let $x(0) = x'(0) = 0$. The forcing function does not oscillate and in fact tends to 0 as $t$ becomes large. Find the solution of this initial value problem and describe its qualitative behavior. Consider questions such as the following: Is the solution periodic? If so, what is the period? What is the amplitude? Does the solution approach a limiting value as $t$ becomes large? Does the solution approach another function as $t$ becomes large? If so, what is it?

(b) Is there an initial condition for which any oscillations in the solution die out as $t$ becomes large? If so, what is it?

**5.** Find the general solution of each of the following differential equations and determine whether it is periodic. If so, find the period.

(a) $x'' + 9x = \cos 5t$

(b) $x'' + 5x = \cos 9t$

(c) $x'' + 4x = \cos t$

(d) $x'' + 4x = \cos 2t$

(e) $x'' + 2x = \cos \sqrt{2} t$

(f) $x'' + 2x = \cos 2\sqrt{2} t$

(g) $x'' + 9x = \sin \pi t$

**6.** For each initial value problem in the following list,

• find the solution;

• determine whether the solution is periodic, and, if so, find the period;

• determine the envelope of the solution.

(a) $x'' + x = \sin \frac{5}{4}t$, $x(0) = 0$, $x'(0) = -4$

(b) $x'' + 4x = 7 \sin \frac{3}{2}t$, $x(0) = 0$, $x'(0) = -2$

(c) $x'' + 3x = 8 \cos \sqrt{27}\, t$, $x(0) = 0$, $x'(0) = 0$

**7.** For each initial value problem in the following list,

- find the solution;

- determine whether the solution is periodic, and, if so, find the period;

- determine whether the solution exhibits resonance.

(a) $x'' + 16x = 8 \cos 4t$, $x(0) = 1$, $x'(0) = 0$

(b) $x'' + 4x = 2$, $x(0) = 0$, $x'(0) = 1$

(c) $x'' + 5x = 10 \sin \sqrt{5}\, t$, $x(0) = x'(0) = 0$

**8.** Repeat Exercise 7 for the initial value problems in the following list:

(a) $x'' + 6x = 4 \cos \sqrt{2}\, t$, $x(0) = 1$, $x'(0) = 0$

(b) $x'' + x = \cos t + \sin t$, $x(0) = x'(0) = 0$

(c) $x'' + x = 6(\cos 2t + \sin 3t)$, $x(0) = 1$, $x'(0) = 0$

**9.** Let $x'' + 5x = \cos 2t$.

(a) Find the solution satisfying the initial conditions $x(0) = 1$ and $x'(0) = \sqrt{5}$.

(b) Plot the solution for $0 \le t \le 100$.

(c) Is the solution periodic? Can you characterize the solution as a "beating" solution? Explain.

**10.** Sometimes it is possible to define the envelope of an oscillating solution even when the oscillations are not periodic. Let $x'' + 3x = 6 \sin 3t$, $x(0) = 0$, $x'(0) = \sqrt{3} - 3$.

(a) Find the solution of this initial value problem and plot it.

(b) Explain why the solution that you found in (a) is not periodic (it is quasiperiodic). Discuss how the quasiperiodicity is reflected in the graph.

(c) Use Eq. (5.99) to rewrite the solution that you found in (a). Identify what the envelope might be, then verify that your candidate function in fact is the envelope by adding its graph to your plot in (a). Discuss the features of the graph that make it a suitable envelope function.

**11.** It is interesting to investigate how errors in the parameters affect the solutions of forced oscillators. Consider the solution of the initial value problem

$$x'' + 4x = \sin \pi t, \quad x(0) = x'(0) = 0. \quad (5.114)$$

(a) Find the solution and plot it for $0 \le t \le 10$. Is the solution periodic? If so, what is the period?

(b) Suppose we approximate Eq. (5.114) by $x'' + 4x = \sin 3t$. Find the solution using the same initial conditions as above and add its graph to the plot in (a). How well do the two solution curves coincide? Is the solution of this initial value problem periodic? If so, what is the period?

(c) Let $x'' + 4x = \sin 3.1t$. Find the solution using the same initial conditions as above and plot its graph with that of the solution in (a). How well does this equation approximate the solution in (a)? Determine whether this solution is periodic, and, if so, identify the period.

(d) Next let $x'' + 4x = \sin 3.14t$ and proceed as above. Determine whether this solution is periodic, and, if so, identify the period. How good is this approximation to the initial value problem in (a)? How large can $t$ be before the graph of this solution is noticeably different from the graph in (a)?

**12.** Let $x(t) = \cos 2t - \frac{1}{2} \cos \frac{9}{4}t$.

(a) What is the period of $x$? Express your answer as an appropriate multiple of $\pi$.

(b) Graph $x(t)$ over an interval that shows approximately two full periods.

(c) Rewrite $x$ as

$$x(t) = \tfrac{1}{2} \cos 2t + \tfrac{1}{2} \left( \cos 2t - \cos \tfrac{9}{4} t \right)$$

and express $x$ a sum of the form $A \cos 2t + B \sin \alpha \sin \beta$ for appropriate terms $A$, $B$, $\alpha$, and $\beta$.

(d) What is the envelope of the product of sine terms that you found in (c)? What is its period?

(e) Graph the product of the sine terms that you found in (c). Does the period of the envelope that you found in (d) correspond with the period of the envelope that appears to enclose the graph of $x$ in (a)? Explain.

**13.** Consider an LC circuit (that is, an LRC circuit without a resistor) to which an external constant voltage $V$ is applied. The charge $q$ in the circuit can be modeled by

$$Lq'' + q/C = V.$$

For notational convenience, we often rewrite the model as

$$q'' + \omega^2 q = V/L, \qquad (5.115)$$

where $\omega^2 = 1/(LC)$.

(a) What are the units of $\omega^2$ and $V/L$?

(b) Find the solution of Eq. (5.115) that satisfies the initial conditions $q(0) = 0$ and $q'(0) = 0$, assuming that the homogeneous equation is underdamped. Show that the solution is periodic and find the period.

(c) If the applied voltage is doubled, is the amplitude of the oscillations doubled? Explain.

(d) More generally, does the amplitude of $q(t)$ vary linearly with $V$? Explain.

(e) Can you change the period of $q$ by changing $V$? Explain.

**14.** The standard A corresponds to a frequency of 440 Hz, that is, 440 cycles per second.

(a) A tuning fork tuned to the standard A at 440 Hz makes a sound whose amplitude is proportional to $\sin \omega t$. Find $\omega$.

(b) Suppose a piano generates an A that is 438 Hz; the amplitude of the sound is proportional to $\sin \alpha t$. Find $\alpha$.

(c) What is the frequency of the beating that you hear when the piano string in (b) and the tuning fork in (a) are struck simultaneously?

(d) The tuning fork in (a) and a piano string are struck simultaneously, producing a low-frequency oscillation that oscillates at 5 Hz. What are the possible frequencies of oscillation of the piano string?

**15.** Let $x'' + 9x = f(t)$ with $x(0) = x'(0) = 0$.

(a) Give an example of a function $f$ for which the solution oscillates with period $2\pi/3$.

(b) Give an example of a function $f$ for which the solution oscillates with period $2\pi$.

(c) Give an example of a function $f$ for which the solution oscillates with period $4\pi$.

(d) Give an example of a function $f$ for which $|f(t)| \le 2$ for every $t$ but $|x(t)| \to \infty$ as $t \to \infty$.

**16.** Let $x'' + \omega^2 x = f(t)$ with $x(0) = x'(0) = 0$. The value of $\omega$ constrains the values of the period of the solution, even in cases in which $f$ is a periodic function. Show that if $\omega = 3$, then no function of the form $f(t) = A \cos \alpha t$ can be found such that the period of $x$ is exactly $3\pi$.

**17.** This problem investigates the question of how the frequency of the forcing affects the amplitude of the response. Let

$$x'' + 9x = 4 \cos \alpha t, \qquad (5.116)$$

where $x(0) = x'(0) = 0$.

(a) Find the solution of Eq. (5.116) assuming that $\alpha \ne 3$.

(b) Identify the envelope of the solution in (a).

(c) Plot the solution and the envelope when $\alpha = 1$. What are the period and amplitude of the envelope? Does the amplitude of the envelope reflect the amplitude of the largest oscillations of $x$? Explain.

(d) Repeat part (c) assuming that $\alpha = 2.5$.

(e) Repeat part (c) assuming that $\alpha = 2.9$.

(f) What happens to the amplitude and the period of the envelope as $\alpha \to 3$?

(g) Now let $\alpha = 3$. Find the solution of Eq. (5.116) for $x(0) = x'(0) = 0$ and plot it. What is the envelope? What happens to the amplitude of the solution as time increases?

**18.** Let $x'' + 9x = g(t)$. Give an example of a function $g(t)$ for which $|g(t)| \le 2$ for every $t$ and for which

(a) $x$ oscillates with period $2\pi$;

(b) $x$ oscillates with period $4\pi$;

(c) $x$ is bounded and quasiperiodic;

(d) $x$ tends to infinity as $t \to \infty$.

**19.** Let $x'' + \omega^2 x = \sin \alpha t$ with $x(0) = x'(0) = 0$.

(a) Find the solution assuming that $\alpha \ne \omega$.

(b) Find the limit of your answer in (a) as $\alpha \to \omega$.

(c) Show that your answer in (b) is a solution of the equation $x'' + \omega^2 x = \sin \omega t$.

**20.** To have resonance, the period of the forcing must be exactly the same as the period of the undamped oscillator. In practice, this condition is difficult to

satisfy, but often the two periods can be matched very closely. This exercise illustrates what can happen in such a situation.

(a) Let $x'' + 16x = 8 \sin 4t$. Find the solution satisfying the initial condition $x(0) = x'(0) = 0$.

(b) Let $x'' + 15.8x = 8 \sin 4t$. Show that the solution satisfying the initial condition $x(0) = x'(0) = 0$ can be written in the form

$$x(t) = 40 \left( \sin \sqrt{15.8}\, t - \sin 4t \right) + r(t),$$

where $r$ is a periodic function whose amplitude is small relative to 40.

(c) Graph the solutions in (a) and (b) for $0 \le t \le 50$. How similar are the graphs?

(d) Extend the $t$ axis farther to the right as needed until you see a significant difference between the graphs. How far out do you need to go?

(e) Use Eq. (5.99) to determine the envelope of the leading expression of the solution in (b). What is the period of the envelope? What is its amplitude?

(f) Add the graph the envelope to the plot that you generated in (a). Is the envelope for (b) also a reasonably good envelope for (a), at least for values of $t$ between, say, 0 and 50?

(g) In what sense is it reasonable to say that a system is "nearly in resonance"? On the basis of a graph of the solution, what criteria might you apply to be able to say that a system is "close to" being a resonant system?

## 5.8   FORCED, DAMPED, LINEAR OSCILLATORS

### The Qualitative Properties of Solutions

In this section, we consider the behavior of forced linear oscillators that have a damping term. A forced damped linear oscillator is a second-order differential equation of the form

$$mx'' + cx' + kx = f(t),$$

where $f$ is the forcing and $m$, $c$, and $k$ are positive constants. Periodic functions such as sines and cosines are of particular technological interest. In this section, we consider the important qualitative features of the behavior. In contrast to the case of undamped oscillators, we will see that there is no sustained beating phenomenon when damping is present. However, damped systems can display resonance, although the solutions remain bounded.

✦ *Example 1*   Let

$$x'' + 3x' + 2x = 10 \sin t \qquad\qquad (5.117)$$

with the initial conditions $x(0) = -1$ and $x'(0) = 1$. The solution is

$$x(t) = 4e^{-t} - 2e^{-2t} + \sin t - 3 \cos t. \qquad\qquad (5.118)$$

The qualitative properties of Eq. (5.118) are straightforward to analyze:

- *Equation (5.118) may be regarded as the sum of a transient and a steady-state function.* This is a basic characterization of the solutions of damped linear oscillators with periodic forcing. The transient function consists of those terms that tend to 0 as $t \to \infty$. In this case, the transient function is $4e^{-t} - 2e^{-2t}$. The remaining terms (that is, $\sin t - 3 \cos t$) are the steady-state function. The amplitude of the steady-state function does not decay with time.

  The transient terms are a homogeneous solution of Eq. (5.117). The characteristic equation corresponding to Eq. (5.117) is $\lambda^2 + 3\lambda + 2 = 0$, whose roots are $-1$ and $-2$. Thus, the homogeneous solution has the form

  $$x_h(t) = c_1 e^{-t} + c_2 e^{-2t},$$

  which is identical to the transient in Eq. (5.118) for appropriate constants $c_1$ and $c_2$.

  The steady-state terms are a particular solution of Eq. (5.117). To prove this, we regard the left-hand side of Eq. (5.117) as the linear differential operator $L(x) = x'' + 3x' + 2x$. Then

  $$L(\sin t - 3 \cos t)$$
  $$= L(\sin t) - 3L(\cos t)$$
  $$= (-\sin t + 3 \cos t + 2 \sin t) - 3(-\cos t - 3 \sin t + 2 \cos t)$$
  $$= 10 \sin t.$$

- *The frequency of the steady-state solution is the same as the frequency of the forcing.* In other words, the forcing function and the steady state are periodic with period $2\pi$. In contrast to undamped oscillators with periodic forcing, there is no sustained beating phenomenon. (See Exercise 16.) Figure 5.23(a) shows the graph of Eq. (5.118) together with the graph of the steady state. When $0 \le t \le 3$, the two curves are relatively

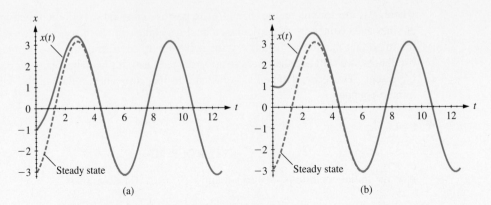

**FIGURE 5.23**    (a) The graph of Eq. (5.118). (b) The graph of Eq. (5.119) with $x_0 = 1$, $v_0 = 0$. The steady-state solution in each case is shown as the dashed curve.

far apart, but they draw closer together as $t$ increases. The two curves are almost indistinguishable for $t > 2\pi$, reflecting the fact that the transient term becomes very small relative to the steady state when $t$ is larger than about 6.

- *The qualitative behavior of the solution is insensitive to changes in the initial conditions.* If we take $x(0) = x_0$ and $x'(0) = v_0$, then the solution of Eq. (5.117) is

$$x(t) = (5 + 2x_0 + v_0)e^{-t} - (2 + x_0 + v_0)e^{-2t} + \sin t - 3\cos t. \qquad (5.119)$$

Hence, the initial conditions are part of the transient. As $t \to \infty$, the transient decays to 0. The solution always approaches the same steady state, regardless of the initial conditions. Figure 5.23(b) shows the graph of the solution corresponding to $x(0) = 1$, $x'(0) = 0$ together with the steady state. The solution curve is different from the one illustrated in Fig. 5.23(a), particularly for $0 \le t \le 1$. However, the solution rapidly approaches the steady state; the two curves are nearly indistinguishable when $t$ is larger than $2\pi$. Thus, when $t$ is large enough, there is little difference between the graphs of the solutions corresponding to the different initial conditions. One practical application of this observation is the following: Solutions of Eq. (5.117) are insensitive to errors in initial conditions, in the sense that for sufficiently large values of $t$, the effect of the errors becomes negligible.                                                    ∎

As is mentioned in the introduction, electrical circuits to which a time-periodic potential is applied have a wide variety of technological uses. Let us examine the mathematical properties of one of the simplest such circuits, an LRC circuit. The differential equation that governs the behavior of the circuit when a time-periodic potential is applied is given by

$$Lq'' + Rq' + \frac{1}{C}q = V \sin \alpha t, \tag{5.120}$$

where $V$ denotes the amplitude of the applied potential. In many applications, Eq. (5.120) does not have convenient coefficients, and it is cumbersome to derive explicit solutions by hand. Nevertheless, the behavior of the solutions can be analyzed in the same way as in the previous example.

◆ *Example 2*    Consider an LRC circuit in which $L = 1$ H, $R = 100\ \Omega$, $C = 100\ \mu$F, $\alpha = 50$ rad/s, and $V = 100$ V. Kirchhoff's law, Eq. (5.120), implies that the behavior of the circuit is governed by the differential equation

$$q'' + 100q' + 10{,}000q = 100 \sin 50t. \tag{5.121}$$

The solution that satisfies initial conditions $q(0) = q_0$ and $q'(0) = I_0$ is

$$q(t) = \tfrac{1}{325}(3 \sin 50t - 2 \cos 50t) + e^{-50t}\left(q_0 + \tfrac{2}{325}\right) \cos 50\sqrt{3}\,t$$
$$+ e^{-50t}\sqrt{3}\left(\tfrac{1}{150}I_0 + \tfrac{1}{3}q_0 - \tfrac{1}{975}\right) \sin 50\sqrt{3}t. \tag{5.122}$$

We have derived Eq. (5.122) with the aid of a computer algebra system. Although the solution (5.122) has a large number of terms, its qualitative properties are similar to those of the previous example:

- *Equation (5.122) may be regarded as the sum of a transient and a steady-state function.* The term $e^{-50t}$ decays very quickly. Figure 5.24 shows a plot of Eq. (5.122) for the case in which $q_0 = I_0 = 0$.

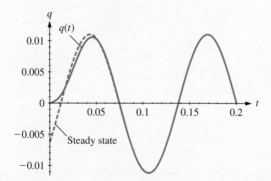

**FIGURE 5.24**    A plot of Eq. (5.122) for the initial conditions $q_0 = I_0 = 0$ (solid curve). The graph of the steady-state solution is shown as the dashed curve.

The solution $q(t)$ and the steady-state function nearly coincide after about 0.1 s. To understand why this is, observe that the transient terms are a homogeneous solution of Eq. (5.121). The characteristic equation corresponding to Eq. (5.121) is $\lambda^2 + 100\lambda + 10000 = 0$, whose roots are $-50 \pm 50\sqrt{3}i$. Thus, the homogeneous solution has the form

$$q_h(t) = e^{-50t}\left(c_1 \cos 50\sqrt{3}\,t + c_2 \sin 50\sqrt{3}\,t\right),$$

which is identical to the transient in Eq. (5.122) for appropriate constants $c_1$ and $c_2$. The amplitude of $q_h$ decreases at a rate proportional to $e^{-50t}$, which implies that the transient decays rapidly to zero.

- *The frequency of the steady-state solution is the same as the frequency of the forcing.* The forcing function and the steady-state are periodic with period $2\pi/50$. The steady-state terms are a particular solution of Eq. (5.121).

- *The solution is insensitive to changes in the initial conditions.* Therefore, the long-term behavior of the circuit does not depend on the initial charge and current. (The meaning of the phrase *long term* depends on the context of the problem. In this case, the transient terms decay exponentially at a rate proportional to $e^{-50t}$ and become negligible after only a few tenths of a second. Here, the phrase *long term* refers to an interval that is at most several hundred milliseconds after the forcing is turned on.) ∎

Although we have examined only two examples in detail, all solutions of Eq. (5.120) have similar properties. We summarize our results in the following theorem.

**Theorem 5.8**   *Let*

$$x'' + cx' + \omega^2 x = f(t), \tag{5.123}$$

*where $\omega^2$ and $c$ are constants and $\omega \neq 0$. The solutions of Eq. (5.123) may be written in the form $x(t) = x_h(t) + x_p(t)$, where $x_h$ is a homogeneous solution and $x_p$ is a particular solution that depends on $f$. Moreover,*

- *if $c > 0$, then $\lim_{t \to \infty} x_h(t) = 0$. That is, the homogeneous solution may be regarded as a transient term. The steady state is a particular solution. (The steady state, denoted $x_{ss}(t)$, is also called the response of the system to the forcing $f$.)*

- *If $f$ is a linear combination of sine and cosine functions and has period $T$, then $x_{ss}(t)$ is periodic with period $T$. The amplitude of $x_{ss}(t)$ depends on $c$, $\omega^2$, and $f$ but does not depend on the initial conditions.*

- *The homogeneous solution $x_h$ depends on the initial conditions, $c$, and $\omega^2$.*

## The Gain Function and Phase Lag

We now consider those aspects of resonant behavior that are preserved to varying degrees in periodically forced, damped oscillators. Theorem 5.8 states that the period of the steady-state solutions of Eq. (5.123) is the same as the period of the forcing $f$, but it says little about the dependence of the amplitude of the steady state as a function of $f$ and the other parameters in the equation. This subsection considers these issues quantitatively. In addition to the questions listed at the beginning of Section 5.7, we also want to know whether phenomena akin to resonance can occur in the presence of damping.

Let

$$x'' + cx' + \omega^2 x = F \sin \alpha t, \qquad (5.124)$$

where $c$, $F$, $\alpha$, and $\omega$ represent positive constants. As an illustration, consider the case in which $c = 1/10$, $\omega = 3$, and $F = 1$. With these constants, Eq. (5.124) becomes

$$x'' + \tfrac{1}{10}x' + 9x = \sin \alpha t. \qquad (5.125)$$

Figure 5.25 shows the solution $x(t)$ of the Eq. (5.125) for various values of $\alpha$, called the *circular frequency*, or simply the *frequency*, of the forcing.[7] (In each case, the solution is plotted for values of $t$ for which the transient terms are very small. Effectively, only the steady state is shown.) In many applications, one is not interested in the transient behavior but rather in the qualitative properties of the steady-state solution (response).

Figure 5.25 illustrates the following features of the response in Eq. (5.125):

- For each value of the forcing frequency $\alpha$, the period of the response is the same as the period of the forcing, as expected in view of Theorem 5.8.

- When $\alpha = 2$ or 2.5, the amplitude of the response is relatively small in comparison to the amplitude of the forcing (that is, the forcing amplitude is 1, but the amplitude of the response is about 0.4).

- The amplitude of the response is much larger at $\alpha = 3$. Notice that $\alpha = \omega = 3$ in this case. (As we will see, the maximum amplitude of the response sometimes occurs when the value of $\alpha$ is close to $\omega$.) In contrast to the undamped case, however, the response remains bounded as $t \to \infty$.

- The amplitude of the response once again is small for $\alpha = 3.5$ or 4.

- When $\alpha = 2$ or 2.5, the maxima, minima, and zeros of the response coincide approximately with those of the forcing.

---

[7]The period of $\sin \alpha t$ is $T = 2\pi/\alpha$. Mathematicians define the frequency as the reciprocal of the period, $1/T = \alpha/2\pi$. However, in engineering contexts, the term *frequency* often refers to $\alpha$, not $1/T$. The constant $\alpha$ is sometimes called the *circular frequency* to distinguish it from the quantity $1/T$.

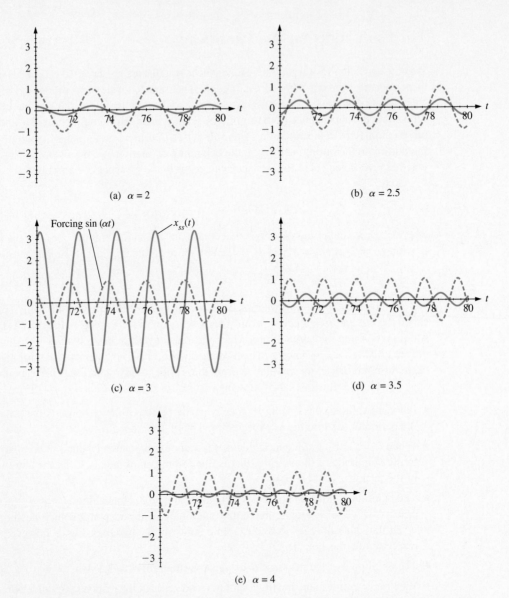

**FIGURE 5.25**  The response $x_{ss}(t)$ (solid curve) of Eq. (5.125) and the corresponding forcing (dashed curve) for the indicated values of $\alpha$.

- When $\alpha = 3$, the zeros and relative extrema of the response no longer coincide with those of the forcing. The relative extrema of the response appear to coincide with the zeros of the forcing, and the relative extrema of the forcing appear to coincide with the zeros of the response.

- When $\alpha = 3.5$ or 4, the response appears to be out of phase with the forcing insofar as a relative maximum of the response appears to coincide with a relative minimum of the forcing, and similarly for the relative minima of the response and the relative maxima of the forcing. However, the zeroes of the response and the forcing do appear to coincide.

This illustration motivates two basic questions about the relationship of the response of the system to the frequency of the applied forcing: (1) How is the amplitude of the response affected, and (2) how can one determine the difference in phase between the response and the forcing?

Both questions can be answered in considerable generality. The solution of Eq. (5.124) can be determined explicitly; the steady state is

$$x_{ss}(t) = \left( \frac{F}{(\omega^2 - \alpha^2)^2 + \alpha^2 c^2} \right) \left( (\omega^2 - \alpha^2) \sin \alpha t - \alpha c \cos \alpha t \right). \tag{5.126}$$

Equation (5.126) can be written in phase-amplitude form as

$$x_{ss}(t) = A \sin (\alpha t + \phi),$$

where

$$A = \frac{F}{\sqrt{\left( \omega^2 - \alpha^2 \right)^2 + \alpha^2 c^2}}$$

and

$$\phi = \tan^{-1} \left( \frac{c\alpha}{\alpha^2 - \omega^2} \right). \tag{5.127}$$

The ratio $A/F$ is called the *gain*, which we denote by $G$. That is,

$$G = \frac{A}{F} = \frac{1}{\sqrt{\left( \omega^2 - \alpha^2 \right)^2 + \alpha^2 c^2}}. \tag{5.128}$$

The gain is a function of the damping $c$, the frequency of the forcing (determined by $\alpha$), and the parameter $\omega^2$.

The angle $\phi$ is called the *phase angle*; by convention, $-\pi < \phi \le 0$. The *phase lag* or *phase shift* is defined as $-\phi/\alpha$. The negative value of $\phi$ reflects the idea that the response "lags behind" the forcing.

✦ *Example 3*     Let

$$x'' + \tfrac{1}{10}x' + 9x = \sin \alpha t, \tag{5.129}$$

for which the response to some forcing functions is shown in Fig. 5.25. The amplitude of the forcing is $F = 1$. Therefore, the gain is

$$G = \frac{1}{\sqrt{\left(9 - \alpha^2\right)^2 + \tfrac{1}{100}\alpha^2}}.$$

Figure 5.26(a) shows a plot of the gain function associated with Eq. (5.129). The plot corroborates the phenomenon observed in Fig. 5.25, namely, that the amplitude of the response of Eq. (5.129) is largest when the forcing frequency $\alpha$ is close to 3.

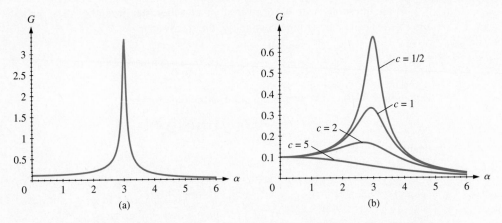

(a)                                                                 (b)

**FIGURE 5.26**     (a) The gain function associated with Eq. (5.129). (b) The gain function associated with Eq. (5.130) with $F = 1$, $\omega = 3$, and various values of $c$. (The vertical axes in the two plots are not the same.)

The value of $\alpha$ for which the gain is maximized occurs when $dG/d\alpha = 0$. A calculation shows that the maximum gain is $G \approx 3.334$ and occurs when

$$\alpha = \sqrt{\frac{1799}{200}} \approx 2.999.$$

Thus, the maximum gain occurs for values of the forcing frequency close to, but not exactly equal to, the parameter $\omega$. In contrast, resonance in undamped oscillators can occur only when the frequency of the forcing exactly equals $\omega$.                    ■

✦ *Example 4*   Let

$$x'' + cx' + 9x = \sin \alpha t, \tag{5.130}$$

which is the same as in the previous example but with the damping set to $c$. Figure 5.26(b) shows the gain function as a function of $\alpha$ for different choices of the damping $c$. As in the case of $c = 1/10$, the maximum gain occurs for value of $\alpha$ close to 3 when $c = 1/2$ and $c = 1$. For $c = 2$, however, the maximum gain occurs for frequencies that have shifted significantly away from $\alpha = 3$. Notice also that the size of the maximum gain decreases as the damping $c$ increases. When $c$ is large enough (such as $c = 5$), the maximum of the gain function occurs when $\alpha = 0$ (that is, when the forcing is constant), and there is no longer any trace of the resonance near $\alpha = 3$.  ∎

In general, if

$$x'' + cx' + \omega^2 x = F \sin \alpha t,$$

then the maximum value of the gain function is

$$G_{\max} = \frac{2}{c\sqrt{4\omega^2 - c^2}} \tag{5.131}$$

and occurs when

$$\alpha = \tfrac{1}{2}\sqrt{4\omega^2 - 2c^2}. \tag{5.132}$$

(See Exercise 12.)

In the limiting case of no damping ($c = 0$), a simple harmonic oscillator exhibits resonance when it is forced with the same frequency as the homogeneous solution. This situation corresponds to an infinite gain, such as would occur in Eq. (5.131) in the limit $c \to 0$. As $c$ increases from 0, the maximum value of the gain function decreases and occurs at smaller values of $\alpha$. When $c = \omega\sqrt{2}$, the maximum of the gain function occurs at $\alpha = 0$—the resonance has disappeared.

We can also see how the phase angle $\phi$ changes as a function of $\alpha$. In general, for fixed values of $c$ and $\omega$, the phase angle changes rapidly as $\alpha \to \omega$. This rapid phase shift is another artifact of resonance in damped linear systems.

✦ *Example 5*   Figure 5.27(a) shows the phase angle $\phi$ associated with the oscillator

$$x'' + \tfrac{1}{10}x' + 9x = \sin \alpha t$$

as a function of $\alpha$. (See Example 3.) The response of the system is nearly in phase with the forcing for $\alpha = 2$ and $\alpha = 2.5$, moves partially out of phase when $\alpha = 3$, and is almost completely out of phase for $\alpha = 3.5$ and $\alpha = 4$. (Compare the graphs of the response in Fig. 5.25.) Clearly, the phase angle shifts very rapidly from near 0 to near

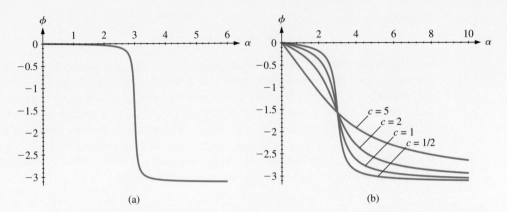

**FIGURE 5.27**    (a) A graph of the phase angle $\phi$ associated with Eq. (5.125) as a function of $\alpha$. (b) A graph of the phase angle associated with Eq. (5.126) for $\omega = 3$. The curves represent $c = 1/2, c = 1, c = 2$, and $c = 5$ as shown.

$-\pi$ as $\alpha$ crosses the value $\omega = 3$. The phase angle shifts more slowly for larger values of the damping $c$, as shown in Fig. 5.27(b). However, once the frequency of the forcing is large enough, the phase angle approaches $-\pi$. This means that the response is almost exactly out of phase with the forcing for sufficiently large values of $\alpha$.     ∎

## Parameter Estimation

The preceding discussion shows that the gain and phase angle depend on $c$, $\omega$, $F$, and $\alpha$. The examples that we have discussed so far vary $\alpha$ and fix the remaining constants. In practice, measured values of the gain, phase angle, and forcing amplitude $F$ can be used to estimate $c$ and $\omega$.

Graphs like those shown in Fig. 5.26 and Fig. 5.27 are called *frequency response diagrams*. They can be used to estimate the unknown parameters in a linear oscillator, such as the LRC circuit. A known signal (forcing) is applied to the circuit, and the output voltage (response) can be measured accurately. The forcing and the response can be used to estimate $c$ and $\omega$ by applying Eqs. (5.131) and (5.132), as follows.

Modern data acquisition equipment makes it possible to record values of the response with high accuracy. Using such data, the phase lag $\phi/\alpha$ can be estimated with good precision. Because the forcing is known, the value of $\phi$ can be determined. Suppose that $\phi$ has been measured from the data. From Eq. (5.127),

$$\alpha^2 - \omega^2 = \frac{\alpha c}{\tan \phi},$$                                                  (5.133)

and we can substitute this expression into Eq. (5.128) to obtain

$$G = \frac{1}{\sqrt{\left(\dfrac{\alpha c}{\tan \phi}\right)^2 + \alpha^2 c^2}},$$

so

$$c = \left(\frac{1}{\alpha G}\right) \sqrt{\frac{\tan^2 \phi}{1 + \tan^2 \phi}}. \tag{5.134}$$

The gain $G$ also can be measured from the output, and $\alpha$ is known because we apply a known forcing. Therefore, $c$ can be computed from Eq. (5.134), and its value can be substituted into Eq. (5.133) to determine $\omega$.

◆ *Example 6*   Let

$$x'' + cx' + \omega^2 x = \sin 2.5t, \tag{5.135}$$

where $c$ and $\omega^2$ are unknown. Here, the forcing is $f(t) = F \sin \alpha t$, where $F = 1$ and $\alpha = 2.5$.

Figure 5.28 shows the response (solid curve) to the applied forcing (dashed curve). Imagine that the data plotted in Fig. 5.28 have been recorded by a computer that has measured the output from the system modeled by Eq. (5.28). Using these data, we find that the amplitude of the response is 0.388. Because $F = 1$, the gain is $G = 0.388$.

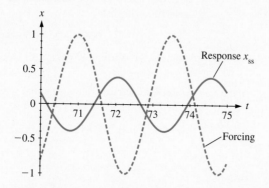

**FIGURE 5.28**   The response (solid curve) and the forcing (dashed curve) for Eq. (5.135). See Example 6.

The forcing crosses 0 when $t \approx 71.63$, and the corresponding value of the response does not cross 0 until $t \approx 72.68$. (It is not possible to determine the 0-crossing times with such precision directly from the graph, but such precision is often possible by inspecting computer-recorded data.) The phase shift is $\phi/\alpha \approx -1.05$. Because $\alpha = 2.5$, we know that $\phi \approx -2.625$, so $\tan \phi \approx 0.568$. Equation (5.134) implies that

$$c = \left(\frac{1}{2.5 \times 0.388}\right) \sqrt{\frac{(0.568)^2}{1 + (0.568)^2}} \approx 0.509.$$

Equation (5.133) implies that

$$(2.5)^2 - \omega^2 = \frac{2.5 \times 0.509}{0.568},$$

or $\omega \approx 2.00$.

In fact, Fig. 5.28 is a plot of the response of Eq. (5.135) with $c = 1/2$ and $\omega^2 = 4$. The relative error in the estimated value of $c$ is less than 2 percent.  ∎

Example 6 is a basic illustration of the problem of parameter estimation. Procedures that are similar in spirit to this one often can be used to calibrate electrical components. The accuracy with which $c$ and $\omega$ can be computed depends on the accuracy with which the gain and phase lag can be measured.

## *Exercises*

1. For each equation in the following list,

   - solve the indicated initial value problem;

   - identify the transient and the steady-state terms;

   - graph the solution and the steady-state terms on the same axes, and determine graphically when the difference between the solution and the steady-state terms becomes negligible (interpret *negligible* in a manner consistent with the resolution of your graph);

   - characterize the steady-state solution by answering the following questions: Is the steady state periodic? If so, what is the period? What is the amplitude? What is the gain?

   (a) $x'' + 4x' + 3x = 20 \cos t$, $x(0) = 2$, $x'(0) = 0$

   (b) $x'' + 4x' + 5x = 8 \sin t$, $x(0) = 1$, $x'(0) = -1$

2. Repeat Exercise 1 for each initial value problem in the following list:

   (a) $x'' + x' + \frac{17}{4}x = 65 \sin 2t$, $x(0) = 4$, $x'(0) = 2$

   (b) $x'' + 2x' + 10x = 85 \cos t$, $x(0) = 1$, $x'(0) = 0$

3. Repeat Exercise 1 for each initial value problem in the following list:

   (a) $16x'' + 8x' + 65x = 257 \sin 2t$, $x(0) = 0$, $x'(0) = 0$

   (b) $9x'' + 6x' + 37x = 145 \sin 2t$, $x(0) = 6$, $x'(0) = 2$

4. For each equation in the following list, determine the solutions that satisfy each pair of initial conditions. Graph each pair of solutions on the same axes. Determine graphically when the difference between the two solutions becomes negligible. (Interpret *neg-*

*ligible* in a manner consistent with the resolution of your graph.)

(a) $x'' + 4x' + 3x = \sin t$. First pair: $x(0) = 0$, $x'(0) = 0$. Second pair: $x(0) = 1$, $x'(0) = 1$.

(b) $x'' + x' + x = \sin t$. First pair: $x(0) = 0$, $x'(0) = 0$. Second pair: $x(0) = 1$, $x'(0) = 0$.

(c) $x'' + 2x' + x = \sin t$. First pair: $x(0) = 1$, $x'(0) = 1$. Second pair: $x(0) = 1/2$, $x'(0) = 1/2$.

**5.** Show that the transient terms in Example 2 are a homogeneous solution of Eq. (5.121).

**6.** Consider a damped forced mass-spring system that is governed by the differential equation

$$x'' + \tfrac{1}{10}x' + 9x = \sin \alpha t.$$

Suppose you want to control the size of the amplitude of the oscillations of the steady state. Assume that the mass and the spring constant are fixed; the only variable in the problem is the frequency of the forcing, which is determined by $\alpha$.

(a) For what values of $\alpha$ will the amplitude of the steady state remain below 2?

(b) For what values of $\alpha$ will the amplitude of the steady state remain above 3?

**7.** Let $x'' + \tfrac{1}{2}x' + 5x = \sin \alpha t$.

(a) Find a general solution, and argue that the steady state is a particular solution.

(b) Determine the gain and the phase angle associated with the steady state in (a) and plot them as functions of $\alpha$.

(c) For which value(s) of $\alpha$ is the maximum value of the gain, given by Eq. (5.131), the largest?

**8.** Let $x'' + cx' + \omega^2 x = f(t)$, where $c$ and $\omega^2$ represent positive constants.

(a) Suppose $f(t) = e^{-kt}$ for some positive constant $k$. Find a particular solution.

(b) Find the general solution assuming that the homogeneous equation is underdamped.

(c) Describe the qualitative behavior of the general solution. Consider questions such as the following: Does it oscillate? If so, what is the period? What happens as $t \to \infty$? Does the solution tend to infinity? To 0?

(d) Repeat (a)–(c) under the assumption that $k < 0$.

**9.** Let $x'' + cx' + \omega^2 x = f(t)$, where $c$ and $\omega$ represent positive constants.

(a) Suppose $f(t) = a_2 t^2 + a_1 t + a_0$ for some constants $a_0$, $a_1$, and $a_2$. Find a particular solution.

(b) Find the general solution assuming that the homogeneous equation is underdamped.

(c) Describe the qualitative nature of the steady state. Consider questions such as the following: Does it oscillate? If so, what is the period? If the steady state is a polynomial, what is its degree?

(d) Describe the qualitative nature of the general solution.

(e) Suppose $f$ is a polynomial of degree $n$. Can you conjecture what the form of a particular solution is for this case? Try to prove your assertion.

**10.** Let $x'' + cx' + \omega^2 x = f(t) = F \sin \alpha t$, where $c$, $\omega$, $F$, and $\alpha$ represent positive constants. Each of the plots in Fig. 5.29 shows a graph of the forcing $f(t)$ as the dashed curve and the response $x_{ss}(t)$ as the solid curve. Each item in the following list gives the values of $F$ and $\alpha$ for the forcing $f(t)$ in the corresponding plot in Fig. 5.29. In each case, estimate the gain from the plot, then use Eqs. (5.133)–(5.134) to estimate $c$ and $\omega$.

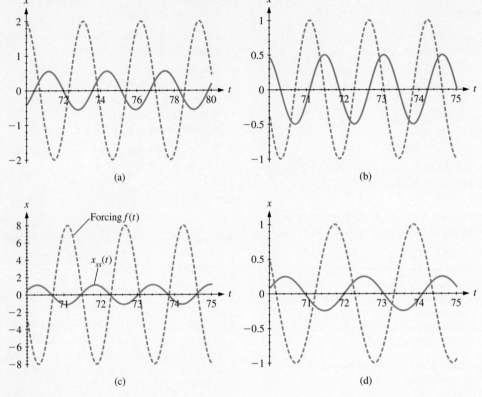

**FIGURE 5.29**   For Exercise 10.

(a) $F = 2, \alpha = 2$      (b) $F = 1, \alpha = 4$

(c) $F = 8, \alpha = 4$      (d) $F = 1, \alpha = 3$

**11.** Let $x'' + cx' + \omega^2 x = f(t) = F \sin \alpha t$ where $c$, $\omega$, $F$, and $\alpha$ represent positive constants.

(a) Let $x_{ss}$ denote the response for a given value of $F$. If $F$ is doubled, how does the amplitude of $x_{ss}$ change?

(b) If $F$ is doubled, how does the gain change?

(c) If the amplitude of the forcing is multiplied by a constant value $\beta$, how does the amplitude of $x_{ss}$ change?

(d) Is the amplitude of the steady state a linear function of $F$? Explain.

(e) Is the gain a linear function of $F$? Explain.

**12.** Let $x'' + cx' + \omega^2 x = F \sin \alpha t$. Show that the maximum value of the gain function is

$$G_{max} = \frac{2}{c\sqrt{4\omega^2 - c^2}}$$

and occurs when

$$\alpha = \tfrac{1}{2}\sqrt{4\omega^2 - 2c^2}.$$

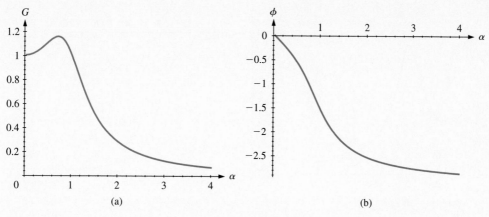

**FIGURE 5.30**    (a) The gain function; (b) the phase angle for Exercise 13.

**13.** Figure 5.30(a) shows a plot of the gain, and Fig. 5.30(b) shows a plot of the phase angle associated with the differential equation

$$x'' + cx' + \omega^2 x = \sin \alpha t$$

as a function of $\alpha$. Estimate $c$ and $\omega$ from the graphs.

**14.** Figure 5.31(a) shows a plot of the gain, and Fig. 5.31(b) shows a plot of the phase angle associated with the differential equation

$$x'' + cx' + \omega^2 x = \sin \alpha t$$

as a function of $\alpha$. Estimate $c$ and $\omega$ from the graphs.

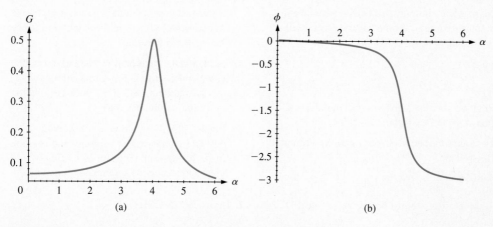

**FIGURE 5.31**    (a) The gain function; (b) the phase angle for Exercise 14.

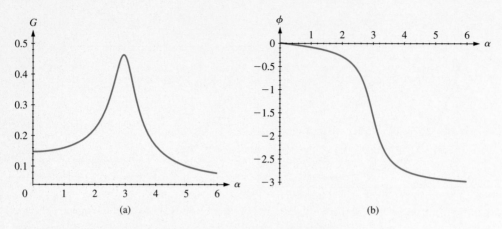

**FIGURE 5.32**    (a) The gain function; (b) the phase angle for Exercise 15.

**15.** Figure 5.32(a) shows a plot of the gain, and Fig. 5.32(b) shows a plot of the phase angle associated with the differential equation

$$x'' + cx' + \omega^2 x = \sin \alpha t$$

as a function of $\alpha$. Estimate $c$ and $\omega$ from the graphs.

**16.** This problem illustrates that beating phenomena exist at most for a limited time when a linear oscillator is damped.

(a) Let

$$x'' + 16x = \sin 5t. \qquad (5.136)$$

Find the solution satisfying the initial conditions $x(0) = 0$ and $x'(0) = -1/9$.

(b) Determine the period of your solution in (a) and its envelope.

(c) Now let

$$x'' + \tfrac{1}{50}x' + 16x = \sin 5t. \qquad (5.137)$$

The numerical values of the coefficients are approximately the same as for Eq. (5.136), but a small amount of damping has been added. Determine the solution using the same initial values as in (a). (Approximate the coefficients as necessary.)

(d) Plot the solution in (c) for $0 \le t \le 30$ and compare it to a plot of the solution in (a) for the same interval of $t$. Can the solution in (c) be said to exhibit beating?

(e) Next, plot the solution in (c) for $200 \le t \le 230$ and compare it to a plot of the solution in (a) for the same interval of $t$. Does the solution in (c) still exhibit beating?

(f) Finally, plot the solution in (c) for $500 \le t \le 530$ and compare it to a plot of the solution in (a) for the same interval of $t$. How would you characterize the solution in (c) now?

**17.** Derive Eq. (5.127).

## 5.9   THE STRUCTURE OF SOLUTIONS

The previous sections discussed the solutions of linear second-order differential equations with *constant coefficients*, that is, differential equations for which the dependent variable and its derivatives are multiplied by simple constants. The purpose of this section is to discuss some of the general theory of linear second-order differential equations. The theory applies also to the case in which the coefficients are not constant. The existence and uniqueness theorem, stated below, allows us to determine when a given linear second-order differential equation has a unique solution. It assures us that a solution exists in principle—even though it usually is not possible to find explicit formulas for the solutions except for special cases. The theory also defines the circumstances under which all solutions of a given linear second-order ordinary differential equation can be written as a linear combination of certain other functions, called *fundamental solutions*.

### An Existence and Uniqueness Theorem

Linear second-order ordinary differential equations can be written most generally as

$$x'' + a(t)x' + b(t)x = g(t), \tag{5.138}$$

where $a$, $b$, and $g$ are functions of the independent variable $t$. If $a$ or $b$ is not constant, then there is no standard procedure to derive explicit solutions of Eq. (5.138). (In practice, however, satisfactory numerical approximations to the solutions usually can be found.) As in the case of first-order equations, there are important scientific questions regarding the solutions of Eq. (5.138). For example:

- How do we know that Eq. (5.138) has any solutions at all? If no solution exists, then the equation cannot be a plausible model for a physical process.

- If there is a solution, is it unique? For instance, if we have found one solution, how do we know that there is not another, different solution that is still consistent with Eq. (5.138)? If solutions are not unique, then Eq. (5.138) has no predictive value.

The following theorem, which we state without proof, gives conditions under which solutions of Eq. (5.138) are guaranteed to exist and be unique.

**Theorem 5.9** *Suppose that a, b, and g are continuous at all t in an open interval I. Then there is a unique solution of Eq. (5.138) satisfying the initial conditions $x(t_0) = x_0$ and $x'(t_0) = v_0$, where $t_0 \in I$. The solution is defined and twice differentiable at every point in I.*

The following examples illustrate the conclusions of Theorem 5.9.

◆ *Example 1*   The solutions of the equation for simple harmonic motion, $x'' + \omega^2 x = 0$, can be written as

$$x(t) = c_1 \cos \omega t + c_2 \sin \omega t. \tag{5.139}$$

There is a unique choice of $c_1$ and $c_2$ that satisfies the initial conditions $x(0) = x_0$ and $x'(0) = v_0$. Theorem 5.9 guarantees that *every* solution of an initial value problem of this form can be written in the form of Eq. (5.139). Moreover, the solution is defined and twice differentiable on the entire real line, because $a$ and $b$ are continuous for every $t$.

If $a$ and $b$ are constant functions, then the roots of the associated characteristic equation can be used to derive solutions of Eq. (5.138). In particular, the roots of the characteristic equation corresponding to $x'' + \omega^2 x = 0$ are $\lambda = \pm \omega i$. Theorem 5.9 implies that the functions given by Eq. (5.139) and the function given by $b_1 e^{i\omega t} + b_2 e^{-i\omega t}$ must represent the same solution when the constants $c_1$, $c_2$, $b_1$, and $b_2$ are chosen to solve the same initial value problem. We used this result in Section 5.4 to derive Euler's formula.                                                                          ■

◆ *Example 2*   A linear, unforced mass-spring system is governed by the second-order equation

$$mx'' + cx' + kx = 0. \tag{5.140}$$

The coefficient functions are $a(t) = c/m$ and $b(t) = k/m$, and $g(t) = 0$. The functions $a, b$, and $g$ are constant and so are continuous at every $t$. Therefore, the solution satisfying the initial conditions $x(t_0) = x_0$ and $x'(t_0) = v_0$ exists and is unique. The interval of existence of the solution is the entire real line. This example shows that Theorem 5.1, discussed in Section 5.1, is a corollary of Theorem 5.9.                                        ■

◆ *Example 3*   The solutions of the differential equation

$$x'' + tx' + t^3 x = 0$$

are twice differentiable for every $t$ and are guaranteed to be unique, because the coefficient functions $a(t) = t$ and $b(t) = t^3$ are continuous at every $t$.                              ■

◆ *Example 4*   The solutions of the differential equation

$$x'' + t^{-1} x' + t^3 x = 0$$

are twice differentiable and are guaranteed to be unique on any interval that does not contain $t = 0$, because the coefficient functions $a(t) = t^{-1}$ and $b(t) = t^3$ are continuous on any interval that excludes $t = 0$. For instance, the solutions satisfying the initial conditions $x(1) = x_0$ and $x'(1) = v_0$ are defined at every positive $t$.                              ■

## The Wronskian and Linear Independence

The existence and uniqueness theorem states when it is possible in principle to find a solution of the linear second-order differential equation

$$x'' + a(t)x' + b(t)x = 0 \tag{5.141}$$

with the specified initial conditions $x(0) = x_0$ and $x'(0) = v_0$. This section considers the circumstances under which the superposition principle can be used to construct *all* solutions of Eq. (5.141).

Suppose that $x_1(t)$ and $x_2(t)$ solve Eq. (5.141). The superposition principle implies that

$$x(t) = c_1 x_1(t) + c_2 x_2(t) \tag{5.142}$$

is also a solution. An important question is whether *all* initial conditions can be satisfied with a suitable choice of the constants $c_1$ and $c_2$. Suppose that the initial conditions are specified as $x(t_0) = \alpha$ and $x'(t_0) = \beta$. The linear combination (5.142) satisfies the initial condition if constants $c_1$ and $c_2$ can be found such that

$$\begin{aligned} c_1 x_1(t_0) + c_2 x_2(t_0) &= \alpha, \\ c_1 x_1'(t_0) + c_2 x_2'(t_0) &= \beta. \end{aligned} \tag{5.143}$$

The solution is

$$c_1 = \frac{x_2'(t_0)\alpha - x_2(t_0)\beta}{x_1(t_0)x_2'(t_0) - x_1'(t_0)x_2(t_0)},$$

$$c_2 = \frac{x_1(t_0)\beta - \alpha x_1'(t_0)}{x_1(t_0)x_2'(t_0) - x_1'(t_0)x_2(t_0)},$$

provided that

$$x_1(t_0)x_2'(t_0) - x_1'(t_0)x_2(t_0) \neq 0. \tag{5.144}$$

The condition (5.144) implies that the determinant

$$\det \begin{pmatrix} x_1(t) & x_2(t) \\ x_1'(t) & x_2'(t) \end{pmatrix}, \tag{5.145}$$

evaluated at $t = t_0$, must be nonzero. The determinant (5.145) is called the *Wronskian determinant* or simply the *Wronskian* of the functions $x_1$ and $x_2$. We summarize these observations in the following theorem.

**Theorem 5.10**   *If the functions $x_1$ and $x_2$ are solutions of the linear, second-order, homogeneous differential equation (5.141) and if the Wronskian determinant (5.145) is nonzero for every $t_0$ in the domain of $x_1$ and $x_2$, then there is a unique choice of the* *constants $c_1$ and $c_2$ such that*

$$x(t) = c_1 x_1(t) + c_2 x_2(t)$$

*solves Eq. (5.141) and satisfies the initial condition specified at $t = t_0$.*

**Definition 5.11**  *Two solutions $x_1$ and $x_2$ of Eq. (5.141) that satisfy the hypotheses of Theorem 5.10 are called* fundamental solutions. *The linear combination $c_1 x_1 + c_2 x_2$ is called a* general solution.

✦ *Example 5*  The functions $x_1(t) = \cos t$ and $x_2(t) = \sin t$ are fundamental solutions of $x'' + x = 0$. Their Wronskian is

$$\det \begin{pmatrix} \cos t & \sin t \\ -\sin t & \cos t \end{pmatrix} = \cos^2 t + \sin^2 t = 1.$$

Since the Wronskian is nonzero for every $t$, Theorem 5.10 implies that it is always possible to find a unique choice of constants $c_1$ and $c_2$ such that $c_1 \cos t + c_2 \sin t$ satisfies the initial condition $x(t_0) = x_0$, $x'(t_0) = v_0$. Therefore, the linear combination $c_1 \cos t + c_2 \sin t$ is a general solution of $x'' + x = 0$.                                          ■

The next result, called Abel's theorem, implies that if the system (5.143) can be solved for one initial condition at $t_0$, then (5.143) can be solved for any initial condition in the interval of existence of the solution of Eq. (5.141).

**Theorem 5.12**  *(Abel's Theorem)*    *Let $x_1$ and $x_2$ be two solutions of Eq. (5.141) whose interval of existence is $I$. Then either the Wronskian determinant of $x_1$ and $x_2$ is zero at every $t \in I$ or the Wronskian of $x_1$ and $x_2$ is never zero for any $t \in I$.*

**Proof**  Since $x_1$ and $x_2$ solve Eq. (5.141), we have

$$x_1'' + a(t)x_1' + b(t)x_1 = 0,$$
$$x_2'' + a(t)x_2' + b(t)x_2 = 0,$$

which implies that

$$x_1'' x_2 + a(t)x_1' x_2 + b(t)x_1 x_2 = 0,$$
$$x_1 x_2'' + a(t)x_1 x_2' + b(t)x_1 x_2 = 0,$$

so

$$a(t)\left(x_1' x_2 - x_1 x_2'\right) = x_1 x_2'' - x_1'' x_2. \tag{5.146}$$

The term on the left-hand side of Eq. (5.146) is just $-a$ times the Wronskian of $x_1$ and $x_2$, which we denote $W(t)$. The right-hand side is $W'(t)$. Therefore, we can rewrite

Eq. (5.146) as

$$W'(t) + a(t)W(t) = 0,$$

which is a separable first-order differential equation for $W(t)$. Its solution is

$$W(t) = W_0 \exp\left(-\int a(t)\,dt\right),$$

where $W_0 = W(t_0)$. As the exponential function is never 0, the Wronskian $W(t)$ is zero if and only if $W_0 = 0$.   ∎

Abel's theorem says that if the system (5.143) can be solved uniquely for an initial condition specified at a convenient time, such as $t_0 = 0$, then (5.143) can be solved if the initial condition is specified at any other time $t_0$ in the interval of existence of the solution.

The case of a zero Wronskian has a special interpretation.

**Theorem 5.13**   *Let $x_1$ and $x_2$ be two solutions of Eq. (5.141). The Wronskian of $x_1$ and $x_2$ is zero at every point in the interval of existence of the solutions if and only if $x_1(t)$ is a constant multiple of $x_2(t)$.*

**Proof**   Suppose that $x_1$ is a constant multiple of $x_2$, that is, $x_1(t) = cx_2(t)$ for every $t$. Then the Wronskian $W(t)$ is

$$W(t) = x_1(t)cx_1'(t) - x_1'(t)cx_1(t) = 0.$$

Conversely, suppose $W(t) = 0$ for every $t$. Then we can write

$$\frac{x_1'(t)}{x_1(t)} = \frac{x_2'(t)}{x_2(t)} \tag{5.147}$$

as long as $x_1(t)$ and $x_2(t)$ are not zero. Equation (5.147) can be integrated to get

$$\ln|x_1(t)| = \ln|x_2(t)| + B,$$

where $B$ is an arbitrary constant of integration. When we exponentiate both sides, we find that

$$x_1(t) = Cx_2(t),$$

where $C = \pm e^B$ depending on the signs of $x_1$ and $x_2$. Hence, $x_1$ and $x_2$ are constant multiples of each other.   ∎

**Definition 5.14**   *The solutions $x_1$ and $x_2$ are linearly dependent on the interval $I$ if they are constant multiples of each other on $I$. Equivalently, $x_1$ and $x_2$ are linearly dependent if there exist constants $c_1$ and $c_2$ that are not both zero for which the linear combination $c_1x_1(t) + c_2x_2(t) = 0$ for every $t \in I$. Two solutions that are not linearly dependent on $I$ are said to be linearly independent on $I$.*

Theorem 5.13 says that if two solutions of the same linear, second-order, homogeneous differential equation are linearly dependent, then their Wronskian is always 0. As a result, two linearly dependent solutions cannot be a fundamental solution, because the system (5.143) cannot be solved uniquely for any given initial condition.

## Consequences of Linear Independence

Abel's theorem (Theorem 5.12) states that the Wronskian of any two fundamental solutions $x_1$ and $x_2$ of Eq. (5.141) either is zero at every $t$ in the interval of existence or else is never zero at any $t$ in the interval of existence. It follows that:

1. $x_1$ and $x_2$ cannot have common zeros. In other words, for a given value of $t$, we cannot have $x_1(t) = x_2(t) = 0$. (It is possible for $x_1(t) = 0$ for one value of $t$ and for $x_2(t) = 0$ for a different value of $t$, however.)

2. $x_1$ and $x_2$ cannot have relative extrema at the same value of $t$. If they did, then $x_1'(t) = x_2'(t) = 0$, which would make their Wronskian zero, contradicting Abel's theorem.

Figure 5.33(a) shows the graph of two linearly dependent solutions—because they are constant multiples of one another, the graphs are identical up to a constant "stretch factor" in the vertical direction. Notice that they must have zeros and extreme values at common values of $t$. Figure 5.33(b) shows two solutions whose graphs are identical except that one is shifted by a constant amount in the horizontal direction. These solutions are linearly independent, because the phase shift depicted in Fig. 5.33(b) does not make them constant multiples of one another. Notice that the solutions would be linearly dependent if the phase shift makes the graphs coincide or if the phase shift makes the graphs mirror images of each other.

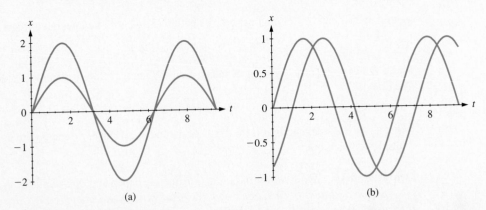

(a)                                                                    (b)

**FIGURE 5.33**    (a) The graph of two linearly dependent solutions; (b) the graph of two linearly independent solutions.

Abel's theorem, together with the existence and uniqueness theorem, can be used to show that the characteristic equation discussed in Section 5.3 generates all possible solutions of second-order linear differential equations with constant coefficients, as the following examples illustrate.

✦ *Example 6* Let

$$x'' + ax' + bx = 0, \qquad (5.148)$$

where $a$ and $b$ are constants. Suppose the corresponding characteristic equation has distinct real roots, $\lambda_1 \neq \lambda_2$. Then every solution of Eq. (5.148) can be written as the linear combination

$$x(t) = c_1 e^{\lambda_1 t} + c_2 e^{\lambda_2 t},$$

where $c_1$ and $c_2$ are determined uniquely by the initial conditions $x(t_0) = x_0$ and $x'(t_0) = v_0$.

To see this, observe that the Wronskian of $x_1$ and $x_2$ is

$$W = \det \begin{pmatrix} e^{\lambda_1 t} & e^{\lambda_2 t} \\ \lambda_1 e^{\lambda_1 t} & \lambda_2 e^{\lambda_2 t} \end{pmatrix} = (\lambda_2 - \lambda_1) e^{(\lambda_2 + \lambda_1)t},$$

which is never 0 for any $t$. Theorem 5.10 implies that every solution can be written as a linear combination of $e^{\lambda_1 t}$ and $e^{\lambda_2 t}$. Thus, $e^{\lambda_1 t}$ and $e^{\lambda_2 t}$ are fundamental solutions of Eq. (5.148); they are also linearly independent. ∎

✦ *Example 7* Suppose the characteristic equation has two complex conjugate roots, $\lambda = \alpha \pm \beta i$. Euler's formula leads to two solutions, $x_1(t) = e^{\alpha t} \cos \beta t$ and $x_2(t) = e^{\alpha t} \sin \beta t$, which are defined on the entire real line. Theorem 5.10 and Abel's theorem imply that if a unique linear combination of $x_1$ and $x_2$ can be found to satisfy an initial condition at $t = 0$, say, then a unique linear combination can be found if the initial condition is given at any other value of $t$. A straightforward calculation shows that the Wronskian of $x_1$ and $x_2$ at $t = 0$ is equal to $\beta$. Thus, $x_1$ and $x_2$ are fundamental solutions; they are also linearly independent. ∎

✦ *Example 8* Suppose the characteristic equation has a double real root $\lambda$. The solutions $x_1(t) = c_1 e^{\lambda t}$ and $x_2(t) = c_2 e^{\lambda t}$ are linearly dependent, because their Wronskian is always zero. Thus, $x_1$ and $x_2$ are not fundamental solutions of Eq. (5.148) when the characteristic equation has a double real root.

However, the solutions $x_1(t) = e^{\lambda t}$ and $x_2(t) = t e^{\lambda t}$ *are* linearly independent, because their Wronskian is $e^{2\lambda t}$, which is never zero. Hence, the linear combination $x(t) = c_1 x_1(t) + c_2 x_2(t)$ is a general solution of Eq. (5.148). ∎

✦ *Example 9* The functions $x_1(t) = \sin t$ and $x_2(t) = \sin 2t$ are not constant multiples of each other. However, their Wronskian is 0 when $t = 0$, which contradicts Theorem 5.12.

Thus, $x_1$ and $x_2$ cannot be solutions of the same second-order, linear, homogeneous differential equation.                                                                                    ∎

## *Exercises*

**1.** Use Theorem 5.9 where possible to determine which of the equations in the following list has a unique solution satisfying the initial conditions $x(0) = x_0$ and $x'(0) = v_0$ in the interval containing $t = 0$.

(a) $t^2 x'' + t x' + x = 0$

(b) $x'' + t x' + t^2 x = 0$

(c) $x'' + t x' + t x^2 = 0$

(d) $x'' + t x' + x = \sin t$

(e) $x'' + t x' + x = \sin x$

**2.** Repeat Exercise 1 but consider an interval containing $t = 1$.

**3.** Which of the following pairs of functions are linearly independent solutions of $x'' = -9x$?

(a) $x_1(t) = \sin 3t,$
    $x_2(t) = \cos (3t + 4)$

(b) $x_1(t) = \cos 3t,$
    $x_2(t) = \cos (3t + 2)$

(c) $x_1(t) = \cos 3t,$
    $x_2(t) = \sin (3t + \pi/2)$

(d) $x_1(t) = \cos 3t,$
    $x_2(t) = \sin (3t - \pi/2)$

**4.** Let $x_1(t) = e^t \sin t$ and $x_2(t) = \sin t$.

(a) Show that $x_1$ and $x_2$ are not constant multiples of each other.

(b) Show that the Wronskian of $x_1$ and $x_2$ is zero for $t = n\pi$, where $n$ is any integer.

(c) Explain how Abel's theorem can be used to determine whether $x_1$ and $x_2$ can be solutions of the same homogeneous, second-order, linear differential equation.

**5.** Let $x_1(t) = e^t$ and $x_2(t) = e^{-t}$.

(a) Observe that $x_1(0) = x_2(0)$. Can $x_1$ and $x_2$ be linearly independent solutions of the same second-order differential equation? If so, exhibit a linear second-order equation for which $x_1$ and $x_2$ are solutions.

(b) Can $x_1$ and $x_2$ be solutions of the same homogeneous, linear, second-order differential equation if the initial condition is specified at $t_0 \neq 0$?

(c) Can $x_1$ and $x_2$ be solutions of the same homogeneous, linear, second-order differential equation if the initial condition is specified at $t_0 = 0$?

**6.** Let $t^2 p'' + t p' + p = 0$.

(a) Show that $p_1(t) = \cos (\ln |t|)$ and $p_2(t) = \sin (\ln |t|)$ are two linearly independent solutions.

(b) Find the solution that satisfies the initial conditions $p(1) = 1$ and $p'(1) = 2$.

(c) What is the interval of existence of the solution in (b)?

(d) Find the solution that satisfies the initial conditions $p(-1) = 1$ and $p'(-1) = 2$.

(e) What is the interval of existence of the solution in (d)?

**7.** Can $\cos t$ and $\cos 2t$ be solutions of the same second-order, homogeneous linear differential equation? Explain.

**8.** If $x_1$ and $x_2$ are two linearly independent solutions of Eq. (5.141), then they cannot both be zero at the same value of $t$ and they cannot have relative extrema at the same $t$. Can they have inflection points at the same $t$? If so, what restrictions must Eq. (5.141) satisfy?

# Chapter 6

---

# THE GEOMETRY OF SYSTEMS OF DIFFERENTIAL EQUATIONS

---

This chapter is an introduction to the subject of coupled systems of differential equations, which arise naturally in a variety of situations. The main goal is to introduce some of the basic geometric tools for analyzing the qualitative behavior of systems of first-order differential equations. We also show how to extend Euler's method for finding numerical approximations of the solutions to such systems.

The principal emphasis is on systems of two equations. We consider two examples in detail:

- A population model of a predator and its prey leads to a coupled system of differential equations. The rate of growth of the predator population depends on the size of the prey population, and conversely.

- The rate at which the position and velocity of the bob of a pendulum change depends on the current position and velocity of the bob.

Additional examples of first-order systems are discussed in Chapters 7 and 8.

Section 6.1 introduces some basic examples and terminology. Section 6.2 shows how to draw phase portraits for systems of two equations and shows how they can be used to determine the qualitative properties of solutions of some simple systems. Section 6.3 describes the basic kinds of equilibrium points that are found in systems of ordinary differential equations. Section 6.4 shows how Euler's method can be extended to find numerical approximations to the solutions of first-order systems of ordinary differential equations.

# 6.1   Examples and Terminology

The previous chapters consider first- and second-order differential equations with a single dependent variable. In many applications, we are interested in the interaction between two or more variables, such as a predator and its prey. A simple model is derived in the next example.

◆ *Example 1*    Let $x(t)$ denote the size of a predator population at time $t$ and let $y(t)$ denote the size of a prey population. (For instance, foxes eat hares, and big fish eat little fish.) In general, it is not easy to devise accurate models for the rate of growth of such populations. Many variables can affect a given population, including the time of year, the weather, food supplies, and the availability of nesting sites.

For the sake of illustration, we make the following idealizations about a predator population and its prey and derive a corresponding model:

1. The predator $x$ and its prey $y$ are the only relevant species in the ecosystem. In particular, $y$ is the only food source for $x$, there is no predation on $x$, and $y$ has no other predators.

2. The prey has an unlimited food supply.

3. Both populations live in a certain fixed region, and there is no immigration or emigration. Moreover, the environment in the region is homogeneous, and the two populations are evenly dispersed throughout the region.

One way to construct a model for these populations is to consider what must happen in certain limiting circumstances. For example, suppose that the prey population becomes extinct ($y = 0$). Since $y$ is the only food source for $x$, the predator population must eventually become extinct.

The simplest model for such a situation is to suppose that the predator population $x$ declines at a rate proportional to its present size. That is, in the absence of prey,

$$x' = -a_1 x, \tag{6.1}$$

where $a_1$ is a positive constant of proportionality that may be viewed as the net death rate for the predator population.

If the predator $x$ becomes extinct and the prey $y$ has an unlimited food supply, then the prey population can increase unchecked. If we suppose that the rate of increase of the population is proportional to its present size, then

$$y' = a_2 y, \tag{6.2}$$

where $a_2$ is a positive constant of proportionality that may be regarded as an intrinsic rate of increase of the prey.

Equations (6.1) and (6.2) are independent differential equations that represent limiting circumstances for the two populations. The interesting case is where both populations are positive and predation occurs.

A simple model of the effects of predation can be derived as follows. Since both populations are assumed to be uniformly dispersed, the likelihood that a predator encounters prey increases with the size of the prey population. Assume that the likelihood of such encounters is proportional to the size of the prey population and that the abundance of prey directly affects the growth rate of the predator population; then

$$\text{rate of growth of predators} = ky,$$

where $k$ is a factor of proportionality.

A moment's thought suggests that $k$ cannot be a fixed constant, even under our simplified assumptions. Given that the prey population is at some fixed level, the likelihood of encounters with predators depends on the size of the predator population. The simplest assumption is that the likelihood of encounters for a fixed $y$ is proportional to the predator population $x$, that is,

$$k = k_1 x,$$

where $k_1$ is another positive constant. Because we suppose that the rate of predation directly affects the reproductive rate of the predators, we have

$$\text{rate of growth of predators} = k_1 xy.$$

Hence,

$$x' = -a_1 x + k_1 xy. \tag{6.3}$$

In a similar way, we assume that predation removes individuals directly from the prey population and that the likelihood of encounters between the prey and predators is proportional to the predator population $x$. Therefore,

$$y' = a_2 y - k_2 xy, \tag{6.4}$$

where $k_2$ is a positive constant of proportionality.    ∎

Equations (6.3) and (6.4) define a system of first-order ordinary differential equations, called the *Lotka-Volterra predator-prey model*. The rate of growth of each population depends in part on the size of the other. We say that the Lotka-Volterra model is a *coupled* system of two differential equations. There is one independent variable, namely $t$ (time), and two dependent variables, $x$ and $y$. The equations are first-order because each equation involves only the first derivative of one of the dependent variables. The equations are ordinary because the derivatives of $x$ and $y$ are taken with respect to a single independent variable.

It is not possible to find explicit formulas for $x(t)$ and $y(t)$ as functions of $t$. Nevertheless, it is possible to describe the qualitative behavior of the solutions and to find numerical approximations of the solutions of Eqs. (6.3) and (6.4). See Section 6.2.

## Higher-Order Equations Are Equivalent to First-Order Systems

This section illustrates the following fact:

*Every second-order scalar ordinary differential equation is equivalent to a system of two first-order ordinary differential equations.*

(It is *not* true that every pair of first-order equations is equivalent to a second-order scalar equation.)

This equivalence has practical consequences, particularly in numerical simulations. Most numerical methods for computing solutions of differential equations are designed for first-order systems (see Euler's method in Section 6.4), but they can be applied to a second- and higher-order scalar differential equation by finding the equivalent first-order system equations. The following examples show how to find an equivalent system of differential equations from a higher-order scalar equation.

✦ *Example 2*    Let

$$x'' = -x, \qquad x(0) = x_0, \qquad x'(0) = v_0. \tag{6.5}$$

Let $x_1(t) = x(t)$, and define $x_2(t) = x_1'(t)$. (You can think of $x_1$ as the angular position at time $t$ of the bob of a linear pendulum and $x_2$ as the angular velocity of the bob.) Notice that $x_2' = x_1'' = x''$. Therefore, the equivalent first-order system is

$$x_1' = x_2,$$
$$x_2' = -x_1,$$

where the second equation follows directly from Eq. (6.5). The initial conditions are equivalent to $x_1(0) = x_0$ and $x_2(0) = v_0$.                                                                            ∎

✦ *Example 3*    Every third-order scalar differential equation is equivalent to a system of three first-order equations. For example, let

$$x''' + \left(x'\right)^2 - x^3 = B \sin t. \tag{6.6}$$

Let $x_1 = x$, $x_2 = x_1'$, and $x_3 = x_2'$. Then $x_3' = x_2'' = x_1''' = x'''$. Equation (6.6) implies that

$$x''' = B \sin t + x^3 - \left(x'\right)^2,$$

so the equivalent system of first-order equations is

$$x_1' = x_2,$$
$$x_2' = x_3,$$
$$x_3' = x_1^3 - x_2^2 + B \sin t.$$

∎

In general, any scalar ordinary differential equation of order $n$ can be rewritten as a system of $n$ first-order differential equations by extending the method in the previous examples (see Exercise 12). For this reason, the theory of first-order systems of equations applies also to second- and higher-order scalar differential equations.

## Notation

Complex phenomena typically involve a large number of interacting variables, and the corresponding models consist of coupled systems of differential equations. A system of $n$ first-order ordinary differential equations can be written as

$$\frac{dx_1}{dt} = f_1(t, x_1, x_2, \ldots, x_n),$$

$$\frac{dx_2}{dt} = f_2(t, x_1, x_2, \ldots, x_n),$$

$$\vdots$$

$$\frac{dx_n}{dt} = f_n(t, x_1, x_2, \ldots, x_n).$$

(6.7)

Equations (6.7) are said to be first-order because they involve only the first derivative of any dependent variable, and they are ordinary because all derivatives are taken with respect to a single independent variable, $t$.

The notation in Eqs. (6.7) is cumbersome. Vector notation provides a more convenient way to express systems of differential equations, as we now demonstrate. The vector-valued function $\mathbf{x}(t)$ from $\mathbf{R}$ to $\mathbf{R}^n$ is written as

$$\mathbf{x}(t) = \begin{pmatrix} x_1(t) \\ x_2(t) \\ \vdots \\ x_n(t) \end{pmatrix},$$

(6.8)

where each of the functions $x_1, \ldots, x_n$ is a function from $\mathbf{R}$ to $\mathbf{R}$, called a *component function* or simply a *component* of **x**. The derivative of **x** with respect to $t$ is

$$\frac{d\mathbf{x}}{dt}(t) = \mathbf{x}'(t) = \begin{pmatrix} x_1'(t) \\ x_2'(t) \\ \vdots \\ x_n'(t) \end{pmatrix}, \tag{6.9}$$

where each component function is differentiated with respect to $t$. We say that **x** is *continuous* on an interval $I$ whenever each of its component functions is continuous on $I$ and that **x** is *differentiable* on $I$ whenever each of its component functions is differentiable on $I$.

Functions from $\mathbf{R}^m$ to $\mathbf{R}^n$ are written in an analogous fashion. The notation $\mathbf{f}(\mathbf{x})$ denotes a function **f** from $\mathbf{R}^m$ to $\mathbf{R}^n$, that is,

$$\mathbf{f}(\mathbf{x}) = \mathbf{f}(x_1, \ldots, x_m) = \begin{pmatrix} f_1(x_1, \ldots, x_m) \\ \vdots \\ f_n(x_1, \ldots, x_m) \end{pmatrix}, \tag{6.10}$$

where each component function in Eq. (6.10) is a map from $\mathbf{R}^m$ to $\mathbf{R}$.

Boldface type is used throughout the text to denote vector-valued quantities, regardless of the number of components. Italic letters are used to denote the components within a vector or vector-valued function. We write the $n$-vector **x** as

$$\mathbf{x} = \begin{pmatrix} x_1 \\ \vdots \\ x_n \end{pmatrix} \qquad \text{or as} \qquad \mathbf{x} = (x_1, \ldots, x_n).$$

The sum of two $n$-vectors is defined as

$$\mathbf{x} + \mathbf{y} = \begin{pmatrix} x_1 + y_1 \\ \vdots \\ x_n + y_n \end{pmatrix},$$

and the scalar multiplication is defined as

$$c\mathbf{x} = \begin{pmatrix} cx_1 \\ \vdots \\ cx_n \end{pmatrix},$$

where $c$ is any constant. (The term *scalar* refers to a number or to a function whose value is a number.)

Whenever confusion might arise, we write $\mathbf{x}(t)$ to emphasize that **x** is a function of $t$ and not a constant vector. The boldface **0** is used to denote both the zero vector and the zero function; the number of components that it represents will always be clear by context.

**Definition 6.1**  *The relation*

$$\mathbf{x}' = \mathbf{f}(t, \mathbf{x}(t)) \tag{6.11}$$

*defines a first-order system of n ordinary differential equations whenever* $\mathbf{x}(t)$ *is a function from* $\mathbf{R}$ *to* $\mathbf{R}^n$ *and* $\mathbf{f}(t, \mathbf{x})$ *is a function from* $\mathbf{R}^{n+1}$ *to* $\mathbf{R}^n$ *. The differential equations are* ordinary *because all derivatives are taken with respect to a single independent variable, namely* $t$ *, and the equations are* first order *because only the first derivative is involved. The* initial condition *is* $\mathbf{x}(t_0) = \mathbf{x}_0$ *and specifies the value of* $\mathbf{x}$ *at some initial time* $t_0$ *. We say that* $\mathbf{x}(t)$ *is a* solution *of Eq. (6.11) if* $\mathbf{x}$ *is differentiable and satisfies Eq. (6.11) for every* $t$ *in an interval containing* $t_0$ *. The component functions* $x_1, \ldots, x_n$ *of* $\mathbf{x}$ *are the* dependent variables, *and* $t$ *is the* independent variable.

**Definition 6.2**  *The system (6.11) is* autonomous *if it can be written equivalently as*

$$\mathbf{x}' = \mathbf{g}(\mathbf{x}), \tag{6.12}$$

*where* $\mathbf{g}$ *is a function from* $\mathbf{R}^n$ *to* $\mathbf{R}^n$ *that does not depend explicitly on* $t$ *.*

The systems (6.11) and (6.12) are often called *vector fields*. A differential equation with a single variable (as discussed in Chapters 1–5) is called a *scalar* differential equation.

✦ *Example 4*  The predator-prey system in Example 1 can be written as $\mathbf{x}' = \mathbf{g}(\mathbf{x})$, where

$$\mathbf{g}(\mathbf{x}) = \begin{pmatrix} -a_1 x + k_1 xy \\ a_2 y - k_2 xy \end{pmatrix},$$

and $\mathbf{x} = (x, y)$ is a vector whose components are the predator and the prey population, respectively.  ∎

✦ *Example 5*  Let

$$\begin{aligned} x_1' &= x_2, \\ x_2' &= -x_1, \end{aligned} \tag{6.13}$$

with the initial conditions $x_1(0) = 1$ and $x_2(0) = 0$. In Example 2, we showed that Eq. (6.13) is equivalent to the scalar, second-order initial value problem

$$x'' = -x, \qquad x(0) = 1, \qquad x'(0) = 0, \tag{6.14}$$

which has the solution $x(t) = \cos t$. We can rewrite the initial value problem as

$$\mathbf{x}' = \mathbf{g}(\mathbf{x}), \qquad \mathbf{x}(0) = \begin{pmatrix} 1 \\ 0 \end{pmatrix},$$

where

$$\mathbf{x}(t) = \begin{pmatrix} x_1(t) \\ x_2(t) \end{pmatrix}$$

and

$$\mathbf{g}(\mathbf{x}) = \begin{pmatrix} x_2 \\ -x_1 \end{pmatrix}.$$

Notice that the system (6.13) is autonomous, as the rate of change of each quantity does not depend explicitly on time. Since we define $x_1 = x$ and $x_2 = x_1'$, the solution of Eq. (6.14) generates a solution of the system (6.13) given by

$$\mathbf{x}(t) = \begin{pmatrix} \cos t \\ -\sin t \end{pmatrix}.$$

We can check by substitution that $\mathbf{x}$ solves Eq. (6.13), as follows:

$$\begin{aligned} \frac{d\mathbf{x}}{dt} &= \frac{d}{dt} \begin{pmatrix} \cos t \\ -\sin t \end{pmatrix} \\ &= \begin{pmatrix} -\sin t \\ -\cos t \end{pmatrix} \\ &= \mathbf{g}(\mathbf{x}), \end{aligned}$$

which is valid for every real number $t$. In particular, $\mathbf{x}(0) = (1, 0)$, so $\mathbf{x}$ satisfies the initial condition. ∎

✦ *Example 6*   Let

$$\begin{aligned} x_1' &= t^2 x_2, \\ x_2' &= x_1 - t. \end{aligned} \qquad (6.15)$$

The right-hand side of the system (6.15) depends explicitly on $t$ as well as on $\mathbf{x}$. Therefore, the system (6.15) is not autonomous. An equivalent formulation is

$$\mathbf{x}' = \mathbf{f}(t, \mathbf{x}) = \begin{pmatrix} t^2 x_2 \\ x_1 - t \end{pmatrix},$$

where $\mathbf{x}(t) = (x_1(t), x_2(t))$. ∎

## Equilibrium Solutions (Fixed Points)

As discussed in Chapter 1, autonomous differential equations often have constant solutions, called *equilibrium solutions*. It is straightforward to extend the definition of equilibrium solutions from autonomous scalar equations to autonomous vector fields.

**Definition 6.3**  *Let* **c** *be a constant vector. The constant function* $\mathbf{x}(t) = \mathbf{c}$ *is an* equilibrium solution *(or* fixed point*) of the autonomous system*

$$\mathbf{x}' = \mathbf{g}(\mathbf{x})$$

*if* $\mathbf{g}(\mathbf{c}) = \mathbf{0}$. *A system of ordinary differential equations is* homogeneous *if the zero function is an equilibrium solution.*

If $\mathbf{g}$ is a function from $\mathbf{R}^n$ to $\mathbf{R}^n$ and $\mathbf{c}$ is a constant vector in $\mathbf{R}^n$, then the relation $\mathbf{g}(\mathbf{c}) = \mathbf{0}$ defines a system of $n$ equations in the $n$ unknowns $c_1, c_2, \ldots, c_n$.

✦ *Example 7*   The function $\mathbf{x}(t) = \mathbf{0}$ is an equilibrium solution of the autonomous system

$$\mathbf{x}'(t) = \begin{pmatrix} x_1' \\ x_2' \end{pmatrix} = \mathbf{g}(\mathbf{x}) = \begin{pmatrix} x_2 \\ -x_1 \end{pmatrix}. \tag{6.16}$$

(See Example 5.)  In this example, $\mathbf{0}$ can be interpreted physically as the rest point of a linear pendulum.  That is, if the pendulum is placed in a straight-down position and given no initial velocity, then the pendulum does not move.  Definition 6.3 implies that the system (6.16) is homogeneous.  ∎

Although the terminology is confusing at first, we often say that the equilibrium solution $\mathbf{x} = \mathbf{0}$ is a *fixed point* or *equilibrium point* of the system (6.16).  The solution curve satisfying the initial condition $(x_1(0), x_2(0)) = (0, 0)$ is a single point, namely $\mathbf{0}$. In other words, if we start at $\mathbf{0}$, then we stay at $\mathbf{0}$ forever under Eqs. (6.16), because $\mathbf{x}' = \mathbf{g}(\mathbf{0}) = \mathbf{0}$.  Therefore, we call the origin a fixed point. Every fixed point corresponds to an equilibrium solution, that is, a solution that is constant in time.

✦ *Example 8*   The constant function $\mathbf{x}(t) = (1, 1, 1)$ is an equilibrium solution (fixed point) of the autonomous system

$$x_1' = x_2 - 1,$$
$$x_2' = x_3 - 1,$$
$$x_3' = x_1^3 - x_3^2.$$

This system is not homogeneous, because $\mathbf{0}$ is not an equilibrium solution.  ∎

## Exercises

**1.** Show by substitution that the function                is a solution of the system

$$\mathbf{x}(t) = \begin{pmatrix} 1 - e^{2t} \\ 1 + e^{2t} \end{pmatrix}$$

$$x_1' = x_1 - x_2,$$
$$x_2' = -x_1 + x_2.$$

**2.** Show by substitution that the function

$$\mathbf{x}(t) = e^t \begin{pmatrix} \sin 2t + \cos 2t \\ \sin 2t - \cos 2t \end{pmatrix}$$

is a solution of the system

$$x_1' = x_1 - 2x_2,$$
$$x_2' = 2x_1 + x_2.$$

**3.** Let

$$x'' + \omega^2 x = 0. \qquad (6.17)$$

(a) Equation (6.17) describes simple harmonic motion. Find the general solution of Eq. (6.17).

(b) Find the first-order system that is equivalent to Eq. (6.17).

(c) Use your answer in (a) to determine the general solution of the system that you found in (b).

**4.** Write each of the following differential equations as an equivalent system of first-order equations.

(a) $x'' + 3x' + 2x = 0$

(b) $x'' + 6x' + 4x = \sin t$

(c) $x'' + \sin x = 0$

(d) $x'' + \frac{1}{10}x + \sin x = 2\cos t$

(e) $x''' + (x')^2 + x(x - 1) = 0$

**5.** Find a constant $B$ so that the system

$$x_1' = 3x_1 + 2x_2 - 7,$$
$$x_2' = 4x_1 + 5x_2 + B$$

has the equilibrium solution $(x_1, x_2) = (1, 2)$.

**6.** Determine which of the systems in the following list is homogeneous.

(a) $x_1' = \sin x_1 - x_2$
    $x_2' = \cos x_1$

(b) $x_1' = x_1^2 - x_2^2$
    $x_2' = 2x_1 - 2x_2$

(c) $x_1' = \sin x_1 - x_2$
    $x_2' = 1 - \cos x_1$

(d) $x_1' = x_2$
    $x_2' = -\sin x_1 - x_2 + \sin t$

(e) $x_1' = t \sin x_1 - x_2$
    $x_2' = \cos x_1$

**7.** Identify the autonomous systems in Exercise 6.

**8.** For each initial value problem in the list below,

- find the equivalent first-order system and determine the appropriate initial condition;

- find the solution $x(t)$ of the second-order equation and use it to determine the solution $\mathbf{x}(t)$ of the equivalent first-order system.

(a) $x'' + 9x = 0, x(0) = 1, x'(0) = -1$

(b) $x'' - 9x = 0, x(0) = 3, x'(0) = -3$

(c) $x'' + 3x' + 2x = 0, x(0) = 5, x'(0) = 1$

(d) $x'' + 2x' + 6x = 0, x(0) = 1, x'(0) = 1$

**9.** Find all equilibrium solutions (fixed points) of each of the following systems.

(a) $x_1' = (\sin x_1) - x_2$      (b) $x_1' = x_1^2 - x_2^2$
    $x_2' = \cos x_1$                  $x_2' = 2x_1 - 2x_2$

**10.** Consider the Lotka-Volterra predator-prey model derived in Example 1.

(a) Is the system autonomous? Explain.

(b) Does the system have any fixed points? If so, try to determine all of them.

**11.** The Lotka-Volterra predator-prey model in Example 1 assumes that the prey has an unlimited food supply, which leads to unlimited exponential growth of the prey in the absence of predation. Suppose instead that the prey obeys a growth rate governed by the logistic equation. (See Exercise 4 in Section 1.1.) Modify the Lotka-Volterra model to reflect this assumption.

**12.** Let $a_0, a_1, \ldots, a_{n-1}$ be functions of $t$. Then the equation

$$x^{(n)} + a_{n-1}x^{(n-1)} + \cdots + a_1 x' + a_0 x = f(t) \quad (6.18)$$

is an $n$th-order differential equation. (The notation $x^{(i)}$ denotes the $i$th derivative of $x$.) Use Examples 2 and 3 as a guide to show that Eq. (6.18) is equivalent to a first-order system of $n$ equations.

## 6.2   GRAPHICAL INTERPRETATION OF SOLUTIONS

### The Phase Plane

Chapter 1 demonstrates that much useful information about the solutions of a given first-order scalar differential equation can be obtained by graphical analysis. Graphical methods are also useful in the analysis of systems of first-order differential equations. We demonstrate this with two examples: the Lotka-Volterra predator-prey model and the linear, undamped pendulum.

The solid curve in Fig. 6.1 is a graph of the solution of the Lotka-Volterra predator-prey model

$$x' = -x + 0.1xy,$$
$$y' = 3y - xy \tag{6.19}$$

satisfying the initial condition $(x(0), y(0)) = (2.2, 5)$. Three axes—$x$, $y$, and $t$—are used to plot the solution curve.

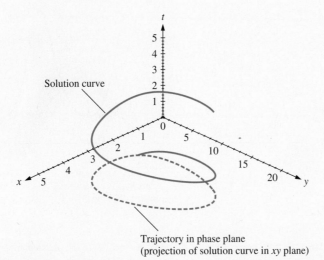

FIGURE 6.1   The graph of the solution of Eqs. (6.19) corresponding to the initial condition $(x(0), y(0)) = (2.2, 5)$.

Despite continuing improvements in computer graphics, it is not easy to represent a three-dimensional curve effectively on a two-dimensional page. Instead, it is often possible to obtain useful information about the solution from a two-dimensional projection of its graph.

The projected solution curve is called a *trajectory*, and the $xy$ plane is called the *phase plane*. The phase plane is particularly useful for illustrating the qualitative behavior of autonomous systems of two equations, because the rate of change of the dependent variables does not depend explicitly on time; therefore, most of the relevant information can be displayed in two dimensions.

The trajectory in Fig. 6.1 is a closed curve and is reproduced in Fig. 6.2(a). The trajectory is a parametric representation of the solution of the form $(x(t), y(t))$. Figure 6.2(b) shows the solutions $x(t)$ and $y(t)$ as functions of time; the graph is often called a *time series plot*. The trajectory and the time series plot are two different ways to view the same information. The trajectory in Fig. 6.2(a) can be sketched by using the time series plot in Fig. 6.2(b) as follows:

- The initial condition is $x = 2.2$ and $y = 5$, specified at $t = 0$. The corresponding point on the trajectory in Fig. 6.2(a) is marked $A$.

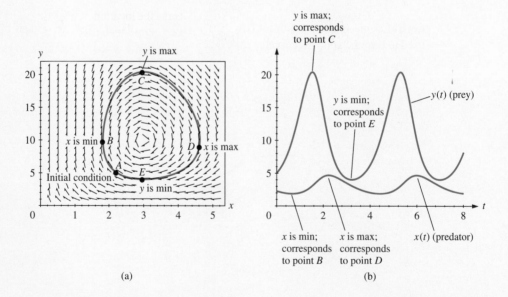

(a)                                                                              (b)

**FIGURE 6.2**   (a) The trajectory corresponding to the system (6.19) in the phase plane. The curve is traversed clockwise as indicated by the small arrows. (b) The corresponding time series plots of $x(t)$ and $y(t)$.

- The predator population ($x$) decreases initially, eventually reaching a minimum value of approximately 1.8 at $t \approx 0.7$. The corresponding point on the trajectory in Fig. 6.2(a) is marked $B$.

- For $0.7 < t < 1.5$, both populations increase until the prey population ($y$) reaches a maximal value of about $y = 20$ at $t \approx 1.5$. The corresponding point on the trajectory in Fig. 6.2(a) is marked $C$.

- For $1.5 < t < 2.2$, the prey population declines and the predator population increases, eventually reaching a maximum value of about $x = 4.5$ at $t \approx 2.2$. The corresponding point on the trajectory in Fig. 6.2(a) is marked $D$.

- For $2.2 < t < 3.2$, both populations decrease. The prey population $y$ reaches its minimal value of about $y = 4$ at $t \approx 3.2$. The corresponding point on the trajectory in Fig. 6.2(a) is marked $E$.

- Both populations return to their original values at $t \approx 3.8$. Therefore, the solutions are periodic with a period of approximately 3.8.

The most notable qualitative feature of the solution of the predator-prey system (6.19) is that it is periodic. The periodicity of the solution is reflected in the trajectory in Fig. 6.2(a), which forms a closed curve.

Another interesting qualitative feature concerns the location of the relative extrema in the populations. For instance, the prey population $y$ reaches a relative maximum value at $t \approx 1.5$, but the corresponding relative maximum value of the predator population $x$ does not occur until $t \approx 2.2$. Thus, there is a lag between the two populations.

Also notice that the prey population $y$ increases rapidly while the predator population declines. Eventually, the predator population increases in response to the increased prey population. As the number of predators grows, the prey population declines, and the lack of prey eventually reduces the predator population. Although the assumptions that are used in Section 6.1 to derive models like (6.19) are oversimplified, the qualitative behavior of the resulting predator-prey model is similar to that of real populations.

## The Direction Field

Consider the system

$$
\begin{aligned}
x_1' &= x_2, \\
x_2' &= -x_1,
\end{aligned}
\tag{6.20}
$$

which is equivalent to the second-order scalar equation

$$x'' + x = 0.$$

The system (6.20) describes simple harmonic motion. A solution of this system is a vector-valued function of the form $\mathbf{x}(t) = (x_1(t), x_2(t))$. A graph of $\mathbf{x}$ can be drawn in three dimensions (using $x$, $y$, and $t$ axes, as in Fig. 6.1) or in two dimensions as a trajectory in the phase plane.

The tangent vector to a trajectory at any time $t$ is the derivative $\mathbf{x}'(t) = (x_1'(t), x_2'(t))$. We can represent the direction of the tangent vector at each point by a *slope line* whose slope is given by

$$\frac{dx_2}{dx_1} = \frac{dx_2/dt}{dx_1/dt} = \frac{x_2'}{x_1'}.$$

If the system is autonomous, then the slope lines are functions only of the points in the phase plane. The *direction field* is the collection of all the slope lines. Figure 6.3 shows the direction field for the system (6.20).

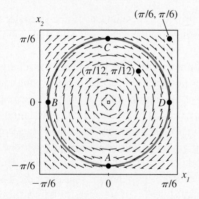

**FIGURE 6.3**   Direction field and a trajectory for Eq. (6.20). The origin is a fixed point and is marked by a small box.

The direction field gives a graphical interpretation of the solutions of autonomous first-order systems:

*Every solution of an autonomous first-order system can be represented parametrically by a trajectory that is tangent at every point to the corresponding slope line.*

The solutions of autonomous second-order differential equations can be represented similarly, because every second-order differential equation is equivalent to a pair of first-order equations.

The direction field is generated by evaluating the derivatives $(x_1', x_2')$ at each point on a grid, then plotting the corresponding slope lines. We demonstrate this in Fig. 6.3:

- Point $A$ is $(0, -\pi/6)$. The corresponding slope line is horizontal and points to the left, because $\mathbf{x}' = (-\pi/6, 0)$ at $A$.

- Point $B$ is the point $(-\pi/6, 0)$, where $(x_1', x_2') = (0, \pi/6)$. The corresponding slope line is vertical and points upward because the second component of the tangent vector is positive.

- Point $C$ is $(0, \pi/6)$, where $(x_1', x_2') = (\pi/6, 0)$. The corresponding slope line is horizontal and points to the right.

- Point $D$ is $(\pi/6, 0)$, where $(x_1', x_2') = (0, -\pi/6)$. The corresponding slope line is vertical and points downward.

In the direction fields that are drawn in this text, the slope lines indicate only the *ratio* $x_2'/x_1'$ at each point. They do not indicate the *length* of the vector that would be tangent to a solution curve at $(x_1, x_2)$. For instance, identical slope lines are drawn in Fig. 6.3 at $(\pi/12, \pi/12)$ and at $(\pi/6, \pi/6)$, even though the length of the tangent vector is twice as great at the latter point.

A direction field lets you sketch the solution curves corresponding to many different initial conditions. The collection of all possible trajectories is called the *phase portrait* of the system of differential equations. Phase portraits typically are illustrated by drawing several representative trajectories. In the case of Fig. 6.3, the phase portrait consists of concentric circles centered at the origin.

◆ *Example 1*   The system (6.20), whose direction field is shown in Fig. 6.3, is a model of a linear undamped pendulum, which exhibits simple harmonic motion.

It is possible to give a physical interpretation of the trajectory shown in Fig. 6.3 as follows. Let $x_1$ be the angular position of the bob of the pendulum, and let $x_2$ be its angular velocity. We adopt the convention that $x_1 = 0$ corresponds to the vertical position in which the bob hangs straight down and that $x_2 > 0$ corresponds to anticlockwise motion.

- Point $A$ in Fig. 6.3 corresponds to the instant at which the bob crosses the vertical $(x_1 = 0)$ position moving clockwise; see graphic $A$ in Fig. 6.4.

- Point $B$ in Fig. 6.3 corresponds to the instant at which the bob has come momentarily to rest at $x_1 = -\pi/6$, which is one extreme of its path; see graphic $B$ in Fig. 6.4.

- Point $C$ in Fig. 6.3 and the corresponding graphic $C$ in Fig. 6.4 show the instant at which the bob crosses the vertical position moving anticlockwise.

- Point $D$ in Fig. 6.3 and the corresponding graphic $D$ in Fig. 6.4 show where the bob comes momentarily to rest at $x_1 = \pi/6$, which is the other extreme of its path.

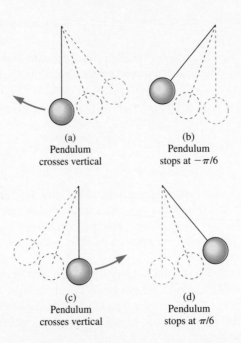

(a)
Pendulum
crosses vertical

(b)
Pendulum
stops at $-\pi/6$

(c)
Pendulum
crosses vertical

(d)
Pendulum
stops at $\pi/6$

**FIGURE 6.4**   Graphics $A$–$D$ show snapshots of the motion of a linear pendulum, modeled by Eq. (6.20). The corresponding points are shown on the trajectory in the direction field in Fig. 6.3.

Notice that points along the bottom half of the trajectory arc in Fig. 6.3 correspond to anticlockwise motion of the bob ($x_2 < 0$) and points along the top half of the trajectory arc correspond to clockwise motion ($x_2 > 0$). Points $B$ and $D$ correspond to times at which the bob changes direction.

The trajectory shown in Fig. 6.3 is a closed curve, indicating that the motion of the pendulum is periodic. Although it is not possible to determine the period of the motion from the trajectory, we can conclude that the same amount of time is always required to traverse the circle. The reason is that the system (6.20) is autonomous: At every instant, the velocity and acceleration depend only on the current position and velocity. Therefore, at any point in a cycle, the state of the pendulum is identical to what it was at the same point in the previous cycle.                                    ∎

♦ *Example 2*   The qualitative behavior of the solution of the predator-prey model

$$x' = -x + 0.1xy,$$
$$y' = 3y - xy$$

(6.21)

satisfying the initial condition $(x(0), y(0)) = (2.2, 5)$ is analyzed in the earlier discussion. It is reasonable to ask whether the solutions of Eq. (6.21) are periodic and have similar lags in the relative extrema of their component functions when other initial conditions are imposed. To answer this question, we must examine trajectories corresponding to other initial conditions.

Figure 6.5 shows a phase portrait and direction field for the system (6.21). Each trajectory corresponds to a different initial condition for the system (6.19). Each trajectory forms a closed curve, implying that the solution is periodic. In addition, each trajectory has a similar qualitative shape, indicating that relative maxima in the prey population precede relative maxima in the predator population, and likewise for the relative minima of the two populations.

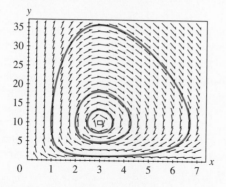

**FIGURE 6.5**   Phase portrait and direction field for the predator-prey system (6.21).

Notice that neither population becomes negative if the initial condition is located in the first quadrant in the phase plane. In fact, if one or the other population becomes extinct, then the trajectory remains on one of the coordinate axes.

Because the trajectories are parametric representations of the solutions, explicit time information is lost. For instance, although it is clear from the trajectories in Fig. 6.5 that the solutions are periodic, the period of each solution cannot be ascertained from Fig. 6.5 alone. Even if the solutions have different periods, the corresponding trajectories are still closed curves. See Exercise 2.   ■

## Fixed Points

By definition, a fixed point occurs where $(x_1', x_2') = (0, 0)$. No slope line is drawn at a fixed point, because the trajectory consists of a single point; therefore, the corresponding derivative is **0**. In many cases, the location of fixed points in a direction field can be inferred by using of the following property:

*Near a fixed point, the slope lines typically point in many different directions.*

This statement holds for *isolated* fixed points, that is, fixed points that are separated from other fixed points.

◆ *Example 3*   In Fig. 6.3, the only fixed point is at the origin, which corresponds to a rest point of the pendulum. The fixed point is marked by a small square in the figure. Notice how the slope lines near the origin point in many different directions.               ■

◆ *Example 4*   The system (6.21) has two fixed points: the origin and the point (3, 10). The latter is marked as the small box in Fig. 6.5. The slope lines near (3, 10) point in all different directions. (The same is true for the origin, though the slope lines on the axes are not shown.)

   These fixed points have a biological interpretation. The origin corresponds to the extinction of both predator and prey. The point (3, 10) corresponds to a situation in which the natural rate of population increase of the prey exactly balances the losses due to predation. Similarly, the intrinsic death rate of the predator population is balanced by the rate of increase due to an adequate food supply. Therefore, the predator and prey populations are in equilibrium; they do not oscillate.               ■

◆ *Example 5*   This example shows that it is sometimes possible to determine a formula for the trajectories in the phase portrait in terms of $x_1$ and $x_2$. Figure 6.6 shows the direction field corresponding to the system

$$\begin{aligned} x_1' &= x_2, \\ x_2' &= -4x_1. \end{aligned} \tag{6.22}$$

Two solution curves are shown. The inner curve corresponds to the initial condition $\mathbf{x}(0) = (0, 0.1)$ and the outer curve to the initial condition $\mathbf{x}(0) = (0.1, 0)$.

   The first-order system (6.22) is equivalent to the second-order scalar equation

$$x'' + 4x = 0.$$

The initial condition $\mathbf{x}_0 = (0.1, 0)$ is equivalent to the initial conditions $x(0) = 0.1$ and $x'(0) = 0$ for the scalar equation. We have

$$x(t) = \tfrac{1}{10} \cos 2t.$$

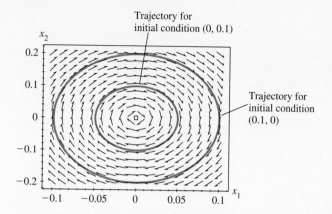

**FIGURE 6.6** The direction field corresponding to Eqs. (6.22). The solution curves correspond to the initial conditions $\mathbf{x}_0 = (0.1, 0)$ and $\mathbf{x}_0 = (0, 0.1)$. The origin (marked by a small box) is at the center.

Since $x_1 = x$ and $x_2 = x_1'$, the solution of Eq. (6.22) satisfying the initial condition $\mathbf{x}_0 = (0.1, 0)$ is

$$x_1(t) = \tfrac{1}{10} \cos 2t,$$
$$x_2(t) = -\tfrac{2}{10} \sin 2t. \tag{6.23}$$

We can determine an analytical representation of the trajectory by eliminating the independent variable $t$. We square both sides of Eqs. (6.23) and use the identity $\cos^2 2t + \sin^2 2t = 1$ to obtain

$$4x_1^2 + x_2^2 = \tfrac{4}{100},$$

which is an ellipse. The trajectory in the phase plane is the outer ellipse in Fig. 6.6. The solutions of Eq. (6.22) are periodic, so the trajectories form closed curves (ellipses). ∎

✦ *Example 6*

$$x'' + \tfrac{1}{5}x' + \sin x = 0 \tag{6.24}$$

is a model of a pendulum with friction. (See Section 5.1.) If the displacement $x$ from the equilibrium (straight-down) position is not too small, then Eq. (6.24) can be replaced by its linear counterpart (we approximate $\sin x$ by $x$). However, if the displacement is not small, as is the case here, then a linear model does not provide an accurate description of the motion.

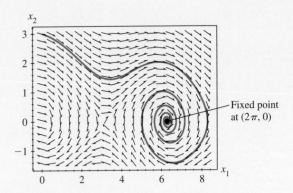

**FIGURE 6.7**   The direction field and a sample trajectory for Eq. (6.25).

Figure 6.7 shows a direction field for the first-order system corresponding to Eq. (6.24), given by

$$\mathbf{x}'(t) = \mathbf{g}(\mathbf{x}) = \begin{pmatrix} x_2 \\ -\frac{1}{5}x_2 - \sin x_1 \end{pmatrix}. \tag{6.25}$$

As in Example 1, $x_1 = 0$ corresponds to the vertical position wherein the bob hangs straight down, and $x_2 > 0$ corresponds to anticlockwise motion. The solution curve corresponds to the initial condition $x(0) = 0$, $x'(0) = 3$. Initially, the pendulum is at the equilibrium position, but it is given a large initial angular velocity.

The solution curve shows that the initial angular velocity causes the pendulum to flip overhead. The straight-up position corresponds to $x_1 = \pi$. The solution trajectory implies that the pendulum crosses the straight-up position at an angular velocity of approximately 1.6 rad/s. The pendulum continues completely around, then oscillates about the straight-down position with decreasing amplitude.

The values $x_1 = 0$ and $x_1 = 2\pi$ refer to the same location—namely, the straight-down position. However, the value $2\pi$ indicates that the pendulum has completed a full revolution.

It is straightforward to verify that $(0, 0)$ and $(2\pi, 0)$ are both fixed points. For instance,

$$\mathbf{g} \begin{pmatrix} 2\pi \\ 0 \end{pmatrix} = \begin{pmatrix} 0 \\ -\sin 2\pi \end{pmatrix} = \begin{pmatrix} 0 \\ 0 \end{pmatrix}.$$

Section 8.4 discusses the nonlinear pendulum equation in more detail.                                 ∎

## Exercises

**1.** Figure 6.8 shows some different direction fields. Determine where the fixed points might be in each picture.

**2.** Using a suitable software package, find and graph the numerically approximated solutions of the predator-prey model (6.21) for four different initial conditions of your choice. Answer the following questions for each initial condition that you choose.

(a) Does the computed solution appear to be periodic? If so, estimate the period.

(b) Does the computed solution have the correct qualitative behavior for predator and prey populations? (Consider the times at which relative minima and maxima of each population occur.) Do the periods of the solutions that you found in (a) appear to depend on the initial condition?

**3.** For each system of equations in the following list, draw the direction field in a box extending from $-1$ to 1 in each coordinate. In each case, the origin is a fixed point. Infer from the direction field whether

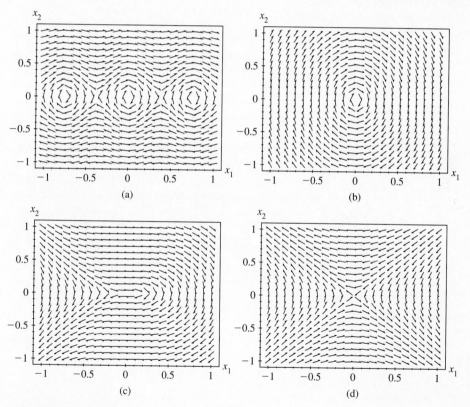

(a)

(b)

(c)

(d)

**FIGURE 6.8**   For Exercise 1.

trajectories eventually approach the origin.

(a) $x_1' = x_1 + x_2$
$\quad x_2' = x_1 - x_2$

(b) $x_1' = -x_1 + x_2$
$\quad x_2' = -x_1 - x_2$

(c) $x_1' = x_1 + x_2$
$\quad x_2' = -x_1 + x_2$

(d) $x_1' = -\frac{3}{2}x_1 + \frac{1}{2}x_2$
$\quad x_2' = \frac{1}{2}x_1 - \frac{3}{2}x_2$

**4.** Show that the two phase portraits shown in Fig. 6.6 represent solutions of the same period by finding explicit solutions of Eq. (6.22).

**5.** The phase portrait in Fig. 6.9 shows the position $x_1$ and velocity $x_2$ of a mass-spring system as a function of each other. The origin is a fixed point. The illustrated curve shows the motion for $t \geq 0$ starting from $\mathbf{x}(0) = (3, -1)$. Answer the following questions.

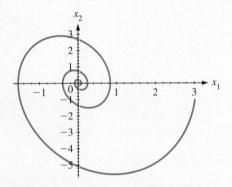

**FIGURE 6.9** For Exercise 5.

(a) What is the maximum displacement of the mass from the equilibrium?

(b) How fast is the mass moving when it crosses the equilibrium the first time?

(c) How fast is the mass moving when it crosses the equilibrium the second time?

(d) When the mass comes to rest the second time, what is its displacement from the equilibrium?

(e) After how many oscillations does the motion of the mass become negligible? (Interpret the

term *negligible* in a manner consistent with the resolution of the graph.)

**6.** Suppose that a pendulum moves according to the following scenario. The pendulum initially is at an angular displacement of 1/2 rad in the anticlockwise direction from the vertical. Its initial velocity is 1 rad/s clockwise. The pendulum crosses the vertical going 0.8 rad/s and momentarily stops at a position of 0.5 rad in the clockwise direction from the vertical. The pendulum then changes direction and crosses the vertical at a rate of 0.3 rad/s. It again reverses direction on the other side of the equilibrium at a position 0.2 rad from the vertical.

Sketch a phase portrait that corresponds to the motion. (That is, plot the position $x_1$ as a function of the velocity $x_2$ of the pendulum.) Draw appropriate numerical labels on each axis.

**7.** Suppose a pendulum starts at position 0 and rotates clockwise without friction, repeatedly flipping overhead. Draw a phase portrait consistent with this description. Use the convention that anticlockwise motion corresponds to positive velocity.

**8.** Find the solution of each of the initial value problems in the following list. Then eliminate the variable $t$ to determine an equation for the corresponding trajectory in the phase plane in terms of $x_1$ and $x_2$ only. (See Example 5.)

(a) $x_1' = x_2$
$\quad x_2' = -4x_1$
$\quad x_1(0) = 1, \quad x_2(0) = 0$

(b) $x_1' = x_2$
$\quad x_2' = -4x_1$
$\quad x_1(0) = 0, \quad x_2(0) = 3$

(c) $x_1' = x_2$
$\quad x_2' = -4x_1$
$\quad x_1(0) = 2, \quad x_2(0) = 3$

(Hint: Find the phase-amplitude form of the solution of the corresponding scalar equation.)

**9.** Let us say that a pendulum is at the 6 o'clock position when it hangs straight down ($x = 0$ rad), at the 12 o'clock position when it is straight up ($x = \pi$ rad), at the 3 o'clock position when $x = \pi/2$ rad, and so on. We adopt the convention that anticlockwise motion corresponds to positive angular velocity. For each of the following descriptions, draw a phase portrait that is consistent with the motion of the pendulum. Label the $x$ axis with appropriate units.

(a) The pendulum starts from the 12 o'clock position, moving anticlockwise rapidly. It flips over once, then changes direction when it reaches the 3 o'clock position again. The pendulum oscillates about the 6 o'clock position with decreasing amplitude.

(b) The pendulum oscillates periodically between the 5 o'clock and 7 o'clock positions without flipping over.

(c) The pendulum gets a push every time it changes direction. The push is strong enough for the pendulum to flip over once before it changes its direction. Draw a phase portrait of a motion that oscillates periodically between the

3 o'clock and 9 o'clock positions and crosses the 12 o'clock position twice during each period.

**10.** Draw a direction field for the system

$$x' = y,$$
$$y' = -\sin x$$

in a box $S = \{-2\pi \leq x \leq 2\pi, -5 \leq y \leq 5\}$. Identify the fixed points.

**11.** Consider the predator-prey model

$$x' = -2x + 0.2xy + 1,$$
$$y' = 2y - xy/2 + 2.$$

(a) What are the slope lines on the positive $x$ axis?

(b) What are the slope lines on the positive $y$ axis?

(c) Show that neither species can ever become extinct.

(d) Discuss how this system reflects immigration into the territory.

## 6.3 SINKS, SOURCES, SADDLES, AND CENTERS

We now consider some of the basic kinds of direction fields associated with first-order systems of differential equations. We are particularly interested in the behavior of trajectories near fixed points. There are four basic kinds of fixed points, called *sinks*, *sources*, *saddles*, and *centers*, which we illustrate in the following discussion. Let

$$mx'' + cx' + kx = 0. \tag{6.26}$$

If there is no friction ($c = 0$), then the solutions are periodic (simple harmonic motion). When there is damping ($c > 0$), the qualitative behavior of the solutions depends on the values of $m$ and $k$; the solutions are described as underdamped, overdamped, or critically damped. Examples 1–4 show the phase portraits corresponding to these different cases.

✦ *Example 1 (A center)* Let $c = 0$, corresponding to a frictionless motion. Then Eq. (6.26) reduces to $x'' + (k/m)x = 0$, which is the equation for simple harmonic motion. For example, if $k = 4$ and $m = 1$, then the general solution is $x(t) = x_0 \cos 2t + v_0/2 \sin 2t$.

The corresponding system is discussed in Example 5 in Section 6.2, and the direction field is shown in Fig. 6.6. The trajectories are ellipses, corresponding to periodic solutions. The ellipses are centered at the origin; we call the origin a *center*.                                                                                  ∎

◆ *Example 2 (A spiral sink)*      Let

$$x'' + x' + 10x = 0, \qquad x(0) = 10, \qquad x'(0) = 0. \tag{6.27}$$

The roots of the corresponding characteristic equation are $-\frac{1}{2}\left(1 \pm i\sqrt{39}\right)$. Equation (6.27) describes an underdamped system, and its solutions oscillate with exponentially decaying amplitude. Figure 6.10(a) shows a time series plot of $x(t)$ and $x'(t)$.

The equivalent first-order system is

$$x_1' = x_2,$$
$$x_2' = -10x_1 - x_2,$$

and the corresponding phase portrait is shown in Fig. 6.10(b). The parametric curve defined by $(x(t), x'(t))$ forms a spiral that winds into the origin. We call the origin a *spiral sink*. The term *sink* refers to the fact that all initial conditions "drain" into the fixed point at the origin as $t \to \infty$.

Sinks are examples of *attracting* fixed points. We say that a fixed point **x** is attracting if trajectories starting from initial conditions near **x** eventually approach **x** as $t \to \infty$. The origin is an attracting fixed point in this example, because each component of the solution curve eventually approaches 0, regardless of the initial condition.                          ∎

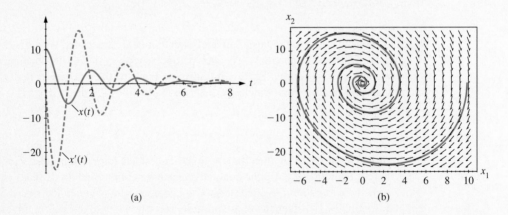

(a)                                                      (b)

**FIGURE 6.10**    (a) A time series plot of $x(t)$ (solid curve) and $x'(t)$ (dashed curve) corresponding to Eq. (6.27). (b) The corresponding trajectory in the phase plane. The origin (marked by a small box) is a spiral sink.

◆ *Example 3 (A sink)*    Let

$$x'' + 7x' + 10x = 0. \tag{6.28}$$

The roots of the corresponding characteristic equation are $-5$ and $-2$, so the solutions of Eq. (6.28) are a linear combination of $e^{-2t}$ and $e^{-5t}$. Equation (6.28) describes an overdamped system, and its solutions decay exponentially, but they do not oscillate.

The equivalent first-order system is

$$x_1' = x_2,$$

$$x_2' = -10x_1 - 7x_2.$$

The phase portrait is shown in Fig. 6.11. Four solution curves are drawn, representing the initial conditions $(x_1(0), x_2(0)) = (10, 0)$, $(x_1(0), x_2(0)) = (0, 10)$, $(x_1(0), x_2(0)) = (-10, 0)$, and $(x_1(0), x_2(0)) = (0, -10)$, respectively. Each solution curve leads to the origin, but the curves are not spirals because the solutions do not oscillate. Notice that solutions cross the $x_1$ axis at most once, which is characteristic of overdamped systems. Here the origin is simply called a *sink*.

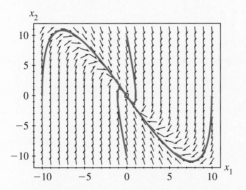

**FIGURE 6.11**    The phase portrait corresponding to Eq. (6.28). The origin (marked by a small box) is a sink.

As in Example 2, the origin is an attracting fixed point. Each component of every solution trajectory approaches 0 as $t \to \infty$.    ■

◆ *Example 4 (A sink)*    Let

$$x'' + 2x' + x = 0. \tag{6.29}$$

The corresponding characteristic equation has $-1$ as a double real root, so the solutions are a linear combination of $e^{-t}$ and $te^{-t}$. Equation (6.29) describes a critically damped system. As in the overdamped case, the solutions decay exponentially but they do not oscillate.

The equivalent first-order system is

$$x_1' = x_2,$$
$$x_2' = -x_1 - 2x_2. \tag{6.30}$$

The phase portrait is shown in Fig. 6.12. Four solution curves are drawn, representing the initial conditions $(x_1(0), x_2(0)) = (10, 0)$, $(x_1(0), x_2(0)) = (0, 10)$, $(x_1(0), x_2(0)) = (-10, 0)$, and $(x_1(0), x_2(0)) = (0, -10)$, respectively. Each solution curve leads to the origin, but the curves are not spirals because the solutions do not oscillate. As in the overdamped case, the origin is a sink, but it is not a spiral sink, and each trajectory crosses the $x_1$ axis at most once. In fact, the phase portrait looks quite similar to that of the overdamped system in Fig. 6.11. The origin is an attracting fixed point.

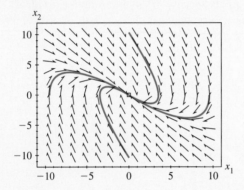

**FIGURE 6.12**   The phase portrait corresponding to Eqs. (6.30). The origin (marked by a small box) is a sink.                                                                              ■

Examples 2–4 illustrate sinks, which arise in unforced, damped systems. (Without forcing, the damping eventually dissipates all the energy from the system, so the system comes to rest at the fixed point.) A natural question is what happens if one or more roots of the characteristic equation are positive or have a positive real part. Such a situation can arise if the damping coefficient is replaced by a negative number. The resulting second-order differential equation is no longer a model of a conventional pendulum, for instance, but such equations do arise in other contexts (see Chapter 8).

◆ *Example 5 (A spiral source)*     Let

$$x'' - x' + 10x = 0 \qquad \text{with} \qquad x(0) = 1/10, \qquad x'(0) = 0. \tag{6.31}$$

The roots of the corresponding characteristic equation are $\frac{1}{2}\left(1 \pm i\sqrt{39}\right)$. Hence, the solution oscillates, and its amplitude increases at a rate proportional to $e^{t/2}$. Figure 6.13(a) shows a time series plot of $x(t)$ and $x'(t)$.

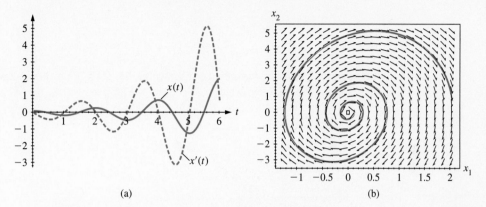

(a)                                          (b)

**FIGURE 6.13**    (a) A time series plot of $x(t)$ (solid curve) and $x'(t)$ (dashed curve) corresponding to Eq. (6.31). (b) The corresponding trajectory in the phase plane. The origin (marked by a small box) is a spiral source.

The equivalent first-order system is

$$x_1' = x_2,$$

$$x_2' = -10x_1 + x_2,$$

and the corresponding phase portrait is shown in Fig. 6.13(b). The parametric curve defined by $(x(t), x'(t))$ forms a spiral that looks similar to that in Fig. 6.10. However, in this case, trajectories move away from the origin as $t$ increases from 0. We call the origin a *spiral source*. The term *source* refers to the fact that all initial conditions lead away from the fixed point at the origin as $t$ increases.                                  ∎

The phase portraits of sources that are not spiral sources look similar to those of sinks. However, the solution curves lead away from the fixed point, not toward it. For this reason, sources are *not* attracting fixed points. We leave the details of how to draw an ordinary source as an exercise. (See Exercise 2.) The last example in this section shows a different kind of fixed point that can arise in systems of equations.

✦ *Example 6 (A saddle)*    A qualitatively different phase portrait arises if the characteristic equation has one positive and one negative real root. Let

$$x'' - x = 0. \tag{6.32}$$

The corresponding characteristic equation is $\lambda^2 - 1$, whose roots are $-1$ and $1$. The solutions of Eq. (6.32) are linear combinations of $e^t$ and $e^{-t}$. For most initial conditions, the solutions tend either to $+\infty$ or to $-\infty$ as $t \to \infty$.

The corresponding first-order system is

$$x_1' = x_2,$$

$$x_2' = x_1,$$

whose phase portrait and direction field are shown in Fig. 6.14. Four different solution curves are drawn. The phase portrait is called a *saddle*, and we call the origin a *saddle fixed point*.

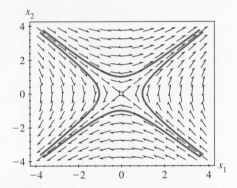

**FIGURE 6.14**    A saddle fixed point at the origin (marked by a small box).

Saddles have some of the characteristics of sinks and sources. For instance, the slope lines along the diagonal line $x_2 = -x_1$ point directly toward the origin. The direction field suggests that initial conditions on or near this diagonal move toward the origin, at least for a while. In this sense, the saddle has an attracting direction.

However, the slope lines along the diagonal line $x_2 = x_1$ point directly away from the origin. Hence, initial conditions that start on or near this diagonal line move more or less directly away from the origin. The saddle has a repelling direction.

The slope lines that do not lie on one of these diagonals have components that point toward the origin and components that point away from the origin. The solution curves in Fig. 6.14 suggest that most solution curves typically approach the origin for a time, then eventually move away. Since most initial conditions do not approach the origin as $t \to \infty$, a saddle fixed point is *not* attracting.     ∎

Table 6.1 displays all the possible phase portraits for first-order systems that correspond to linear second-order differential equations with constant coefficients.

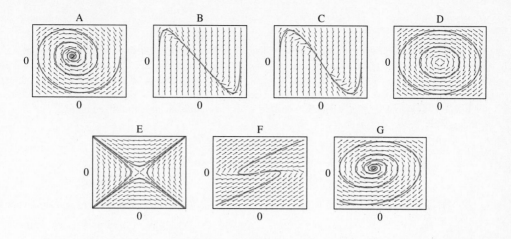

**Differential Equation: $mx'' + cx' + kx = 0$**

| Type of Roots of the Characteristic Equation | Type of Motion | Phase Portrait |
|---|---|---|
| Two complex roots with negative real part | Underdamped | Spiral sink (A) |
| Two real, distinct and negative roots | Overdamped | Sink (B) |
| Two real and equal roots | Critically damped | Sink (C) |
| Two purely imaginary roots | Simple harmonic motion, no damping | Center (D) |
| Two real roots with opposite sign | Not a mass-spring system | Saddle (E) |
| Two real positive roots | Not a mass-spring system | Source (F) |
| Two complex roots with positive real part | Not a mass-spring system | Spiral source (G) |

TABLE 6.1 Typical phase portraits for constant-coefficient linear second-order differential equations.

## *Exercises*

**1.** For each of the equations in the following list, draw the direction field and a phase portrait in a box from $-5$ to $5$. Experiment with different initial conditions. In each case, the origin is a fixed point. Use the direction field to determine whether the origin is a source, sink, center, or saddle.

(a) $x_1' = x_1 + x_2$
    $x_2' = x_1 - x_2$

(b) $x_1' = -x_1 + x_2$
    $x_2' = -x_1 - x_2$

(c) $x_1' = x_1 + x_2$
    $x_2' = -x_1 + x_2$

(d) $x_1' = -\frac{3}{2}x_1 + \frac{1}{2}x_2$
    $x_2' = \frac{1}{2}x_1 - \frac{3}{2}x_2$

**2.** Let $x'' - \frac{3}{2}x' + x = 0$, and consider the two sets of initial conditions $x(0) = 1/10, x'(0) = 0$ and $x(0) = 0, x'(0) = 1/10$.

(a) Determine the corresponding first-order system.

(b) Draw a direction field and argue that the origin is a source.

(c) Sketch the solution curves corresponding to the initial conditions listed above.

**3.** Let $x'' + cx' + 16x = 0$. For each of the following values of the damping constant $c$,

• determine the type of the fixed point at the origin;

• generate a direction field and include the trajectories of the solutions satisfying the initial conditions $(x(0), x'(0)) = (10, 0)$ and $(x(0), x'(0)) = (0, 10)$.

(a) $c = 1$

(b) $c = 2$

(c) $c = 4$

How do the phase portraits differ as you increase the damping?

**4.** Two of the initial conditions in Example 3 cause the mass to overshoot the equilibrium position exactly once. Which ones are they?

**5.** Each time series plot in Fig. 6.15 shows $x(t)$ as a solid curve and $x'(t)$ as a dashed curve. Define $x_1 = x$ and $x_2 = x_1'$. Sketch the corresponding phase portrait in the $x_1 x_2$ plane. Determine whether the origin is an attracting fixed point.

**6.** Consider the system

$$x' = -y,$$
$$y' = 4x$$

(a) Generate a phase portrait and a direction field.

(b) What type of fixed point is the origin?

(c) How is this phase portrait different from Fig. 6.6?

**7.** Consider a mass-spring system modeled by the equations

$$x' = y,$$
$$y' = -y - 25x.$$

Generate a direction field in the box $S = \{-3 \le x \le 3, -3 \le y \le 3\}$ and use it to argue that:

(a) when the mass is released from a position away from the equilibrium with zero initial velocity, the mass initially approaches the equilibrium;

(b) when the mass is released from the equilibrium position with a nonzero initial velocity, then the mass initially moves away from the equilibrium in the same direction as the initial velocity.

**8.** Consider the predator-prey model

$$x' = -2x + 0.2xy,$$
$$y' = 2y - xy/2.$$

(a) Determine all of the fixed points.

(b) Draw a direction field together with some representative trajectories and use them to classify each fixed point in (a) as a sink, source, saddle, or center.

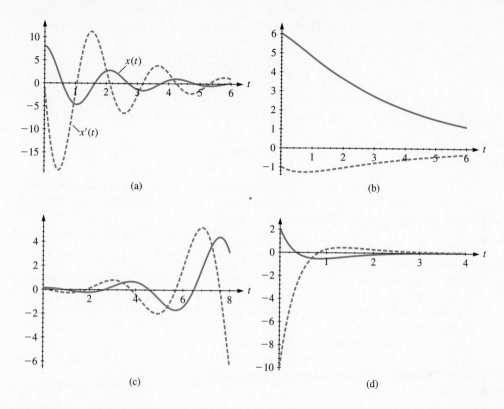

**FIGURE 6.15**   For Exercise 5.

## 6.4   EULER'S METHOD

Euler's method provides a simple way to approximate solutions of first-order systems of differential equations. We discuss the method here because it provides a straightforward, quantitative way to approximate solutions directly from the direction field. Euler's method has important limitations, as we will see, but it is a useful starting point for understanding the nature of solutions of systems of differential equations. To illustrate the method graphically in the plane, we restrict attention to systems of two first-order equations. However, the method works for systems of arbitrary dimension.

The basic idea of Euler's method is the same in two (or more) dimensions as it is in one dimension. Let

$$\frac{d\mathbf{x}}{dt} = \mathbf{f}(t, \mathbf{x}). \tag{6.33}$$

Equation (6.33) determines the tangent vector to the solution curve at any point $\mathbf{x}$. Euler's method begins at a starting point (initial condition) given by $\mathbf{x}(t_0) = \mathbf{x}_0$. The tangent vector

$$\frac{d\mathbf{x}_0}{dt} = \mathbf{f}(t_0, \mathbf{x}_0)$$

is evaluated by using Eq. (6.33). The next point on the solution curve is approximated by moving along the tangent vector by an amount $h\mathbf{f}(t_0, \mathbf{x}_0)$, where $h$ is called the *step size*. After one time step, the point

$$\mathbf{x}_1 = \mathbf{x}_0 + h\frac{d\mathbf{x}_0}{dt}$$

is an approximation to the true solution $\mathbf{x}(t_1)$, where $t_1 = t_0 + h$. The process is repeated for as long as desired.

Figure 6.16 shows a schematic illustration of two steps of the procedure. The dashed curve shows the true solution curve $\mathbf{x}(t)$. In the figure, the points $\mathbf{x}_1$ and $\mathbf{x}_2$ are approximations to $\mathbf{x}(t_1)$ and $\mathbf{x}(t_2)$, respectively. The length of the segment from $\mathbf{x}_0$ to $\mathbf{x}_1$ is determined by $h\mathbf{f}(t_0, \mathbf{x}_0)$, which is the product of the derivative at $t = t_0$ and the step size $h$. Similarly, the length of the segment from $\mathbf{x}_1$ to $\mathbf{x}_2$ is $h\mathbf{f}(t_1, \mathbf{x}_1)$.

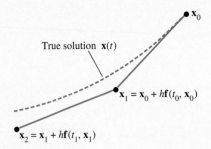

**FIGURE 6.16**   Schematic illustration of two Euler steps for a system of differential equations in the plane. The dashed curve indicates the true solution $\mathbf{x}(t)$. The points $\mathbf{x}_1$ and $\mathbf{x}_2$ are approximations to $\mathbf{x}(t_1)$ and $\mathbf{x}(t_2)$, respectively.

Euler's method can be summarized in the following steps:

**Step 0.** Select the initial condition $\mathbf{x}(t_0) = \mathbf{x}_0$ and the step size $h$. Set $n = 0$.

**Step 1.** Set $\mathbf{x}_{n+1} = \mathbf{x}_n + h\mathbf{f}(t_n, \mathbf{x}_n)$ and $t_{n+1} = t_n + h$.

**Step 2.** If $t_{n+1}$ equals or exceeds some prespecified stopping time, then halt. Otherwise, increment $n$ by 1 and go to Step 1.

Figure 6.17 shows a Fortran program that implements Euler's method for the system of equations to be discussed in Example 1. Any number of equations can be handled by changing the value of the parameter NEQ in the main routine in the left-hand column. The number of time steps is given by the value of the parameter NSTEPS, and the initial condition and step size are specified in the subsequent statements. The subroutine deriv evaluates the derivatives. The array dx holds the components of $\mathbf{x}'$ after each call. Of course, deriv must be modified to integrate a system of equations that is different from system (6.34).

```
parameter(NEQ=2,NSTEPS=3)          subroutine deriv(t,x,dx)
real x(NEQ),dx(NEQ),t,h            real t,x(2),dx(2)
x(1)=1.0                           dx(1) = x(2)
x(2)=0.0                           dx(2) = -x(1)
h=0.01                             return
t=0.0                              end
do j=1,NSTEPS
   call deriv(t,x,dx)
   do k=1,NEQ
      x(k) = x(k) + h*dx(k)
   enddo
   t = t+h
   print *, t,x
enddo
end
```

**FIGURE 6.17** Fortran program to integrate Eqs. (6.34) from $t = 0$ to $t = 0.03$ using a step size of 0.01. The initial condition is $\mathbf{x}_0 = (1, 0)$.

◆ *Example 1*  Let $x'' + x = 0$. Euler's method can be applied to this second-order differential equation by writing it as the equivalent first-order system

$$x_1' = x_2,$$
$$x_2' = -x_1. \tag{6.34}$$

The right-hand side of Eqs. (6.34) defines $\mathbf{f}(t, \mathbf{x}) = (x_2, -x_1)$. Notice that the system (6.34) is autonomous, because $\mathbf{f}$ does not depend explicitly on the independent variable $t$.

Suppose $\mathbf{x}(0) = (x_1(0), x_2(0)) = (1, 0)$. Euler's method can be used to approximate the solution at $t = 0.03$ using a step size of 0.01 as follows. The computations are rounded to four places after the decimal point.

Set $\mathbf{x}_1 = \mathbf{x}_0 + h\mathbf{f}(0, \mathbf{x}_0)$. This yields

$$\mathbf{x}_1 = \begin{pmatrix} 1 \\ 0 \end{pmatrix} + 0.01 \begin{pmatrix} 0 \\ -1 \end{pmatrix}$$

$$= \begin{pmatrix} 1 \\ -0.01 \end{pmatrix}.$$

Set $\mathbf{x}_2 = \mathbf{x}_1 + h\mathbf{f}(0.01, \mathbf{x}_1)$. This yields

$$\mathbf{x}_2 = \begin{pmatrix} 1 \\ -0.01 \end{pmatrix} + 0.01 \begin{pmatrix} -0.01 \\ -1 \end{pmatrix}$$

$$= \begin{pmatrix} 0.9999 \\ -0.02 \end{pmatrix}.$$

Set $\mathbf{x}_3 = \mathbf{x}_2 + h\mathbf{f}(0.02, \mathbf{x}_2)$. This yields

$$\mathbf{x}_3 = \begin{pmatrix} 0.9999 \\ -0.02 \end{pmatrix} + 0.01 \begin{pmatrix} -0.02 \\ -0.9999 \end{pmatrix}$$

$$= \begin{pmatrix} 0.9997 \\ -0.0300 \end{pmatrix}.$$

Here $\mathbf{x}_3$ is an approximation to $\mathbf{x}(0.03)$. The program in Fig. 6.17 carries out these calculations, but with more significant digits. ∎

Figure 6.18 shows a plot of the approximation generated in Example 1. The numerical approximation of the solution is given by the indicated line segments. Notice

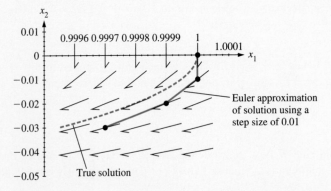

**FIGURE 6.18**   Direction field corresponding to Eqs. (6.34). The dashed curve shows the true solution. The solid line segments show the numerical approximation to the solution generated by Euler's method in Example 1. The arrows indicate the direction field.

how Euler's method simply follows a given slope line for a fixed amount of time at each step. The graph of the true solution is denoted by the dashed curve.

## How Accurate Is Euler's Method?

Euler's method is only an approximation to a given solution curve. As illustrated in Fig. 6.18, each step is subject to some error. While the Euler approximation tracks the true solution reasonably closely in the region displayed in Fig. 6.18, the errors that are introduced in each step accumulate with time. An important question, therefore, is how the accuracy of the numerical approximation deteriorates as Euler's method is iterated for many steps.

As in Example 1, consider the system

$$
\begin{aligned}
x_1' &= x_2, \\
x_2' &= -x_1.
\end{aligned}
\tag{6.35}
$$

The solution is straightforward to derive from the corresponding second-order differential equation. In this case, the initial condition $\mathbf{x}_0 = (1, 0)$ implies that $\mathbf{x}(t) = (x_1(t), x_2(t)) = (\cos t, -\sin t)$.

Figure 6.19(a) shows a plot of the numerical approximation of $x_1(t) = \cos t$ generated by Euler's method when the step size is $2\pi/50$. (The step size is 1/50 of the period of the solution.) The approximation is reasonably good for $0 < t < 2$, corresponding to the first 15 time steps. However, the accuracy deteriorates rapidly thereafter. Figure 6.19(b) shows a phase portrait of the approximated solution curve as the solid line and the true solution as the dashed line.

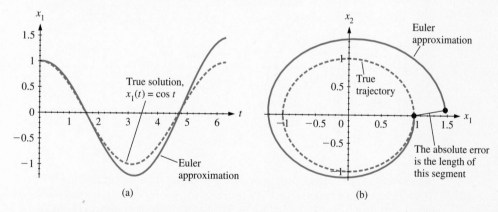

**FIGURE 6.19**  (a) A plot of the Euler approximation of $x_1(t) = \cos t$ (solid curve) and the graph of $\cos t$ (dashed curve). (b) The corresponding phase portrait showing the computed solution for $0 \le t \le 2\pi$.

We define the *absolute error* between the true solution $\mathbf{x}(t) = (x_1(t), x_2(t))$ and the numerically approximated solution $\hat{\mathbf{x}}(t) = (\hat{x}_1(t), \hat{x}_2(t))$ as

$$\|\mathbf{x}(t) - \hat{\mathbf{x}}(t)\| = \sqrt{(x_1(t) - \hat{x}_1(t))^2 + (x_2(t) - \hat{x}_2(t))^2}. \qquad (6.36)$$

Equation (6.36) is the euclidean distance between $\mathbf{x}(t)$ and $\hat{\mathbf{x}}(t)$. The absolute error in the numerically approximated solution at $t = 2\pi$ is simply the length of the indicated segment. After one full period, Euler's method leads to the estimate $\hat{\mathbf{x}}(2\pi) \approx (1.479, 0.0485)$. The true solution is $\mathbf{x}(2\pi) = (1, 0)$. Therefore, the absolute error is

$$\|\mathbf{x}(2\pi) - \hat{\mathbf{x}}(2\pi)\| \approx 0.48.$$

Greater accuracy can be obtained by cutting the step size. Figure 6.20(a) shows the phase portrait for the computed solution of Eqs. (6.35) when the step size is cut in half to $2\pi/100$. After one full period, Euler's method yields

$$\hat{\mathbf{x}}(2\pi) \approx (1.218, 0.0100),$$

which implies that

$$\|\mathbf{x}(2\pi) - \hat{\mathbf{x}}(2\pi)\| \approx 0.22.$$

Thus, by cutting the step size in half, the absolute error in the solution has been cut roughly in half.

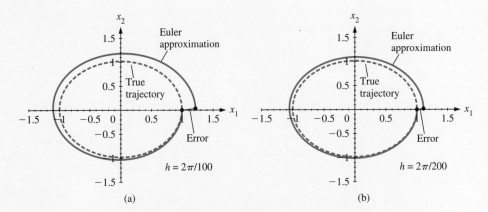

**FIGURE 6.20**  (a) The phase portrait showing the computed solution of Eqs. (6.35) for $0 \le t \le 2\pi$ with a step size of $2\pi/100$. (b) The phase portrait showing the computed solution with a step size of $2\pi/200$. Notice that the Euler approximation is twice as close to the true trajectory in (b) as in (a).

Figure 6.20(b) shows the phase portrait when the step size is cut in half again, to $2\pi/200$. After one full period, Euler's method yields

$$\hat{\mathbf{x}}(2\pi) \approx (1.104, 0.00228),$$

which implies that

$$\|\mathbf{x}(2\pi) - \hat{\mathbf{x}}(2\pi)\| \approx 0.10.$$

This value is about half of the absolute error when $h = 2\pi/100$ and about one-fourth of the absolute error when $h = 2\pi/50$.

This numerical example illustrates the following property of Euler's method, provided that the step size $h$ is sufficiently small:

*The absolute error in the numerical solution generated by Euler's method is reduced by approximately a factor of 2 whenever $h$ is cut in half.*

An identical result holds for Euler's method for scalar equations (see Section 1.4).

This discussion shows that Euler's method sometimes generates poor approximations to the actual solution. Greater accuracy can be obtained by reducing the step size further. In general, very small steps are needed to obtain accurate solutions from Euler's method.[1]

Note also that Euler's method is particularly problematic for finding numerical approximations to periodic solutions. The small errors that arise at each step of the method accumulate over time and often produce a numerical solution that is not periodic, as shown in Fig. 6.20.

## *Exercises*

**1.** For each second-order initial value problem in the following list, apply Euler's method to the equivalent first-order system with a step size of 0.01 to approximate the solution at $t = 0.04$.

(a) $x'' + 4x = 0$, $x(0) = 1$, $x'(0) = 1$

(b) $x'' + 2x' + 2x = 0$, $x(0) = 1$, $x'(0) = -2$

**2.** Let $x'' + 2x' + 2x = 0$ with $x(0) = 1$, $x'(0) = -2$.

(a) Find the solution of the initial value problem and plot it. Suppose that the equation models a mass-spring system. When would you say that the motion of the mass becomes negligible? (Interpret the term *negligible* in a manner that is consistent with the resolution of your plot.)

(b) Find the corresponding first-order system and apply Euler's method to it, using a step size of 0.1. Integrate from $t = 0$ to a value of $t$ at which the motion of the mass becomes negligible (i.e., the upper time limit for Euler's method is your answer to (a)).

---

[1] Although it is easy to understand conceptually and to implement in a computer program, Euler's method is not the numerical method of choice for most differential equations. Better numerical algorithms are described in Chapter 9.

(c) Repeat (b) using a step size of 0.01.

(d) Plot the trajectories corresponding to the true solution and the two Euler approximations that you found in (b) and (c). Does Euler's method reproduce the true solution with reasonable accuracy? Which solution curve follows the true solution curve more closely?

**3.** Let $x'' + 4x = 0$ with $x(0) = 1$, $x'(0) = 1$.

(a) Find the solution of the initial value problem.

(b) Find the solution of the corresponding first-order system.

(c) Let $T$ be the period of the solution that you found in (a). Use Euler's method to approximate the solution at $t = T$ using a step size of $T/100$. (In other words, set $h = T/100$ and iterate Euler's method 100 times. The last iteration will be an approximation of $\mathbf{x}(T)$.)

(d) Repeat (c) for step sizes of $T/200$, $T/400$, and $T/800$.

(e) Use your results in (a), (c), and (d) to complete the entries in Table 6.2. The notation $\hat{\mathbf{x}}(t)$ refers to the numerical approximation of $\mathbf{x}(T)$ that you found in (c) and (d). In the last two columns, compute the absolute error in the solution at time $T$ as defined by Eq. (6.36).

| Step Size | $\hat{x}_1(T)$ | $\hat{x}_2(T)$ | $\|\mathbf{x}(T) - \hat{\mathbf{x}}(T)\|$ |
|-----------|----------------|----------------|--------------------------------------------|
| $T/100$   |                |                |                                            |
| $T/200$   |                |                |                                            |
| $T/400$   |                |                |                                            |
| $T/800$   |                |                |                                            |

**TABLE 6.2**   For Exercise 3.

(f) Do the errors listed in the last column of the table appear to decrease linearly with the step size? Explain.

(g) Suppose that the numerical approximation of the solution is acceptably accurate if, after one period, the absolute error is less than 0.07. Which step sizes satisfy this criterion?

(h) How many periods can you iterate Euler's method with a step size of $T/800$ and still satisfy this error tolerance?

**4.** The nonlinear pendulum is described by the differential equation

$$x'' + \sin x = 0, \qquad (6.37)$$

but the solution does not have a convenient analytical representation.

Equation (6.37) can be investigated numerically, provided that the numerical method is applied carefully. In this exercise, we will use Euler's method to investigate the behavior of the solutions for certain initial conditions.

(a) The linear pendulum is described by the equation $x'' + x = 0$. Let $x(0) = 0$. Find an explicit solution for each of the following initial values for $x'(0)$: $x'(0) = 0.5$, $x'(0) = 1$, $x'(0) = 1.5$. (That is, you will find a solution for each choice of $x'(0)$ in this list.) What is the period of each solution?

(b) Use Euler's method with a step size of 0.01 to solve Eq. (6.37) using the same initial conditions as in (a). Plot the solution curves that you obtain on the same set of axes. What is the approximate period of each solution?

(c) Repeat (b) using a time step of 0.005. Do you obtain similar results?

(d) If you have access to a Runge-Kutta or other ODE solver (see Chapter 9), use it to repeat the calculations in (b).

(e) Discuss how your numerical investigations support or refute the contention that the period of the solutions of Eq. (6.37) varies with the initial condition. How do your findings compare to your answer in (a)?

# Chapter 7

---

# THE ANALYSIS OF LINEAR SYSTEMS

---

This chapter develops the theory of systems of first-order, linear, autonomous differential equations with constant coefficients. Such systems have the form

$$\mathbf{x}' = \mathbf{A}\mathbf{x},$$

where $\mathbf{A}$ is a matrix of constants. We will see that explicit formulas for the solutions of this type of system can be found. We focus mostly on systems of two first-order differential equations; the basic ideas carry over with few changes to systems with three or more equations.

## 7.1  BASIC LINEAR ALGEBRA

In this section, we discuss the basics of linear algebra, which is concerned with the study of matrices and vectors. The presentation is only a brief introduction to a much larger subject. (If you have already studied linear algebra in another course, then you may wish to skim most of the material in this section.) We have chosen only those topics from linear algebra that have the greatest relevance to the study of differential equations.

### Linear Transformations and Matrix-Vector Multiplication

A linear function $\mathbf{f}$ is called a *linear transformation* if for all vectors $\mathbf{x}$ and $\mathbf{y}$ and all constants $c$,

- $\mathbf{f}(\mathbf{x} + \mathbf{y}) = \mathbf{f}(\mathbf{x}) + \mathbf{f}(\mathbf{y})$, and
- $\mathbf{f}(c\mathbf{x}) = c\mathbf{f}(\mathbf{x})$.

Linear transformations are usually written in matrix notation. The notation $a_{ij}$ denotes the entry in the $i$th row and $j$th column of the matrix $\mathbf{A}$. The matrix $\mathbf{A}$ is a *constant matrix* if and only if each $a_{ij}$ is a constant. If the entries $a_{ij}$ are functions of $t$, then $\mathbf{A}(t)$ is a *matrix function*. An $m \times n$ matrix is one that has $m$ rows and $n$ columns. The notation $\mathbf{a}_k$ denotes the $k$th column of $\mathbf{A}$, that is,

$$\mathbf{a}_k = \begin{pmatrix} a_{1k} \\ a_{2k} \\ \vdots \\ a_{mk} \end{pmatrix}$$

if $\mathbf{A}$ has $m$ rows. We write the $2 \times 2$ matrix $\mathbf{A}$ as

$$\begin{pmatrix} a_{11} & a_{12} \\ a_{21} & a_{22} \end{pmatrix} \quad \text{or} \quad (\mathbf{a}_1 \quad \mathbf{a}_2) \quad \text{or} \quad \begin{pmatrix} a & b \\ c & d \end{pmatrix},$$

depending on the application.

**Definition 7.1**   *Let $\mathbf{A}$ be a $2 \times 2$ matrix and let $\mathbf{x}$ be a 2-vector. The* matrix-vector product $\mathbf{A}\mathbf{x}$ *is defined as*

$$\mathbf{A}\mathbf{x} = \begin{pmatrix} a_{11} & a_{12} \\ a_{21} & a_{22} \end{pmatrix} \begin{pmatrix} x_1 \\ x_2 \end{pmatrix}$$

$$= x_1 \begin{pmatrix} a_{11} \\ a_{21} \end{pmatrix} + x_2 \begin{pmatrix} a_{12} \\ a_{22} \end{pmatrix}$$

$$= x_1 \mathbf{a}_1 + x_2 \mathbf{a}_2.$$

*In other words, the matrix-vector product is a linear combination of the columns of $\mathbf{A}$. An analogous definition applies when $\mathbf{A}$ is an $m \times n$ matrix and $\mathbf{x}$ is an $n$-vector:*

$$\mathbf{A}\mathbf{x} = x_1 \mathbf{a}_1 + x_2 \mathbf{a}_2 + \cdots + x_n \mathbf{a}_n,$$

*where each vector $\mathbf{a}_i$, and hence the product $\mathbf{A}\mathbf{x}$, has $m$ components.*

✦ *Example 1*

$$\begin{pmatrix} 1 & 2 \\ 3 & 4 \end{pmatrix} \begin{pmatrix} 5 \\ 6 \end{pmatrix} = 5 \begin{pmatrix} 1 \\ 3 \end{pmatrix} + 6 \begin{pmatrix} 2 \\ 4 \end{pmatrix} = \begin{pmatrix} 17 \\ 39 \end{pmatrix}. \qquad \blacksquare$$

✦ *Example 2*   The $2 \times 2$ *identity matrix* is the matrix

$$\mathbf{I} = \begin{pmatrix} 1 & 0 \\ 0 & 1 \end{pmatrix}.$$

The term reflects the property that $\mathbf{Ix} = \mathbf{x}$ for every 2-vector $\mathbf{x}$.

The $n \times n$ identity matrix has 1's down its diagonal and 0's elsewhere, that is,

$$\mathbf{I} = \begin{pmatrix} 1 & 0 & \cdots & 0 \\ 0 & 1 & \cdots & 0 \\ & & \ddots & \\ 0 & 0 & \cdots & 1 \end{pmatrix}.$$   ■

✦ *Example 3*   The linear system

$$ax_1 + bx_2 = y_1,$$
$$cx_1 + dx_2 = y_2$$

can be rewritten as the matrix-vector product

$$\begin{pmatrix} a & b \\ c & d \end{pmatrix} \begin{pmatrix} x_1 \\ x_2 \end{pmatrix} = \begin{pmatrix} y_1 \\ y_2 \end{pmatrix}$$

or more compactly as

$$\mathbf{Ax} = \mathbf{y}.$$   ■

✦ *Example 4*   The product of a $2 \times 3$ matrix and a 3-vector is a 2-vector, as the following calculation illustrates:

$$\begin{pmatrix} 1 & 2 & 3 \\ 4 & 5 & 6 \end{pmatrix} \begin{pmatrix} -1 \\ 2 \\ 3 \end{pmatrix} = (-1)\begin{pmatrix} 1 \\ 4 \end{pmatrix} + 2\begin{pmatrix} 2 \\ 5 \end{pmatrix} + 3\begin{pmatrix} 3 \\ 6 \end{pmatrix} = \begin{pmatrix} 12 \\ 24 \end{pmatrix}.$$   ■

✦ *Example 5*   The matrix-vector product

$$\mathbf{Ax} = \begin{pmatrix} 1 & 2 & 3 \\ 4 & 5 & 6 \end{pmatrix} \begin{pmatrix} 1 \\ 2 \end{pmatrix}$$

is not defined, because $\mathbf{x}$ has fewer components than $\mathbf{A}$ has columns. Definition 7.1 requires that the number of components of $\mathbf{x}$ must equal the number of columns of $\mathbf{A}$.   ■

Let $\mathbf{A}$ and $\mathbf{B}$ be $m \times n$ matrices, $\mathbf{x}$ and $\mathbf{y}$ be $n$-vectors, and $c$ be a constant (scalar). Then the definition of matrix-vector multiplication implies that the following properties hold:

$$\mathbf{A}(\mathbf{x} + \mathbf{y}) = \mathbf{A}\mathbf{x} + \mathbf{A}\mathbf{y}, \tag{7.1}$$

$$\mathbf{A}(c\mathbf{x}) = c(\mathbf{A}\mathbf{x}), \tag{7.2}$$

$$\mathbf{A}\mathbf{x} + \mathbf{B}\mathbf{x} = (\mathbf{A} + \mathbf{B})\mathbf{x}, \tag{7.3}$$

$$c(\mathbf{A}\mathbf{x}) = (c\mathbf{A})\mathbf{x}. \tag{7.4}$$

The proof of Eqs. (7.1)–(7.4) is left as an exercise. (See Exercise 11.)

## Matrix Multiplication

The product of two matrices is defined as follows.

**Definition 7.2**   *Let* $\mathbf{A}$ *and* $\mathbf{B}$ *be* $2 \times 2$ *matrices. The matrix product* $\mathbf{AB}$ *is defined as*

$$\mathbf{AB} = \begin{pmatrix} \mathbf{A}\mathbf{b}_1 & \mathbf{A}\mathbf{b}_2 \end{pmatrix},$$

*where each column of* $\mathbf{AB}$ *is the matrix-vector product of* $\mathbf{A}$ *with the corresponding column of* $\mathbf{B}$*. In general, if* $\mathbf{A}$ *is an* $m \times n$ *matrix and* $\mathbf{B}$ *is an* $n \times p$ *matrix, then the product* $\mathbf{AB}$ *is defined as*

$$\mathbf{AB} = \begin{pmatrix} \mathbf{A}\mathbf{b}_1 & \mathbf{A}\mathbf{b}_2 & \cdots & \mathbf{A}\mathbf{b}_p \end{pmatrix},$$

*which is an* $m \times p$ *matrix.*

Notice that the product $\mathbf{AB}$ is defined if and only if $\mathbf{B}$ has as many rows as $\mathbf{A}$ has columns.

✦ *Example 6*   Let

$$\mathbf{A} = \begin{pmatrix} 1 & 2 \\ 3 & 4 \end{pmatrix} \quad \text{and} \quad \mathbf{B} = \begin{pmatrix} -1 & 2 \\ -3 & 4 \end{pmatrix}.$$

Then

$$\mathbf{AB} = \begin{pmatrix} \mathbf{A}\begin{pmatrix} -1 \\ -3 \end{pmatrix} & \mathbf{A}\begin{pmatrix} 2 \\ 4 \end{pmatrix} \end{pmatrix} = \begin{pmatrix} -7 & 10 \\ -15 & 22 \end{pmatrix}$$

and

$$\mathbf{BA} = \begin{pmatrix} \mathbf{B}\begin{pmatrix} 1 \\ 3 \end{pmatrix} & \mathbf{B}\begin{pmatrix} 2 \\ 4 \end{pmatrix} \end{pmatrix} = \begin{pmatrix} 5 & 6 \\ 9 & 10 \end{pmatrix}.$$

In general, $\mathbf{AB} \neq \mathbf{BA}$. In other words, *matrix multiplication is not a commutative operation.* Only in some special cases is it true that $\mathbf{AB} = \mathbf{BA}$. ∎

✦ *Example 7* Let

$$\mathbf{A} = \begin{pmatrix} 1 & 2 & 3 \\ 4 & 5 & 6 \end{pmatrix} \quad \text{and} \quad \mathbf{B} = \begin{pmatrix} -1 & 2 \\ -3 & 4 \\ -5 & 6 \end{pmatrix}.$$

Then

$$\mathbf{AB} = \left( \mathbf{A} \begin{pmatrix} -1 \\ -3 \\ -5 \end{pmatrix} \ \mathbf{A} \begin{pmatrix} 2 \\ 4 \\ 6 \end{pmatrix} \right) = \begin{pmatrix} -22 & 28 \\ -49 & 64 \end{pmatrix}$$

and

$$\mathbf{BA} = \begin{pmatrix} 7 & 8 & 9 \\ 13 & 14 & 15 \\ 19 & 20 & 21 \end{pmatrix}.$$

Once again, $\mathbf{AB} \neq \mathbf{BA}$. In fact, $\mathbf{AB}$ is a $2 \times 2$ matrix, and $\mathbf{BA}$ is a $3 \times 3$ matrix. ∎

✦ *Example 8* The product

$$\mathbf{AB} = \begin{pmatrix} 1 \\ 2 \end{pmatrix} \begin{pmatrix} 1 & 2 \\ 3 & 4 \end{pmatrix}$$

is not defined, because $\mathbf{A}$ has only one column and $\mathbf{B}$ has two rows. ∎

The following two properties of matrix multiplication are very important.

**Property 1** **Matrix multiplication in general is not a commutative operation.** *That is, $\mathbf{AB} \neq \mathbf{BA}$ in general. There are some special cases in which it is possible to interchange the order in which the matrices are multiplied. Two special cases are the following:*

- *If $\mathbf{A}$ is any $n \times n$ matrix and $\mathbf{I}$ is the $n \times n$ identity matrix, then*

$$\mathbf{AI} = \mathbf{IA} = \mathbf{A}. \tag{7.5}$$

- *If the inverse $\mathbf{A}^{-1}$ is defined (see Definition 7.4), then*

$$\mathbf{AA}^{-1} = \mathbf{A}^{-1}\mathbf{A} = \mathbf{I}. \tag{7.6}$$

*(See also Exercise 10.)*

**Property 2   Matrix multiplication is an associative operation.** *Suppose* $A$ *is an* $m \times n$ *matrix,* $B$ *is an* $n \times p$ *matrix, and* $C$ *is a* $p \times q$ *matrix. Then*

$$ABC = (AB)C = A(BC). \tag{7.7}$$

*The product is an* $m \times q$ *matrix.*

## Determinants and Inverse Matrices

**Definition 7.3**   *Let*

$$A = \begin{pmatrix} a & b \\ c & d \end{pmatrix}.$$

*The determinant of* $A$ *is*

$$\det A = \det \begin{pmatrix} a & b \\ c & d \end{pmatrix} = ad - bc.$$

The determinant plays an important role in the solution of linear systems of equations. If $ad - bc \neq 0$, then the linear system

$$ax_1 + bx_2 = y_1, \\ cx_1 + dx_2 = y_2 \tag{7.8}$$

has a unique solution, given by

$$x_1 = \frac{dy_1 - by_2}{ad - bc} \tag{7.9}$$

and

$$x_2 = \frac{ay_2 - cy_1}{ad - bc}. \tag{7.10}$$

(See Exercise 4.)

As was shown in Example 3, we can write the system (7.8) more compactly as $Ax = y$, where

$$A = \begin{pmatrix} a & b \\ c & d \end{pmatrix}.$$

The solution can be expressed as

$$\begin{pmatrix} x_1 \\ x_2 \end{pmatrix} = \frac{1}{\det A} \begin{pmatrix} dy_1 - by_2 \\ ay_2 - cy_1 \end{pmatrix} = \frac{1}{\det A} \begin{pmatrix} d & -b \\ -c & a \end{pmatrix} \begin{pmatrix} y_1 \\ y_2 \end{pmatrix}. \tag{7.11}$$

Thus, the system (7.8) has a unique solution provided that $\det A \neq 0$.

**Definition 7.4**  *Let* **A** *be an n × n square matrix. We say that* **A** *is* nonsingular *or* invertible *if there exists an n × n matrix* **B** *such that*

$$\mathbf{AB} = \mathbf{BA} = \mathbf{I},$$

*where* **I** *is the n × n identity matrix. The matrix* **B** *is called the* inverse *of* **A** *and is usually denoted as* $\mathbf{A}^{-1}$.

The inverse of a 2 × 2 matrix can be calculated using the determinant.

**Theorem 7.5**  *Let*

$$\mathbf{A} = \begin{pmatrix} a & b \\ c & d \end{pmatrix}.$$

*If* $\det \mathbf{A} \neq 0$, *then*

$$\mathbf{A}^{-1} = \frac{1}{\det \mathbf{A}} \begin{pmatrix} d & -b \\ -c & a \end{pmatrix}.$$

Note that **A** is nonsingular (invertible) if and only if $\det \mathbf{A} \neq 0$. Otherwise, **A** is singular (noninvertible).

◆ *Example 9*   The system

$$\begin{aligned} x_1 + 2x_2 &= 4, \\ 3x_1 + 4x_2 &= 6 \end{aligned} \qquad (7.12)$$

is equivalent to the matrix-vector product $\mathbf{Ax} = \mathbf{y}$ given by

$$\begin{pmatrix} 1 & 2 \\ 3 & 4 \end{pmatrix} \mathbf{x} = \begin{pmatrix} 4 \\ 6 \end{pmatrix}.$$

The determinant of **A** is

$$\det \mathbf{A} = (1)(4) - (2)(3) = -2,$$

so

$$\mathbf{A}^{-1} = -\frac{1}{2} \begin{pmatrix} 4 & -2 \\ -3 & 1 \end{pmatrix} = \begin{pmatrix} -2 & 1 \\ \frac{3}{2} & -\frac{1}{2} \end{pmatrix}.$$

The solution of the system (7.12) is

$$\mathbf{x} = \mathbf{A}^{-1}\mathbf{y} = \begin{pmatrix} -2 & 1 \\ \frac{3}{2} & -\frac{1}{2} \end{pmatrix} \begin{pmatrix} 4 \\ 6 \end{pmatrix} = \begin{pmatrix} -2 \\ 3 \end{pmatrix}. \qquad \blacksquare$$

Some linear systems of two equations do not have a unique solution. The equations may be inconsistent, in which case there is no solution, or the equations may be redundant, in which case there are infinitely many solutions. Both kinds of systems have a singular matrix associated with them, as the next examples illustrate.

✦ *Example 10*  Let

$$2x_1 + 4x_2 = 0,$$
$$2x_1 + 4x_2 = 1. \tag{7.13}$$

The system (7.13) is inconsistent, because the left-hand side of each equation cannot simultaneously be equal to both 0 and 1. Therefore, there is no solution. We can rewrite Eqs. (7.13) as

$$\mathbf{Ax} = \begin{pmatrix} 2 & 4 \\ 2 & 4 \end{pmatrix} \begin{pmatrix} x_1 \\ x_2 \end{pmatrix} = \begin{pmatrix} 0 \\ 1 \end{pmatrix}.$$

Here

$$\det \mathbf{A} = (2)(4) - (2)(4) = 0,$$

and $\mathbf{A}^{-1}$ is undefined.  ∎

✦ *Example 11*  Let

$$2x_1 + 4x_2 = 1,$$
$$4x_2 + 8x_2 = 2. \tag{7.14}$$

The system (7.14) is redundant, because the second equation is a multiple of the first. Therefore, there are infinitely many solutions. We can rewrite Eqs. (7.14) as

$$\mathbf{Ax} = \begin{pmatrix} 2 & 4 \\ 4 & 8 \end{pmatrix} \begin{pmatrix} x_1 \\ x_2 \end{pmatrix} = \begin{pmatrix} 1 \\ 2 \end{pmatrix}.$$

Here,

$$\det \mathbf{A} = (2)(8) - (4)(4) = 0,$$

and $\mathbf{A}^{-1}$ is undefined.  ∎

In general, the matrix $\mathbf{A}$ is singular whenever one row of $\mathbf{A}$ is a multiple of the other or one column of $\mathbf{A}$ is a multiple of the other. (See Exercise 12.)

The inverse is defined for any square ($n \times n$) matrix whose determinant is nonzero. However, there are no convenient formulas for the inverse for $n$ greater than 2.

## Important Properties of Nonsingular Matrices

We now turn to some important properties of singular and nonsingular matrices that will be used repeatedly in the remainder of the chapter. The following terminology is useful.

**Definition 7.6**  *Let $\mathbf{x}_1$ and $\mathbf{x}_2$ be 2-vectors. We say that $\mathbf{x}_1$ and $\mathbf{x}_2$ are linearly independent if they are not constant multiples of each other or, equivalently, if the relation*

$$c_1\mathbf{x}_1 + c_2\mathbf{x}_2 = \mathbf{0}$$

*is satisfied only when $c_1 = c_2 = 0$. Otherwise, the vectors $\mathbf{x}_1$ and $\mathbf{x}_2$ are linearly dependent.*

*More generally, the $n$-vectors $\mathbf{x}_1, \mathbf{x}_2, \ldots, \mathbf{x}_m$ are linearly independent if it is impossible to express any one of the vectors as a linear combination of the others. Equivalently, the vectors are linearly independent if the relation*

$$c_1\mathbf{x}_1 + c_2\mathbf{x}_2 + \cdots + c_m\mathbf{x}_m = \mathbf{0}$$

*is satisfied only when $c_1 = c_2 = \cdots = c_m = 0$. If the vectors $\mathbf{x}_1, \mathbf{x}_2, \ldots, \mathbf{x}_m$ are not linearly independent, then they are linearly dependent.*

An important consequence of linear independence is the following theorem, whose proof is left as an exercise. (See Exercise 1.6.)

**Theorem 7.7**  *Let $\mathbf{x}_1, \mathbf{x}_2, \ldots, \mathbf{x}_n$ be a collection of linearly independent $n$-vectors. For any $n$-vector $\mathbf{y}$, there is a unique set of constants $c_1, c_2, \ldots, c_n$ such that*

$$c_1\mathbf{x}_1 + c_2\mathbf{x}_2 + \cdots + c_n\mathbf{x}_n = \mathbf{y}.$$

Figure 7.1 illustrates the concept of linear independence for two 2-vectors. Two 2-vectors are linearly independent if they are not collinear.

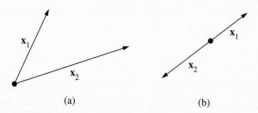

(a)                              (b)

**FIGURE 7.1**  (a) Two linearly independent vectors. (b) Two linearly dependent vectors.

◆ *Example 12*    The vectors

$$\mathbf{x}_1 = \begin{pmatrix} 1 \\ 2 \end{pmatrix} \quad \text{and} \quad \mathbf{x}_2 = \begin{pmatrix} 3 \\ 4 \end{pmatrix}$$

are linearly independent, because the relation

$$c_1\mathbf{x}_1 + c_2\mathbf{x}_2 = \mathbf{0}$$

is equivalent to the linear system

$$c_1 + 3c_2 = 0,$$
$$2c_1 + 4c_2 = 0,$$

the only solution of which is $c_1 = c_2 = 0$.                                      ∎

◆ *Example 13*    The vectors

$$\mathbf{x}_1 = \begin{pmatrix} 1 \\ 2 \end{pmatrix} \quad \text{and} \quad \mathbf{x}_2 = \begin{pmatrix} 2 \\ 4 \end{pmatrix}$$

are linearly dependent, because the relation

$$c_1\mathbf{x}_1 + c_2\mathbf{x}_2 = \mathbf{0}$$

is equivalent to the linear system

$$c_1 + 2c_2 = 0,$$
$$2c_1 + 4c_2 = 0,$$

which has infinitely many solutions, such as $c_1 = 2$ and $c_2 = -1$. (It is straightforward to check that $2\mathbf{x}_1 - \mathbf{x}_2 = \mathbf{0}$.) Notice that $\mathbf{x}_1$ and $\mathbf{x}_2$ are collinear.        ∎

The following theorem summarizes some important properties of singular and non-singular matrices that we will use repeatedly in the remainder of this chapter. Although the two statements are true for any $n \times n$ matrix, for simplicity the proof of part (a) is demonstrated only for $2 \times 2$ matrices.

**Theorem 7.8**    *Let* $\mathbf{A}$ *be an* $n \times n$ *matrix, and let* $\mathbf{x}$ *and* $\mathbf{y}$ *be* $n$-*vectors.*

*(a)* $\mathbf{A}$ *is singular if and only if the equation* $\mathbf{A}\mathbf{x} = \mathbf{0}$ *has more than one solution.*

*(b) If* $\mathbf{A}$ *is nonsingular, then the equation* $\mathbf{A}\mathbf{x} = \mathbf{y}$ *has exactly one solution for any vector* $\mathbf{y}$.

*Proof*

(a) The equation $\mathbf{Ax} = \mathbf{0}$ is equivalent to the system

$$ax_1 + bx_2 = 0,$$
$$cx_1 + dx_2 = 0,$$

which implies that $(ad - bc)x_1 = 0$ and $(ad - bc)x_2 = 0$. This situation arises only if the two equations in the system are redundant (that is, one is a multiple of the other). In this case, there are infinitely many pairs $(x_1, x_2)$ that satisfy both equations. Conversely, if there is more than one solution of $\mathbf{Ax} = \mathbf{0}$, then we can find constants $c_1$ and $c_2$, not both zero, such that $c_1\mathbf{a}_1 + c_2\mathbf{a}_2 = \mathbf{0}$, where $\mathbf{a}_1$ and $\mathbf{a}_2$ are the columns of $\mathbf{A}$. This implies that $\mathbf{a}_1 = k\mathbf{a}_2$ for some constant $k$. Therefore, $\mathbf{A}$ is a matrix of the form

$$\begin{pmatrix} \alpha_1 & k\alpha_1 \\ \alpha_2 & k\alpha_2 \end{pmatrix},$$

whose determinant is 0. Hence, $\mathbf{A}$ is singular.

(b) Let $\mathbf{A}$ be nonsingular. Suppose that there were two distinct solutions $\mathbf{x}_1$ and $\mathbf{x}_2$ such that $\mathbf{Ax}_1 = \mathbf{Ax}_2 = \mathbf{y}$. Then

$$\mathbf{0} = \mathbf{Ax}_1 - \mathbf{Ax}_2 = \mathbf{A}(\mathbf{x}_1 - \mathbf{x}_2).$$

Since $\mathbf{x}_1$ and $\mathbf{x}_2$ are distinct, $\mathbf{x}_1 - \mathbf{x}_2 \neq \mathbf{0}$, which implies that $\mathbf{A}$ is singular, a contradiction. Hence, $\mathbf{x}_1 = \mathbf{x}_2$. ∎

The proof of the following useful result is left as an exercise. (See Exercise 12.)

**Theorem 7.9** *Two 2-vectors $\mathbf{a}_1$ and $\mathbf{a}_2$ are linearly dependent if and only if $\mathbf{A} = (\mathbf{a}_1 \ \mathbf{a}_2)$ is singular.*

## The Euclidean Norm

A *norm* is a function that measures the length of a vector. The *euclidean norm* of the 2-vector $\mathbf{x} = (x_1, x_2)$ is

$$\|\mathbf{x}\| = \sqrt{x_1^2 + x_2^2}. \tag{7.15}$$

Equation (7.15) simply expresses the length of the vector $\mathbf{x}$ as the euclidean distance of the point $(x_1, x_2)$ from the origin. Figure 7.2 shows that the euclidean norm of a

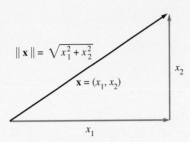

$$\|\mathbf{x}\| = \sqrt{x_1^2 + x_2^2}$$

$$\mathbf{x} = (x_1, x_2)$$

$$x_2$$

$$x_1$$

**FIGURE 7.2**  The euclidean norm measures the length of a 2-vector in terms of the hypotenuse of a right triangle consisting of the vector and its components.

2-vector can be regarded as the length of the hypotenuse of a right triangle whose sides are composed of the components of $\mathbf{x}$ along the horizontal and vertical axes.

✦ *Example 14*   Let $\mathbf{x}(t) = (1, t)$ be a function from $\mathbf{R}$ to $\mathbf{R}^2$. Then for any $t$,

$$\|\mathbf{x}(t)\| = \sqrt{t^2 + 1}.$$    ∎

## *Exercises*

**1.** Determine whether each of the matrix-vector products in the following list is defined, and if so, compute it.

(a) $\begin{pmatrix} -1 & 1 \\ 2 & 3 \end{pmatrix}\begin{pmatrix} 1 \\ 3 \end{pmatrix}$        (b) $\begin{pmatrix} 3 & 0 \\ 0 & -2 \end{pmatrix}\begin{pmatrix} 1 \\ 1 \end{pmatrix}$

(c) $\begin{pmatrix} 1 & 6 & 0 \\ 4 & 3 & 1 \end{pmatrix}\begin{pmatrix} -2 \\ 3 \\ 4 \end{pmatrix}$        (d) $\begin{pmatrix} 2 \\ 4 \end{pmatrix}\begin{pmatrix} 1 & 2 \\ 3 & 4 \end{pmatrix}$

(e) $\begin{pmatrix} 1 & 4 & -2 \\ 6 & 2 & 3 \end{pmatrix}\begin{pmatrix} 0 \\ 1 \end{pmatrix}$        (f) $\begin{pmatrix} 2 & 5 \\ -3 & 2 \end{pmatrix}\begin{pmatrix} 1 \\ 2 \\ 3 \end{pmatrix}$

**2.** Find the inverse of each matrix in the following list. If the inverse is not defined, then so state.

(a) $\begin{pmatrix} 2 & 0 \\ 0 & 4 \end{pmatrix}$        (b) $\begin{pmatrix} 0 & 2 \\ 4 & 0 \end{pmatrix}$

(c) $\begin{pmatrix} 2 & 1 \\ 0 & 4 \end{pmatrix}$        (d) $\begin{pmatrix} 0 & 2 \\ 4 & 1 \end{pmatrix}$

(e) $\begin{pmatrix} 2 & 0 \\ 4 & 0 \end{pmatrix}$        (f) $\begin{pmatrix} 0 & 0 \\ 2 & 4 \end{pmatrix}$

**3.** For each pair of matrices in the following list, compute $\mathbf{AB}$ and $\mathbf{BA}$.

(a) $\mathbf{A} = \begin{pmatrix} 1 & 2 \\ 3 & 4 \end{pmatrix}$        $\mathbf{B} = \begin{pmatrix} 1 & 0 \\ 0 & 10 \end{pmatrix}$

(b) $\mathbf{A} = \begin{pmatrix} -1 & 2 \\ 1 & -1 \end{pmatrix}$        $\mathbf{B} = \begin{pmatrix} 1 & 2 \\ 1 & 1 \end{pmatrix}$

(c) $\mathbf{A} = \begin{pmatrix} a & b \\ b & a \end{pmatrix}$        $\mathbf{B} = \begin{pmatrix} c & d \\ d & c \end{pmatrix}$

(d) $\mathbf{A} = \begin{pmatrix} a & -b \\ b & a \end{pmatrix}$        $\mathbf{B} = \begin{pmatrix} c & -d \\ d & c \end{pmatrix}$

(e) $\mathbf{A} = \begin{pmatrix} a & -b \\ b & a \end{pmatrix}$        $\mathbf{B} = \begin{pmatrix} c & d \\ e & f \end{pmatrix}$

**4.** Derive Eqs. (7.9) and (7.10).

**5.** Let $\mathbf{A}$ and $\mathbf{B}$ be $2 \times 2$ matrices and let $c$ be a scalar.

(a) Show that

$$\mathbf{A}(c\mathbf{B}) = (c\mathbf{A})\mathbf{B} = c(\mathbf{A}\mathbf{B}). \qquad (7.16)$$

Equation (7.16) holds for any two matrices for which the product $\mathbf{A}\mathbf{B}$ is defined. Stated informally, Eq. (7.16) implies that scalars can be factored out of any matrix product.

(b) Compute

$$\begin{pmatrix} 3 & 1 \\ 2 & -1 \end{pmatrix} \begin{pmatrix} 1/7 & -1/7 \\ 1/7 & 1/7 \end{pmatrix}.$$

(c) Compute

$$\begin{pmatrix} 1/3 & 2/3 \\ -1 & 4/3 \end{pmatrix} \begin{pmatrix} 5/8 & -1/8 \\ 3/8 & 1/8 \end{pmatrix}.$$

(d) Compute

$$\begin{pmatrix} 3\cos t & 2\cos t \\ \cos t & 6\cos t \end{pmatrix} \begin{pmatrix} e^{-t} & 0 \\ 2e^{-t} & 3e^{-t} \end{pmatrix}.$$

**6.** Suppose that $\mathbf{A}$ is a $2 \times 2$ matrix for which

$$\mathbf{A}\begin{pmatrix} -1 \\ 2 \end{pmatrix} = \begin{pmatrix} 1 \\ 2 \end{pmatrix}$$

and

$$\mathbf{A}\begin{pmatrix} 4 \\ 6 \end{pmatrix} = \begin{pmatrix} 3 \\ 4 \end{pmatrix}.$$

What is the matrix $\mathbf{B}$ for which

$$\mathbf{A}\mathbf{B} = \begin{pmatrix} 1 & 3 \\ 2 & 4 \end{pmatrix}?$$

**7.** True or false: If $\mathbf{A}$ is not the zero matrix, then $\mathbf{A}\mathbf{x} = \mathbf{0}$ if and only if $\mathbf{x} = \mathbf{0}$. (If you say "false," then give a counterexample.)

**8.** Let $\mathbf{A}$ and $\mathbf{B}$ be $2 \times 2$ matrices. Give two different examples of a nonzero matrix $\mathbf{A}$ and a nonzero matrix $\mathbf{B}$ such that $\mathbf{A}\mathbf{B}$ is the zero matrix. (In other words, at least one entry in $\mathbf{A}$ and at least one entry in $\mathbf{B}$ is not zero, but their product is a $2 \times 2$ matrix of zeros.)

**9.** Find all constants $a$ for which the matrix

$$\begin{pmatrix} 8 - a & -2 \\ -1 & 7 - a \end{pmatrix}$$

is singular.

**10.** A *diagonal* matrix is one whose elements are all zero, except possibly on the diagonal from the upper left to lower right. In the $2 \times 2$ case, a diagonal matrix has the form

$$\mathbf{D} = \begin{pmatrix} a_{11} & 0 \\ 0 & a_{22} \end{pmatrix}.$$

(a) Show that $\mathbf{D}$ is nonsingular if and only if all of its diagonal elements are nonzero.

(b) Assuming that $\mathbf{D}$ is nonsingular, find a formula for $\mathbf{D}^{-1}$.

(c) Prove that if $\mathbf{D}_1$ and $\mathbf{D}_2$ are diagonal matrices, then

$$\mathbf{D}_1\mathbf{D}_2 = \mathbf{D}_2\mathbf{D}_1.$$

Diagonal matrices are another special case in which matrix multiplication is commutative.

(d) Find the diagonal matrix $\mathbf{D}$ such that

$$\mathbf{D}\begin{pmatrix} 2 \\ -3 \end{pmatrix} = \begin{pmatrix} 3 \\ 6 \end{pmatrix}.$$

**11.** Suppose $\mathbf{A}$ and $\mathbf{B}$ are $2 \times 2$ matrices and $\mathbf{x}$ and $\mathbf{y}$ are 2-vectors.

(a) Use the definition of matrix-vector multiplication to show that $\mathbf{f}(\mathbf{x}) = \mathbf{A}\mathbf{x}$ is a linear function of $\mathbf{x}$. This establishes Eqs. (7.1) and (7.2).

(b) Establish Eq. (7.3).

(c) Let $c$ be a constant. The matrix $c\mathbf{A}$ is the matrix obtained by multiplying every entry of $\mathbf{A}$ by $c$. Establish Eq. (7.4).

(Note: The properties illustrated in (a) and (b) hold for any $m \times n$ matrix $\mathbf{A}$ and any $n$-vectors $\mathbf{x}$ and $\mathbf{y}$. The proofs are similar to those for the $2 \times 2$ case.)

**12.** Let $A$ be a $2 \times 2$ matrix.

  (a) Prove that $\det A = 0$ if and only if the rows of $A$ are linearly dependent.

  (b) Prove that $\det A = 0$ if and only if the columns of $A$ are linearly dependent. (See Theorem 7.9.)

**13.** Let $A$ and $B$ be $n \times n$ matrices, and suppose that $A^{-1}$ is defined.

  (a) Explain why $ABA^{-1} \neq B$ in general, assuming that $A$ is nonsingular and the matrix product is defined.

  (b) Suppose $D$ is an $n \times n$ diagonal matrix. Is it true that $ADA^{-1} = D$? Explain why or why not.

**14.** Verify Eqs. (7.5) and (7.6) for any $2 \times 2$ matrices by expanding both sides.

**15.** Linear systems of algebraic equations of the form

$$ax_1 + bx_2 = e,$$
$$cx_1 + dx_2 = f$$

can be described as having homogeneous and particular solutions. New solutions can be formed by using the properties of linearity in a manner that is entirely analogous to the solution of linear differential equations discussed in Chapter 4. This exercise illustrates an example. Let

$$A = \begin{pmatrix} 1 & 2 \\ 1 & 2 \end{pmatrix}.$$

  (a) Show that there are infinitely many solutions of the equation

$$Ax = 0. \qquad (7.17)$$

  (b) A solution of Eq. (7.17) that is not the zero vector is called a *homogeneous solution*, denoted $x_h$. Find a homogeneous solution of Eq. (7.17).

  (c) A solution of the equation

$$Ax = \begin{pmatrix} 11 \\ 11 \end{pmatrix} \qquad (7.18)$$

is called a *particular solution*, denoted $x_p$. Find a particular solution of Eq. (7.18).

  (d) Use the linearity of matrix-vector multiplication to show that any vector of the form $x = cx_h + x_p$ solves Eq. (7.18), where $c$ is any constant.

  (e) True or false: The set of all possible solutions of the equation in (c) forms a line. Give a rationale for your answer.

**16.** Consider Theorem 7.7.

  (a) Why would you expect the theorem not to hold for three 2-vectors or two 3-vectors?

  (b) Prove Theorem 7.7. (Suggestion: Suppose that there were two sets of constants $b_1, \ldots, b_n$ and $c_1, \ldots, c_n$ such that the respective linear combinations yielded $y$.)

## 7.2 DIRECTION FIELDS FOR LINEAR SYSTEMS

This section considers some of the special properties of the direction fields of linear systems.

**Definition 7.10** *Let $A(t)$ be an $n \times n$ matrix function, let $x(t)$ be a differentiable function from $\mathbf{R}$ to $\mathbf{R}^n$, and let $f(t)$ be a continuous function from $\mathbf{R}$ to $\mathbf{R}^n$. The relation*

$$\frac{dx(t)}{dt} = A(t)x(t) + f(t)$$

*defines a linear, first-order system of ordinary differential equations. If* $\mathbf{f}(t) = \mathbf{0}$ *for all* $t$*, then the system is homogeneous. If* $\mathbf{A}$ *is a constant matrix, then the system is said to have constant coefficients. A homogeneous, linear, first-order system with constant coefficients is written as*

$$\frac{d\mathbf{x}}{dt} = \mathbf{A}\mathbf{x}. \tag{7.19}$$

In this chapter, we are primarily concerned with first-order systems like Eq. (7.19) that have two equations (that is, $\mathbf{A}$ is a $2 \times 2$ constant matrix). In such a case, it is possible to draw the direction field associated with Eq. (7.19) in a manner similar to that outlined in Chapter 6. For any fixed value of $\mathbf{x}$, the slope lines are determined by the right-hand side of Eq. (7.19).

The first-order system defined by Eq. (7.19) has the following special properties.

1. Equation (7.19) is linear and homogeneous. This means that if $\mathbf{x}(t)$ is a solution, then $c\mathbf{x}(t)$ is also a solution for any constant $c$. In particular, we can choose $c = -1$, so $-\mathbf{x}(t)$ is also a solution. Therefore, the phase portraits associated with any linear homogeneous system are symmetric with respect to a reflection about the origin. This graphical interpretation of linearity is similar to that in Chapter 4, where we saw that the direction fields of linear scalar differential equations have a reflection symmetry with respect to the $t$ axis.

2. The length of the tangent vector to the solution at $c\mathbf{x}$ is $c$ times as great as the length of the tangent vector to the solution at $\mathbf{x}$—regardless of $\mathbf{x}$. That is, the size of the derivative $\mathbf{x}'(t)$ grows proportionately with $\mathbf{x}(t)$. If $\mathbf{x}$ were a scalar function, we could conclude immediately that $\mathbf{x}$ must be an exponential function. In fact, we will show that even when $\mathbf{x}$ is a function from $\mathbf{R}$ to $\mathbf{R}^2$, the components of $\mathbf{x}$ must be a linear combination of exponential functions.

3. As was discussed in Chapter 6, a *fixed point* (or *equilibrium point*) of a system of autonomous differential equations is a point where $\mathbf{x}' = \mathbf{0}$. The origin is the only fixed point if $\mathbf{A}$ is nonsingular. (See Exercise 3.)

✦ *Example 1*   Figure 7.3 shows the direction field associated with the differential equation

$$\mathbf{x}' = \mathbf{I}\mathbf{x}, \tag{7.20}$$

where $\mathbf{I}$ is the $2 \times 2$ identity matrix. Another way to write Eq. (7.20) is

$$x_1' = x_1,$$
$$x_2' = x_2.$$

For instance, at $\mathbf{x} = (1/2, 1/2)$, we have $\mathbf{x}' = (1/2, 1/2)$. The corresponding slope line has slope 1 and points away from the origin. No slope line is drawn at the origin, because the trajectory corresponding to an initial condition at the origin is a single point.

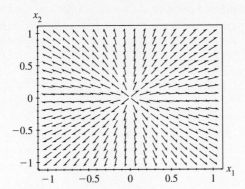

**FIGURE 7.3**    The direction field associated with $\mathbf{x}' = \mathbf{Ix}$.

The slope lines suggest that the phase portrait consists of trajectories leading away from the origin in straight lines. In other words, if $x_2(0) \neq 0$, then the ratio $x_1(t)/x_2(t)$ remains constant, *even though $x_1(t)$ and $x_2(t)$ themselves may not be linear functions of $t$.* (See Exercise 5.)                                                        ■

◆ *Example 2*    Figures 7.4(a) and 7.4(c) show the direction fields associated with the differential equation

$$\mathbf{x}' = \begin{pmatrix} 0 & 1 \\ 1 & 0 \end{pmatrix} \mathbf{x},$$

and Figs. 7.4(b) and 7.4(d) show the direction field for

$$\mathbf{x}' = \begin{pmatrix} 0 & -1 \\ 1 & 0 \end{pmatrix} \mathbf{x}.$$

Each differential equation is linear. Figure 7.4(a) and Fig. 7.4(c) show four pairs of solution curves. With the exception of the rays extending along the $\pm 45$-degree diagonals, any ray starting from the origin intersects two of the solution curves; the first intersection point is half as far from the origin as the second. Figures 7.4(b) and 7.4(d) show three solution curves, each of which is a circle. The radius of the middle circle is twice as large as the radius of the innermost circle, and the radius of the outer circle is four times as large. The linearity of the underlying differential equation in each picture is reflected in the fact that any multiple (magnification) of a solution curve is again a solution curve.                                                                                       ■

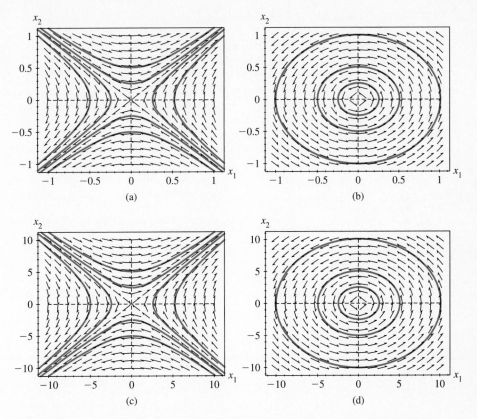

**FIGURE 7.4** (a) A saddle. (b) A center. Each phase portrait and direction field is symmetric about the origin. The dashed lines cross at the origin and are intended only to guide the eye. (c)–(d) The phase portraits in (a) and (b), respectively, on a larger scale.

These examples, together with the three properties discussed before them, imply that the direction field of a linear nonsingular homogeneous $2 \times 2$ system is *scale-invariant*. That is, the direction field looks the same in any box centered at the origin, regardless of the size of the box.

✦ *Example 3* Figures 7.4(c)–(d) show that the phase portraits corresponding to the systems discussed in Examples 1 and 2 look exactly the same when they are drawn over a larger region. ∎

## Exercises

**1.** Which of the direction fields in Fig. 7.5 corresponds to a system of the form $\mathbf{x}' = \mathbf{D}\mathbf{x}$, where $\mathbf{D}$ is a diagonal matrix? (Hint: What must the slope of the slope lines be on the coordinate axes if $\mathbf{D}$ is diagonal?)

**2.** Sketch the direction field associated with each first-order system in the following list. Draw enough of the vector field so that you can discern whether the origin is a source, sink, or saddle.

(a) $\mathbf{x}' = \begin{pmatrix} 2 & -3 \\ 1 & 4 \end{pmatrix} \mathbf{x}$    (b) $\mathbf{x}' = \begin{pmatrix} 2 & 0 \\ -1 & 3 \end{pmatrix} \mathbf{x}$

(c) $\mathbf{x}' = \begin{pmatrix} 1 & -2 \\ 5 & -2 \end{pmatrix} \mathbf{x}$    (d) $\mathbf{x}' = \begin{pmatrix} 3 & 1 \\ 1 & -4 \end{pmatrix} \mathbf{x}$

**3.** Let $\mathbf{x}' = \mathbf{A}\mathbf{x}$, where $\mathbf{A}$ is a constant matrix. Show that the origin is the only fixed point if $\mathbf{A}$ is nonsingular.

**4.** Which of the direction fields in Fig. 7.6 appear to correspond to linear systems of differential equations? Briefly explain how you classify each direction field.

**5.** This exercise discusses the concluding statement in Example 1 in detail.

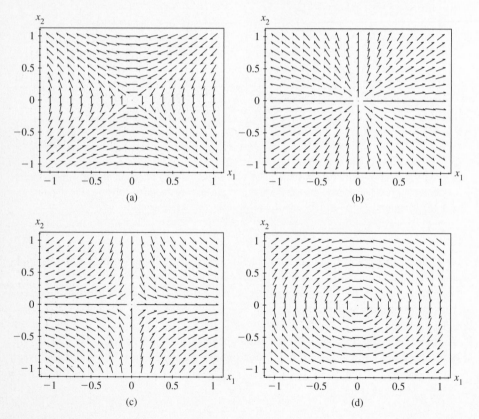

(a)    (b)

(c)    (d)

**Figure 7.5**    For Exercise 1.

(a) Find the general solution of Eq. (7.20). (Hint: The two equations are independent.)

(b) Show that $x_1(t)$ and $x_2(t)$ are not linear functions of $t$.

(c) Show that the ratio $x_1(t)/x_2(t)$ is constant.

(d) Show that, in general, the parametric curve $(x_1(t), x_2(t))$ forms a straight line in the $x_1x_2$ plane.

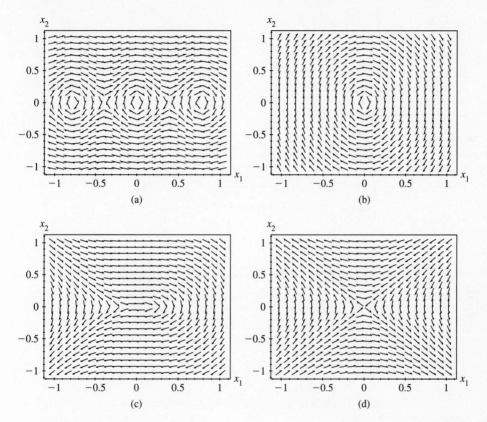

**FIGURE 7.6** For Exercise 4.

## 7.3 EIGENVALUES, EIGENVECTORS, AND DIRECTION FIELDS

In this section, we study some special vectors, called eigenvectors, that are associated with square ($n \times n$) matrices. The nonzero vector **x** is called an *eigenvector* of **A** if the product **Ax** is vector that is a scalar multiple of **x**. Eigenvectors play a crucial role in the solutions of linear systems of first-order differential equations.

## The Geometry of Eigenvectors

Figure 7.7 shows the direction field associated with the system

$$\mathbf{x}' = \begin{pmatrix} 0 & 1 \\ 1 & 0 \end{pmatrix} \mathbf{x}. \tag{7.21}$$

We now describe some patterns in the direction field to gain more intuition about the qualitative properties of the solution curves. Consider points along the same diagonal as the point $(-1, 1)$. Suppose that $\mathbf{x}(t) = (-a, a)$ for some value of $t$ and some constant $a$. Then $\mathbf{x}'(t) = (a, -a)$. The matrix in Eq. (7.21) associates with each point along the $-45$-degree diagonal through the origin a tangent vector that points directly toward the origin.

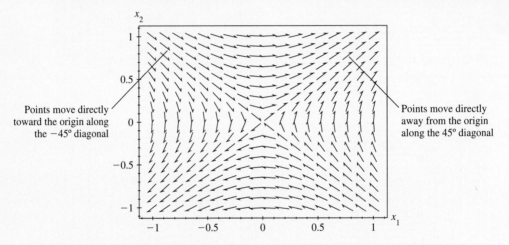

**Figure 7.7** The direction field associated with Eq. (7.21).

Let us use Euler's method to approximate the solution curve starting at $\mathbf{x}_0 = (-1, 1)$ with a step size $h$. Then

$$\mathbf{x}_1 = \mathbf{x}_0 + h\mathbf{x}_0'$$

$$= \begin{pmatrix} -1 \\ 1 \end{pmatrix} + h \begin{pmatrix} 1 \\ -1 \end{pmatrix}$$

$$= \begin{pmatrix} -(1-h) \\ 1-h \end{pmatrix}.$$

Notice that $\mathbf{x}_1$ lies along the same diagonal as $\mathbf{x}_0$. In fact, every iterate lies along the same diagonal as $\mathbf{x}_0$. The direction field implies that every solution curve starting at

$\mathbf{x}_0 = (-a, a)$ moves along the diagonal toward the origin. That is, if $\mathbf{x}(0) = (x_1(0), x_2(0))$ $= (-a, a)$, then $x_1(t)/x_2(t) = -1$ for all values of $t$.

In contrast, the slope line associated with the point $(1, 1)$ points directly away from the origin. The tangent line to any solution curve at a point of the form $(a, a)$ points directly away from the origin. Thus, solution curves starting at initial conditions along the 45-degree diagonal through the origin move directly away from the origin. Similar to the case of the other diagonal, $x_1(t)/x_2(t) = 1$ for all $t$. These observations imply that the origin is a saddle fixed point.

The direction field in Fig. 7.7 suggests that the diagonals are special. Consider how the matrix

$$\mathbf{A} = \begin{pmatrix} 0 & 1 \\ 1 & 0 \end{pmatrix}$$

maps each of the following vectors:

(a)  $\mathbf{A}\begin{pmatrix} 2 \\ 2 \end{pmatrix} = \begin{pmatrix} 2 \\ 2 \end{pmatrix}$    (b)  $\mathbf{A}\begin{pmatrix} -2 \\ 2 \end{pmatrix} = \begin{pmatrix} 2 \\ -2 \end{pmatrix}$

(c)  $\mathbf{A}\begin{pmatrix} 3 \\ 1 \end{pmatrix} = \begin{pmatrix} 1 \\ 3 \end{pmatrix}$    (d)  $\mathbf{A}\begin{pmatrix} 0 \\ 1 \end{pmatrix} = \begin{pmatrix} 1 \\ 0 \end{pmatrix}$

In case (a), the vector $(2, 2)$ is mapped to itself, and in case (b), the vector $(-2, 2)$ is mapped to the negative of itself. In cases (a) and (b), $\mathbf{A}$ maps the vector to a *multiple* of itself. Such vectors are called *eigenvectors*, and the corresponding multiple is called an *eigenvalue*.

**Definition 7.11**  *Let $\mathbf{A}$ be a square $(n \times n)$ matrix. We say that the scalar $\lambda$ is an* eigenvalue *of $\mathbf{A}$ if there exists an $n$-vector $\mathbf{x}$, different from $\mathbf{0}$, such that*

$$\mathbf{Ax} = \lambda\mathbf{x}. \tag{7.22}$$

*The vector $\mathbf{x}$ is called an* eigenvector. *More precisely, $\mathbf{x}$ is an eigenvector corresponding to $\lambda$.*

The terms *characteristic value* and *characteristic vector* are synonyms for *eigenvalue* and *eigenvector*, respectively. Equation (7.22) implies that *any multiple of an eigenvector is also an eigenvector*. In case (a), the vector $(2, 2)$ is an eigenvector corresponding to the eigenvalue 1; and in case (b), the vector $(-2, 2)$ is an eigenvector corresponding to the eigenvalue $-1$.

Not every vector is an eigenvector. As cases (c) and (d) show, the vectors $(3, 1)$ and $(0, 1)$ are not mapped to multiples of themselves. Therefore, $(3, 1)$ and $(0, 1)$ are not eigenvectors.

✦ *Example 1*    One eigenvalue of the matrix **A** in Eq. (7.21) is $\lambda_1 = 1$ with corresponding eigenvector $\mathbf{p}_1 = (1, 1)$, and another eigenvalue is $\lambda_2 = -1$ with corresponding eigenvector $\mathbf{p}_2 = (-1, 1)$. The diagonals in Fig. 7.7 are the lines that contain $\mathbf{p}_1$ and $\mathbf{p}_2$.                                                                  ∎

We discuss the relationship between eigenvalues, eigenvectors, and direction fields in more detail at the end of this section.

## The Algebra of Eigenvectors

We now consider more generally the relationship between **A**, $\lambda$, and **x** when Eq. (7.22) holds and **A** is a $2 \times 2$ matrix. Equation (7.22) implies that

$$\mathbf{A}\mathbf{x} - \lambda\mathbf{x} = \mathbf{0},$$

which holds whenever

$$\mathbf{A}\mathbf{x} - \lambda\mathbf{I}\mathbf{x} = \mathbf{0},$$

by the definition of the identity matrix **I**. The distributive law of matrix multiplication, Eq. (7.3), implies that

$$(\mathbf{A} - \lambda\mathbf{I})\mathbf{x} = \mathbf{0}. \tag{7.23}$$

We are interested only in *nonzero* vectors that satisfy Eq. (7.23). Therefore, $\mathbf{A} - \lambda\mathbf{I}$ must be a singular matrix, and its determinant must be 0. In the $2 \times 2$ case, Eq. (7.23) implies that

$$\begin{aligned}
\det(\mathbf{A} - \lambda\mathbf{I}) &= \det\left[ \begin{pmatrix} a & b \\ c & d \end{pmatrix} - \begin{pmatrix} \lambda & 0 \\ 0 & \lambda \end{pmatrix} \right] \\
&= \det \begin{pmatrix} a - \lambda & b \\ c & d - \lambda \end{pmatrix} \\
&= (a - \lambda)(d - \lambda) - bc \\
&= \lambda^2 - (a + d)\lambda + (ad - bc) \\
&= 0.
\end{aligned}$$

**Definition 7.12**    *Let*

$$\mathbf{A} = \begin{pmatrix} a & b \\ c & d \end{pmatrix}.$$

*The* characteristic equation *of* **A** *is*

$$\lambda^2 - (a + d)\lambda + (ad - bc) = 0. \tag{7.24}$$

As shown by the preceding computation, the eigenvalues of **A** are the roots of the characteristic equation of **A**. A corresponding eigenvector is any nonzero solution of the equation $(\mathbf{A} - \lambda \mathbf{I})\mathbf{x} = \mathbf{0}$. Notice that if **x** is one solution, then so is $c\mathbf{x}$, where $c$ is any constant.

The characteristic equation (7.24) of a $2 \times 2$ matrix is a quadratic polynomial. If the matrix **A** has real entries, then the polynomial has real coefficients. The roots of the characteristic equation (and therefore the eigenvalues of **A**) may be real and distinct, real and equal, or complex conjugate.

## Real, Distinct Eigenvalues

We first consider the case in which both eigenvalues of **A** are real and distinct.

✦ *Example 2*   The eigenvalues and eigenvectors of the matrix

$$\mathbf{A} = \begin{pmatrix} 2 & 2 \\ 2 & -1 \end{pmatrix}$$

can be found as follows.

**Step 1.** Solve the characteristic equation to find the eigenvalues. The characteristic equation is

$$\begin{aligned} \det(\mathbf{A} - \lambda \mathbf{I}) &= \det \begin{pmatrix} 2 - \lambda & 2 \\ 2 & -1 - \lambda \end{pmatrix} \\ &= (2 - \lambda)(-1 - \lambda) - (2)(2) \\ &= \lambda^2 - \lambda - 6 \\ &= (\lambda - 3)(\lambda + 2). \end{aligned}$$

The determinant is zero if and only if $\lambda = 3$ or $\lambda = -2$. Therefore, the eigenvalues of **A** are $\lambda_1 = 3$ and $\lambda_2 = -2$.

**Step 2.** Substitute each eigenvalue in turn into Eq. (7.23) to determine the corresponding eigenvectors. In the case of $\lambda_1 = 3$, Eq. (7.23) becomes

$$(\mathbf{A} - 3\mathbf{I})\mathbf{x} = \begin{pmatrix} -1 & 2 \\ 2 & -4 \end{pmatrix} \begin{pmatrix} x_1 \\ x_2 \end{pmatrix} = \begin{pmatrix} 0 \\ 0 \end{pmatrix},$$

which is equivalent to the system

$$\begin{aligned} -x_1 + 2x_2 &= 0, \\ 2x_1 - 4x_2 &= 0. \end{aligned} \tag{7.25}$$

The system (7.25) is redundant, because the corresponding matrix $\mathbf{A} - 3\mathbf{I}$ is singular. There are infinitely many solutions (see Theorem 7.8). Any nonzero solution yields an eigenvector. For example, we can take $x_1 = 2$ and $x_2 = 1$. We verify that $(2, 1)$ is an eigenvector corresponding to $\lambda_1 = 3$ by multiplying out:

$$\mathbf{A} \begin{pmatrix} 2 \\ 1 \end{pmatrix} = \begin{pmatrix} 2 & 2 \\ 2 & -1 \end{pmatrix} \begin{pmatrix} 2 \\ 1 \end{pmatrix} = \begin{pmatrix} 6 \\ 3 \end{pmatrix} = 3 \begin{pmatrix} 2 \\ 1 \end{pmatrix}.$$

We could have chosen other values for $x_1$ and $x_2$, as long as they solve Eq. (7.25). For instance, the choice $x_1 = 1$ and $x_2 = 1/2$ also produces an eigenvector, as the following calculation verifies:

$$\begin{pmatrix} 2 & 2 \\ 2 & -1 \end{pmatrix} \begin{pmatrix} 1 \\ \frac{1}{2} \end{pmatrix} = \begin{pmatrix} 3 \\ \frac{3}{2} \end{pmatrix} = 3 \begin{pmatrix} 1 \\ \frac{1}{2} \end{pmatrix}.$$

We proceed similarly to find an eigenvector corresponding to $\lambda_2 = -2$. Equation (7.23) gives

$$(\mathbf{A} + 2\mathbf{I})\mathbf{x} = \begin{pmatrix} 4 & 2 \\ 2 & 1 \end{pmatrix} \begin{pmatrix} x_1 \\ x_2 \end{pmatrix} = \begin{pmatrix} 0 \\ 0 \end{pmatrix},$$

for which one solution is $\mathbf{x} = (1, -2)$. Hence, $(1, -2)$ is an eigenvector corresponding to $\lambda_2 = -2$.                                                                                  ∎

## The Eigenvalues of Singular Matrices

If $\mathbf{A}$ is singular, then one eigenvalue is 0, because there is a solution of $\mathbf{A}\mathbf{x} = \mathbf{0}$ for which $\mathbf{x} \neq \mathbf{0}$. The next example illustrates this idea for one specific matrix.

✦ *Example 3*   Let

$$\mathbf{A} = \begin{pmatrix} 2 & 0 \\ 2 & 0 \end{pmatrix}.$$

The equation $\mathbf{A}\mathbf{x} = \mathbf{0}$ is equivalent to the condition $2x_1 = 0$. Hence, any vector of the form $(0, a)$ is a solution, which means that $(0, a)$ is an eigenvector corresponding to the eigenvalue 0.

The characteristic equation of $\mathbf{A}$ is

$$\det(\mathbf{A} - \lambda\mathbf{I}) = \det \begin{pmatrix} 2 - \lambda & 0 \\ 2 & -\lambda \end{pmatrix} = (2 - \lambda)(-\lambda),$$

for which the roots are $\lambda_1 = 0$ and $\lambda_2 = 2$. For $\lambda_1 = 0$, Eq. (7.23) reduces to $\mathbf{A}\mathbf{x} = \mathbf{0}$.

A corresponding eigenvector is $(0, 1)$. For $\lambda_2 = 2$, we have

$$(\mathbf{A} - 2\mathbf{I})\mathbf{x} = \begin{pmatrix} 0 & 0 \\ 2 & -2 \end{pmatrix} \begin{pmatrix} x_1 \\ x_2 \end{pmatrix} = \begin{pmatrix} 0 \\ 0 \end{pmatrix},$$

which implies that $2x_1 - 2x_2 = 0$. A corresponding eigenvector is $(1, 1)$. Exercise 3 considers some of the properties of the corresponding direction field.   ∎

## Double Real Eigenvalues

In most cases, if the characteristic equation has a double real root, then the corresponding matrix has only one eigenvector. (See also Exercise 4.)

◆ *Example 4*   Let

$$\mathbf{A} = \begin{pmatrix} -1 & -1 \\ 1 & -3 \end{pmatrix}.$$

The characteristic equation of $\mathbf{A}$ is

$$\det(\mathbf{A} - \lambda\mathbf{I}) = \det \begin{pmatrix} -1-\lambda & -1 \\ 1 & -3-\lambda \end{pmatrix} = \lambda^2 + 4\lambda + 4,$$

for which $\lambda = -2$ is a double root. An eigenvector corresponding to $-2$ solves

$$(\mathbf{A} + 2\mathbf{I})\mathbf{x} = \begin{pmatrix} 1 & -1 \\ 1 & -1 \end{pmatrix} \begin{pmatrix} x_1 \\ x_2 \end{pmatrix} = \begin{pmatrix} 0 \\ 0 \end{pmatrix},$$

which implies that $x_1 - x_2 = 0$. Hence, an eigenvector corresponding to $\lambda = -2$ is $\mathbf{x} = (1, 1)$.   ∎

## Complex Conjugate Eigenvalues

The eigenvectors corresponding to complex conjugate eigenvalues can be calculated in precisely the same manner as for real eigenvalues.

◆ *Example 5*   Let

$$\mathbf{A} = \begin{pmatrix} -\frac{1}{4} & 2 \\ -2 & -\frac{1}{4} \end{pmatrix}.$$

The characteristic equation is

$$\det(\mathbf{A} - \lambda \mathbf{I}) = \det \begin{pmatrix} -\frac{1}{4} - \lambda & 2 \\ -2 & -\frac{1}{4} - \lambda \end{pmatrix} = \lambda^2 + \frac{1}{2}\lambda + \frac{65}{16}.$$

The quadratic formula implies that the roots of the characteristic equation, and therefore the eigenvalues of $\mathbf{A}$, are

$$\lambda = -\frac{1}{4} \pm 2i.$$

An eigenvector corresponding to $\lambda_1 = -\frac{1}{4} + 2i$ is a nonzero solution of

$$\left(\mathbf{A} - \left(-\frac{1}{4} + 2i\right)\mathbf{I}\right)\mathbf{x} = \begin{pmatrix} -2i & 2 \\ -2 & -2i \end{pmatrix} \begin{pmatrix} x_1 \\ x_2 \end{pmatrix} = \begin{pmatrix} 0 \\ 0 \end{pmatrix},$$

which is equivalent to the system

$$\begin{aligned} -2ix_1 + 2x_2 &= 0, \\ -2x_1 - 2ix_2 &= 0. \end{aligned} \tag{7.26}$$

The system (7.26) is redundant, because the first equation is $i$ times the second. One solution, and hence an eigenvector corresponding to $\lambda_1$, is $\mathbf{p}_1 = (1, i)$. ∎

A similar procedure can be used to determine the other eigenvalue and corresponding eigenvector. However, having found $\lambda_1$ and $\mathbf{p}_1$, the theorem below allows us to determine $\lambda_2$ and $\mathbf{p}_2$ very easily. Although the proof is restricted to the case of $2 \times 2$ matrices, the theorem is true in general.

**Theorem 7.13** *If* $\mathbf{A}$ *is a real matrix with a complex eigenvalue* $\lambda_1$ *and corresponding eigenvector* $\mathbf{p}_1$*, then* $\lambda_2 = \overline{\lambda}_1$ *is also an eigenvalue of* $\mathbf{A}$*, and the corresponding eigenvector is* $\mathbf{p}_2 = \overline{\mathbf{p}}_1$*.*

**Proof**   If $\mathbf{A}$ is a $2 \times 2$ matrix, then the characteristic equation is a quadratic polynomial. Any complex roots come in complex conjugate pairs. (See Exercise 5.) If $w$ and $z$ are any complex numbers, then $\overline{w + z} = \overline{w} + \overline{z}$ and $\overline{wz} = \overline{w}\,\overline{z}$.

Suppose that $\lambda$ is a complex eigenvalue with corresponding eigenvector $\mathbf{p}$. Then $\mathbf{A}\mathbf{p} = \lambda\mathbf{p}$, so

$$\overline{\mathbf{A}\mathbf{p}} = \overline{\lambda\mathbf{p}} = \overline{\lambda}\,\overline{\mathbf{p}}. \tag{7.27}$$

On the other hand,

$$\overline{\mathbf{A}\mathbf{p}} = \overline{\mathbf{A}}\,\overline{\mathbf{p}} = \mathbf{A}\overline{\mathbf{p}}, \tag{7.28}$$

because **A** is a real matrix. Equations (7.27) and (7.28) therefore imply that $\overline{\lambda}$ is an eigenvalue of **A** with corresponding eigenvector $\overline{\mathbf{p}}$.   ∎

✦ *Example 6*   Example 5 shows that one eigenvalue of

$$\mathbf{A} = \begin{pmatrix} -\frac{1}{4} & 2 \\ -2 & -\frac{1}{4} \end{pmatrix}$$

is $\lambda_1 = -\frac{1}{4} + 2i$ with corresponding eigenvector $\mathbf{p}_1 = (1, i)$. Since **A** is real, a second eigenvalue is

$$\lambda_2 = \overline{\lambda}_1 = -\frac{1}{4} - 2i.$$

An eigenvector corresponding to $\lambda_2$ is $\mathbf{p}_2 = \overline{\mathbf{p}}_1 = (1, -i)$.   ∎

## The Linear Independence of Eigenvectors

The following theorem addresses an important property of eigenvectors that is particularly useful in the analysis of linear first-order systems of differential equations.

**Theorem 7.14**   *Let **A** have distinct eigenvalues $\lambda_1$ and $\lambda_2$ with corresponding eigenvectors $\mathbf{p}_1$ and $\mathbf{p}_2$. Then $\mathbf{p}_1$ and $\mathbf{p}_2$ are linearly independent.*

**Proof**   This proof assumes that **A** is a $2 \times 2$ matrix, but the theorem holds for $n \times n$ matrices as well. Suppose $\mathbf{p}_1$ and $\mathbf{p}_2$ were linearly dependent. Then for some nonzero constant $c_1$, $c_1\mathbf{p}_1 = \mathbf{p}_2 \neq \mathbf{0}$ (because eigenvectors by definition are nonzero). Therefore, $\mathbf{A}(c_1\mathbf{p}_1) = \mathbf{A}\mathbf{p}_2$, which in turn implies that

$$\lambda_1 c_1 \mathbf{p}_1 = \lambda_2 \mathbf{p}_2. \tag{7.29}$$

On the other hand, since $c_1\mathbf{p}_1 = \mathbf{p}_2$, we have

$$\lambda_2 c_1 \mathbf{p}_1 = \lambda_2 \mathbf{p}_2. \tag{7.30}$$

Equations (7.29) and (7.30) imply that

$$c_1(\lambda_1 - \lambda_2)\mathbf{p}_1 = \mathbf{0}.$$

Since $\lambda_1 \neq \lambda_2$, we have $c_1 = 0$, which implies that $\mathbf{p}_2 = \mathbf{0}$, a contradiction. Therefore, $\mathbf{p}_1$ and $\mathbf{p}_2$ must be linearly independent.   ∎

## Phase Portraits of Linear Systems with Constant Coefficients

As was discussed in Section 7.4, the eigenvectors of the matrix $\mathbf{A}$ are the building blocks of solutions of first-order linear systems with constant coefficients. The eigenvalues, however, determine the qualitative nature of the phase portrait near the fixed point at the origin, as we now show.

### Real Distinct Eigenvalues

Figure 7.8 shows the direction field corresponding to the system

$$\mathbf{x}' = \mathbf{A}\mathbf{x} = \begin{pmatrix} 2 & 2 \\ 2 & -1 \end{pmatrix} \mathbf{x}, \tag{7.31}$$

where $\mathbf{A}$ is the matrix analyzed in Example 2. The indicated lines contain the eigenvector $\mathbf{p}_1 = (2, 1)$, corresponding to the eigenvalue $\lambda_1 = 3$, and the eigenvector $\mathbf{p}_2 = (1, -2)$, corresponding to the eigenvalue $\lambda_2 = -2$. These lines are called the *eigendirections*. Notice the following aspects of Fig. 7.8.

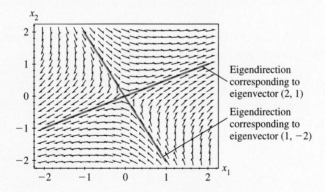

**FIGURE 7.8**   The direction field corresponding to Eq. (7.31).

- The lines comprising the eigendirections appear to be consistent with the direction field. That is, they appear to be solution trajectories for appropriate initial conditions.
- The slope lines along the eigendirection containing $\mathbf{p}_1$ point directly away from the origin. The corresponding eigenvalue is $\lambda_1 = 3$, which is positive.

- The slope lines along the eigendirection containing $\mathbf{p}_2$ point directly toward the origin. The corresponding eigenvalue is $\lambda_1 = -2$, which is negative.
- The origin is a saddle fixed point.

Figure 7.9(a) shows the direction field for the equation

$$\frac{d\mathbf{x}}{dt} = \mathbf{A}\mathbf{x} = \frac{1}{5} \begin{pmatrix} -14 & -2 \\ -2 & -11 \end{pmatrix} \mathbf{x}. \tag{7.32}$$

The methods discussed previously can be used to show that the eigenvalues of $\mathbf{A}$ are $\lambda_1 = -3$ and $\lambda_2 = -2$ with eigenvectors $(2, 1)$ and $(1, -2)$, respectively. The eigendirections are shown as solid lines. The slope lines along both eigendirections point directly toward the origin.

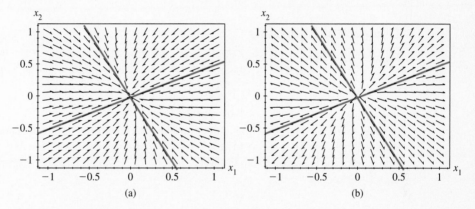

(a)                         (b)

**FIGURE 7.9**   The direction field corresponding to (a) Eq. (7.32); (b) Eq. (7.33). The solid lines indicate the eigendirections.

Suppose we take any initial condition in the region illustrated in Fig. 7.9(a). Every solution trajectory must be consistent with the slope lines; it is evident from the direction field that all trajectories eventually approach the origin. Hence, the origin is a sink.

Figure 7.9(b) shows the direction field for the equation

$$\frac{d\mathbf{x}}{dt} = \mathbf{B}\mathbf{x} = \frac{1}{5} \begin{pmatrix} 11 & -2 \\ -2 & 14 \end{pmatrix} \mathbf{x}. \tag{7.33}$$

The eigenvectors of $\mathbf{B}$ are the same as those of $\mathbf{A}$ in Eq. (7.32), but the eigenvalues of $\mathbf{B}$ are $\lambda_1 = +3$ and $\lambda_2 = +2$. The direction field is similar to the one in Fig. 7.9(a), except that corresponding slope lines point in opposite directions. Hence, the origin is a source.

The direction fields that we have just discussed illustrate the following general facts about the $2 \times 2$ system $\mathbf{x}' = \mathbf{A}\mathbf{x}$ when the eigenvalues of $\mathbf{A}$ are real, nonzero, and distinct.

- There are exactly two trajectories that lie on straight lines through the origin. The direction of these lines is given by the eigenvectors of $\mathbf{A}$.

- Eigenvectors corresponding to *negative* eigenvalues are associated with solution curves that *approach* the origin as $t \to \infty$. Therefore, if all the eigenvalues of $\mathbf{A}$ are negative, then the origin is a sink.

- Eigenvectors corresponding to *positive* eigenvalues are associated with solution curves that *move away from* the origin as $t \to \infty$. (Alternatively, one can say that the solution curves approach the origin as $t \to -\infty$.) Therefore, if all of the eigenvalues of $\mathbf{A}$ are positive, then the origin is a source.

- If the matrix $\mathbf{A}$ has one positive and one negative eigenvalue, then the origin is a saddle.

## Double Real Eigenvalues

Figure 7.10 shows the direction field corresponding to the system

$$\frac{d\mathbf{x}}{dt} = \begin{pmatrix} -1 & -1 \\ 1 & -3 \end{pmatrix} \mathbf{x}. \tag{7.34}$$

(See Example 4.) The solid line in the figure is the eigendirection (the line contains the eigenvector). The line appears to be a trajectory that is consistent with the direction field. The eigenvalue is negative, and the slope lines point directly toward the origin along the eigendirection.

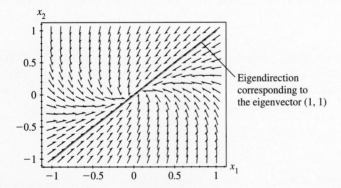

**FIGURE 7.10** The direction field corresponding to Eq. (7.34). See Example 4.

Notice that the origin is attracting. If we take an initial condition anywhere in the region shown in Fig. 7.10 and draw a trajectory that is consistent with the slope lines, then we approach the origin. Therefore, the origin is a sink. In this case, we sometimes call the origin a *node*, because there is only one eigendirection.

If **A** is a $2 \times 2$ matrix with a double real eigenvalue, and if the eigenvalue is positive, then the origin is a source. If the eigenvalue is negative, then the origin is a sink.

### Complex Conjugate Eigenvalues

Since components of the eigenvectors are complex numbers, it is not possible to graph complex eigenvectors in $\mathbf{R}^2$. However, the presence of complex eigenvalues and eigenvectors is easy to detect from the direction field.

Figure 7.11 shows the direction field and a solution trajectory corresponding to the system

$$\mathbf{x}' = \mathbf{A}\mathbf{x} = \begin{pmatrix} -\frac{1}{4} & 2 \\ -2 & -\frac{1}{4} \end{pmatrix} \mathbf{x}. \tag{7.35}$$

The matrix **A** has eigenvalues $-\frac{1}{4} \pm 2i$, as illustrated in Examples 5 and 6. The trajectory is a spiral; the direction field suggests that its shape is typical of other trajectories in the phase portrait. The spiral shape implies that each component of the solution $\mathbf{x}(t)$ changes sign repeatedly.

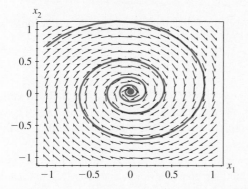

**FIGURE 7.11**    The direction field and a solution trajectory corresponding to Eq. (7.35). Spiral trajectories are typical of linear systems with complex conjugate eigenvalues.

The situation depicted in Fig. 7.11 is typical of first-order systems with complex eigenvalues. As we traverse the solution trajectory, we see that the sign of $x_1(t)$ and the sign of $x_2(t)$ change repeatedly. Therefore, $x_1(t)$ and $x_2(t)$ are oscillating functions.

Figure 7.12(a) shows the direction field corresponding to the first-order system

$$\frac{d\mathbf{x}}{dt} = \begin{pmatrix} 1/4 & -1/2 \\ 1/2 & 1/4 \end{pmatrix} \mathbf{x}, \tag{7.36}$$

for which the eigenvalues of the corresponding matrix are $\frac{1}{4} \pm \frac{1}{2}i$. The trajectory starts from the initial condition $\mathbf{x}_0 = (0.01, 0)$. The origin is a spiral source.

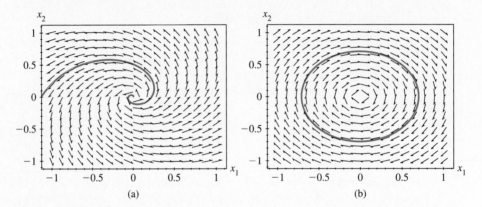

**FIGURE 7.12**    The direction field and a typical solution curve corresponding to (a) Eq. (7.36); (b) Eq. (7.37).

Figure 7.12(b) shows the direction field corresponding to the first-order system

$$\frac{d\mathbf{x}}{dt} = \begin{pmatrix} 0 & 1 \\ -1 & 0 \end{pmatrix} \mathbf{x}, \tag{7.37}$$

for which the eigenvalues of the corresponding matrix are $\pm i$. (The system (7.37) is equivalent to the second-order scalar equation $x'' + x = 0$.) The trajectory starts from the initial condition $\mathbf{x}_0 = (0.7, 0)$. In this case, the trajectory is a circle, indicating that the solutions are periodic.

The direction fields that we have just discussed illustrate the following general facts about the system $\mathbf{x}' = \mathbf{A}\mathbf{x}$ when the eigenvalues of $\mathbf{A}$ are complex conjugates. (These properties follow from the analytical solutions, which are derived in Section 7.4.)

- If the real part of the eigenvalues is negative, then the origin is a sink.
- If the real part of the eigenvalues is positive, then the origin is a source.
- If the real part of the eigenvalues is zero, then the origin is a center.
- In all cases, the solutions spiral around the origin, meaning that the solutions oscillate when the eigenvalues are complex. If the origin is a center, then the solutions are periodic.

There is an interesting parallel between the eigenvalues of first-order systems, the roots of the second-order scalar equations discussed in Chapter 5, and the corresponding solutions. Exercises 10 and 11 consider this correspondence in more detail.

## Exercises

**1.** Determine the eigenvalues and corresponding eigenvectors of each matrix in the following list:

(a) $\mathbf{A} = \begin{pmatrix} -8 & 18 \\ -3 & 7 \end{pmatrix}$      (b) $\mathbf{A} = \begin{pmatrix} -2 & 1 \\ 1 & -2 \end{pmatrix}$

(c) $\mathbf{A} = \begin{pmatrix} 0 & \frac{1}{2} \\ -2 & 2 \end{pmatrix}$      (d) $\mathbf{A} = \begin{pmatrix} -2 & \frac{1}{3} \\ -3 & -2 \end{pmatrix}$

(e) $\mathbf{A} = \begin{pmatrix} -1 & 1 \\ -5 & 1 \end{pmatrix}$      (f) $\mathbf{A} = \begin{pmatrix} 3 & 2 \\ 2 & 3 \end{pmatrix}$

(g) $\mathbf{A} = \begin{pmatrix} 1 & 1 \\ -1 & 1 \end{pmatrix}$      (h) $\mathbf{A} = \begin{pmatrix} 0 & \frac{3}{2} \\ 6 & 0 \end{pmatrix}$

(i) $\mathbf{A} = \begin{pmatrix} -1 & 1 \\ -6 & 4 \end{pmatrix}$      (j) $\mathbf{A} = \begin{pmatrix} 1 & -1 \\ 1 & 3 \end{pmatrix}$

**2.** Sketch the direction field for the system $\mathbf{x}' = \mathbf{A}\mathbf{x}$ for each matrix in Exercise 1 and determine whether the origin is a sink, source, saddle, or center.

**3.** Consider Example 3.

(a) Draw the direction field for the system $\mathbf{x}' = \mathbf{A}\mathbf{x}$, where $\mathbf{A}$ is the matrix in Example 3.

(b) Show that the direction field in (a) contains infinitely many fixed points. (Systems with infinitely many fixed points are sometimes called *degenerate*.)

(c) Prove or disprove: If $\mathbf{B}$ is a singular matrix, then the equation $\mathbf{x}' = \mathbf{B}\mathbf{x}$ has an infinite number of fixed points.

**4.** Let

$$\mathbf{A} = \begin{pmatrix} a & 0 \\ 0 & a \end{pmatrix}.$$

(a) What are the eigenvalues of $\mathbf{A}$?

(b) The matrix $\mathbf{A}$ has two linearly independent eigenvectors. What are they?

**5.** Suppose that $\lambda$ is a complex-valued root of a quadratic polynomial with real coefficients.

(a) Show that the complex conjugate, $\bar{\lambda}$, is also a root.

(b) Why does the result in (a) imply that complex eigenvalues of a real $2 \times 2$ matrix $\mathbf{A}$ always come in complex conjugate pairs?

**6.** Prove that if $\mathbf{x}$ is an eigenvector of $\mathbf{A}$ with eigenvalue $\lambda$, then any nonzero multiple of $\mathbf{x}$ is also an eigenvector of $\mathbf{A}$ corresponding to $\lambda$.

**7.** A $2 \times 2$ matrix is *symmetric* if it has the form

$$\mathbf{A} = \begin{pmatrix} a & b \\ b & a \end{pmatrix}.$$

(a) Show that every $2 \times 2$ symmetric matrix has real eigenvalues.

(b) What conditions must $a$ and $b$ satisfy for the eigenvalues of $\mathbf{A}$ to be negative? Positive? Distinct?

**8.** Let

$$x_1' = -x_1 + bx_2,$$
$$x_2' = -2x_2,$$

where $b$ is a constant.

(a) Show that the origin is a sink regardless of the value of $b$.

(b) Suppose $\mathbf{x}(0) = (0, 1/2)$. Sketch the direction field for different values of $b$ and describe how the solution trajectory for $t \geq 0$ depends on the value of $b$.

**9.** Let

$$x_1' = ax_1,$$
$$x_2' = bx_1 + cx_2,$$

where $a$, $b$, and $c$ are constants.

(a) Determine the values of $a$, $b$, and $c$ for which the origin is a sink; a saddle; a source.

(b) Do your answers depend on the value of $b$? Explain.

**10.** Let

$$mx'' + cx' + kx = 0, \qquad (7.38)$$

where $m$, $c$, and $k$ are constants. Equation (7.38) is the standard model for a linear pendulum or linear mass-spring system.

(a) What is the characteristic equation for Eq. (7.38)?

(b) Write Eq. (7.38) in the form $\mathbf{x}' = \mathbf{Ax}$ for an appropriate $2 \times 2$ matrix $\mathbf{A}$. Notice that if $x$ is a solution of Eq. (7.38), then $\mathbf{x} = (x, x')$.

(c) Find the characteristic equation of $\mathbf{A}$ and compare it to your answer in (a).

(d) Suppose Eq. (7.38) is overdamped. Characterize the eigenvalues of $\mathbf{A}$. Are they real or complex? Is the real part positive or negative?

(e) In (d), is the phase portrait of the corresponding first-order system a sink, source, or saddle?

Do solutions oscillate, forming either a spiral sink or a spiral source?

(f) Repeat (d)–(e), assuming that Eq. (7.38) is underdamped.

(g) Repeat (d)–(e), assuming that Eq. (7.38) is critically damped.

**11.** Let

$$x'' + ax' + bx = 0, \qquad (7.39)$$

where $a$ and $b$ are constants.

(a) What is the characteristic equation for Eq. (7.39)?

(b) Write Eq. (7.39) as a corresponding first-order system of equations

$$\mathbf{x}' = \mathbf{Ax} \qquad (7.40)$$

and determine the eigenvalues of $\mathbf{A}$.

(c) Suppose $a > 0$ and $b > 0$, and let $x$ be the general solution of Eq. (7.39). What is $\lim_{t \to \infty} |x(t)|$? (Hint: See Exercise 10(e).)

(d) Let $\mathbf{x}(t)$ be the corresponding solution of Eq. (7.40), and let $\|\mathbf{x}\| = \sqrt{x_1^2 + x_2^2}$. What is $\lim_{t \to \infty} \|\mathbf{x}(t)\|$?

(e) Repeat (c)–(d) assuming that $a < 0$ and $b > 0$.

(f) Repeat (c)–(d) assuming that $a < 0$ and $b < 0$.

## 7.4   SOLUTIONS OF HOMOGENEOUS SYSTEMS WITH CONSTANT COEFFICIENTS

So far, we have considered graphical and numerical representations of solutions of first-order systems of the form

$$\mathbf{x}' = \mathbf{Ax}.$$

In this section, we show how to determine explicit solutions when $\mathbf{A}$ is a constant $2 \times 2$ matrix. The ideas carry over in a natural way when $\mathbf{A}$ is a $3 \times 3$ or larger constant matrix, although the computations can be cumbersome to do by hand. (Computer software packages are very useful for solving large systems of differential equations.)

## Motivation

The discussion in Chapter 5 shows that the solutions of second-order differential equations of the form

$$x'' + ax' + bx = 0$$

are exponential functions. Every second-order scalar equation is equivalent to a first-order system of two equations; therefore, the components of the solutions of the corresponding first-order system are exponential functions.

More generally, if $\mathbf{A}$ is a constant matrix and $c$ is any constant, then the system

$$\mathbf{x}' = \mathbf{A}\mathbf{x} \qquad (7.41)$$

has the property that $\|\mathbf{x}'\|$ changes by a factor of $c$ whenever $\mathbf{x}$ is multiplied by a factor of $c$. This property implies that the components of $\mathbf{x}$ are exponential functions. The eigenvalues and eigenvectors of $\mathbf{A}$ determine the solutions, as is shown below.

## The General $2 \times 2$ Case

We restrict attention to the case in which $\mathbf{A}$ is a $2 \times 2$ matrix of real coefficients. The characteristic polynomial of $\mathbf{A}$ is quadratic, and its solutions can be real and distinct, complex conjugates, or real and equal.

As a motivating example, suppose that $\mathbf{A}$ has real, distinct eigenvalues $\lambda_1$ and $\lambda_2$. Let $\mathbf{p}_1$ be the eigenvector corresponding to $\lambda_1$, and let $\mathbf{x}(0) = \mathbf{p}_1$ be the initial condition. Now apply Euler's method to approximate the solution curve of Eq. (7.41). At $t = 0$, we have

$$\mathbf{x}'(0) = \mathbf{A}\mathbf{p}_1 = \lambda_1 \mathbf{p}_1 = \lambda_1 \mathbf{x}(0),$$

which says that the tangent vector is just a multiple of $\mathbf{x}(0) = \mathbf{p}_1$. Hence, the tangent vector lies along the eigendirection determined by $\mathbf{p}_1$. As we apply Euler's method, we advance the numerical approximation to the solution curve along the eigenvector; the distance traveled at each step is determined by the step size. At subsequent time steps, the derivative $\mathbf{x}'$, and hence the solution curve, still lie along the eigendirection determined by $\mathbf{p}_1$.

A similar sequence of Euler steps would be generated to approximate the solution of the uncoupled system

$$x_1' = \lambda_1 x_1,$$
$$x_2' = \lambda_1 x_2, \tag{7.42}$$

whose solution is

$$\mathbf{x}(t) = \begin{pmatrix} x_1(t) \\ x_2(t) \end{pmatrix} = \begin{pmatrix} a_1 e^{\lambda_1 t} \\ a_2 e^{\lambda_1 t} \end{pmatrix} = e^{\lambda_1 t} \begin{pmatrix} a_1 \\ a_2 \end{pmatrix}.$$

Since $\mathbf{x}(0) = (a_1, a_2)$, the initial condition $\mathbf{x}(0) = \mathbf{p}_1$ implies that $\mathbf{p}_1 = (a_1, a_2)$. Therefore,

$$\mathbf{x}(t) = e^{\lambda_1 t} \mathbf{p}_1. \tag{7.43}$$

It is straightforward to check that Eq. (7.43) solves Eq. (7.41). The right-hand side yields

$$\begin{aligned} \mathbf{A}\mathbf{x} &= \mathbf{A}\left(e^{\lambda_1 t} \mathbf{p}_1\right) \\ &= e^{\lambda_1 t}(\mathbf{A}\mathbf{p}_1) & \text{because } e^{\lambda_1 t} \text{ is a scalar} \\ &= \lambda_1 e^{\lambda_1 t} \mathbf{p}_1 & \text{because } \mathbf{p}_1 \text{ is an eigenvector} \\ &= \lambda_1 \mathbf{x}. \end{aligned}$$

The left-hand side yields

$$\begin{aligned} \mathbf{x}'(t) &= \frac{d}{dt}\left(e^{\lambda_1 t} \mathbf{p}_1\right) \\ &= \lambda_1 e^{\lambda_1 t} \mathbf{p}_1 \\ &= \lambda_1 \mathbf{x}, \end{aligned}$$

so Eq. (7.43) is a solution.

Let $\mathbf{p}_2$ be the eigenvector corresponding to $\lambda_2$. A similar argument shows that if $\mathbf{x}(0) = \mathbf{p}_2$, then the solution of Eq. (7.41) is

$$\mathbf{x}(t) = e^{\lambda_2 t} \mathbf{p}_2.$$

Equation (7.41) therefore has two solutions:

$$\mathbf{x}_1(t) = e^{\lambda_1 t} \mathbf{p}_1 \qquad \text{and} \qquad \mathbf{x}_2(t) = e^{\lambda_2 t} \mathbf{p}_2.$$

Because Eq. (7.41) is linear, the superposition principle implies that the function

$$\mathbf{x}(t) = c_1 e^{\lambda_1 t} \mathbf{p}_1 + c_2 e^{\lambda_2 t} \mathbf{p}_2 \tag{7.44}$$

also is a solution. As we show later in this section, the constants $c_1$ and $c_2$ can be chosen to satisfy any initial condition. Hence, the function in (7.44) is a *general solution* of Eq. (7.41).

## Distinct Eigenvalues

These observations give us a procedure for finding the solutions of systems of the form $x' = Ax$, provided that $A$ is a matrix with distinct eigenvalues. We begin with the case in which the eigenvalues are real and distinct.

✦ *Example 1*   Let

$$x' = Ax = \begin{pmatrix} 2 & 2 \\ 2 & -1 \end{pmatrix} x. \tag{7.45}$$

The eigenvalues of $A$ are $\lambda_1 = 3$, with corresponding eigenvector $p_1 = (2, 1)$, and $\lambda_2 = -2$, with corresponding eigenvector $p_2 = (1, -2)$. (See Example 2 in Section 7.3.)

One solution of Eq. (7.45) is

$$x_1(t) = e^{3t} p_1 = e^{3t} \begin{pmatrix} 2 \\ 1 \end{pmatrix};$$

the other one is

$$x_2(t) = e^{-2t} p_2 = e^{-2t} \begin{pmatrix} 1 \\ -2 \end{pmatrix},$$

as can easily be checked by substitution. The superposition principle implies that

$$x(t) = c_1 e^{3t} p_1 + c_2 e^{-2t} p_2 \tag{7.46}$$

also is a solution of Eq. (7.45).

The constants $c_1$ and $c_2$ can be chosen to satisfy any initial condition of the form $x(0) = x_0 = (a_1, a_2)$. For example, let $x_0 = (a_1, a_2) = (2, 6)$. When $t = 0$, Eq. (7.46) is equivalent to the linear system

$$\begin{aligned} 2c_1 + c_2 &= 2, \\ c_1 - 2c_2 &= 6, \end{aligned} \tag{7.47}$$

whose solution is $c_1 = 2$ and $c_2 = -2$. Thus, the solution of Eq. (7.45) satisfying the initial condition $x_0 = (2, 6)$ is

$$x(t) = 2e^{3t} \begin{pmatrix} 2 \\ 1 \end{pmatrix} - 2e^{-2t} \begin{pmatrix} 1 \\ -2 \end{pmatrix}. \tag{7.48}$$

The solution trajectory moves away from the origin as $t$ increases from 0.

The solution (7.46) implies that the origin is a saddle fixed point. For instance, if the initial condition is a multiple of $p_2$, then $c_1 = 0$. (Why?) In this case, $x \to 0$ as $t \to \infty$: The solution trajectory goes to the origin as $t$ increases. If the initial condition is a multiple of $p_1$, then $c_2 = 0$, and the solution trajectory leads directly away from the origin as $t$ increases. ∎

The analytical ideas described previously carry over in the same way for complex eigenvalues, as the next example illustrates.

✦ *Example 2*  Let

$$\mathbf{x}' = \mathbf{A}\mathbf{x} = \begin{pmatrix} -1 & 2 \\ -2 & -1 \end{pmatrix} \mathbf{x}. \tag{7.49}$$

One eigenvalue of $\mathbf{A}$ is $\lambda_1 = -1 + 2i$ with corresponding eigenvector $\mathbf{p}_1 = (1, i)$. The other eigenvalue of $\mathbf{A}$ is $\lambda_2 = -1 - 2i$ with corresponding eigenvector $\mathbf{p}_2 = (1, -i)$. Notice that $\lambda_1 = \overline{\lambda}_2$ and $\mathbf{p}_1 = \overline{\mathbf{p}}_2$, as discussed in Section 7.3.

One solution of Eq. (7.49) is

$$\mathbf{x}_1(t) = e^{\lambda_1 t}\mathbf{p}_1 = e^{(-1+2i)t} \begin{pmatrix} 1 \\ i \end{pmatrix},$$

and another solution is

$$\mathbf{x}_2(t) = e^{\lambda_2 t}\mathbf{p}_2 = e^{(-1-2i)t} \begin{pmatrix} 1 \\ -i \end{pmatrix}.$$

The general solution of Eq. (7.49) is

$$\mathbf{x}(t) = c_1 e^{(-1+2i)t} \begin{pmatrix} 1 \\ i \end{pmatrix} + c_2 e^{(-1-2i)t} \begin{pmatrix} 1 \\ -i \end{pmatrix}. \tag{7.50}$$

The constants $c_1$ and $c_2$ can be chosen to satisfy any initial condition of the form $\mathbf{x}_0 = (a_1, a_2)$. For instance, if $\mathbf{x}_0 = (a_1, a_2) = (1, 1)$, then Eq. (7.46) implies that

$$c_1 + c_2 = 1,$$
$$ic_1 - ic_2 = 1,$$

which yields the unique solution

$$c_1 = \tfrac{1}{2} - \tfrac{1}{2}i \quad \text{and} \quad c_2 = \tfrac{1}{2} + \tfrac{1}{2}i.$$

Hence, the solution of Eq. (7.49) satisfying the initial condition $\mathbf{x}(0) = (1, 1)$ can be written as

$$\mathbf{x}(t) = \left(\tfrac{1}{2} - \tfrac{1}{2}i\right) e^{(-1+2i)t} \begin{pmatrix} 1 \\ i \end{pmatrix} + \left(\tfrac{1}{2} + \tfrac{1}{2}i\right) e^{(-1-2i)t} \begin{pmatrix} 1 \\ -i \end{pmatrix}. \tag{7.51}$$

Equation (7.51) is the complex form of the solution.

Euler's formula can be used to determine the real form of the solution (7.51). The computation, though tedious, is routine. We start with Eq. (7.51), apply the identity

$$e^{(-1\pm 2i)t} = e^{-t}(\cos 2t \pm i \sin 2t),$$

collect real and imaginary parts, and simplify to obtain

$$\mathbf{x}(t) = e^{-t} \begin{pmatrix} \cos 2t + \sin 2t \\ \cos 2t - \sin 2t \end{pmatrix}. \tag{7.52}$$

Equation (7.51) and Equation (7.52) represent the *same* solution $\mathbf{x}(t)$. The solution implies that the origin is a sink, because $\|\mathbf{x}(t)\| \to 0$ as $t \to \infty$. ∎

Example 2 illustrates the following general fact:

*When the eigenvalues of* $\mathbf{A}$ *are complex conjugates, the solutions of* $\mathbf{x}' = \mathbf{A}\mathbf{x}$ *oscillate. If the real part of the eigenvalue is positive, then the amplitude of the solution increases with time; if it is negative, then the amplitude decreases with time. The imaginary part of the eigenvalue determines the frequency of the oscillation.*

There is a close relationship between Euler's formula and complex eigenvalues. The complex form of the solution given by Eq. (7.51) is entirely analogous to the form of the solution when the eigenvalues are real and distinct. Euler's formula relates the complex exponentials to the sine and cosine functions that comprise the real form of the solution.

The real part of the eigenvalues determines whether the origin is a sink or a source. In the case of Eq. (7.49), the real part of the eigenvalue is $-1$. Hence, the amplitude of each component of the solution decays at a rate proportional to $e^{-t}$. Therefore, the origin is a sink (in fact, a spiral sink).

## Linear Independence and the Wronskian

We have shown that if $\lambda_1$ and $\lambda_2$ are distinct eigenvalues of the $2 \times 2$ matrix $\mathbf{A}$ with corresponding eigenvectors $\mathbf{p}_1$ and $\mathbf{p}_2$, respectively, then a solution of the system $\mathbf{x}' = \mathbf{A}\mathbf{x}$ is

$$\mathbf{x}(t) = c_1 e^{\lambda_1 t} \mathbf{p}_1 + c_2 e^{\lambda_2 t} \mathbf{p}_2. \tag{7.53}$$

An important theoretical question is whether Eq. (7.53) is a *general* solution, that is, whether a *unique* choice of constants $c_1$ and $c_2$ can always be found that satisfies any initial condition of the form $\mathbf{x}(t_0) = (a_1, a_2)$.

When $t = t_0$, Eq. (7.53) can be expressed equivalently as

$$(\mathbf{p}_1 \quad \mathbf{p}_2) \begin{pmatrix} e^{\lambda_1 t_0} c_1 \\ e^{\lambda_2 t_0} c_2 \end{pmatrix} = \begin{pmatrix} a_1 \\ a_2 \end{pmatrix}, \tag{7.54}$$

where $\mathbf{P} = (\mathbf{p}_1 \ \mathbf{p}_2)$ is a $2 \times 2$ matrix whose columns are the eigenvectors of $\mathbf{A}$. Since $\mathbf{p}_1$ and $\mathbf{p}_2$ correspond to distinct eigenvalues, they are linearly independent by Theorem 7.14.

Therefore, **P** is nonsingular, by Theorem 7.9. As a result, Eq. (7.54) has a unique solution. The quantities $e^{\lambda_1 t_0}$ and $e^{\lambda_2 t_0}$ are never zero, so there is a unique choice of $c_1$ and $c_2$ that satisfies Eq. (7.53) for any given initial condition. These considerations establish the following theorem.

**Theorem 7.15**  *Let* **A** *be a constant* $2 \times 2$ *matrix with distinct eigenvalues* $\lambda_1$ *and* $\lambda_2$ *and corresponding eigenvectors* $\mathbf{p}_1$ *and* $\mathbf{p}_2$. *Then Eq. (7.53) is the general solution of the first-order system* $\mathbf{x}' = \mathbf{A}\mathbf{x}$.

The Wronskian is introduced in Chapter 5 to assess the linear independence of solutions of second-order scalar equations. The idea can be generalized to systems of two first-order equations (and to systems of $n$ first-order equations).

**Definition 7.16**  *Let* $\mathbf{x}_1(t)$ *and* $\mathbf{x}_2(t)$ *be two continuous functions from* $\mathbf{R}$ *to* $\mathbf{R}^2$. *The Wronskian of* $\mathbf{x}_1$ *and* $\mathbf{x}_2$ *is*

$$W(\mathbf{x}_1(t), \mathbf{x}_2(t)) = \det \left( \mathbf{x}_1(t) \quad \mathbf{x}_2(t) \right). \tag{7.55}$$

The following result, which is an extension of Abel's theorem for systems, is a corollary of Theorem 7.15. It also motivates Definition 7.18.

**Theorem 7.17**  *Let* **A** *be as in Theorem 7.15. Then either*

$$W \left( e^{\lambda_1 t} \mathbf{p}_1, e^{\lambda_2 t} \mathbf{p}_2 \right)$$

*is never zero for any* $t$ *or it is always zero for every* $t$.

**Definition 7.18**  *Let* **A** *be a* $2 \times 2$ *matrix, and let* $\mathbf{x}_1(t)$ *and* $\mathbf{x}_2(t)$ *be two solutions of the system* $\mathbf{x}' = \mathbf{A}\mathbf{x}$. *If* $W(\mathbf{x}_1(t), \mathbf{x}_2(t)) \neq 0$ *for every* $t$, *then* $\mathbf{x}_1$ *and* $\mathbf{x}_2$ *are* linearly independent *solutions.*

## A Note on Complex Eigenvalues

The Wronskian can be used to simplify the computations involved when the matrix **A** has complex conjugate eigenvalues. Let

$$\mathbf{x}' = \mathbf{A}\mathbf{x}, \tag{7.56}$$

and suppose that $\lambda = a + bi$ is an eigenvalue of **A** with corresponding eigenvector **p**. Then

$$\mathbf{x}_1(t) = e^{\lambda t} \mathbf{p} \tag{7.57}$$

is one solution of Eq. (7.56). The other solution is the complex conjugate,

$$\mathbf{x}_2(t) = \overline{\mathbf{x}}_1(t) = \overline{e^{\lambda t} \mathbf{p}} = e^{\overline{\lambda} t} \overline{\mathbf{p}}.$$

(See Exercise 4.) Because Eq. (7.56) is linear, the functions

$$\mathbf{x}_{\text{re}} = \tfrac{1}{2}(\mathbf{x}_1 + \mathbf{x}_2) \tag{7.58}$$

and

$$\mathbf{x}_{\text{im}} = \frac{1}{2i}(\mathbf{x}_1 - \mathbf{x}_2) \tag{7.59}$$

are solutions. The properties of complex arithmetic imply[1] that $\mathbf{x}_{\text{re}}$ is the real part and $\mathbf{x}_{\text{im}}$ is the imaginary part of the solution $\mathbf{x}_1$. Both $\mathbf{x}_{\text{re}}$ and $\mathbf{x}_{\text{im}}$ are real-valued functions.

The Wronskian can be used to show that $\mathbf{x}_{\text{re}}$ and $\mathbf{x}_{\text{im}}$ are linearly independent. (See Exercise 8.) The following theorem summarizes this discussion.

**Theorem 7.19**  *Let* $\mathbf{A}$ *be a real* $2 \times 2$ *matrix with complex conjugate eigenvalues. Let* $\lambda = a + bi$ *be one of the eigenvalues, and let* $\mathbf{p}$ *be a corresponding eigenvector. The general solution of the system* $\mathbf{x}' = \mathbf{A}\mathbf{x}$ *is*

$$\mathbf{x}(t) = c_1 \operatorname{Re}\left(e^{\lambda t}\mathbf{p}\right) + c_2 \operatorname{Im}\left(e^{\lambda t}\mathbf{p}\right), \tag{7.60}$$

*where the constants* $c_1$ *and* $c_2$ *are real if the initial condition is real.*

Equation (7.60) is called the *real form* of the general solution.

✦ *Example 3*   Let

$$\mathbf{x}' = \mathbf{A}\mathbf{x} = \begin{pmatrix} -1 & 2 \\ -2 & -1 \end{pmatrix} \mathbf{x}. \tag{7.61}$$

The real form of the solution was derived from the complex form of the solution in Example 2. It is possible to use Euler's formula to derive the real form of the solution directly, as follows.

One eigenvalue of the matrix

$$\mathbf{A} = \begin{pmatrix} -1 & 2 \\ -2 & -1 \end{pmatrix}$$

---

[1]For any complex number $z = a + bi$, where $a$ and $b$ are real, we have $\operatorname{Re} z = a = (z + \overline{z})/2$ and $\operatorname{Im} z = b = (z - \overline{z})/(2i)$.

is $\lambda = -1 + 2i$ with corresponding eigenvector $\mathbf{p} = (1, i)$. Therefore, one solution is

$$\mathbf{x}_1(t) = e^{(-1+2i)t} \begin{pmatrix} 1 \\ i \end{pmatrix}.$$

Euler's formula implies that $e^{(-1+2i)t} = e^{-t}(\cos 2t + i \sin 2t)$. Therefore, the real part of $\mathbf{x}_1(t)$ is

$$\mathbf{x}_{\text{re}}(t) = e^{-t} \begin{pmatrix} \cos 2t \\ -\sin 2t \end{pmatrix},$$

and the imaginary part is

$$\mathbf{x}_{\text{im}}(t) = e^{-t} \begin{pmatrix} \sin 2t \\ \cos 2t \end{pmatrix}.$$

The corresponding Wronskian is

$$W(\mathbf{x}_{\text{re}}, \mathbf{x}_{\text{im}}) = \det \left[ e^{-t} \begin{pmatrix} \cos 2t & \sin 2t \\ -\sin 2t & \cos 2t \end{pmatrix} \right] = e^{-t},$$

which is never 0 for any $t$. Therefore, $\mathbf{x}_{\text{re}}$ and $\mathbf{x}_{\text{im}}$ are linearly independent for every $t$, which implies that

$$\mathbf{x}(t) = c_1\mathbf{x}_{\text{re}}(t) + c_2\mathbf{x}_{\text{im}}(t) = c_1 e^{-t} \begin{pmatrix} \cos 2t \\ -\sin 2t \end{pmatrix} + c_2 e^{-t} \begin{pmatrix} \sin 2t \\ \cos 2t \end{pmatrix}$$

is a general solution of Eq. (7.61). The initial condition $\mathbf{x}(0) = (1, 1)$ implies that

$$c_1 \begin{pmatrix} 1 \\ 0 \end{pmatrix} + c_2 \begin{pmatrix} 0 \\ 1 \end{pmatrix} = \begin{pmatrix} 1 \\ 1 \end{pmatrix},$$

so $c_1 = c_2 = 1$. Therefore, the solution satisfying this initial condition is

$$\mathbf{x}(t) = e^{-t} \begin{pmatrix} \cos 2t + \sin 2t \\ -\sin 2t + \cos 2t \end{pmatrix}. \qquad \blacksquare$$

## Double Real Eigenvalues

If the characteristic equation of the $2 \times 2$ matrix $\mathbf{A}$ has a double root, then $\mathbf{A}$ has only one eigenvalue and typically only one eigenvector. Suppose $\mathbf{p}$ is an eigenvector corresponding to $\lambda$. The function $\mathbf{x}(t) = e^{\lambda t}\mathbf{p}$ is a solution, because

$$\mathbf{x}'(t) = \lambda e^{\lambda t}\mathbf{p} = e^{\lambda t}\mathbf{A}\mathbf{p}.$$

However, $\mathbf{x}(t)$ is not a general solution, for the following reason. Suppose that the initial condition $\mathbf{x}_0$ is not a multiple of the eigenvector $\mathbf{p}$. Then there is no constant $c$ such that $c\mathbf{x}(t) = ce^{\lambda t}\mathbf{p}$ satisfies the initial condition. Since the initial condition can be any point in the plane, we must find a second, linearly independent solution.

It is useful to recall the situation for the second-order scalar equation

$$x'' + 2rx' + r^2 x = 0, \tag{7.62}$$

where $r$ is a constant. Because the characteristic equation has a double root (namely, $\lambda = -r$), the general solution of Eq. (7.62) is

$$x(t) = c_1 e^{-rt} + c_2 t e^{-rt}. \tag{7.63}$$

(See Section 5.5.) The equivalent first-order system is

$$\mathbf{x}' = \mathbf{A}\mathbf{x} = \begin{pmatrix} 0 & 1 \\ -r^2 & -2r \end{pmatrix} \mathbf{x},$$

so the solutions of this system must also have terms involving $e^{\lambda t}$ and $te^{\lambda t}$.

This consideration suggests a procedure for finding solutions in the general $2 \times 2$ case. Let

$$\mathbf{x}' = \mathbf{A}\mathbf{x}, \tag{7.64}$$

where $\mathbf{A}$ has a double real eigenvalue $\lambda$. If $\mathbf{p}$ is a corresponding eigenvector, then one solution of Eq. (7.64) is

$$\mathbf{x}_1(t) = e^{\lambda t}\mathbf{p}.$$

Let us suppose that a second solution of Eq. (7.64) has the form

$$\mathbf{x}_2(t) = e^{\lambda t}\mathbf{q}_1 + te^{\lambda t}\mathbf{q}_2, \tag{7.65}$$

for some vectors $\mathbf{q}_1$ and $\mathbf{q}_2$. We substitute $\mathbf{x}_2$ into Eq. (7.64) to determine what $\mathbf{q}_1$ and $\mathbf{q}_2$ must be. The left-hand side yields

$$\mathbf{x}_2'(t) = \lambda e^{\lambda t}\mathbf{q}_1 + (1 + \lambda t)e^{\lambda t}\mathbf{q}_2$$
$$= e^{\lambda t}(\lambda\mathbf{q}_1 + t\lambda\mathbf{q}_2 + \mathbf{q}_2). \tag{7.66}$$

The right-hand side yields

$$\mathbf{x}_2'(t) = \mathbf{A}\mathbf{x}_2(t)$$
$$= e^{\lambda t}\mathbf{A}(\mathbf{q}_1 + t\mathbf{q}_2). \tag{7.67}$$

We equate coefficients for $e^{\lambda t}$ and $te^{\lambda t}$ in Eqs. (7.66) and (7.67) to obtain

$$\mathbf{A}\mathbf{q}_2 = \lambda\mathbf{q}_2 \tag{7.68}$$

and

$$\mathbf{q}_2 = (\mathbf{A} - \lambda\mathbf{I})\mathbf{q}_1. \tag{7.69}$$

Equation (7.68) implies that $\mathbf{q}_2$ is an eigenvector corresponding to the eigenvalue $\lambda$. (That is, $\mathbf{q}_2 = \mathbf{p}$.) Equation (7.69) expresses a different kind of relationship: We say that $\mathbf{q}_1$ is a *generalized eigenvector*. It is not obvious that Eq. (7.69) even has a solution, because $\mathbf{A} - \lambda\mathbf{I}$ is singular. Although the details are omitted here, it can be shown that Eq. (7.69) always has a solution when $\mathbf{q}_2$ is an eigenvector of $\mathbf{A}$. It can also be shown that $\mathbf{q}_1$ and $\mathbf{q}_2$ are linearly independent. (See Exercise 12.)

We can summarize the results as follows. Let $\mathbf{A}$ have an eigenvalue $\lambda$ with only one corresponding eigenvector $\mathbf{p}$. Let $\mathbf{q}$ be a generalized eigenvector, that is, a solution of the equation $(\mathbf{A} - \lambda\mathbf{I})\mathbf{q} = \mathbf{p}$. Then the general solution of Eq. (7.64) is

$$\mathbf{x}(t) = c_1 e^{\lambda t}\mathbf{p} + c_2 e^{\lambda t}(\mathbf{q} + t\mathbf{p}).$$

◆ *Example 4*   The solution of the initial value problem

$$\mathbf{x}' = \mathbf{A}\mathbf{x} = \begin{pmatrix} -3 & 1 \\ -1 & -1 \end{pmatrix}\mathbf{x}, \qquad \mathbf{x}_0 = \begin{pmatrix} 1 \\ 3 \end{pmatrix}$$

can be derived as follows. The characteristic equation of $\mathbf{A}$ is $\lambda^2 + 4\lambda + 4 = 0$; $\lambda = -2$ is a double root and hence is the only eigenvalue of $\mathbf{A}$. A corresponding eigenvector is $\mathbf{p} = (1, 1)$. The generalized eigenvector satisfies Eq. (7.69). Here,

$$(\mathbf{A} - \lambda\mathbf{I})\mathbf{q} = \begin{pmatrix} -1 & 1 \\ -1 & 1 \end{pmatrix}\mathbf{q} = \mathbf{p} = \begin{pmatrix} 1 \\ 1 \end{pmatrix},$$

for which one solution is $\mathbf{q} = (0, 1)$. The general solution of the system (7.64) is

$$\mathbf{x}(t) = c_1\mathbf{x}_1(t) + c_2\mathbf{x}_2(t)$$

$$= c_1 e^{-2t}\begin{pmatrix} 1 \\ 1 \end{pmatrix} + c_2 e^{-2t}\left[\begin{pmatrix} 0 \\ 1 \end{pmatrix} + t\begin{pmatrix} 1 \\ 1 \end{pmatrix}\right].$$

The initial condition $\mathbf{x}_0 = (1, 3)$ implies that $c_1 = 1$ and $c_2 = 2$. Hence, the solution of the initial value problem is

$$\mathbf{x}(t) = \begin{pmatrix} e^{-2t} + 2te^{-2t} \\ 3e^{-2t} + 2te^{-2t} \end{pmatrix}. \qquad \blacksquare$$

## Exercises

**1.** Let $\mathbf{x}' = \mathbf{Ax}$. For each matrix $\mathbf{A}$ in the following list,

- find the general solution;
- find the solution of the indicated initial value problem;
- classify the origin as a sink, saddle, or source.

(a) $\mathbf{A} = \begin{pmatrix} -8 & 18 \\ -3 & 7 \end{pmatrix}$, $\mathbf{x}(0) = \begin{pmatrix} 4 \\ 1 \end{pmatrix}$

(b) $\mathbf{A} = \begin{pmatrix} -2 & 1 \\ 1 & -2 \end{pmatrix}$, $\mathbf{x}(0) = \begin{pmatrix} 4 \\ 1 \end{pmatrix}$

(c) $\mathbf{A} = \begin{pmatrix} 4 & -2 \\ 1 & 1 \end{pmatrix}$, $\mathbf{x}(0) = \begin{pmatrix} 1 \\ 0 \end{pmatrix}$

(d) $\mathbf{A} = \begin{pmatrix} -5 & 1 \\ -6 & 0 \end{pmatrix}$, $\mathbf{x}(0) = \begin{pmatrix} 3 \\ 2 \end{pmatrix}$

**2.** Let $\mathbf{x}' = \mathbf{Ax}$. For each matrix $\mathbf{A}$ in the following list,

- find the general solution;
- find the real solution of the indicated initial value problem;
- classify the origin as a sink, source, saddle, or center.

(a) $\mathbf{A} = \begin{pmatrix} 1 & 1 \\ -1 & 1 \end{pmatrix}$, $\mathbf{x}(0) = \begin{pmatrix} 1 \\ 1 \end{pmatrix}$

(b) $\mathbf{A} = \begin{pmatrix} -3 & 1 \\ -5 & 1 \end{pmatrix}$, $\mathbf{x}(0) = \begin{pmatrix} 5 \\ 5 \end{pmatrix}$

(c) $\mathbf{A} = \begin{pmatrix} -1 & 1 \\ -5 & 1 \end{pmatrix}$, $\mathbf{x}(0) = \begin{pmatrix} 1 \\ 1 \end{pmatrix}$

(d) $\mathbf{A} = \begin{pmatrix} -2 & \frac{1}{3} \\ -3 & -2 \end{pmatrix}$, $\mathbf{x}(0) = \begin{pmatrix} 2 \\ 4 \end{pmatrix}$

**3.** Use Euler's formula to derive Eq. (7.52) directly from Eq. (7.51).

**4.** Let $\mathbf{A}$ be a constant, real matrix with complex eigenvalue $\lambda$ and corresponding eigenvector $\mathbf{p}$. Show

by substitution that $\mathbf{x}(t) = e^{\lambda t}\mathbf{p}$ is a solution of $\mathbf{x}' = \mathbf{Ax}$. Then argue that $\overline{\mathbf{x}}(t)$ is also a solution.

**5.** Let $\mathbf{x}' = \mathbf{Ax}$. Find the complex form and the real form of the solutions of each of the initial value problems in the following list:

(a) $\mathbf{A} = \begin{pmatrix} -2 & 1 \\ -4 & -2 \end{pmatrix}$, $\mathbf{x}(0) = \begin{pmatrix} 2 \\ 4 \end{pmatrix}$

(b) $\mathbf{A} = \begin{pmatrix} -5 & 3 \\ -15 & 7 \end{pmatrix}$, $\mathbf{x}(0) = \begin{pmatrix} 2 \\ -2 \end{pmatrix}$

**6.** This exercise considers some basic properties of the Wronskian.

(a) Let $\mathbf{v}$, $\mathbf{x}$, and $\mathbf{y}$ be any 2-vectors, and let $a$ and $b$ be any constants. Establish the following statements:

  (i) $W(a\mathbf{v} + b\mathbf{x}, \mathbf{y}) = aW(\mathbf{v}, \mathbf{y}) + bW(\mathbf{x}, \mathbf{y})$
  (ii) $W(\mathbf{x}, a\mathbf{y} + b\mathbf{v}) = aW(\mathbf{x}, \mathbf{y}) + bW(\mathbf{x}, \mathbf{v})$
  (iii) $W(\mathbf{x}, \mathbf{y}) = -W(\mathbf{y}, \mathbf{x})$
  (iv) $W(\mathbf{x}, \mathbf{x}) = 0$.

(b) Let $\mathbf{A}$ be a $2 \times 2$ real matrix with a complex eigenvalue $\lambda$ and corresponding eigenvector $\mathbf{p}$. Use the results in (a) to prove that $\mathrm{Re}\,\mathbf{p}$ and $\mathrm{Im}\,\mathbf{p}$ are linearly independent.

**7.** Let $\lambda = a + bi$ and $\mathbf{p} = \mathbf{r} + i\mathbf{s}$, where $a$ and $b$ are real numbers and $\mathbf{r}$ and $\mathbf{s}$ are real vectors. Use these definitions to express Eq. (7.60) in terms of $\mathrm{Re}\,\mathbf{p}$ and $\mathrm{Im}\,\mathbf{p}$.

**8.** Show that the functions defined by Eqs. (7.58) and (7.59) are linearly independent.

**9.** The origin is a *center* if the phase portraits starting at points near the origin form closed curves. Prove that if $\mathbf{A}$ is a $2 \times 2$ matrix whose eigenvalues are purely imaginary (i.e., the real part is 0), then the origin is a center for the system $\mathbf{x}' = \mathbf{Ax}$.

**10.** For each of the initial value problems in the following list,

- find the solution;
- classify the origin as a source, sink, or saddle.

(a) $\mathbf{x}' = \begin{pmatrix} 1 & 1 \\ -1 & 3 \end{pmatrix} \mathbf{x}, \ \mathbf{x}_0 = \begin{pmatrix} 1 \\ 2 \end{pmatrix}$

(b) $\mathbf{x}' = \begin{pmatrix} -1 & 1 \\ -4 & 3 \end{pmatrix} \mathbf{x}, \ \mathbf{x}_0 = \begin{pmatrix} -2 \\ 3 \end{pmatrix}$

(c) $\mathbf{x}' = \begin{pmatrix} 0 & 1 \\ -1 & 2 \end{pmatrix} \mathbf{x}, \ \mathbf{x}_0 = \begin{pmatrix} 0 \\ 1 \end{pmatrix}$

(d) $\mathbf{x}' = \begin{pmatrix} -6 & 9 \\ -1 & 0 \end{pmatrix} \mathbf{x}, \ \mathbf{x}_0 = \begin{pmatrix} -1 \\ 1 \end{pmatrix}$

**11.** In Example 4, we let $\mathbf{q} = (0, 1)$ be the generalized eigenvector. Show that we could just as well have let $\mathbf{q} = (-1, 0)$ and that this choice leads to the same solution.

**12.** Let $\mathbf{A}$ be a $2 \times 2$ matrix with a double real eigenvalue $\lambda$ and one corresponding eigenvector $\mathbf{p}$. Suppose that $\mathbf{q}$ is a generalized eigenvector.

(a) Show that $\mathbf{p}$ and $\mathbf{q}$ are linearly independent.

(b) Use the Wronskian to show that $\mathbf{x}_1(t) = e^{\lambda t}\mathbf{p}$ and $\mathbf{x}_2(t) = e^{\lambda t}\mathbf{q} + te^{\lambda t}\mathbf{p}$ form a linearly independent pair of functions for all values of $t$.

**13.** Let

$$x'' + ax' + bx = 0, \qquad (7.70)$$

where $a$ and $b$ are constants. Suppose that Eq. (7.70) represents a critically damped system.

(a) What is the characteristic equation corresponding to Eq. (7.70), and what must be the relationship between $a$ and $b$ if Eq. (7.70) is critically damped?

(b) Find the system $\mathbf{x}' = \mathbf{A}\mathbf{x}$ that is equivalent to Eq. (7.70).

(c) Find the characteristic equation of $\mathbf{A}$. Comment on how it is related to the characteristic equation that you found in (a).

(d) What must be the relationship between $a$ and $b$ if $\mathbf{A}$ has a double real eigenvalue?

## 7.5   STABILITY, PREDICTABILITY, AND UNCERTAINTY

We have classified fixed points as sinks, sources, saddles, and centers. In the case of linear systems of the form $\mathbf{x}' = \mathbf{A}\mathbf{x}$, where $\mathbf{A}$ is a constant matrix, the fixed point at the origin can be classified simply by finding the eigenvalues of $\mathbf{A}$.

We have characterized sinks as *attracting* fixed points, because trajectories starting from points near the sink approach the sink as $t \to \infty$. We can think of sinks as *stable*, because trajectories that start at initial conditions slightly away from the sink eventually return to the sink as $t$ tends to infinity. In this section, we explore the notion of stability in more detail. We also consider the role of eigenvalues in quantifying uncertainties in the solutions of linear systems when the initial condition is not known precisely.

### What Is a Stable Fixed Point?

In Chapter 1, we considered the notion of stability in the context of scalar, first-order differential equations. Suppose some system (say, a population) is in equilibrium: Its rate of change is zero. Now suppose that some initial disturbance changes the population so that it is no longer at its equilibrium value. A stable population is one that returns to its previous value, provided that the initial disturbance is not too large. The following definition makes this idea precise for systems of differential equations.

**Definition 7.20**  *Let* $x' = f(x)$ *be an autonomous system of $n$ first-order equations, and let $0$ be an equilibrium point. Suppose there is a ball $U$, centered at $0$, such that if* $x(0) \in U$, *then* $\lim_{t \to \infty} x(t) = 0$. *Then $0$ is an* asymptotically stable *equilibrium point.*

Stated informally, an asymptotically stable fixed point is one to which the system eventually returns, provided that the initial disturbance is not too large. If the disturbance corresponds to an initial condition in the circle $U$, then the system returns to its previous equilibrium as $t \to \infty$. An attracting fixed point is asymptotically stable. We say that the origin is *globally asymptotically stable* if *all* solutions eventually approach the origin, regardless of the initial condition.

Figure 7.13 illustrates the idea behind Definition 7.20 for the case of an asymptotically stable fixed point at the origin. Here, $x_0$ is an initial condition in the set $U$. The origin is asymptotically stable as long as the trajectory starting from $x_0$ approaches $0$ as $t \to \infty$; it does not matter whether the trajectory leaves $U$ for a time. In the case of a linear system of the form $x' = Ax$, the eigenvalues of $A$ completely determine whether the origin is asymptotically stable. The following theorem summarizes the main result.

**FIGURE 7.13**  An asymptotically stable equilibrium point (attracting fixed point) at the origin.

**Theorem 7.21**  *Let* $x' = Ax$, *where $A$ is a real $2 \times 2$ matrix whose eigenvalues are $\lambda_1$ and $\lambda_2$. Suppose that $\lambda_1$ and $\lambda_2$ satisfy any one of the following three criteria:*

1. $\lambda_1 < \lambda_2 < 0$; *or*
2. $\lambda_1 = \lambda_2 < 0$; *or*
3. $\lambda_1 = a + ib$ *and* $\lambda_2 = a - ib$, *where* $a < 0$.

*Then the origin is an asymptotically stable equilibrium point.*

*Proof*  We prove the first conclusion assuming that the eigenvalues are real, negative, and distinct; the remaining cases are left as exercises. (See Exercises 11 and 12.)

Let the eigenvectors corresponding to $\lambda_1$ and $\lambda_2$ be $\mathbf{p}_1$ and $\mathbf{p}_2$, respectively. Then

$$\mathbf{x}(t) = c_1 e^{\lambda_1 t}\mathbf{p}_1 + c_2 e^{\lambda_2 t}\mathbf{p}_2, \tag{7.71}$$

where $c_1$ and $c_2$ depend on the initial condition $\mathbf{x}(0)$. If $\lambda_1 < \lambda_2 < 0$, then $e^{\lambda_1 t} \to 0$ and $e^{\lambda_2 t} \to 0$ as $t \to \infty$. Hence, $\mathbf{x}(t) \to \mathbf{0}$ as $t \to \infty$.    ∎

Notice that the conclusion of the proof holds regardless of the initial condition, because the initial condition determines only $c_1$ and $c_2$. Thus, the origin is globally asymptotically stable.

✦ *Example 1*  Let

$$\mathbf{x}' = \mathbf{A}\mathbf{x} = \begin{pmatrix} -5 & 2 \\ -6 & 2 \end{pmatrix}\mathbf{x}.$$

The eigenvalues of $\mathbf{A}$ are $\lambda_1 = -1$ and $\lambda_2 = -2$ with corresponding eigenvectors $\mathbf{p}_1 = (1, 2)$ and $\mathbf{p}_2 = (2, 3)$. Therefore, the general solution is

$$\mathbf{x}(t) = c_1 e^{-t}\begin{pmatrix} 1 \\ 2 \end{pmatrix} + c_2 e^{-2t}\begin{pmatrix} 2 \\ 3 \end{pmatrix},$$

and $\mathbf{x}(t) \to \mathbf{0}$ as $t \to \infty$, regardless of the constants $c_1$ and $c_2$. Hence, the origin is asymptotically stable (in fact, globally asymptotically stable).    ∎

✦ *Example 2*  The general solution of the system

$$\mathbf{x}' = \mathbf{A}\mathbf{x} = \begin{pmatrix} 2 & 2 \\ 2 & -1 \end{pmatrix}\mathbf{x}$$

is

$$\mathbf{x}(t) = c_1 e^{3t}\begin{pmatrix} 2 \\ 1 \end{pmatrix} + c_2 e^{-2t}\begin{pmatrix} 1 \\ -2 \end{pmatrix}, \tag{7.72}$$

where $c_1$ and $c_2$ depend on the initial conditions. If $c_1 \neq 0$, then the term $c_1 e^{3t}$ tends to infinity. Therefore, the origin is *not* an asymptotically stable fixed point, because the solution does *not* in general tend to $\mathbf{0}$ as $t \to \infty$. It can be shown that any circle centered at the origin contains initial conditions for which $c_1 \neq 0$. (See Exercise 5.)    ∎

◆ *Example 3* Let

$$\mathbf{x}' = \mathbf{A}\mathbf{x} = \begin{pmatrix} -1 & -1 \\ 1 & -1 \end{pmatrix} \mathbf{x}.$$

The eigenvalues of $\mathbf{A}$ are $\lambda_1 = -1 + i$ and $\lambda_2 = -1 - i$ with corresponding eigenvectors $\mathbf{p}_1 = (-1, i)$ and $\mathbf{p}_2 = (-1, -i)$. Therefore, the general solution is

$$\mathbf{x}(t) = c_1 e^{-t} \begin{pmatrix} -\cos t \\ -\sin t \end{pmatrix} + c_2 e^{-t} \begin{pmatrix} -\sin t \\ \cos t \end{pmatrix}.$$

As $t \to \infty$, the term $e^{-t}$ goes to zero, so $\mathbf{x}(t) \to \mathbf{0}$ irrespective of the initial conditions. Hence, the origin is globally asymptotically stable. ∎

What happens when the real part of the eigenvalues is not negative? Euler's formula implies that the imaginary parts of the eigenvalues determine the frequency of the oscillations and that the real parts determine the amplitude. It can be shown that

*if $\mathbf{x}_0 \neq \mathbf{0}$ and the real parts of the eigenvalues of $\mathbf{A}$ are positive, then the solution* $\mathbf{x}(t)$ *satisfying the initial condition* $\mathbf{x}(0) = \mathbf{x}_0$ *tends to infinity as* $t \to \infty$.

The details are left as an exercise. (See Exercise 7.)

◆ *Example 4* This example illustrates the role that the eigenvalues and eigenvectors play in a phase portrait in which the origin is a sink. Let

$$\mathbf{x}' = \mathbf{A}\mathbf{x} = \frac{1}{5} \begin{pmatrix} -21 & -8 \\ -8 & -9 \end{pmatrix} \mathbf{x}. \tag{7.73}$$

The eigenvalues of $\mathbf{A}$ are $\lambda_1 = -1$ and $\lambda_2 = -5$ with corresponding eigenvectors $\mathbf{p}_1 = (1, -2)$ and $\mathbf{p}_2 = (2, 1)$. The general solution is

$$\mathbf{x}(t) = c_1 e^{-t} \mathbf{p}_1 + c_2 e^{-5t} \mathbf{p}_2,$$

which can be rewritten as

$$\mathbf{x}(t) = e^{-t} \left( c_1 \mathbf{p}_1 + c_2 e^{-4t} \mathbf{p}_2 \right). \tag{7.74}$$

The term in parentheses in Eq. (7.74) has an exponentially decaying component. Therefore, as $t$ becomes large, solution trajectories approach the line containing the eigenvector $\mathbf{p}_1$, because $\mathbf{x}(t) \to c_1 e^{-t} \mathbf{p}_1$.

Figure 7.14 shows the phase portrait corresponding to Eq. (7.73). The solution trajectories all approach the origin as $t \to \infty$ along the line containing $\mathbf{p}_1$, which is the eigenvector corresponding to the larger of the two eigenvalues of $\mathbf{A}$. (See Exercises 8–10.)

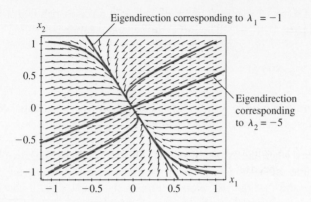

**FIGURE 7.14**    The phase portrait corresponding to Eq. (7.73).                    ■

## Predictability and Uncertainty

Suppose that some physical system, such as an electrical circuit, can be described by a linear system of first-order differential equations of the form $\mathbf{x}' = \mathbf{A}\mathbf{x}$. If the initial state of the system is known precisely, then the state of the system can be predicted at any future time.[2] In other words, if $\mathbf{x}(0)$ is known precisely, then $\mathbf{x}(t)$ is completely determined for any $t$.

In practice, however, the initial state of the system is known only approximately. The "true" initial state might be $\mathbf{x}(0)$, but its measured value, $\hat{\mathbf{x}}(0)$, usually is slightly different. How large is the error in a prediction of the state of the system at some future time $t$? We are interested in the difference between a predicted state, say $\hat{\mathbf{x}}(t)$, and the "true" state, given by $\mathbf{x}(t)$, that arises from errors in measuring a given initial condition. In this context, the *absolute error in the solution* or the *absolute difference between the solutions* at time $t$ is defined as the *root mean square error*, which for functions from $\mathbf{R}$ to $\mathbf{R}^2$ is

$$\epsilon_a(t) = \|\mathbf{x}(t) - \hat{\mathbf{x}}(t)\|$$

$$= \sqrt{(x_1(t) - \hat{x}_1(t))^2 + (x_2(t) - \hat{x}_2(t))^2}. \tag{7.75}$$

If $\mathbf{x}(t) \neq \mathbf{0}$, then the *relative error in the solution* or the *relative difference between the*

---

[2]This statement assumes that initial value problems of the form $\mathbf{x}' = \mathbf{A}\mathbf{x}$, $\mathbf{x}(0) = \mathbf{x}_0$, have unique solutions. In fact they do, as discussed in Section 7.7.

*solutions* at time $t$ is defined as

$$\epsilon_r(t) = \frac{\|\mathbf{x}(t) - \hat{\mathbf{x}}(t)\|}{\|\mathbf{x}(t)\|}. \tag{7.76}$$

The eigenvalues of the matrix $\mathbf{A}$ determine the stability of the origin with respect to solutions of systems like $\mathbf{x}' = \mathbf{A}\mathbf{x}$. If the origin is a sink, then solutions that tend to $\mathbf{0}$ with time show little sensitivity to errors in the initial conditions, as illustrated in the next example.

◆ *Example 5*   Let

$$\mathbf{x}' = \mathbf{A}\mathbf{x} = \begin{pmatrix} -5 & 2 \\ -6 & 2 \end{pmatrix} \mathbf{x}.$$

(See Example 1.)   The solutions satisfying the initial conditions $\mathbf{x}(0) = (1, 1)$ and $\hat{\mathbf{x}}(0) = (1.1, 0.9)$ are respectively

$$\mathbf{x}(t) = -e^{-t} \begin{pmatrix} 1 \\ 2 \end{pmatrix} + e^{-2t} \begin{pmatrix} 2 \\ 3 \end{pmatrix}$$

and

$$\hat{\mathbf{x}}(t) = -1.5e^{-t} \begin{pmatrix} 1 \\ 2 \end{pmatrix} + 1.3e^{-2t} \begin{pmatrix} 2 \\ 3 \end{pmatrix}.$$

The absolute difference between the solutions is

$$\epsilon_a(t) = \|\mathbf{x}(t) - \hat{\mathbf{x}}(t)\| = \left\| 0.5e^{-t} \begin{pmatrix} 1 \\ 2 \end{pmatrix} - 0.3e^{-2t} \begin{pmatrix} 2 \\ 3 \end{pmatrix} \right\|.$$

Each component of each term in the rightmost expression tends to 0 as $t \to \infty$. Therefore, $\epsilon_a \to 0$ as $t \to \infty$. In the long run, differences between solutions that arise from changes in initial conditions eventually disappear.                                                ■

The result in Example 5 can be generalized as follows:

*If the origin is asymptotically stable, then the absolute difference between two solutions arising from differences in the initial conditions goes to 0 as* $t \to \infty$.

By virtue of Theorem 7.21, this statement applies to all systems of the form $\mathbf{x}' = \mathbf{A}\mathbf{x}$ where the eigenvalues of $\mathbf{A}$ are negative or have negative real part.

Now suppose that all eigenvalues of $\mathbf{A}$ are positive or have positive real part. Then the origin is a source, and in fact, $\|\mathbf{x}(t)\| \to \infty$ as $t \to \infty$, regardless of the initial condition. (See Exercises 6 and 7.) Therefore, $\|\mathbf{x}(t) - \hat{\mathbf{x}}(t)\| \to \infty$ as $t \to \infty$.

Although the *absolute* error in the solution becomes large as $t$ grows, we can ask whether the *relative* error becomes large compared to the size of the "true" solution $\mathbf{x}(t)$. The answer in general is "no," in a sense that is made precise by the following example.

✦ *Example 6*  Let

$$\mathbf{x}' = \mathbf{A}\mathbf{x} = \begin{pmatrix} 5 & -1 \\ 6 & 0 \end{pmatrix} \mathbf{x}, \qquad \mathbf{x}_0 = \begin{pmatrix} x_1(0) \\ x_2(0) \end{pmatrix}. \tag{7.77}$$

The eigenvalues of $\mathbf{A}$ are $\lambda_1 = 2$ and $\lambda_2 = 3$ with corresponding eigenvectors $\mathbf{p}_1 = (1, 3)$ and $\mathbf{p}_2 = (1, 2)$. The solution satisfying the initial condition $\mathbf{x}_0 = (x_0, y_0)$ is

$$\mathbf{x}(t) = \begin{pmatrix} (3x_0 - y_0)e^{3t} + (y_0 - 2x_0)e^{2t} \\ (6x_0 - 2y_0)e^{3t} + (3y_0 - 6x_0)e^{2t} \end{pmatrix}.$$

Let $\mathbf{x}(t)$ denote the "true" solution satisfying the initial condition $\mathbf{x}_0 = (1, 1)$, and let $\hat{\mathbf{x}}(t)$ denote the solution satisfying the initial condition $\hat{\mathbf{x}}(t) = (1.1, 1)$.

Figure 7.15(a) shows a plot of the absolute error in the solution,

$$\epsilon_a(t) = \|\mathbf{x}(t) - \hat{\mathbf{x}}(t)\| = \sqrt{\left(-0.3e^{3t} + 0.2e^{2t}\right)^2 + \left(-0.6e^{3t} + 0.6e^{2t}\right)^2}.$$

Figure 7.15(b) shows a plot of the relative error in the solution,

$$\epsilon_r(t) = \frac{\|\mathbf{x}(t) - \hat{\mathbf{x}}(t)\|}{\|\mathbf{x}(t)\|}.$$

The origin is a source, and each term in the solution grows exponentially with time. Figure 7.15(a) reflects the exponential growth in the absolute error. Although the absolute error becomes large, the relative error approaches a constant value—about 15 percent in this example.

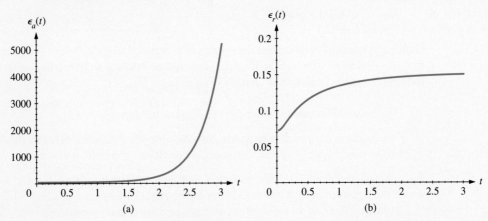

**FIGURE 7.15**   A plot of (a) the absolute error and (b) the relative error in the solution of Eq. (7.77).

This example illustrates the following general statement:

*Errors in the initial conditions of linear systems lead in almost all cases to relative errors in the solutions that are bounded for all $t$.*

There are three important things to notice about this statement:

- Exercise 14 explores what is meant by the qualifier "almost all."

- The relative errors in most cases are bounded, regardless of whether the eigenvalues of $\mathbf{A}$ are positive, negative, or zero. If one or more of the eigenvalues is positive, then the *absolute* error in the solution may tend to infinity as $t \to \infty$, but the *relative* error usually remains bounded.

- The statement should be contrasted with the result derived in Section 2.4 for the scalar equation $x' = ax$. In the scalar case, the relative error in the solution does not vary with time if there is an error in the initial condition; it remains constant.

## Exercises

**1.** Let $\mathbf{x}' = \mathbf{Ax}$, where

$$\mathbf{A} = \begin{pmatrix} -3 & 1 \\ -5 & 1 \end{pmatrix}.$$

(a) Find the eigenvalues of $\mathbf{A}$.

(b) Plot the solutions satisfying the initial conditions $\mathbf{x}_0 = (1, 0)$ and $\hat{\mathbf{x}}_0 = (1.1, 0)$.

(c) On the same axes, plot the solutions satisfying the initial conditions $\mathbf{x}_0 = (-1, -1)$ and $\hat{\mathbf{x}}_0 = (-0.8, -1)$.

(d) Pick several other pairs of nearby initial points and plot the trajectories. What happens to the distance between the various trajectories as $t$ becomes large? If $\hat{\mathbf{x}}_0$ represents a slightly erroneous measurement of the "true" initial condition $\mathbf{x}_0$, then what can you say about the absolute error in the solution $\mathbf{x}(t)$?

**2.** Repeat Exercise 1 when

$$\mathbf{A} = \begin{pmatrix} -6 & -1 \\ 9 & 0 \end{pmatrix}.$$

**3.** Repeat Exercise 1 when

$$\mathbf{A} = \begin{pmatrix} -2 & 1 \\ 1 & -2 \end{pmatrix}.$$

**4.** Let

$$x_1' = -x_1 + bx_2,$$
$$x_2' = -2x_2,$$

where $b$ is a constant.

(a) Argue that the origin is an asymptotically stable equilibrium point regardless of the value of $b$.

(b) Suppose $\mathbf{x}(0) = (0, 1/2)$. Sketch the direction field for different values of $b$ and describe how the phase portrait of the solution curve for $t \geq 0$ depends on the value of $b$.

**5.** This exercise asks you to prove that if the origin is a saddle fixed point for the linear system $\mathbf{x}' = \mathbf{Ax}$ in the plane, then typical solutions tend to infinity as $t \to \infty$.

(a) Let

$$\mathbf{x}(t) = c_1 e^{3t} \begin{pmatrix} 2 \\ 1 \end{pmatrix} + c_2 e^{-2t} \begin{pmatrix} 1 \\ -2 \end{pmatrix}.$$

(See Examples 1 and 2.) Show that in any circle centered at the origin, there are initial conditions for which $c_1 \neq 0$. Conclude that for such initial conditions, $\|\mathbf{x}(t)\|$ tends to infinity as $t \to \infty$.

(b) Prove the following theorem. *Let* $\mathbf{x}' = \mathbf{Ax}$, *where* $\mathbf{A}$ *is a real* $2 \times 2$ *matrix. Suppose that the eigenvalues of* $\mathbf{A}$ *are* $\lambda_1 > 0$ *and* $\lambda_2 < 0$ *with corresponding eigenvectors* $\mathbf{p}_1$ *and* $\mathbf{p}_2$. *If* $\mathbf{x}(0)$ *is not a multiple of* $\mathbf{p}_2$, *then* $\|\mathbf{x}(t)\|$ *tends to infinity as* $t \to \infty$.

(c) Explain what happens in (b) if $\mathbf{x}(0)$ is a multiple of $\mathbf{p}_2$. We sometimes call $\mathbf{p}_1$ the *unstable* eigenvector and $\mathbf{p}_2$ the *stable* eigenvector. Why is this terminology justified?

**6.** The origin is an equilibrium point for every linear system of the form $\mathbf{x}' = \mathbf{Ax}$. The origin is a *source* if there exists a positive constant $K$ such that $\lim_{t \to -\infty} \|\mathbf{x}(t)\| = 0$ whenever $\|\mathbf{x}(0)\| \leq K$. Let $\mathbf{A}$ be a $2 \times 2$ matrix. Prove that the origin is a source if both eigenvalues of $\mathbf{A}$ are real and positive.

**7.** Let $\mathbf{x}' = \mathbf{Ax}$, where $\mathbf{A}$ is a $2 \times 2$ constant matrix with complex conjugate eigenvalues $\lambda_1 = a + bi$ and $\lambda_2 = a - bi$. Suppose $a > 0$. Show that if $\mathbf{x}(0) \neq \mathbf{0}$, then $\|\mathbf{x}(t)\| \to \infty$ as $t \to \infty$.

**8.** Let $\mathbf{A}$ be a constant $2 \times 2$ matrix, and let $\mathbf{x}' = \mathbf{Ax}$. Suppose that $\mathbf{A}$ has two eigenvalues such that $\lambda_1 < \lambda_2 < 0$. Prove that trajectories in the phase plane approach the origin along the eigendirection corresponding to $\lambda_2$ as $t \to \infty$, provided that the initial condition is not on the eigendirection corresponding to $\lambda_1$.

**9.** Let

$$\mathbf{x}' = \mathbf{Ax} = \begin{pmatrix} 3 & 2 \\ 2 & 3 \end{pmatrix} \mathbf{x}.$$

(a) Find the general solution. Is the origin a source, a sink, or a saddle?

(b) Find the solutions that satisfy the following initial conditions: $\mathbf{x}_0 = (1, 0)$, $\mathbf{x}_0 = (-1, 0)$, $\mathbf{x}_0 = (0, -1)$, and $\mathbf{x}_0 = (0, 1)$.

(c) Plot the solution curves in the phase plane corresponding to each of the solutions that you found in (b) on the same axes for $0 \leq t \leq 0.5$. Also indicate the eigendirections on your plot.

(d) Make another plot, analogous to the one in (c) but for $0 \leq t \leq 1$. You will need to change the scaling on the axes.

(e) Make another plot, analogous to the one in (d) but for $0 \leq t \leq 2$.

(f) What can you conclude about the relationship between typical solution trajectories and the eigenvectors of $\mathbf{A}$ as $t$ becomes large?

**10.** This problem generalizes the result in Exercise 8. Let $\mathbf{A}$ be a real $2 \times 2$ matrix whose eigenvalues are real and distinct. Prove that solution trajectories of the system $\mathbf{x}' = \mathbf{Ax}$ in the phase plane approach the eigendirection corresponding to the largest eigenvalue of $\mathbf{A}$, provided that the initial condition does not lie on the eigendirection corresponding to the smaller eigenvalue. (Notice that this statement is independent of whether the origin is a sink, a source, or a saddle.)

**11.** Let $\mathbf{x}' = \mathbf{Ax}$, where $\mathbf{A}$ is a $2 \times 2$ matrix with a double real eigenvalue $\lambda$ and one corresponding eigenvector.

(a) Show that if $\lambda < 0$, then the origin is asymptotically stable (i.e., a sink).

(b) Show that if $\lambda > 0$ and $\mathbf{x}(t) \neq \mathbf{0}$, then $\|\mathbf{x}(t)\| \to \infty$ as $t \to \infty$. (Hint: See Exercise 7.)

(c) Describe the solutions if $\lambda = 0$.

**12.** Let $\mathbf{A}$ be a $2 \times 2$ constant matrix with complex conjugate eigenvalues $\lambda$ and $\bar{\lambda}$. Prove that the origin is an asymptotically stable fixed point for the equation $\mathbf{x}' = \mathbf{Ax}$ if Re $\lambda < 0$. (Hint: Use the strategy of the proof of Theorem 7.21.)

**13.** Consider the system $\mathbf{x}' = \mathbf{Ax}$, where

$$\mathbf{A} = \begin{pmatrix} 0 & 1 \\ 1 & 0 \end{pmatrix}.$$

(a) Find a solution that satisfies the initial condition $\mathbf{x}(0) = (-2, 1)$.

(b) Find a solution that satisfies the initial condition $\hat{\mathbf{x}}(0) = (-2.1, 1.3)$.

(c) Generate a phase portrait for the two initial conditions for the time interval $0 \leq t \leq 2.5$. What happens to the trajectories of the solution curves?

(d) Show that the solutions in (a) and (b) have a component that decays exponentially with time.

(e) Plot the relative difference between the two solutions.

(f) Argue that the relative difference between the two solutions approaches a constant value as $t$ becomes large.

**14.** Let $\mathbf{x}' = \mathbf{Ax}$, where

$$\mathbf{A} = \begin{pmatrix} 4 & -2 \\ 1 & 1 \end{pmatrix}.$$

(a) Find the eigenvalues of $\mathbf{A}$ and use them to classify the origin as a sink, a source, or a saddle.

(b) Find the solution $\mathbf{x}(t)$ satisfying the initial condition $\mathbf{x}(0) = (2, 1)$ and the solution $\hat{\mathbf{x}}(t)$ satisfying the initial condition $\hat{\mathbf{x}}(0) = (2.1, 1)$.

(c) Compute the absolute difference $\|\mathbf{x}(t) - \hat{\mathbf{x}}(t)\|$ and plot it for $0 \leq t \leq 3$. Characterize the graph. (Consider questions such as the following: Does the graph increase or decrease with $t$? Does the graph appear to be linear or exponential? Does the curve approach a constant value?)

(d) Compute the relative difference

$$\frac{\|\mathbf{x}(t) - \hat{\mathbf{x}}(t)\|}{\|\mathbf{x}(t)\|}$$

and plot it for $0 \leq t \leq 3$. Characterize the graph.

**15.** Redo parts (b)–(d) of Exercise 14 for the initial conditions $\mathbf{x}(0) = (1, 1)$ and $\hat{\mathbf{x}}(0) = (1.1, 1)$. Explain why the relative error does not approach a constant value.

**16.** Repeat Exercise 14 when

$$\mathbf{A} = \begin{pmatrix} 1 & 1 \\ -1 & 1 \end{pmatrix}.$$

**17.** Repeat Exercise 14 when

$$\mathbf{A} = \begin{pmatrix} -1 & 1 \\ -4 & 3 \end{pmatrix}.$$

# 7.6   HIGHER-DIMENSIONAL LINEAR SYSTEMS

The analysis described in the previous sections carries over with little change to higher-dimensional systems. The role of eigenvalues and eigenvectors in determining the solution is essentially the same as in the $2 \times 2$ case. In this section, we concentrate on first-order systems of three differential equations with constant coefficients.

## Eigenvalues and Eigenvectors

Let $\mathbf{A}$ be a $3 \times 3$ matrix. The determinant of $\mathbf{A}$ is defined as

$$\det \mathbf{A} = a_{11} \det \begin{pmatrix} a_{22} & a_{23} \\ a_{32} & a_{33} \end{pmatrix} - a_{21} \det \begin{pmatrix} a_{12} & a_{13} \\ a_{32} & a_{33} \end{pmatrix} + a_{31} \det \begin{pmatrix} a_{12} & a_{13} \\ a_{22} & a_{23} \end{pmatrix}. \quad (7.78)$$

The determinants on the right-hand side of Eq. (7.78) are called the *minors* of $\mathbf{A}$. The

$(i, j)$th minor of $\mathbf{A}$, denoted $M_{ij}$, is the determinant of the $2 \times 2$ matrix formed by deleting the $i$th row and $j$th column of $\mathbf{A}$. For instance,

$$M_{11} = \det \begin{pmatrix} a_{22} & a_{23} \\ a_{32} & a_{33} \end{pmatrix},$$

the determinant of the $2 \times 2$ matrix that results when the first row and first column of $\mathbf{A}$ are deleted. Notice that

$$\det \mathbf{A} = a_{11}M_{11} - a_{21}M_{21} + a_{31}M_{31},$$

which makes the formula easier to remember. (See Exercise 9.)

The eigenvalues of $\mathbf{A}$ are the values $\lambda$ for which

$$\det (\mathbf{A} - \lambda \mathbf{I}) = 0.$$

This relation, which is the characteristic equation of $\mathbf{A}$, defines a cubic polynomial in $\lambda$. In general, a $3 \times 3$ matrix $\mathbf{A}$ has three eigenvalues. If $\mathbf{A}$ is real, then either the eigenvalues are all real or two of the eigenvalues are complex conjugates and one is real. The computation of eigenvectors is entirely similar to the $2 \times 2$ case, as the following examples illustrate.

✦ *Example 1*   Let

$$\mathbf{A} = \begin{pmatrix} 7 & -10 & 12 \\ 5 & -8 & 9 \\ 0 & 0 & -1 \end{pmatrix}.$$

The eigenvalues are the roots of the characteristic equation, given by

$$
\begin{aligned}
0 &= \det (\mathbf{A} - \lambda \mathbf{I}) \\
&= \det \begin{pmatrix} 7 - \lambda & -10 & 12 \\ 5 & -8 - \lambda & 9 \\ 0 & 0 & -1 - \lambda \end{pmatrix} \\
&= (7 - \lambda) \det \begin{pmatrix} -8 - \lambda & 9 \\ 0 & -1 - \lambda \end{pmatrix} - 5 \det \begin{pmatrix} -10 & 12 \\ 0 & -1 - \lambda \end{pmatrix} + 0 \\
&= (7 - \lambda)(8 + \lambda)(1 + \lambda) - 50(1 + \lambda) \\
&= -\lambda^3 - 2\lambda^2 + 5\lambda + 6 \\
&= -(\lambda + 1)(\lambda - 2)(\lambda + 3).
\end{aligned}
$$

Therefore, the eigenvalues are $\lambda_1 = -1$, $\lambda_2 = 2$, and $\lambda_3 = -3$.

An eigenvector corresponding to $\lambda$ is a nonzero vector $\mathbf{p}$ for which $(\mathbf{A} - \lambda\mathbf{I})\mathbf{p} = \mathbf{0}$. For example, an eigenvector corresponding to $\lambda_1 = -1$ is a solution of the system

$$(\mathbf{A} + \mathbf{I})\mathbf{p} = \begin{pmatrix} 8 & -10 & 12 \\ 5 & -7 & 9 \\ 0 & 0 & 0 \end{pmatrix} \mathbf{p} = \mathbf{0}.$$

The system is underdetermined because the matrix $\mathbf{A} + \mathbf{I}$ is singular. One solution, and hence an eigenvector corresponding to $\lambda_1 = -1$, is $\mathbf{p}_1 = (1, 2, 1)$.

In a similar way, an eigenvector $\mathbf{p}_2$ corresponding to $\lambda_2 = 2$ is a solution of

$$(\mathbf{A} - 2\mathbf{I})\mathbf{p}_2 = \begin{pmatrix} 5 & -10 & 12 \\ 5 & -10 & 9 \\ 0 & 0 & -3 \end{pmatrix} \mathbf{p}_2 = \mathbf{0},$$

one of which is $\mathbf{p}_2 = (2, 1, 0)$. An eigenvector $\mathbf{p}_3$ corresponding to $\lambda_3 = -3$ is a solution of

$$(\mathbf{A} + 3\mathbf{I})\mathbf{p}_3 = \begin{pmatrix} 10 & -10 & 12 \\ 5 & -5 & 9 \\ 0 & 0 & 2 \end{pmatrix} \mathbf{p}_3 = \mathbf{0},$$

one of which is $\mathbf{p}_3 = (1, 1, 0)$.   ∎

✦ *Example 2*   Let

$$\mathbf{A} = \begin{pmatrix} -1 & 1 & 0 \\ 5 & -2 & 5 \\ 5 & -3 & 4 \end{pmatrix}.$$

The characteristic equation is

$$\begin{aligned} 0 &= \det(\mathbf{A} - \lambda\mathbf{I}) \\ &= -(\lambda^3 - \lambda^2 + 2) \\ &= -(\lambda + 1)(\lambda^2 - 2\lambda + 2). \end{aligned}$$

Therefore, one eigenvalue is $\lambda_1 = -1$, and a computation similar to one in Example 1 shows that a corresponding eigenvector is $\mathbf{p}_1 = (-1, 0, 1)$.

The quadratic formula implies that the remaining eigenvalues are $\lambda_2 = 1 + i$ and $\lambda_3 = 1 - i$. We compute corresponding eigenvectors in exactly the same manner as in Example 1, except that the calculations involve complex arithmetic. An eigenvector $\mathbf{p}_2$ corresponding to $\lambda_2 = 1 + i$ is a solution of $[\mathbf{A} - (1 + i)\mathbf{I}]\mathbf{p}_2 = \mathbf{0}$, which implies that

$$\begin{pmatrix} -2 - i & 1 & 0 \\ 5 & -3 - i & 5 \\ 5 & -3 & 3 - i \end{pmatrix} \mathbf{p}_2 = \mathbf{0}.$$

One solution is $p_2 = (1, 2 + i, i)$. Since $A$ is real, an eigenvector $p_3$ corresponding to $\lambda_3$ is just $p_3 = \bar{p}_2 = (1, 2 - i, -i)$. (Why?)    ∎

## The Solutions of Linear Systems

The solutions of first-order systems of three equations have the same form as described in Section 7.4. That is, if $\lambda$ is an eigenvalue of $A$ with corresponding eigenvector $p$, then $x(t) = e^{\lambda t} p$ is a solution of the system $x' = Ax$.

✦ *Example 3*   Let

$$x' = Ax = \begin{pmatrix} 7 & -10 & 12 \\ 5 & -8 & 9 \\ 0 & 0 & -1 \end{pmatrix} x. \tag{7.79}$$

As shown in Example 1, the eigenvalues of $A$ are $\lambda_1 = -1$, $\lambda_2 = 2$, and $\lambda_3 = -3$ with corresponding eigenvectors $p_1 = (1, 2, 1)$, $p_2 = (2, 1, 0)$, and $p_3 = (1, 1, 0)$. Therefore, the general solution of Eq. (7.79) is

$$x(t) = c_1 e^{\lambda_1 t} p_1 + c_2 e^{\lambda_2 t} p_2 + c_3 e^{\lambda_3 t} p_3$$

$$= c_1 e^{-t} \begin{pmatrix} 1 \\ 2 \\ 1 \end{pmatrix} + c_2 e^{2t} \begin{pmatrix} 2 \\ 1 \\ 0 \end{pmatrix} + c_3 e^{-3t} \begin{pmatrix} 1 \\ 1 \\ 0 \end{pmatrix}.$$

Because the vectors $p_1$, $p_2$, and $p_3$ are linearly independent, it is possible to find a unique set of the constants $c_1$, $c_2$, and $c_3$ to satisfy any initial condition. For example, the initial condition $x(0) = (1, 2, 3)$ implies that

$$\begin{pmatrix} 1 \\ 2 \\ 3 \end{pmatrix} = c_1 \begin{pmatrix} 1 \\ 2 \\ 1 \end{pmatrix} + c_2 \begin{pmatrix} 2 \\ 1 \\ 0 \end{pmatrix} + c_3 \begin{pmatrix} 1 \\ 1 \\ 0 \end{pmatrix}.$$

We solve the resulting $3 \times 3$ system of linear equations for the constants $c_1$, $c_2$, and $c_3$ to show that the solution of the initial value problem is

$$x(t) = \begin{pmatrix} 3e^{-t} + 4e^{2t} - 6e^{-3t} \\ 6e^{-t} + 2e^{2t} - 6e^{-3t} \\ 3e^{-t} \end{pmatrix}. \qquad ∎$$

✦ *Example 4* Let

$$\mathbf{x}' = \mathbf{A}\mathbf{x} = \begin{pmatrix} -1 & 1 & 0 \\ 5 & -2 & 5 \\ 5 & -3 & 4 \end{pmatrix} \mathbf{x}.$$

As shown in Example 2, the eigenvalues of $\mathbf{A}$ are $\lambda_1 = -1$, $\lambda_2 = 1 + i$, and $\lambda_3 = 1 - i$, with corresponding eigenvectors $\mathbf{p}_1 = (-1, 0, 1)$, $\mathbf{p}_2 = (1, 2 + i, i)$, and $\mathbf{p}_3 = (1, 2 - i, -i)$. Therefore, the general solution has the form

$$\mathbf{x}(t) = c_1 e^{-t} \begin{pmatrix} -1 \\ 0 \\ 1 \end{pmatrix} + c_2 e^{(1+i)t} \begin{pmatrix} 1 \\ 2 + i \\ i \end{pmatrix} + c_3 e^{(1-i)t} \begin{pmatrix} 1 \\ 2 - i \\ -i \end{pmatrix}. \tag{7.80}$$

Equation (7.80) is the *complex form* of the general solution. As in the $2 \times 2$ case, it can be shown that if the complex form of the solution is

$$\mathbf{x}(t) = c_1 e^{\lambda_1 t} \mathbf{p}_1 + c_2 e^{\lambda_2 t} \mathbf{p}_2 + c_3 e^{\bar{\lambda}_2 t} \bar{\mathbf{p}}_2,$$

where $\lambda_3 = \bar{\lambda}_2$ and $\mathbf{p}_3 = \bar{\mathbf{p}}_2$, then the *real form* of the general solution is

$$\mathbf{x}(t) = b_1 e^{\lambda_1 t} \mathbf{p}_1 + b_2 \operatorname{Re}\left(e^{\lambda_2 t} \mathbf{p}_2\right) + b_3 \operatorname{Im}\left(e^{\lambda_2 t} \mathbf{p}_2\right).$$

Euler's formula can be used to find the real form of the general solution. We have

$$\operatorname{Re}\left[ e^{(1+i)t} \begin{pmatrix} 1 \\ 2 + i \\ i \end{pmatrix} \right] = e^t \begin{pmatrix} \cos t \\ 2 \cos t - \sin t \\ -\sin t \end{pmatrix}$$

and

$$\operatorname{Im}\left[ e^{(1+i)t} \begin{pmatrix} 1 \\ 2 + i \\ i \end{pmatrix} \right] = e^t \begin{pmatrix} \sin t \\ \cos t + 2 \sin t \\ \cos t \end{pmatrix}.$$

Therefore, the real form of the general solution is

$$\mathbf{x}(t) = b_1 e^{-t} \begin{pmatrix} -1 \\ 0 \\ 1 \end{pmatrix} + b_2 e^t \begin{pmatrix} \cos t \\ 2 \cos t - \sin t \\ -\sin t \end{pmatrix} + b_3 e^t \begin{pmatrix} \sin t \\ \cos t + 2 \sin t \\ \cos t \end{pmatrix}. \tag{7.81}$$

We have relabeled the arbitrary constants, because $b_2 \neq c_2$ and $b_3 \neq c_3$ in general.

It is possible to determine unique values of $b_1$, $b_2$, and $b_3$ (or $c_1$, $c_2$, and $c_3$) to satisfy any initial condition. For example, given $\mathbf{x}(0) = (1, -1, 2)$, we can substitute into Eq. (7.81) to obtain

$$\mathbf{x}(0) = b_1 \begin{pmatrix} -1 \\ 0 \\ 1 \end{pmatrix} + b_2 \begin{pmatrix} 1 \\ 2 \\ 0 \end{pmatrix} + b_3 \begin{pmatrix} 0 \\ 1 \\ 1 \end{pmatrix} = \begin{pmatrix} 1 \\ -1 \\ 2 \end{pmatrix}.$$

We solve the resulting system of three equations for $b_1$, $b_2$, and $b_3$ to find that the solution of the initial value problem is

$$\mathbf{x}(t) = \begin{pmatrix} 5e^{-t} + e^t(7\sin t - 4\cos t) \\ e^t(18\sin t - \cos t) \\ -5e^{-t} + e^t(4\sin t + 7\cos t) \end{pmatrix}. \qquad \blacksquare$$

## Sources, Sinks, and Saddles

The origin is always a fixed point for any system of the form $\mathbf{x}' = \mathbf{Ax}$. Typical curves move toward the origin if all components have exponentially decaying terms and move away from the origin if one or more components have an exponentially growing term. The eigenvalues of $\mathbf{A}$ determine the nature of the origin.

- If all the eigenvalues of $\mathbf{A}$ are negative or have negative real part, then the origin is a sink.
- If all the eigenvalues of $\mathbf{A}$ are positive or have positive real part, then the origin is a source.
- If at least one eigenvalue of $\mathbf{A}$ is positive or has positive real part, and if at least one eigenvalue is negative or has negative real part, then the origin is a saddle.

The geometry of saddle fixed points is particularly interesting in three-dimensional systems. There are two qualitatively different types of saddles, according to whether there are one or two eigenvalues of $\mathbf{A}$ with positive real parts.

✦ *Example 5*   Let

$$\mathbf{x}(t) = \mathbf{Ax} = \begin{pmatrix} 0.1 & 0 & 0 \\ 0 & -0.05 & 1 \\ 0 & -1 & -0.05 \end{pmatrix} \mathbf{x}. \qquad (7.82)$$

The eigenvalues of $\mathbf{A}$ are $\lambda_1 = 0.1$, $\lambda_2 = -0.05 + i$, and $\lambda_3 = -0.05 - i$. The general solution has the form

$$\mathbf{x}(t) = \begin{pmatrix} c_1 e^{0.1t} \\ e^{-0.05t}(c_2 \cos t + c_3 \sin t) \\ e^{-0.05t}(-c_2 \sin t + c_3 \cos t) \end{pmatrix}.$$

The solution trajectory grows exponentially in one direction only. This situation reflects the fact that only one eigenvalue of $\mathbf{A}$ is positive or has positive real part. In this example, the projection of the solution curve onto the $x_2 x_3$ plane tends to $\mathbf{0}$ as $t$ becomes large. However, the solution curve moves out toward infinity along the $x_1$ axis. Figure 7.16(a) shows the solution curve corresponding to a typical initial condition.

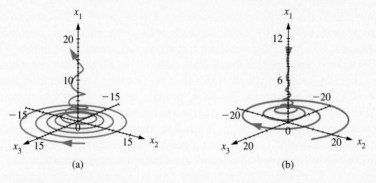

(a)                                             (b)

**FIGURE 7.16**   Typical solution curves for (a) Eq. (7.82); (b) Eq. (7.83).

◆ *Example 6*   Let

$$\mathbf{x}' = \mathbf{A}\mathbf{x} = \begin{pmatrix} -0.05 & 0 & 0 \\ 0 & 0.1 & 1 \\ 0 & -1 & 0.1 \end{pmatrix} \mathbf{x}. \tag{7.83}$$

The eigenvalues of $\mathbf{A}$ are $\lambda_1 = -0.05$, $\lambda_2 = 0.1 + i$, and $\lambda_3 = 0.1 - i$. The general solution has the form

$$\mathbf{x}(t) = \begin{pmatrix} c_1 e^{-0.05t} \\ e^{0.1t}(c_2 \cos t + c_3 \sin t) \\ e^{0.1t}(-c_2 \sin t + c_3 \cos t) \end{pmatrix}.$$

Figure 7.16(b) shows a typical solution curve. Notice that the projection of the curve onto the $x_2 x_3$ plane forms a spiral that grows with time. As $t$ increases, typical solution curves spiral away from the origin, but they move in along the $x_1$ axis. In other words, typical solution curves flatten out onto the $x_2 x_3$ plane. ∎

In general, the eigendirections do not correspond to the coordinate axes or the coordinate planes. Instead, typical trajectories move along directions that are determined by the eigenvectors corresponding to the eigenvalues that have positive real parts.

✦ *Example 7*   Let

$$\mathbf{x'} = \mathbf{A}\mathbf{x} = \begin{pmatrix} 6.75 & -3.25 & 7.25 \\ 20 & -11.75 & 20 \\ 5.5 & -4.75 & 5 \end{pmatrix} \mathbf{x}. \tag{7.84}$$

A computation shows that one eigenvalue of $\mathbf{A}$ is $\lambda_1 = -\frac{1}{2}$ with corresponding eigenvector $\mathbf{p}_1 = (1, 0, -1)$. The remaining eigenvalues of $\mathbf{A}$ are complex conjugates: $\lambda_2 = \frac{1}{4} + 4i$ and $\lambda_3 = \frac{1}{4} - 4i$. The general solution contains a term of the form $e^{-t/2}\mathbf{p}_1$, which implies that the component of the solution along the vector $\mathbf{p}_1$ decays exponentially with time. The complex conjugate eigenvalues imply that the general solution also has terms that oscillate with exponentially growing amplitudes. Therefore, the origin is a saddle.

Figure 7.17 shows a typical solution trajectory for Eq. (7.84). Because the component along one direction in space (that is, along $\mathbf{p}_1$) damps out exponentially, the solution curve eventually approaches a plane. The situation is qualitatively similar to that described in Example 6. The curve approaches the origin for a time, but eventually, the exponential growth of the amplitude of the oscillating terms in the solution causes the curve to move away from the origin after a sufficiently long interval. The only solutions that approach the origin as $t \to \infty$ are those whose initial conditions lie along the eigendirection determined by $\mathbf{p}_1$.

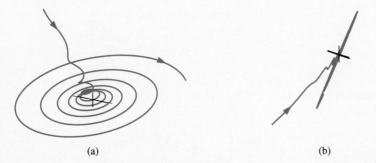

(a)                                                                 (b)

**FIGURE 7.17**   A solution trajectory corresponding to Eq. (7.84). Graphic (b) shows the same curve as (a) from a different viewing angle, as indicated by the cross in the middle of each graphic. Most of the spiral curve lies close to a plane in 3-space. ∎

## Multiple Eigenvalues

If $\mathbf{A}$ is a $3 \times 3$ matrix, then the characteristic equation of $\mathbf{A}$ is a cubic polynomial whose roots can have multiplicity two or three; that is, $\mathbf{A}$ can have a double or triple eigenvalue. We discuss these possibilities in turn.

### Case 1: A Double Eigenvalue

Suppose that $\mathbf{A}$ is a $3 \times 3$ matrix with a repeated eigenvalue, that is, $\lambda_1 \neq \lambda_2$ but $\lambda_2 = \lambda_3$. If there are two linearly independent eigenvectors $\mathbf{p}_2$ and $\mathbf{p}_3$ corresponding to $\lambda_2$, then the general solution of the system $\mathbf{x}' = \mathbf{Ax}$ can be written in the form

$$\mathbf{x}(t) = c_1 e^{\lambda_1 t} \mathbf{p}_1 + c_2 e^{\lambda_2 t} \mathbf{p}_2 + c_3 e^{\lambda_2 t} \mathbf{p}_3,$$

similar to the case in which all the eigenvalues are distinct. If $\lambda_2$ has only one associated eigenvector, then the general solution has the form

$$\mathbf{x}(t) = c_1 e^{\lambda_1 t} \mathbf{p}_1 + c_2 e^{\lambda_2 t} \mathbf{p}_2 + c_3 e^{\lambda_2 t} (\mathbf{q} + t\mathbf{p}_2),$$

where $\mathbf{q}$ is a generalized eigenvector corresponding to $\lambda_2$. The following examples illustrate the possibilities.

✦ *Example 8*   Let

$$\mathbf{A} = \begin{pmatrix} 2 & -3 & 0 \\ 0 & -1 & 0 \\ 0 & 0 & -1 \end{pmatrix}.$$

The characteristic equation is

$$\det(\mathbf{A} - \lambda \mathbf{I}) = (2 - \lambda)(-1 - \lambda)^2 = 0,$$

which implies that $\lambda_1 = 2$ and $\lambda_2 = \lambda_3 = -1$. An eigenvector corresponding to $\lambda_1$ is $\mathbf{p}_1 = (1, 0, 0)$; the computation is similar to those in the previous examples.

An eigenvector corresponding to $\lambda_2 = -1$ is a solution of

$$(\mathbf{A} + \mathbf{I})\mathbf{p} = \begin{pmatrix} 3 & -3 & 0 \\ 0 & 0 & 0 \\ 0 & 0 & 0 \end{pmatrix} \mathbf{p} = \mathbf{0}.$$

This relation is equivalent to a linear equation in three unknowns; thus, there are two different sets of solutions. One solution is $\mathbf{p}_2 = (1, 1, 0)$. Another solution is $\mathbf{p}_3 = (0, 0, 1)$. The vectors $\mathbf{p}_2$ and $\mathbf{p}_3$ are linearly independent.

The general solution of the system $\mathbf{x}' = \mathbf{A}\mathbf{x}$ is

$$\mathbf{x}(t) = c_1 e^{2t} \begin{pmatrix} 1 \\ 0 \\ 0 \end{pmatrix} + c_2 e^{-t} \begin{pmatrix} 1 \\ 1 \\ 0 \end{pmatrix} + c_3 e^{-t} \begin{pmatrix} 0 \\ 0 \\ 1 \end{pmatrix}.$$

The constants $c_1$, $c_2$, and $c_3$ can be chosen to satisfy any initial condition. For instance, if $\mathbf{x}(0) = (3, 1, 2)$, then we have

$$\begin{pmatrix} 3 \\ 1 \\ 2 \end{pmatrix} = c_1 \begin{pmatrix} 1 \\ 0 \\ 0 \end{pmatrix} + c_2 \begin{pmatrix} 1 \\ 1 \\ 0 \end{pmatrix} + c_3 \begin{pmatrix} 0 \\ 0 \\ 1 \end{pmatrix},$$

which implies that the solution is

$$\mathbf{x}(t) = \begin{pmatrix} 2e^{2t} + e^{-t} \\ e^{-t} \\ 2e^{-t} \end{pmatrix}.$$

Notice that the origin is a saddle.                                                        ∎

✦ *Example 9*   Let

$$\mathbf{A} = \begin{pmatrix} 2 & -3 & 1 \\ 0 & -1 & 1 \\ 0 & 0 & -1 \end{pmatrix}.$$

As in the previous example, the characteristic equation is

$$\det (\mathbf{A} - \lambda \mathbf{I}) = (2 - \lambda)(-1 - \lambda)^2 = 0,$$

so $\lambda_1 = 2$ and $\lambda_2 = \lambda_3 = -1$. An eigenvector corresponding to $\lambda_1$ is $\mathbf{p}_1 = (1, 0, 0)$. An eigenvector corresponding to $\lambda_2 = -1$ is a solution of

$$(\mathbf{A} + \mathbf{I})\mathbf{p} = \begin{pmatrix} 3 & -3 & 1 \\ 0 & 0 & 1 \\ 0 & 0 & 0 \end{pmatrix} \mathbf{p} = \mathbf{0}.$$

As before, one solution is $\mathbf{p}_2 = (1, 1, 0)$, but there is no other linearly independent eigenvector. (In particular, $(0, 0, 1)$ is not a solution.)

As in the $2 \times 2$ case, it is possible to determine a generalized eigenvector $\mathbf{q}$ such that $(A - \lambda_2 I)\mathbf{q} = \mathbf{p}_2$. In this case, we have

$$(\mathbf{A} + \mathbf{I})\mathbf{q} = \begin{pmatrix} 3 & -3 & 1 \\ 0 & 0 & 1 \\ 0 & 0 & 0 \end{pmatrix} \mathbf{q} = \begin{pmatrix} 1 \\ 1 \\ 0 \end{pmatrix},$$

which implies that $\mathbf{q} = (0, 0, 1)$. (There are other choices for $\mathbf{q}$ because $\mathbf{A} + \mathbf{I}$ is a singular matrix. It does not matter which solution is chosen, as long as $\mathbf{q} \neq \mathbf{0}$.)

The general solution of the system $\mathbf{x}' = \mathbf{Ax}$ is

$$\mathbf{x}(t) = c_1 e^{2t} \begin{pmatrix} 1 \\ 0 \\ 0 \end{pmatrix} + c_2 e^{-t} \begin{pmatrix} 1 \\ 1 \\ 0 \end{pmatrix} + c_3 e^{-t} \left[ \begin{pmatrix} 0 \\ 0 \\ 1 \end{pmatrix} + t \begin{pmatrix} 1 \\ 1 \\ 0 \end{pmatrix} \right].$$

The constants $c_1$, $c_2$, and $c_3$ can be chosen to satisfy any initial condition. For instance, if $\mathbf{x}(0) = (3, 1, 2)$, then

$$\begin{pmatrix} 3 \\ 1 \\ 2 \end{pmatrix} = c_1 \begin{pmatrix} 1 \\ 0 \\ 0 \end{pmatrix} + c_2 \begin{pmatrix} 1 \\ 1 \\ 0 \end{pmatrix} + c_3 \begin{pmatrix} 0 \\ 0 \\ 1 \end{pmatrix},$$

which implies that the solution satisfying the initial condition is

$$\mathbf{x}(t) = \begin{pmatrix} 2e^{2t} + e^{-t}(1 + 2t) \\ e^{-t} + 2te^{-t} \\ 2e^{-t} \end{pmatrix}.$$

As in the previous example, the origin is a saddle.                                      ■

## Case 2: A Triple Eigenvalue

If $\mathbf{A}$ is a $3 \times 3$ matrix whose characteristic equation has a triple root, then there may be only one eigenvector $\mathbf{p}$ associated with the triple eigenvalue $\lambda$. In this case, it is possible to determine two generalized eigenvectors $\mathbf{q}_1$ and $\mathbf{q}_2$ such that

$$(\mathbf{A} - \lambda \mathbf{I})\mathbf{q}_1 = \mathbf{p} \qquad \text{and} \qquad (\mathbf{A} - \lambda \mathbf{I})\mathbf{q}_2 = \mathbf{q}_1.$$

It can be shown that the general solution of the system $\mathbf{x}' = \mathbf{Ax}$ is

$$\mathbf{x}(t) = c_1 e^{\lambda t} \mathbf{p} + c_2 e^{\lambda t} (\mathbf{q}_1 + t\mathbf{p}) + c_3 e^{\lambda t} \left( \mathbf{q}_2 + t\mathbf{q}_1 + \tfrac{1}{2}t^2 \mathbf{p} \right). \qquad (7.85)$$

♦ *Example 10*   Let

$$\mathbf{A} = \begin{pmatrix} -1 & 0 & 1 \\ 0 & -1 & 1 \\ 1 & -1 & -1 \end{pmatrix}.$$

The characteristic equation is

$$
\begin{aligned}
0 &= \det\left(\mathbf{A} - \lambda\mathbf{I}\right) \\
&= \det \begin{pmatrix} -1-\lambda & 0 & 1 \\ 0 & -1-\lambda & 1 \\ 1 & -1 & -1-\lambda \end{pmatrix} \\
&= (-1-\lambda)\left[(-1-\lambda)^2 + 1\right] + 1 + \lambda \\
&= -\lambda^3 - 3\lambda^2 - 3\lambda - 1 \\
&= -(\lambda+1)^3.
\end{aligned}
$$

Thus, $\lambda = -1$ is a triple root. The associated eigenvalue is a solution of the equation

$$(\mathbf{A}+\mathbf{I})\mathbf{p} = \begin{pmatrix} 0 & 0 & 1 \\ 0 & 0 & 1 \\ 1 & -1 & 0 \end{pmatrix}\mathbf{p} = \mathbf{0}.$$

One solution is $\mathbf{p} = (1, 1, 0)$. However, there is no other linearly independent eigenvector.

Instead, we consider two generalized eigenvectors. The first, $\mathbf{q}_1$, is a solution of the equation $(\mathbf{A}+\mathbf{I})\mathbf{q}_1 = \mathbf{p}$. We have

$$\begin{pmatrix} 0 & 0 & 1 \\ 0 & 0 & 1 \\ 1 & -1 & 0 \end{pmatrix}\mathbf{q}_1 = \begin{pmatrix} 1 \\ 1 \\ 0 \end{pmatrix},$$

which implies that $\mathbf{q}_1 = (0, 0, 1)$. (There are other choices for $\mathbf{q}_1$ because $\mathbf{A}+\mathbf{I}$ is singular. Any nonzero choice is satisfactory.) The second generalized eigenvector $\mathbf{q}_2$ is a solution of the equation $(\mathbf{A}+\mathbf{I})\mathbf{q}_2 = \mathbf{q}_1$. We have

$$\begin{pmatrix} 0 & 0 & 1 \\ 0 & 0 & 1 \\ 1 & -1 & 0 \end{pmatrix}\mathbf{q}_2 = \begin{pmatrix} 0 \\ 0 \\ 1 \end{pmatrix};$$

one solution is $\mathbf{q}_2 = (0, -1, 0)$.

By Eq. (7.85), the general solution of the system $\mathbf{x}' = \mathbf{A}\mathbf{x}$ is

$$\mathbf{x}(t) = c_1 e^{-t} \begin{pmatrix} 1 \\ 1 \\ 0 \end{pmatrix} + c_2 e^{-t} \left[ \begin{pmatrix} 0 \\ 0 \\ 1 \end{pmatrix} + t \begin{pmatrix} 1 \\ 1 \\ 0 \end{pmatrix} \right]$$

$$+ c_3 e^{-t} \left[ \begin{pmatrix} 0 \\ -1 \\ 0 \end{pmatrix} + t \begin{pmatrix} 0 \\ 0 \\ 1 \end{pmatrix} + \frac{t^2}{2} \begin{pmatrix} 1 \\ 1 \\ 0 \end{pmatrix} \right].$$

The origin is a sink, because $\mathbf{x}(t) \to \mathbf{0}$ as $t \to \infty$, regardless of the constants $c_1$, $c_2$, and $c_3$.

The constants $c_1$, $c_2$, and $c_3$ can be chosen to satisfy any initial condition. For instance, if $\mathbf{x}(0) = (2, 1, 1)$, then

$$\begin{pmatrix} 2 \\ 1 \\ 1 \end{pmatrix} = c_1 \begin{pmatrix} 1 \\ 1 \\ 0 \end{pmatrix} + c_2 \begin{pmatrix} 0 \\ 0 \\ 1 \end{pmatrix} + c_3 \begin{pmatrix} 0 \\ -1 \\ 0 \end{pmatrix}.$$

We can solve this system of equations to obtain the solution of the initial value problem, which is

$$\mathbf{x}(1) = \begin{pmatrix} e^{-t} \left(2 + t + t^2/2\right) \\ e^{-t} \left(1 + t + t^2/2\right) \\ e^{-t}(1 + t) \end{pmatrix}. \qquad \blacksquare$$

## Exercises

**1.** For each of the matrices in the following list,

- Find the solution of the system $\mathbf{x}' = \mathbf{A}\mathbf{x}$ that satisfies the indicated initial condition. (Determine the real form of the solution as applicable.)

- Classify the origin as a source, a sink, or a saddle.

(a) $\mathbf{A} = \begin{pmatrix} 20 & -12 & -6 \\ 23 & -14 & -7 \\ 21 & -12 & -7 \end{pmatrix}$, $\mathbf{x}(0) = (1, -1, 2)$

(b) $\mathbf{A} = \begin{pmatrix} -5 & 0 & 2 \\ -11 & 2 & 3 \\ -24 & 0 & 9 \end{pmatrix}$, $\mathbf{x}(0) = (-1, 1, 2)$

(c) $\mathbf{A} = \begin{pmatrix} 15 & -50 & -14 \\ -1 & 2 & 1 \\ 20 & -62 & -19 \end{pmatrix}$, $\mathbf{x}(0) = (1, 2, 1)$

(d) $\mathbf{A} = \begin{pmatrix} 0 & -1 & 0 \\ -1 & 4 & -4 \\ -3 & 9 & -7 \end{pmatrix}$, $\mathbf{x}(0) = (-2, 0, 2)$

**2.** Repeat Exercise 1 for each of the following matrices:

(a) $\mathbf{A} = \begin{pmatrix} 0 & -1 & -1 \\ -6 & -8 & 6 \\ -10 & -16 & 9 \end{pmatrix}$, $\mathbf{x}(0) = (3, 2, 1)$

(b) $\mathbf{A} = \begin{pmatrix} 2 & 0 & 0 \\ 7 & -1 & -4 \\ -3 & 2 & 3 \end{pmatrix}$,  $\mathbf{x}(0) = (1, -1, 2)$

(c) $\mathbf{A} = \begin{pmatrix} -5 & 2 & 4 \\ 24 & -7 & -16 \\ -18 & 6 & 13 \end{pmatrix}$,  $\mathbf{x}(0) = (0, 1, 1)$

(d) $\mathbf{A} = \begin{pmatrix} 1 & 6 & -3 \\ 2 & 5 & -3 \\ 6 & 18 & -10 \end{pmatrix}$,  $\mathbf{x}(0) = (-1, 2, 0)$

**3.** Repeat Exercise 1 for each of the following matrices:

(a) $\mathbf{A} = \begin{pmatrix} 5 & -4 & 4 \\ 3 & -1 & 2 \\ -3 & 4 & -3 \end{pmatrix}$,  $\mathbf{x}(0) = (1, 1, 1)$

(b) $\mathbf{A} = \begin{pmatrix} -1 & -1 & 0 \\ 0 & -3 & 1 \\ 1 & -1 & -2 \end{pmatrix}$,  $\mathbf{x}(0) = (0, 1, 1)$

(c) $\mathbf{A} = \begin{pmatrix} -6 & 1 & 1 \\ -13 & 1 & 3 \\ -15 & 3 & 2 \end{pmatrix}$,  $\mathbf{x}(0) = (1, 0, 1)$

**4.** Let

$$\mathbf{x}' = \begin{pmatrix} -1 & 0 & 0 \\ 0 & \frac{1}{10} & 0 \\ 0 & 0 & 1 \end{pmatrix} \mathbf{x}.$$

(a) Find the general solution. Is the origin a saddle or a source?

(b) Let $\mathbf{x}(t)$ be the solution that satisfies the initial condition $\mathbf{x}(0) = (10, 0.1, 0.1)$. Calculate $\mathbf{x}(1), \mathbf{x}(5)$, and $\mathbf{x}(10)$. Discuss the relative magnitudes of the components of $\mathbf{x}$ for the different values of $t$.

(c) Draw the solution trajectory corresponding to (b). In what direction does the trajectory leave the origin as $t$ increases from 0?

**5.** Repeat Exercise 4 for the system

$$\mathbf{x}' = \begin{pmatrix} -3 & 2 & 2 \\ -1 & 0 & 2 \\ -3 & 3 & 1 \end{pmatrix} \mathbf{x}.$$

**6.** Let

$$\mathbf{x}' = \begin{pmatrix} -1 & 0 & 0 \\ 0 & -0.1 & 0 \\ 0 & 0 & -2 \end{pmatrix} \mathbf{x}.$$

(a) Find the general solution. Is the origin a saddle or a sink?

(b) Let $\mathbf{x}(0) = (10, 9, 8)$. Compute $\mathbf{x}(1), \mathbf{x}(5)$, and $\mathbf{x}(10)$. Discuss the relative magnitudes of the components of $\mathbf{x}$ for the different values of $t$.

(c) Draw the solution trajectory corresponding to (b). In what direction does the trajectory approach the origin as $t$ increases from 0?

**7.** Repeat Exercise 6 for the system

$$\mathbf{x}'(t) = \begin{pmatrix} -3 & 1 & 1 \\ 0 & -2 & 1 \\ -2 & 2 & -1 \end{pmatrix} \mathbf{x}.$$

**8.** Let

$$\mathbf{x}' = \begin{pmatrix} 5 & 2 & -4 \\ -6 & -3 & 4 \\ 1 & -1 & -1 \end{pmatrix} \mathbf{x}.$$

(a) Is the origin a source or a saddle?

(b) Find the solution that satisfies the initial condition $\mathbf{x}(0) = (10, 2, 2)$.

(c) Argue that the trajectory approaches a plane as $t$ becomes large.

**9.** The $(i, j)th$ *cofactor* of the matrix $\mathbf{A}$ is

$$C_{ij} = (-1)^{i+j} M_{ij},$$

where $M_{ij}$ is the $(i, j)$th minor. Equation (7.78) represents an evaluation of the determinant of a $3 \times 3$ matrix $\mathbf{A}$ by an expansion of cofactors along the first column of $\mathbf{A}$. Show that the determinant can be evaluated equivalently by a cofactor expansion along the first row of $\mathbf{A}$. (In fact, the determinant can be evaluated by a cofactor expansion along any column or any row of $\mathbf{A}$.)

**10.** An upper triangular $3 \times 3$ matrix $\mathbf{A}$ is one whose only nonzero entries lie along and to the right of the diagonal; that is, $a_{21} = a_{31} = a_{32} = 0$. State and prove a theorem that gives an easy way to determine the eigenvalues of an upper triangular matrix. (In fact, your theorem holds for an arbitrary $n \times n$ matrix, but you need only state the proof for the $3 \times 3$ case.)

**11.** The determinant of the $4 \times 4$ matrix $\mathbf{A}$ is defined as

$$\det \mathbf{A} = a_{11} \det \begin{pmatrix} a_{22} & a_{23} & a_{24} \\ a_{32} & a_{33} & a_{34} \\ a_{42} & a_{43} & a_{44} \end{pmatrix}$$

$$- a_{21} \det \begin{pmatrix} a_{12} & a_{13} & a_{14} \\ a_{32} & a_{33} & a_{34} \\ a_{42} & a_{43} & a_{44} \end{pmatrix}$$

$$+ a_{31} \det \begin{pmatrix} a_{12} & a_{13} & a_{14} \\ a_{22} & a_{23} & a_{24} \\ a_{42} & a_{43} & a_{44} \end{pmatrix}$$

$$- a_{41} \det \begin{pmatrix} a_{12} & a_{13} & a_{14} \\ a_{22} & a_{23} & a_{24} \\ a_{32} & a_{33} & a_{34} \end{pmatrix}.$$

The characteristic equation of $\mathbf{A}$ is $\det(\mathbf{A} - \lambda\mathbf{I}) = 0$. The eigenvalues and eigenvectors of $\mathbf{A}$, as well as the general solution of the system $\mathbf{x}' = \mathbf{A}\mathbf{x}$, are determined in a manner analogous to that described in Example 1.

For each matrix $\mathbf{A}$ in the following list,

- determine the characteristic equation of $\mathbf{A}$;

- find the eigenvalues and eigenvectors of $\mathbf{A}$;

- find the general solution of the system $\mathbf{x}' = \mathbf{A}\mathbf{x}$.

(a) $\mathbf{A} = \begin{pmatrix} 2 & -1 & 1 & -4 \\ 4 & -3 & 1 & -5 \\ 4 & -4 & 2 & -7 \\ 0 & 0 & 0 & -1 \end{pmatrix}$

(b) $\mathbf{A} = \begin{pmatrix} 14 & -5 & -26 & 21 \\ 18 & -7 & -36 & 27 \\ -15 & 5 & 31 & -24 \\ -24 & 8 & 48 & -38 \end{pmatrix}$

(c) $\mathbf{A} = \begin{pmatrix} 7 & 3 & 6 & -3 \\ -2 & -7 & -4 & 6 \\ -14 & -8 & -13 & 8 \\ -12 & -15 & -14 & 14 \end{pmatrix}$

# 7.7   THE FUNDAMENTAL MATRIX

This section discusses some theoretical concepts that unify many of the details of the analysis of linear systems. In particular, we investigate the definition of the exponential of a constant matrix and discuss its relationship to the solutions of first-order systems of the form $\mathbf{x}' = \mathbf{A}\mathbf{x}$.

## An Existence and Uniqueness Theorem

The previous sections have discussed ways to derive explicit solutions of $\mathbf{x}' = \mathbf{A}\mathbf{x}$, where $\mathbf{A}$ is a constant matrix. The following theorem, which we state without proof, states that linear systems of this form have unique solutions.

**Theorem 7.22**   *Let $\mathbf{A}$ be a constant $n \times n$ matrix. The initial value problem $\mathbf{x}' = \mathbf{A}\mathbf{x}$, $\mathbf{x}(t_0) = \mathbf{x}_0$, has a unique solution whose interval of existence is the entire real number line.*

The uniqueness property is very important. In particular, Theorem 7.22 guarantees that the solutions found in Sections 7.4 and 7.6 are the only solutions of $2 \times 2$ and $3 \times 3$ linear systems of the form $\mathbf{x}' = \mathbf{Ax}$.

Another important consequence of the uniqueness theorem is reflected in the phase portrait of the system $\mathbf{x}' = \mathbf{Ax}$:

> *The trajectories corresponding to two different solutions of the autonomous equation* $\mathbf{x}' = \mathbf{Ax}$ *cannot cross.*

If two different solution curves intersected at a common point $\mathbf{y}_0$, then there would not be a unique solution satisfying the initial condition $\mathbf{x}(t_0) = \mathbf{y}_0$, because there would be two possible trajectories starting from the common point $\mathbf{y}_0$.

## A Matrix Formulation of Solutions

Let $a$ be a constant; then the scalar first-order equation $x' = ax$ has solutions of the form $x(t) = e^{at}x_0$. The objective of this section is to explore the parallels between scalar first-order equations with constant coefficients and systems of the form $\mathbf{x}' = \mathbf{Ax}$, where $\mathbf{A}$ is a constant matrix. In particular, we will discuss how the quantity $e^{\mathbf{A}t}$ can be defined for the case of a $2 \times 2$ matrix $\mathbf{A}$.

To motivate the basic idea, consider the system

$$x_1' = \lambda_1 x_1,$$
$$x_2' = \lambda_2 x_2,$$

which can be expressed equivalently in matrix notation as

$$\mathbf{x}' = \begin{pmatrix} \lambda_1 & 0 \\ 0 & \lambda_2 \end{pmatrix} \mathbf{x} = \mathbf{\Lambda x}.$$

The solution is

$$x_1(t) = e^{\lambda_1 t} x_1(0),$$
$$x_2(t) = e^{\lambda_2 t} x_2(0),$$

which can be expressed equivalently as

$$\mathbf{x}(t) = \begin{pmatrix} e^{\lambda_1 t} & 0 \\ 0 & e^{\lambda_2 t} \end{pmatrix} \mathbf{x}(0).$$

This example suggests that we can *define*

$$e^{\Lambda t} = \exp \begin{pmatrix} \lambda_1 t & 0 \\ 0 & \lambda_2 t \end{pmatrix} = \begin{pmatrix} e^{\lambda_1 t} & 0 \\ 0 & e^{\lambda_2 t} \end{pmatrix} \tag{7.86}$$

for a $2 \times 2$ diagonal matrix $\Lambda$.

Further investigation shows that Eq. (7.86) has properties that are analogous to those for the conventional exponential function. For example, $e^0 = 1$. If we substitute $t = 0$ into Eq. (7.86), we find

$$e^0 = \begin{pmatrix} 1 & 0 \\ 0 & 1 \end{pmatrix} = \mathbf{I}, \tag{7.87}$$

which is the identity matrix. Thus, $e^0$ is the identity element with respect to matrix multiplication in the same way that $e^0$ is the identity element with respect to multiplication of real and complex numbers.

Another law of exponents is

$$e^{t+s} = e^t e^s = e^s e^t = e^{s+t}$$

for all numbers $t$ and $s$. It is straightforward to check that Eq. (7.86) satisfies an analogous property:

$$e^{\Lambda(t+s)} = \begin{pmatrix} e^{\lambda_1(t+s)} & 0 \\ 0 & e^{\lambda_2(t+s)} \end{pmatrix}$$

$$= \begin{pmatrix} e^{\lambda_1 t} e^{\lambda_1 s} & 0 \\ 0 & e^{\lambda_2 t} e^{\lambda_2 s} \end{pmatrix}$$

$$= e^{\Lambda t} e^{\Lambda s}. \tag{7.88}$$

A similar computation shows that $e^{\Lambda t} e^{\Lambda s} = e^{\Lambda s} e^{\Lambda t}$.

A corollary of the basic law of exponents is that $e^{-t} = 1/e^t$. It can be shown (see Exercise 3) that

$$e^{-\Lambda t} = \left( e^{\Lambda t} \right)^{-1}.$$

Thus, $e^{-\Lambda t}$ is the inverse of $e^{\Lambda t}$ in the same way that $e^{-t}$ is the reciprocal of $e^t$.

We also know that $d\left( e^{at} \right)/dt = a e^{at}$. Notice that

$$\frac{d}{dt} \left( e^{\Lambda t} \right) = \frac{d}{dt} \begin{pmatrix} e^{\lambda_1 t} & 0 \\ 0 & e^{\lambda_2 t} \end{pmatrix}$$

$$= \begin{pmatrix} \lambda_1 e^{\lambda_1 t} & 0 \\ 0 & \lambda_2 e^{\lambda_2 t} \end{pmatrix}$$

$$= \Lambda e^{\Lambda t}. \tag{7.89}$$

Thus, the first derivative of $e^{\Lambda t}$ is analogous to that of the scalar function $e^{at}$. (It is straightforward to check that $\Lambda e^{\Lambda t} = e^{\Lambda t} \Lambda$ as well. See Exercise 12.)

These considerations suggest that the definition (7.86) is sensible, at least for a diagonal matrix $\Lambda$. In the rest of this section, we show that the definition can be extended to more general $2 \times 2$ matrices. (In fact, the ideas carry over to $n \times n$ matrices, but some of the details require some background in linear algebra that is beyond the scope of this book.) We address the cases of distinct eigenvalues, complex conjugate eigenvalues, and double real eigenvalues in turn.

## Case 1: Distinct Eigenvalues

Let $\mathbf{A}$ be a $2 \times 2$ matrix with real, distinct eigenvalues $\lambda_1$ and $\lambda_2$ and corresponding eigenvectors $\mathbf{p}_1$ and $\mathbf{p}_2$. Let $\mathbf{P}$ be the matrix whose columns are $\mathbf{p}_1$ and $\mathbf{p}_2$, that is,

$$\mathbf{P} = \begin{pmatrix} \mathbf{p}_1 & \mathbf{p}_2 \end{pmatrix}.$$

Then the definition of matrix-vector multiplication implies that

$$
\begin{aligned}
\mathbf{AP} &= \mathbf{A}\begin{pmatrix} \mathbf{p}_1 & \mathbf{p}_2 \end{pmatrix} \\
&= \begin{pmatrix} \mathbf{Ap}_1 & \mathbf{Ap}_2 \end{pmatrix} \\
&= \begin{pmatrix} \lambda_1 \mathbf{p}_1 & \lambda_2 \mathbf{p}_2 \end{pmatrix} \\
&= \begin{pmatrix} \mathbf{p}_1 & \mathbf{p}_2 \end{pmatrix} \begin{pmatrix} \lambda_1 & 0 \\ 0 & \lambda_2 \end{pmatrix}.
\end{aligned}
$$

We express this relation as

$$\mathbf{AP} = \mathbf{P\Lambda}, \tag{7.90}$$

where $\Lambda$ is the diagonal matrix

$$\Lambda = \begin{pmatrix} \lambda_1 & 0 \\ 0 & \lambda_2 \end{pmatrix}.$$

Since $\mathbf{p}_1$ and $\mathbf{p}_2$ are linearly independent by Theorem 7.14, the matrix $\mathbf{P}$ is nonsingular; therefore, Eq. (7.90) may be expressed equivalently as

$$\mathbf{P}^{-1}\mathbf{AP} = \Lambda. \tag{7.91}$$

Now consider the general solution of the system $\mathbf{x}' = \mathbf{Ax}$, given by

$$\mathbf{x}(t) = c_1 e^{\lambda_1 t} \mathbf{p}_1 + c_2 e^{\lambda_2 t} \mathbf{p}_2. \tag{7.92}$$

Equation (7.92) is equivalent to the matrix-vector product

$$\mathbf{x}(t) = \mathbf{P}e^{\Lambda t}\mathbf{c},$$

where $\mathbf{c} = (c_1, c_2)$. When $t = 0$,

$$\mathbf{x}(0) = \mathbf{x}_0 = \mathbf{Pc}.$$

Since $\mathbf{P}$ is nonsingular, $\mathbf{c} = \mathbf{P}^{-1}\mathbf{x}_0$. Therefore, the general solution (7.92) can be written as

$$\mathbf{x}(t) = \mathbf{P}e^{\Lambda t}\mathbf{P}^{-1}\mathbf{x}_0. \tag{7.93}$$

**Definition 7.23** *If* $\mathbf{A}$ *is a* $2 \times 2$ *matrix with distinct eigenvalues* $\lambda_1$ *and* $\lambda_2$ *and corresponding eigenvectors* $\mathbf{p}_1$ *and* $\mathbf{p}_2$, *then the matrix*

$$\boldsymbol{\Psi}(t) = \mathbf{P}\begin{pmatrix} e^{\lambda_1 t} & 0 \\ 0 & e^{\lambda_2 t} \end{pmatrix}\mathbf{P}^{-1}$$

*is called the* fundamental matrix *associated with* $\mathbf{A}$. *Here,* $\mathbf{P} = \begin{pmatrix} \mathbf{p}_1 & \mathbf{p}_2 \end{pmatrix}$ *is the matrix whose columns are the eigenvectors of* $\mathbf{A}$.

At the end of this section, we show that it is sensible to regard $\boldsymbol{\Psi}(t)$ as a matrix exponential function.

✦ *Example 1* The general solution of the system

$$\mathbf{x}' = \mathbf{A}\mathbf{x} = \begin{pmatrix} -5 & 2 \\ -6 & 2 \end{pmatrix}\mathbf{x}$$

is

$$\mathbf{x}(t) = c_1 e^{-t}\begin{pmatrix} 1 \\ 2 \end{pmatrix} + c_2 e^{-2t}\begin{pmatrix} 2 \\ 3 \end{pmatrix}.$$

(See Example 1 in Section 7.5.) A matrix representation of the general solution is

$$\mathbf{x}(t) = \begin{pmatrix} 1 & 2 \\ 2 & 3 \end{pmatrix}\begin{pmatrix} e^{-t} & 0 \\ 0 & e^{-2t} \end{pmatrix}\begin{pmatrix} c_1 \\ c_2 \end{pmatrix}.$$

The eigenvectors of $\mathbf{A}$ are linearly independent, so $\mathbf{P}^{-1}$ is defined and is given by

$$\mathbf{P}^{-1} = \begin{pmatrix} -3 & 2 \\ 2 & -1 \end{pmatrix}.$$

The initial condition $\mathbf{x}(0) = \mathbf{x}_0$ implies that $\mathbf{x}_0 = \mathbf{P}\mathbf{c}$, so $\mathbf{c} = \mathbf{P}^{-1}\mathbf{x}_0$. Therefore,

$$\mathbf{x}(t) = \begin{pmatrix} 1 & 2 \\ 2 & 3 \end{pmatrix} \begin{pmatrix} e^{-t} & 0 \\ 0 & e^{-2t} \end{pmatrix} \begin{pmatrix} -3 & 2 \\ 2 & -1 \end{pmatrix} \mathbf{x}_0$$

$$= \begin{pmatrix} -3e^{-t} + 4e^{-2t} & 2e^{-t} - 2e^{-2t} \\ -6e^{-t} + 6e^{-2t} & 4e^{-t} - 3e^{-2t} \end{pmatrix} \mathbf{x}_0$$

$$= \mathbf{\Psi}(t)\mathbf{x}_0.$$

The fundamental matrix is

$$\mathbf{\Psi}(t) = \begin{pmatrix} -3e^{-t} + 4e^{-2t} & 2e^{-t} - 2e^{-2t} \\ -6e^{-t} + 6e^{-2t} & 4e^{-t} - 3e^{-2t} \end{pmatrix}. \qquad \blacksquare$$

## Case 2: Complex Conjugate Eigenvalues

Suppose that $\mathbf{A}$ has complex conjugate eigenvalues. If we use complex arithmetic, then the fundamental matrix associated with $\mathbf{A}$ can be derived in the manner just discussed. Let $\mathbf{x}' = \mathbf{A}\mathbf{x}$, where the eigenvalues of $\mathbf{A}$ are $\lambda_1 = a + bi$ and $\lambda_2 = \bar{\lambda}_1 = a - bi$ with corresponding eigenvectors $\mathbf{p}_1$ and $\mathbf{p}_2$. Then

$$\mathbf{A}\mathbf{P} = \mathbf{P}\mathbf{\Lambda},$$

where $\mathbf{P} = \begin{pmatrix} \mathbf{p}_1 & \mathbf{p}_2 \end{pmatrix}$. The fundamental matrix, expressed in complex arithmetic, is

$$\mathbf{\Psi}(t) = \mathbf{P} \begin{pmatrix} e^{(a+bi)t} & 0 \\ 0 & e^{(a-bi)t} \end{pmatrix} \mathbf{P}^{-1}.$$

If we restrict ourselves to real arithmetic, then the fundamental matrix has a different form, which we now describe. The real form of the general solution is

$$\mathbf{x}(t) = c_1 \operatorname{Re}\left(e^{\lambda t}\mathbf{p}\right) + c_2 \operatorname{Im}\left(e^{\lambda t}\mathbf{p}\right), \tag{7.94}$$

where $\mathbf{p}$ is an eigenvector corresponding to $\lambda = a + bi$. Euler's formula can be applied to Eq. (7.94) to show that it is equivalent to

$$\mathbf{x}(t) = c_1 e^{at}(\cos bt \operatorname{Re}\mathbf{p} - \sin bt \operatorname{Im}\mathbf{p}) + c_2 e^{at}(\sin bt \operatorname{Re}\mathbf{p} + \cos bt \operatorname{Im}\mathbf{p}). \tag{7.95}$$

Equation (7.95) can be expressed as the equivalent matrix-vector product

$$\mathbf{x}(t) = e^{at}\begin{pmatrix} \operatorname{Re}\mathbf{p} & \operatorname{Im}\mathbf{p} \end{pmatrix} \begin{pmatrix} \cos bt & \sin bt \\ -\sin bt & \cos bt \end{pmatrix} \mathbf{c},$$

where $\mathbf{c} = (c_1, c_2)$.

Let $\mathbf{P} = (\text{Re}\,\mathbf{p}\ \text{Im}\,\mathbf{p})$, which is an invertible matrix. (See Exercise 6 in Section 7.4.) Notice that $\mathbf{x}(0) = \mathbf{Pc}$, so $\mathbf{c} = \mathbf{P}^{-1}\mathbf{x}(0)$. Therefore,

$$\mathbf{x}(t) = e^{at}\mathbf{PR}(t)\mathbf{P}^{-1}\mathbf{x}_0,$$

where $\mathbf{R}(t)$ is the rotation matrix

$$\mathbf{R}(t) = \begin{pmatrix} \cos bt & \sin bt \\ -\sin bt & \cos bt \end{pmatrix}. \tag{7.96}$$

(See Section 4.1.)

**Definition 7.24**   *Let* $\mathbf{A}$ *be a real* $2 \times 2$ *matrix with an eigenvalue* $\lambda = a + bi$ *and corresponding eigenvector* $\mathbf{p}$. *The matrix*

$$\mathbf{\Psi}(t) = e^{at}\mathbf{PR}(t)\mathbf{P}^{-1}$$

*is the fundamental matrix associated with* $\mathbf{A}$, *where* $\mathbf{P} = (\text{Re}\,\mathbf{p}\ \text{Im}\,\mathbf{p})$ *and* $\mathbf{R}$ *is the rotation matrix given by Eq. (7.96).*

There is a nice geometric parallel between the phase portrait and the rotation matrix $\mathbf{R}(t)$. When $\mathbf{A}$ has complex conjugate eigenvalues, solution trajectories spiral about the origin. The matrix $\mathbf{R}(t)$ rotates points in the plane about the origin in a time-dependent manner. The term $e^{at}$ determines whether the trajectories spiral into the origin, spiral away from the origin, or form closed curves, according to whether the origin is a sink, a source, or a center.

◆ *Example 2*   Let

$$\mathbf{x}' = \mathbf{Ax} = \begin{pmatrix} -\frac{59}{10} & -10 \\ 4 & \frac{61}{10} \end{pmatrix}\mathbf{x}. \tag{7.97}$$

It is straightforward to verify that one eigenvalue of $\mathbf{A}$ is $\lambda = \frac{1}{10} + 2i$ and that $\mathbf{p} = \left(-\frac{3}{2} + \frac{1}{2}i, 1\right)$ is a corresponding eigenvector. Let

$$\mathbf{P} = (\text{Re}\,\mathbf{p}\ \ \text{Im}\,\mathbf{p}) = \begin{pmatrix} -\frac{3}{2} & \frac{1}{2} \\ 1 & 0 \end{pmatrix}.$$

As was discussed in Section 7.4, the general solution of Eq. (7.97) has the form

$$\mathbf{x}(t) = c_1 \text{Re}\left(e^{\lambda t}\mathbf{p}\right) + c_2 \text{Im}\left(e^{\lambda t}\mathbf{p}\right).$$

(See Theorem 7.19.) By Eq. (7.95), we may express the general solution equivalently in the form

$$\mathbf{x}(t) = e^{t/10}\mathbf{PR}(t)\mathbf{P}^{-1}\mathbf{x}_0$$

$$= e^{t/10} \begin{pmatrix} -\frac{3}{2} & \frac{1}{2} \\ 1 & 0 \end{pmatrix} \begin{pmatrix} \cos 2t & \sin 2t \\ -\sin 2t & \cos 2t \end{pmatrix} \begin{pmatrix} -\frac{3}{2} & \frac{1}{2} \\ 1 & 0 \end{pmatrix}^{-1} \mathbf{x}_0$$

$$= e^{t/10} \begin{pmatrix} \cos 2t - 3\sin 2t & -5\sin 2t \\ 2\sin 2t & 3\sin 2t + \cos 2t \end{pmatrix} \mathbf{x}_0.$$

The fundamental matrix associated with $\mathbf{A}$ is

$$\mathbf{\Psi}(t) = e^{t/10}\mathbf{PR}(t)\mathbf{P}^{-1}.$$

Figure 7.18 shows the phase portrait associated with Eq. (7.97). The solution curve spirals away from the origin as $t$ increases, reflecting the oscillatory nature of the components of $\mathbf{x}(t)$, the action of the rotation matrix $\mathbf{R}(t)$, and the growth of the term $e^{t/10}$.

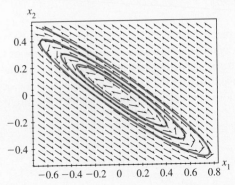

**FIGURE 7.18**    The phase portrait associated with Eq. (7.97).

## Case 3: Double Real Eigenvalues

If $\mathbf{A}$ has a double real eigenvalue $\lambda$ with only one corresponding eigenvector $\mathbf{p}$, then the general solution of the system $\mathbf{x}' = \mathbf{Ax}$ is

$$\mathbf{x}(t) = c_1 e^{\lambda t}\mathbf{p} + c_2 e^{\lambda t}(\mathbf{q} + t\mathbf{p}), \tag{7.98}$$

where $\mathbf{q}$ is a generalized eigenvector. (See Section 7.4.)

Equation (7.98) can be expressed as the corresponding matrix-vector product

$$\mathbf{x}(t) = \mathbf{P} \begin{pmatrix} e^{\lambda t} & te^{\lambda t} \\ 0 & e^{\lambda t} \end{pmatrix} \mathbf{c},$$

where $\mathbf{P} = \begin{pmatrix} \mathbf{p} & \mathbf{q} \end{pmatrix}$ and $\mathbf{c} = (c_1, c_2)$. As in the previous cases, $\mathbf{x}(0) = \mathbf{Pc}$, and $\mathbf{P}$ is nonsingular. Therefore, Eq. (7.98) is equivalent to

$$\mathbf{x}(t) = \mathbf{P} \begin{pmatrix} e^{\lambda t} & t e^{\lambda t} \\ 0 & e^{\lambda t} \end{pmatrix} \mathbf{P}^{-1} \mathbf{x}(0).$$

**Definition 7.25**   *Let $\mathbf{A}$ be a $2 \times 2$ real matrix with a double real eigenvalue $\lambda$, one corresponding eigenvector $\mathbf{p}$, and a generalized eigenvector $\mathbf{q}$. The fundamental matrix associated with $\mathbf{A}$ is*

$$\mathbf{\Psi}(t) = \mathbf{P} \begin{pmatrix} e^{\lambda t} & t e^{\lambda t} \\ 0 & e^{\lambda t} \end{pmatrix} \mathbf{P}^{-1},$$

*where $\mathbf{P} = \begin{pmatrix} \mathbf{p} & \mathbf{q} \end{pmatrix}$.*

✦ *Example 3*   Let

$$\mathbf{x}' = \mathbf{Ax} = \begin{pmatrix} -3 & 1 \\ -1 & -1 \end{pmatrix} \mathbf{x}.$$

The general solution is

$$\mathbf{x}(t) = c_1 e^{-2t} \begin{pmatrix} 1 \\ 1 \end{pmatrix} + c_2 e^{-2t} \left[ \begin{pmatrix} 0 \\ 1 \end{pmatrix} + t \begin{pmatrix} 1 \\ 1 \end{pmatrix} \right], \tag{7.99}$$

where $\mathbf{p} = (1, 1)$ is the eigenvector corresponding to the eigenvalue $-2$ and $\mathbf{q} = (0, 1)$ is the generalized eigenvector. (See Example 4 in Section 7.4.)

Equation (7.99) can be expressed as the equivalent matrix-vector product

$$\mathbf{x}(t) = \mathbf{P} \begin{pmatrix} e^{-2t} & t e^{-2t} \\ 0 & e^{-2t} \end{pmatrix} \mathbf{P}^{-1} \mathbf{x}_0,$$

where $\mathbf{P}$ is the nonsingular matrix

$$\mathbf{P} = \begin{pmatrix} \mathbf{p} & \mathbf{q} \end{pmatrix} = \begin{pmatrix} 1 & 0 \\ 1 & 1 \end{pmatrix}.$$

Therefore, the fundamental matrix associated with $\mathbf{A}$ is

$$\mathbf{\Psi}(t) = \mathbf{P} \begin{pmatrix} e^{-2t} & t e^{-2t} \\ 0 & e^{-2t} \end{pmatrix} \mathbf{P}^{-1}$$

$$= \begin{pmatrix} 1 & 0 \\ 1 & 1 \end{pmatrix} \begin{pmatrix} e^{-2t} & t e^{-2t} \\ 0 & e^{-2t} \end{pmatrix} \begin{pmatrix} 1 & 0 \\ -1 & 1 \end{pmatrix}$$

$$= e^{-2t} \begin{pmatrix} 1 - t & t \\ -t & t + 1 \end{pmatrix}. \qquad ∎$$

## Analytical Properties of the Fundamental Matrix

The fundamental matrix has important analytical properties, which are summarized in the following theorem.

**Theorem 7.26**  *Let* $x' = Ax$, *where* $A$ *is a constant matrix and* $\Psi(t)$ *is the associated fundamental matrix. Then*

1. $\Psi(0) = I$;
2. *for all real numbers* $s$ *and* $t$, $\Psi(s + t) = \Psi(s)\Psi(t)$;
3. $\dfrac{d}{dt}\Psi(t) = A\Psi(t)$;
4. $\Psi(-t) = \Psi(t)^{-1}$.

Theorem 7.26 says that the fundamental matrix has properties that are analogous to those of the ordinary exponential function. For this reason, we make the following definition.

**Definition 7.27**  *Let* $A$ *be a constant matrix, and let* $\Psi(t)$ *be the fundamental matrix associated with the first-order differential equation* $x' = Ax$. *The exponential of the matrix* $At$ *is* $e^{At} = \Psi(t)$.

Theorem 7.26 and Definition 7.27 allow us to express the solution of the initial value problem $x' = Ax$, $x(0) = x_0$, as $x(t) = e^{At}x_0$. In this way, the general solution of first-order linear systems of differential equations with constant coefficients can be written in a manner that is analogous to that of the scalar first-order equation $x' = ax$.

We now consider the proof of Theorem 7.26 when $A$ is a $2 \times 2$ matrix. We treat the case in which $A$ has distinct real eigenvalues here and leave the cases of complex conjugate and double real eigenvalues as exercises.

If $A$ has distinct real eigenvalues $\lambda_1$ and $\lambda_2$, then it follows from Eq. (7.91) that

$$A = P\Lambda P^{-1}, \tag{7.100}$$

where $P$ is the matrix of corresponding eigenvectors. The fundamental matrix associated with $A$ can be expressed as the product

$$\Psi(t) = Pe^{\Lambda t}P^{-1},$$

where

$$e^{\Lambda t} = \begin{pmatrix} e^{\lambda_1 t} & 0 \\ 0 & e^{\lambda_2 t} \end{pmatrix}.$$

To prove property (1), observe that

$$\boldsymbol{\Psi}(0) = \mathbf{P}e^{\mathbf{0}}\mathbf{P}^{-1}$$
$$= \mathbf{PIP}^{-1} \qquad \text{by Eq. (7.87)}$$
$$= \mathbf{I}.$$

To prove property (2), observe that

$$\boldsymbol{\Psi}(s + t) = \mathbf{P}e^{\boldsymbol{\Lambda}(s+t)}\mathbf{P}^{-1}$$
$$= \mathbf{P}e^{\boldsymbol{\Lambda}s}e^{\boldsymbol{\Lambda}t}\mathbf{P}^{-1} \qquad \text{by Eq. (7.88)}$$
$$= \left(\mathbf{P}e^{\boldsymbol{\Lambda}s}\mathbf{P}^{-1}\right)\left(\mathbf{P}e^{\boldsymbol{\Lambda}t}\mathbf{P}^{-1}\right)$$
$$= \boldsymbol{\Psi}(s)\boldsymbol{\Psi}(t).$$

To prove property (3), observe that

$$\frac{d}{dt}\boldsymbol{\Psi}(t) = \mathbf{P}\left(\frac{d}{dt}e^{\boldsymbol{\Lambda}t}\right)\mathbf{P}^{-1} \qquad \text{because } \mathbf{P} \text{ is constant; see Exercise 16}$$
$$= \mathbf{P}\left(\boldsymbol{\Lambda}e^{\boldsymbol{\Lambda}t}\right)\mathbf{P}^{-1} \qquad \text{by Eq. (7.89)}$$
$$= (\mathbf{AP})e^{\boldsymbol{\Lambda}t}\mathbf{P}^{-1} \qquad \text{by Eq. (7.100)}$$
$$= \mathbf{A}\boldsymbol{\Psi}(t)$$

by the definition of $\boldsymbol{\Psi}(t)$. Property (4) follows from properties (1) and (2) (see Exercise 13).

## What Is a Flow?

Theorem 7.26 implies the following important fact about the solutions of homogeneous first-order systems with constant coefficients:

*The solution is a linear function of the initial condition.*

The following theorem makes this idea precise.

**Theorem 7.28**   *Let* $\mathbf{A}$ *be a constant* $2 \times 2$ *matrix, and let* $c$ *be any scalar. Let* $\mathbf{x}(t)$ *be a solution of* $\mathbf{x}' = \mathbf{Ax}$ *that satisfies the initial condition* $\mathbf{x}(0) = \mathbf{x}_0$*; that is,* $\mathbf{x}(t) = \boldsymbol{\Psi}(t)\mathbf{x}_0$*, where* $\boldsymbol{\Psi}(t)$ *is the fundamental matrix associated with* $\mathbf{A}$*. Then the solution that satisfies* *the initial condition* $z_0 = cx_0$ *is*

$$z(t) = cx(t) = c\Psi(t)x_0,$$

*and the solution that satisfies the initial condition* $z_0 = x_0 + y_0$ *is*

$$z(t) = \Psi(t)(x_0 + y_0).$$

Theorem 7.28 can be extended to the case in which $A$ is an $n \times n$ matrix, but the details of defining the associated fundamental matrix are beyond the scope of this book.

The discussion in this section shows that the $2 \times 2$ linear system of the form

$$x' = Ax \tag{7.101}$$

defines a rule for moving points in the phase plane. The point $x_0$ is moved at time $t$ to the point $\Psi(t)x_0$, where $\Psi(t)$ is the fundamental matrix associated with $A$. We say that Eq. (7.101) defines a *flow*. Every initial condition "flows" along a trajectory that is consistent with the slope lines in the phase plane. By expressing solutions in the form $x(t) = \Psi(t)x_0$, we emphasize the importance of the initial condition.

In practice, we are often not as interested in the fate of a particular initial condition as we are in the fate of an ensemble of initial conditions. It is natural to ask whether there are sets of initial conditions for which the system tends to an equilibrium solution, to a periodic solution, to infinity, and so on. (For instance, a set of initial conditions that tend to infinity might correspond to a set of dangerous operating conditions for some apparatus whose behavior is modeled by the system of differential equations.)

The notion of a flow allows us to ask about the fate of ensembles of initial conditions. In the case of linear systems like Eq. (7.101), the fundamental matrix allows us to analyze the behavior of many initial conditions at once.

◆ *Example 4*  Let

$$x' = Ax = \frac{1}{4}\begin{pmatrix} -3 & -1 \\ -1 & -3 \end{pmatrix} x. \tag{7.102}$$

The eigenvalues of $A$ are $\lambda_1 = -1$ and $\lambda_2 = -1/2$ with corresponding eigenvectors $p_1 = (1, 1)$ and $p_2 = (-1, 1)$. The fundamental matrix associated with $A$ is

$$\Psi(t) = \frac{1}{2}\begin{pmatrix} e^{-t/2} + e^{-t} & e^{-t} - e^{-t/2} \\ e^{-t} - e^{-t/2} & e^{-t/2} + e^{-t} \end{pmatrix}.$$

The solution of Eq. (7.102) can be written as

$$x(t) = \Psi(t)x_0, \tag{7.103}$$

where $\mathbf{x}_0 = \mathbf{x}(0)$ is the initial condition. Equation (7.103) lets us investigate how sets of initial points are mapped with time.

Figure 7.19 shows a box $B$ in the phase plane extending from $-1$ to $1$ in each coordinate direction. The quadrilateral denoted $\mathbf{\Psi}(1)(B)$ shows the image of $B$ under the flow at $t = 1$. The flow shrinks the overall box, but its relative length is greater in one direction. The quadrilateral denoted $\mathbf{\Psi}(2)(B)$ shows the image of $B$ at $t = 2$. The original box has shrunk still more, and its relative length is even greater in one direction.

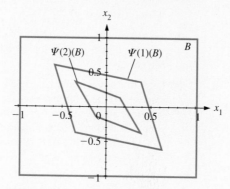

**FIGURE 7.19**  Successive images of the box $B$ under the flow defined by Eqs. (7.102) and (7.103).

The box $B$ is contracted faster along the $+45$-degree direction than along the $-45$-degree direction. The $+45$-degree direction is the eigendirection corresponding to $\lambda_1 = -1$; points along this direction move toward the origin at a rate proportional to $e^{-t}$. In contrast, points along the $-45$-degree direction move toward the origin at a slower rate, which is proportional to $e^{-t/2}$. Thus, $\mathbf{\Psi}(t)(B)$ appears to be relatively longer in the $-45$-degree direction. ∎

## Exercises

**1.** Determine the fundamental matrix associated with each matrix in the following list:

(a) $\mathbf{A} = \begin{pmatrix} -2 & 3 \\ 0 & -1 \end{pmatrix}$  (b) $\mathbf{A} = \begin{pmatrix} -3 & -1 \\ 1 & -3 \end{pmatrix}$

**2.** Determine the fundamental matrix associated with each matrix in the following list:

(a) $\mathbf{A} = \begin{pmatrix} -3 & 5 \\ -1 & -3 \end{pmatrix}$  (b) $\mathbf{A} = \begin{pmatrix} -3 & -5 \\ -1 & -3 \end{pmatrix}$

**3.** Use Eq. (7.88) to show that $e^{-\mathbf{A}t} = \left(e^{\mathbf{A}t}\right)^{-1}$.

**4.** Let $B$ be the box whose vertices are $(1, 1)$, $(-1, 1)$, $(-1, -1)$, and $(1, -1)$. Use the fundamental matrices that you found in Exercise 1 to plot $\Psi\left(\frac{1}{4}\right)(B)$, $\Psi\left(\frac{1}{2}\right)(B)$, and $\Psi\left(\frac{3}{4}\right)(B)$. Compare and contrast how the box is mapped, and relate your findings to the eigenvalues of the original matrix. Also explain what happens to the images as $t$ becomes large.

**5.** The determinant of the matrix $\mathbf{A}$ is related to the way in which the flow generated by the system $\mathbf{x}' = \mathbf{Ax}$ changes the area enclosed by a box centered at the origin. The next few exercises explore the connection. We begin with the following theorems:

- If $\mathbf{A}$ and $\mathbf{B}$ are $n \times n$ matrices, then $\det(\mathbf{AB}) = (\det \mathbf{A})(\det \mathbf{B})$.

- Let $c$ be any scalar, and let $\mathbf{A}$ be an $n \times n$ matrix. Then
$$\det(c\mathbf{A}) = c^n \det \mathbf{A}.$$

(The proofs are omitted here but can be found in most linear algebra texts.)

(a) Find the determinant of the $2 \times 2$ and $3 \times 3$ identity matrices.

(b) Use the preceding theorems and your result in (a) to prove that if $\mathbf{A}$ is a nonsingular $2 \times 2$ or $3 \times 3$ matrix, then $\det(\mathbf{A}^{-1}) = 1/\det \mathbf{A}$. (This result holds for nonsingular $n \times n$ matrices as well.)

(c) Let $\mathbf{A}$ be a constant $2 \times 2$ matrix with distinct real eigenvalues $\lambda_1$ and $\lambda_2$, and let $\Psi(t)$ be the associated fundamental matrix. Find a formula for $\det \Psi(t)$. (Hint: See Definition 7.23.)

**6.** Let $\mathbf{A}$ be a real $2 \times 2$ matrix with real, distinct eigenvalues, and let $\Psi(t)$ be the associated fundamental matrix. Let $B$ be a box centered at the origin, and let
$$\mathbf{A}(B) = \{\mathbf{y} : \mathbf{y} = \mathbf{Ax} \text{ where } \mathbf{x} \in B\}.$$

The set $\mathbf{A}(B)$ is the image of $B$, that is, the set of points that results when each point in the box $B$ is multiplied by the matrix $\mathbf{A}$. It can be shown that area $\mathbf{A}(B) = |\det \mathbf{A}|$ area $B$. Use this fact and the

results of Exercise 5 to state and prove a theorem that gives conditions under which area $\Psi(t)(B)$ is an increasing function of $t$.

**7.** The results of Exercise 6 can be extended to the case in which $\mathbf{A}$ is a $2 \times 2$ matrix with complex conjugate eigenvalues.

(a) Let $\mathbf{R}(t)$ be a rotation matrix. (See Eq. (7.96).) What is $\det \mathbf{R}(t)$?

(b) Let $\Psi(t)$ be the fundamental matrix associated with $\mathbf{A}$. Find a formula for $\det \Psi(t)$. (Hint: See Definition 7.24.)

(c) Let $B$ be a box centered at the origin. Use your results in (a) and (b) and Exercise 5 to state and prove a theorem that gives conditions under which area $\Psi(t)(B)$ is an increasing function of $t$.

**8.** Suppose $\mathbf{A}$ is a $2 \times 2$ matrix with a double real eigenvalue. Let $\Psi(t)$ be the associated fundamental matrix, and let $B$ be a box centered at the origin. Under what circumstances is area $\Psi(t)(B)$ an increasing function of $t$? (See Exercises 5–7.)

**9.** Let $\mathbf{A}$ be a $2 \times 2$ matrix, and let $\Psi(t)$ be the associated fundamental matrix. Let $B$ be a box centered at the origin. Use the results of Exercises 5–8 to state and prove a theorem that gives conditions under which area $\Psi(t)(B)$ is a decreasing function of $t$.

**10.** Let $\mathbf{A}$ be a $2 \times 2$ constant matrix with associated fundamental matrix $\Psi(t)$, and let $B$ be a box centered at the origin. The resulting flow is said to be *area preserving* if area $\Psi(t)(B) = $ area $B$ for every $t$.

(a) Show that if the origin is a center, then the flow generated by the system $\mathbf{x}' = \mathbf{Ax}$ is area preserving.

(b) State and prove a theorem that gives conditions under which the flow generated by the system $\mathbf{x}' = \mathbf{Ax}$ is area preserving.

(c) If the flow is area preserving, is it true that the origin is a center?

**11.** Let $\Lambda = \begin{pmatrix} \lambda_1 & 0 \\ 0 & \lambda_2 \end{pmatrix}$. Prove that

$$\frac{d}{dt}e^{\Lambda t} = \Lambda e^{\Lambda t} = e^{\Lambda t}\Lambda.$$

**12.** Show that property (4) in Theorem 7.26 follows from properties (1) and (2).

**13.** Given $A = \begin{pmatrix} a & b \\ c & d \end{pmatrix}$, one might be tempted to define

$$e^{At} = \begin{pmatrix} e^{at} & e^{bt} \\ e^{ct} & e^{dt} \end{pmatrix},$$

analogous to Eq. (7.86). Suggest some reasons why this definition is not especially useful.

**14.** Use Euler's formula to derive Eq. (7.95) from Eq. (7.94).

**15.** If $f(t)$ is a differentiable function and $c$ is a constant, then

$$\frac{d}{dt}(cf(t)) = c\frac{df}{dt}.$$

This exercise asks you to prove that the analogous property holds for constant matrices and vector functions.

(a) Let $C$ be a constant $n \times n$ matrix, and let $\mathbf{f}(t)$ be a differentiable function from $\mathbf{R}$ to $\mathbf{R}^n$. Show that

$$\frac{d}{dt}(C\mathbf{f}(t)) = C\frac{d\mathbf{f}}{dt}.$$

(b) Let $C$ be a constant $n \times n$ matrix, and let $\mathbf{F}(t)$ be a differentiable $n \times n$ matrix function of $t$. Show that

$$\frac{d}{dt}(C\mathbf{F}(t)) = C\frac{d\mathbf{F}}{dt}.$$

**16.** Let $\Lambda = \begin{pmatrix} a & b \\ -b & a \end{pmatrix}$. This exercise asks you to show that it is reasonable to define

$$e^{\Lambda t} = e^{at}\begin{pmatrix} \cos bt & \sin bt \\ -\sin bt & \cos bt \end{pmatrix}.$$

(a) Show that the eigenvalues of $\Lambda$ are $a \pm bi$.

(b) Let $e^{\Lambda t}$ be defined as indicated above. Show that $e^{\Lambda \cdot 0} = I$, the identity matrix.

(c) Show that if $s$ and $t$ are any real numbers, then $e^{\Lambda(s+t)} = e^{\Lambda s}e^{\Lambda t}$.

(d) Show that

$$\frac{d}{dt}\left(e^{\Lambda t}\right) = \Lambda e^{\Lambda t}$$

and that $\Lambda e^{\Lambda t} = e^{\Lambda t}\Lambda$.

(e) Discuss how the above properties are analogous to those of the ordinary exponential function.

**17.** Use the results of Exercise 16 to prove Theorem 7.26 for the case in which $A$ is a real $2 \times 2$ matrix with complex conjugate eigenvalues $a \pm bi$.

**18.** Let $\Lambda = \begin{pmatrix} \lambda & 1 \\ 0 & \lambda \end{pmatrix}$. This exercise asks you to show that it is reasonable to define

$$e^{\Lambda t} = e^{\lambda t}\begin{pmatrix} 1 & t \\ 0 & 1 \end{pmatrix}.$$

(a) Show that $e^{\Lambda \cdot 0} = I$, the identity matrix.

(b) Show that if $s$ and $t$ are any real numbers, then $e^{\Lambda(s+t)} = e^{\Lambda s}e^{\Lambda t}$.

(c) Show that

$$\frac{d}{dt}\left(e^{\Lambda t}\right) = \Lambda e^{\Lambda t}$$

and that $\Lambda e^{\Lambda t} = e^{\Lambda t}\Lambda$.

(d) Discuss how the above properties are analogous to those of the ordinary exponential function.

**19.** Use the results of Exercise 18 to prove Theorem 7.26 for the case in which $A$ is a matrix with a double real eigenvalue $\lambda$ and only one corresponding eigenvector.

# 7.8   NONHOMOGENEOUS LINEAR SYSTEMS

This section discusses the solutions of first-order systems of the form

$$\mathbf{x}' = \mathbf{A}\mathbf{x} + \mathbf{f}(t), \tag{7.104}$$

where $\mathbf{A}$ is a constant matrix and $\mathbf{f}$ is a nonzero function of $t$. Equation (7.104) defines a nonhomogeneous system of equations, because $\mathbf{0}$ (the zero vector function) is not a solution. In this section, we describe an important theoretical result:

> *Every solution of Eq. (7.104) can be expressed in terms of an integral that involves the fundamental matrix.*

## Preliminaries

Before discussing Eq. (7.104) in more detail, we first consider the extension of some basic calculus results to vector functions.

**Definition 7.29**   *Let* $\mathbf{f}(t) = (f_1(t), f_2(t), \ldots, f_n(t))$ *be a continuous function from* $\mathbf{R}$ *to* $\mathbf{R}^n$. *Then*

$$\int \mathbf{f}(t)\, dt = \begin{pmatrix} \int f_1(t)\, dt \\ \int f_2(t)\, dt \\ \vdots \\ \int f_n(t)\, dt \end{pmatrix}.$$

*The function* $\mathbf{F}(t) = \int \mathbf{f}(t)\, dt$ *is called an antiderivative of* $\mathbf{f}$.

In other words, the integral of a vector-valued function is simply the integral of its component functions.

**Theorem 7.30**   *Let* $\mathbf{f}(t)$ *be a continuous function from* $\mathbf{R}$ *to* $\mathbf{R}^n$, *and let* $\mathbf{F}(t)$ *be an antiderivative of* $\mathbf{f}$. *Then*

$$\int_a^b \mathbf{f}(t)\, dt = \mathbf{F}(b) - \mathbf{F}(a). \tag{7.105}$$

Theorem 7.30 is just a statement of the fundamental theorem of calculus for vector-valued functions.

✦ *Example 1*  Let $\mathbf{f}(t) = (t, \cos t)$. An antiderivative of $\mathbf{f}$ is

$$\int \mathbf{f}(t)\, dt = \begin{pmatrix} \int t\, dt \\ \int \cos t\, dt \end{pmatrix} = \begin{pmatrix} \frac{1}{2}t^2 \\ \sin t \end{pmatrix}.$$

Theorem 7.30 implies that

$$\int_0^{\pi/2} \mathbf{f}(t)\, dt = \begin{pmatrix} \frac{1}{2}t^2 \\ \sin t \end{pmatrix}\Bigg|_{t=\pi/2} - \begin{pmatrix} \frac{1}{2}t^2 \\ \sin t \end{pmatrix}\Bigg|_{t=0} = \begin{pmatrix} \frac{1}{8}\pi^2 \\ 1 \end{pmatrix} - \begin{pmatrix} 0 \\ 0 \end{pmatrix}.$$  ∎

The product rule for differentiation can be extended to matrix- and vector-valued functions in a manner that is analogous to that for scalar functions.

**Theorem 7.31**  *Let $\mathbf{A}(t)$ and $\mathbf{x}(t)$ be a differentiable matrix- and vector-valued function of $t$, respectively, such that the product $\mathbf{y}(t) = \mathbf{A}(t)\mathbf{x}(t)$ is defined. Then*

$$\mathbf{y}'(t) = (\mathbf{A}(t)\mathbf{x}(t))' = \mathbf{A}'(t)\mathbf{x}(t) + \mathbf{A}(t)\mathbf{x}'(t),$$

*where the prime denotes differentiation with respect to $t$.*

## The Solutions of $\mathbf{x}' = \mathbf{A}\mathbf{x} + \mathbf{f}(t)$

We now turn to the solution of the equation

$$\mathbf{x}' = \mathbf{A}\mathbf{x} + \mathbf{f}(t), \tag{7.106}$$

where $\mathbf{A}$ is a constant matrix. Since Eq. (7.106) is a linear system, the superposition principle implies that the general solution of Eq. (7.106) has the form

$$\mathbf{x}(t) = \mathbf{x}_h(t) + \mathbf{x}_p(t).$$

The homogeneous solution has the form

$$\mathbf{x}_h(t) = \mathbf{\Psi}(t)\mathbf{c},$$

where $\mathbf{c}$ is a constant vector that depends on the initial condition. We will show that a particular solution has the form

$$\mathbf{x}_p(t) = \mathbf{\Psi}(t)\mathbf{u}(t), \tag{7.107}$$

where $\mathbf{u}$ is a function that depends on $\mathbf{f}$.

Let $\boldsymbol{\Psi}(t)$ be the fundamental matrix associated with $\mathbf{A}$. Then

$$(\boldsymbol{\Psi}(t)\mathbf{u}(t))' = \boldsymbol{\Psi}'(t)\mathbf{u}(t) + \boldsymbol{\Psi}(t)\mathbf{u}'(t) \qquad \text{by Theorem 7.31}$$
$$= \mathbf{A}\boldsymbol{\Psi}(t)\mathbf{u}(t) + \boldsymbol{\Psi}(t)\mathbf{u}'(t), \qquad\qquad (7.108)$$

by Theorem 7.26. We substitute Eq. (7.107) into Eq. (7.106) to obtain

$$(\boldsymbol{\Psi}(t)\mathbf{u}(t))' = \mathbf{A}\boldsymbol{\Psi}(t)\mathbf{u}(t) + \mathbf{f}(t).$$

However, this relation is equivalent to

$$\boldsymbol{\Psi}(t)\mathbf{u}'(t) + \mathbf{A}\boldsymbol{\Psi}(t)\mathbf{u}(t) = \mathbf{A}\boldsymbol{\Psi}(t)\mathbf{u}(t) + \mathbf{f}(t),$$

by Eq. (7.108). Therefore, $\boldsymbol{\Psi}(t)\mathbf{u}'(t) = \mathbf{f}(t)$, which implies that

$$\mathbf{u}(t) = \int_0^t \boldsymbol{\Psi}(s)^{-1}\mathbf{f}(s)\, ds. \qquad\qquad (7.109)$$

The integrand in Eq. (7.109) is well defined, because $\boldsymbol{\Psi}(s)$ is a fundamental matrix, which is always invertible. A particular solution of Eq. (7.106) is

$$\mathbf{x}_p(t) = \boldsymbol{\Psi}(t)\int_0^t \boldsymbol{\Psi}(s)^{-1}\mathbf{f}(s)\, ds. \qquad\qquad (7.110)$$

The following theorem summarizes the main result.

**Theorem 7.32** *Let $\mathbf{A}$ be a constant $n \times n$ matrix, and let $\boldsymbol{\Psi}(t)$ be the fundamental matrix associated with $\mathbf{A}$. Then the general solution of the system*

$$\mathbf{x}' = \mathbf{A}\mathbf{x} + \mathbf{f}(t)$$

*is*

$$\mathbf{x}(t) = \underset{e^{At}}{\boldsymbol{\Psi}(t)}\mathbf{x}_0 + \underset{e^{At}}{\boldsymbol{\Psi}(t)}\int_0^t \boldsymbol{\Psi}(s)^{-1}\mathbf{f}(s)\, ds, \qquad\qquad (7.111)$$

*where the initial condition is $\mathbf{x}_0 = \mathbf{x}(0)$.*

Equation (7.111) is of more theoretical than practical interest, because the integral may be difficult or impossible to evaluate in closed form. Nevertheless, Theorem 7.32 shows that an explicit solution of Eq. (7.106) exists, at least in principle, and it can be expressed in terms of an integral.

It is straightforward to verify that Eq. (7.111) satisfies the initial condition. We have $\boldsymbol{\Psi}(0) = \mathbf{I}$, so

$$\mathbf{x}(0) = \boldsymbol{\Psi}(0)\mathbf{x}_0 + \boldsymbol{\Psi}(0) \int_0^0 \boldsymbol{\Psi}(s)^{-1}\mathbf{f}(s)\,ds$$
$$= \mathbf{x}_0 + \mathbf{0}.$$

Theorem 7.26 implies that $\boldsymbol{\Psi}(t)^{-1} = \boldsymbol{\Psi}(-t)$. This special property makes the inverse of the fundamental matrix easy to calculate, provided that $\boldsymbol{\Psi}(t)$ is known. Equation (7.111) is called the *variation of parameters* or the *variation of constants* for linear systems, and it can be used directly to find explicit solutions in simple cases.

✦ *Example 2*   Find the solution of the initial value problem

$$\mathbf{x}' = \mathbf{A}\mathbf{x} + \mathbf{f}(t) = \begin{pmatrix} 3 & 2 \\ 2 & 3 \end{pmatrix}\mathbf{x} + \begin{pmatrix} 2e^{-2t} \\ 2e^{-t} \end{pmatrix}, \tag{7.112}$$

$\mathbf{x}(0) = (4, 2)$.

*Solution*   We first determine the fundamental matrix associated with $\mathbf{A}$. The eigenvalues of $\mathbf{A}$ are $\lambda_1 = 1$ and $\lambda_2 = 5$ with corresponding eigenvectors $\mathbf{p}_1 = (-1, 1)$ and $\mathbf{p}_2 = (1, 1)$. Let

$$\mathbf{P} = (\mathbf{p}_1 \quad \mathbf{p}_2) = \begin{pmatrix} -1 & 1 \\ 1 & 1 \end{pmatrix}.$$

Then the fundamental matrix is

$$\boldsymbol{\Psi}(t) = \begin{pmatrix} -1 & 1 \\ 1 & 1 \end{pmatrix} \begin{pmatrix} e^t & 0 \\ 0 & e^{5t} \end{pmatrix} \begin{pmatrix} -1 & 1 \\ 1 & 1 \end{pmatrix}^{-1}$$
$$= \frac{1}{2} \begin{pmatrix} e^{5t} + e^t & e^{5t} - e^t \\ e^{5t} - e^t & e^{5t} + e^t \end{pmatrix}.$$

The particular solution has the form $\mathbf{x}_p(t) = \boldsymbol{\Psi}(t)\mathbf{u}(t)$, where $\mathbf{u}(t)$ is given by Eq. (7.109). Theorem 7.26 implies that $\boldsymbol{\Psi}(s)^{-1} = \boldsymbol{\Psi}(-s)$, so

$$\mathbf{u}(t) = \int_0^t \boldsymbol{\Psi}(-s)\mathbf{f}(s)\,ds$$
$$= \int_0^t \frac{1}{2} \begin{pmatrix} e^{-5s} + e^{-s} & -e^{-s} + e^{-5s} \\ -e^{-s} + e^{-5s} & e^{-5s} + e^{-s} \end{pmatrix} \begin{pmatrix} 2e^{-2s} \\ 2e^{-s} \end{pmatrix} ds$$
$$= \int_0^t \begin{pmatrix} e^{-7s} + e^{-3s} - e^{-2s} + e^{-6s} \\ -e^{-3s} + e^{-7s} + e^{-6s} + e^{-2s} \end{pmatrix} ds$$
$$= \begin{pmatrix} -e^{-7t}/7 - e^{-3t}/3 + e^{-2t}/2 - e^{-6t}/6 + 1/7 \\ e^{-3t}/3 - e^{-7t}/7 - e^{-6t}/6 - e^{-2t}/2 + 10/21 \end{pmatrix}.$$

A particular solution of Eq. (7.112) is

$$\mathbf{x}_p(t) = \mathbf{\Psi}(t)\mathbf{u}(t) \tag{7.113}$$

$$= \frac{1}{2}\begin{pmatrix} e^{5t}+e^t & e^{5t}-e^t \\ e^{5t}-e^t & e^{5t}+e^t \end{pmatrix}\begin{pmatrix} -e^{-7t}/7 - e^{-3t}/3 + e^{-2t}/2 - e^{-6t}/6 + 1/7 \\ e^{-3t}/3 - e^{-7t}/7 - e^{-6t}/6 - e^{-2t}/2 + 10/21 \end{pmatrix}$$

$$= \begin{pmatrix} -10e^{-2t}/21 + e^{-t}/3 + 13e^{5t}/42 - e^t/6 \\ 4e^{-2t}/21 - 2e^{-t}/3 + e^t/6 + 13e^{5t}/42 \end{pmatrix}. \tag{7.114}$$

The general solution of Eq. (7.112) has the form

$$\mathbf{x}(t) = \mathbf{\Psi}(t)\mathbf{x}_0 + \mathbf{x}_p(t),$$

where $\mathbf{x}_p(t)$ is given by Eq. (7.114). Notice that $\mathbf{x}_p(0) = \mathbf{0}$ and $\mathbf{\Psi}(0) = \mathbf{I}$. (Why?) Therefore, the solution of the initial value problem is

$$\mathbf{x}(t) = \mathbf{\Psi}(t)\begin{pmatrix} 1 \\ 1 \end{pmatrix} + \mathbf{x}_p(t)$$

$$= \begin{pmatrix} 3e^{5t}+e^t \\ 3e^{5t}-e^t \end{pmatrix} + \begin{pmatrix} -10e^{-2t}/21 + e^{-t}/3 + 13e^{5t}/42 - e^t/6 \\ 4e^{-2t}/21 - 2e^{-t}/3 + e^t/6 + 13e^{5t}/42 \end{pmatrix}.$$

The solution of Eq. (7.112) is a linear combination of terms from $\mathbf{f}(t)$ and terms involving $e^t$ and $e^{5t}$. The latter functions are part of the solution of the homogeneous equation $\mathbf{x}' = \mathbf{A}\mathbf{x}$ and reflect the fact that the eigenvalues of $\mathbf{A}$ are 1 and 5.     ∎

The solution of nonhomogeneous linear systems usually requires lengthy calculations. Computer algebra systems are often helpful for deriving the solutions.

## *Exercises*

**1.** Let

$$\mathbf{x}' = \begin{pmatrix} 3 & -2 \\ 10 & -6 \end{pmatrix}\mathbf{x} + \begin{pmatrix} e^t \\ e^{-t} \end{pmatrix}.$$

(a) Find the solution that satisfies the initial condition $\mathbf{x}_0 = (1, 0)$.

(b) Describe the qualitative properties of the solution. Consider whether the solution oscillates,

goes to 0, or tends to infinity as $t$ becomes large.

**2.** Let

$$\mathbf{x}' = \begin{pmatrix} 3 & -2 \\ 10 & -6 \end{pmatrix}\mathbf{x} + \begin{pmatrix} \cos t \\ \sin t \end{pmatrix}.$$

(a) Find the solution that satisfies the initial condition $\mathbf{x}_0 = (2, 1)$.

(b) Describe the qualitative properties of the solution. Consider whether the solution oscillates, goes to 0, or tends to infinity as $t$ becomes large.

**3.** Let

$$\mathbf{x}' = \begin{pmatrix} -3 & 4 \\ -2 & 3 \end{pmatrix}\mathbf{x} + \begin{pmatrix} e^{-t} \\ t \end{pmatrix}.$$

(a) Find the solution that satisfies the initial condition $\mathbf{x}_0 = (-1, 1)$.

(b) Describe the qualitative properties of the solution. Consider whether the solution oscillates, goes to 0, or tends to infinity as $t$ becomes large.

**4.** Let

$$\mathbf{x}' = \begin{pmatrix} -1 & 3 \\ -2 & 4 \end{pmatrix}\mathbf{x} + \begin{pmatrix} 5\sin t \\ 5\cos t \end{pmatrix}.$$

(a) Find the solution that satisfies the initial condition $\mathbf{x}_0 = (-1, 1)$.

(b) Describe the qualitative properties of the solution. Consider whether the solution oscillates, goes to 0, or tends to infinity as $t$ becomes large.

**5.** Prove Theorem 7.30.

**6.** Prove Theorem 7.31.

# Chapter 8

# Nonlinear Differential Equations

Most of the discussion in the previous chapters has been concerned with the study of linear differential equations. Many aspects of exponential growth and oscillatory motion can be modeled accurately with linear differential equations, and the solutions of linear systems of first-order equations with constant coefficients can be characterized completely (see Section 7.8).

In contrast, nonlinear differential equations are much harder to analyze. It is usually not possible to find explicit solutions of nonlinear equations; the best that one can do is to compute numerical approximations of the solutions. As we will see later in this chapter, the solutions of nonlinear differential equations can exhibit extremely complex behavior. In fact, although nonlinear systems of equations can be very accurate models of deterministic phenomena, it may be difficult or impossible to make long-term predictions of the state of the system. The difficulty arises not because of poor modeling, but rather because errors in initial conditions have an intrinsic tendency to grow with time in many nonlinear differential equations.

Section 8.1 discusses some of the basic theory of nonlinear systems of differential equations. Sections 8.2–8.4 describe three examples of nonlinear systems that we call case studies. Each example incorporates progressively more complex behavior, as follows:

- The Lotka-Volterra predator-prey model shows that multiple fixed points can exist in nonlinear differential equations. Moreover, the stability of each fixed point can change as parameters in the model are varied.

- The van der Pol oscillator shows that periodic solutions can exist in nonlinear systems that are not periodically forced.

- The linear and nonlinear pendulum equations illustrate some differences between the solutions of linear and nonlinear systems.

Section 8.5 discusses the stability of fixed points in nonlinear systems of differential equations. Sections 8.6–8.7 are a brief introduction to the subject of bifurcation theory, which addresses the question of how the number and stability of fixed points changes with the parameters in a differential equation. Section 8.8 is an introduction to the subject of sensitive dependence on initial conditions. It shows that even simple deterministic systems can defy long-term prediction.

## 8.1 BASIC THEORY

In this chapter, we are interested in systems of differential equations of the form

$$\mathbf{x}' = \mathbf{f}(t, \mathbf{x}), \tag{8.1}$$

where $\mathbf{x}$ denotes a vector of dependent variables and $t$ (time) is the independent variable. Chapter 3 considers some of the basic scientific questions regarding the solutions of scalar differential equations. The same questions apply to systems of the form (8.1):

1. Under what circumstances does a solution of Eq. (8.1) exist that satisfies the initial condition $\mathbf{x}(t_0) = \mathbf{x}_0$?

2. If the solution exists, is it unique?

3. What is the interval of existence of the solution?

4. How do solutions depend on the initial condition? For instance, if $\mathbf{x}_0$ represents the "true" initial condition and $\hat{\mathbf{x}}_0$ represents a measured value, what can be said about the difference between the actual solution $\mathbf{x}(t)$ and the "predicted" solution $\hat{\mathbf{x}}(t)$ as a function of $t$?

These questions can be answered in varying degrees of detail, although much of the theory is beyond the scope of this text. The next subsection states a basic theorem that answers the questions about the existence and uniqueness of solutions.

### An Existence and Uniqueness Theorem

#### Two Motivating Examples

Example 1 shows that, in general, solutions of an initial value problem may not exist. Example 2 shows that a given initial value problem can have more than one solution in certain cases.

✦ *Example 1* Let

$$x_1' = 1/x_2,$$
$$x_2' = x_1^2 + \tan t. \tag{8.2}$$

Here,

$$\mathbf{x}' = \mathbf{f}(t, \mathbf{x}) = \begin{pmatrix} f_1(t, x_1, x_2) \\ f_2(t, x_1, x_2) \end{pmatrix} = \begin{pmatrix} 1/x_2 \\ x_1 + \tan t \end{pmatrix}.$$

It is not obvious how one can find an explicit solution of Eq. (8.2). Nevertheless, it is possible to determine some initial conditions for which one may be unable to find a solution, even in principle. For example, $\mathbf{f}(t, \mathbf{0})$ is undefined. Therefore, it may not be possible to find a function $\mathbf{x}(t)$ that satisfies Eq. (8.2) for all $t$, that satisfies the initial condition $\mathbf{x}(0) = \mathbf{0}$, and whose derivative $d\mathbf{x}/dt$ is defined at the origin. Similarly, $\mathbf{f}$ is undefined when $t = \pi/2$, so similar difficulties may exist for initial conditions of the form $\mathbf{x}(\pi/2) = \mathbf{x}_0$. ∎

✦ *Example 2* Let

$$x_1' = \sqrt{x_1},$$
$$x_2' = x_1 + 2x_2. \tag{8.3}$$

Let $\mathbf{x}(t_0) = \mathbf{0}$. The first equation in the system is uncoupled from the second. The discussion in Section 3.2 shows that $x_1(t) = 0$ and $x_1(t) = t^2/4$ are two different functions that satisfy the first equation in the system (8.3) and the initial condition $x_1(0) = 0$. The system (8.3) cannot have a unique solution, because there is more than one function $x_1(t)$ that satisfies the first equation in the list, and each choice of $x_1$ that satisfies the first equation also satisfies the second. ∎

## The Jacobian Matrix

Let $\mathbf{x} \in \mathbf{R}^n$ and let $\mathbf{f}(t, \mathbf{x})$ be a function from $\mathbf{R}^{n+1}$ to $\mathbf{R}^n$. The *Jacobian* matrix of partial derivatives of $\mathbf{f}$ with respect to $\mathbf{x}$ is the matrix $\mathbf{D_x f}$ whose $(i, j)$th entry is $\partial f_i/\partial x_j$. For instance, if $n = 2$, then

$$\mathbf{D_x f} = \begin{pmatrix} \partial f_1/\partial x_1 & \partial f_1/\partial x_2 \\ \partial f_2/\partial x_1 & \partial f_2/\partial x_2 \end{pmatrix}.$$

We say that the Jacobian matrix $\mathbf{D_x f}$ is continuous on a region $S$ if each of its entries is defined and continuous on the region $S$.

✦ *Example 3* The system in Example 2 can be written as

$$\mathbf{f}(t, \mathbf{x}) = \begin{pmatrix} f_1(t, x_1, x_2) \\ f_2(t, x_1, x_2) \end{pmatrix} = \begin{pmatrix} \sqrt{x_1} \\ x_1 + 2x_2 \end{pmatrix}.$$

The Jacobian matrix of $\mathbf{f}$ with respect to $\mathbf{x}$ is

$$\mathbf{D_x f} = \begin{pmatrix} \frac{1}{2}x_1^{-1/2} & 0 \\ 1 & 2 \end{pmatrix}.$$

Let $S$ be a cube centered at the origin. Then $\mathbf{D_x f}$ is *not* continuous on $S$, because $\partial f_1/\partial x_1 = 1/\left(2\sqrt{x_1}\right)$ is not continuous at $x_1 = 0$.

Now suppose that $S$ is a cube that does not contain the line $x_1 = 0$. Then $\mathbf{D_x f}$ is continuous on $S$, because the only points where any element of $\mathbf{D_x f}$ is discontinuous must lie on the line $x_1 = 0$. ∎

## An Existence and Uniqueness Theorem

**Theorem 8.1**  *Let* $\mathbf{x}' = \mathbf{f}(t, \mathbf{x})$. *Let $S$ be a cube centered at the initial condition* $(t_0, \mathbf{x}_0)$, *and suppose that* $\mathbf{f}$ *and* $\mathbf{D_x f}$ *are defined and continuous at each point in $S$. Then there exists a solution* $\mathbf{x}(t)$ *that satisfies the initial condition. It is unique as long as the graph of* $\mathbf{x}(t)$ *stays in $S$.*

Theorem 8.1 is a basic result. It says that a unique solution of the initial value problem exists provided that $\mathbf{f}$ and its Jacobian matrix are defined and continuous in some cube $S$ around the initial condition. Theorem 8.1 does not say *how* to find a solution, but it does say that exactly one solution exists when the hypotheses are satisfied.

Mathematical models of many phenomena consist of large systems of first-order ordinary differential equations that satisfy the hypotheses of Theorem 8.1. The theorem implies that in principle, such systems are predictable: If the initial condition is known, then the state of the system at future times can be forecast.

✦ *Example 4*  Consider the system (8.3). The solution of the initial value problem $\mathbf{x}(1) = (1, 1)$ is unique. We can regard the initial condition as a 3-vector of the form

$$(t_0, \mathbf{x}_0) = (t_0, x_1(t_0), x_2(t_0)) = (1, 1, 1).$$

It is possible to center a cube on the initial condition on which $\mathbf{f}$ and $\mathbf{D_x f}$ are continuous, as discussed in Example 3.

In contrast, a unique solution satisfying the initial condition $\mathbf{x}(0) = \mathbf{0}$ is *not* guaranteed, because the initial point is equivalent to

$$(t_0, \mathbf{x}_0)_1 = (t_0, x_1(t_0), x_2(t_0)) = (0, 0, 0),$$

and $\mathbf{D_x f}$ is not continuous at the origin. ∎

✦ *Example 5*   Consider the system (8.2). A unique solution of the initial value problem $(t_0, \mathbf{x}_0) = (\pi/2, 0, 0)$ is not guaranteed to exist, because the function $\mathbf{f}$ is not continuous at $t = \pi/2$. Similarly, a unique solution satisfying the initial condition $\mathbf{x}(t_0) = \mathbf{0}$ is not guaranteed to exist either, because $\mathbf{f}$ is not continuous at the origin. However, a unique solution is guaranteed for the initial condition $\mathbf{x}(0) = (1, 1)$, because there is a cube centered at the point $(0, 1, 1)$ on which $\mathbf{f}$ and $\mathbf{D_x f}$ are continuous.                         ∎

## Continuous Dependence on Initial Conditions

Suppose $\mathbf{x}' = \mathbf{f}(t, \mathbf{x})$, and suppose that $\mathbf{f}$ and $\mathbf{D_x f}$ are continuous functions. Theorem 8.1 guarantees the existence of a unique solution satisfying the initial condition $\mathbf{x}(t_0) = \mathbf{x}_0$. Suppose that the value of $\mathbf{x}_0$ is known only approximately. If $\hat{\mathbf{x}}(t)$ is the solution satisfying the initial condition $\hat{\mathbf{x}}(t_0) = \hat{\mathbf{x}}_0$, where $\hat{\mathbf{x}}_0$ is close to $\mathbf{x}_0$, then what can be said about the difference between the "predicted" solution $\hat{\mathbf{x}}(t)$ and the "true" solution $\mathbf{x}(t)$?

Chapter 3 discusses Gronwall's inequality, which states how quickly two solutions of a first-order scalar equation can separate. It is possible to state a version of Gronwall's inequality for systems of differential equations. However, we omit the details. The important point is the following:

*If $\mathbf{x}_0$ and $\hat{\mathbf{x}}_0$ are close together, then the corresponding solutions $\mathbf{x}(t)$ and $\hat{\mathbf{x}}(t)$ are close together for values of $t$ near $t_0$.*

Figure 8.1 illustrates the idea. Two trajectories are shown: one through the initial point $\mathbf{x}_0$ and another through the initial point $\hat{\mathbf{x}}_0$. The two trajectories stay close to each other for a while. They need not remain close forever.

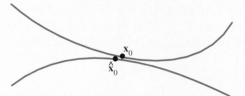

**FIGURE 8.1**   Two solution trajectories through nearby initial points.

A practical interpretation of this result is the following. Suppose we have an accurate mathematical model of some phenomenon, but the initial condition cannot be measured with complete precision. Provided that the measured initial condition is close to the "true" initial condition, then *predictions based on the measured initial condition are accurate in the short term.* Long-term predictions may not be accurate, as we will see later.

## Exercises

**1.** Use Theorem 8.1 to identify the systems in the following list for which a unique solution is guaranteed if $\mathbf{x}(0) = \mathbf{0}$. Also identify those systems for which the hypotheses of the theorem are not satisfied, and explain precisely why the hypotheses are not satisfied.

(a) $x_1' = x_1 + x_2, \quad x_2' = \sin x_1$

(b) $x_1' = x_1^{1/3}, \quad x_2' = x_1 + x_2$

(c) $x_1' = t^{1/3} + x_2, \quad x_2' = x_1 + t$

(d) $x_1' = x_1/t, \quad x_2' = x_1 \sin x_2$

**2.** For each system in the following list, give an example of an initial condition for which Theorem 8.1 guarantees a unique solution, and give an example of an initial condition for which Theorem 8.1 does not guarantee a unique solution.

(a) $x_1' = 1/x_1^2, \quad x_2' = x_1 + x_2$

(b) $x_1' = \sqrt{x_2}, \quad x_2' = \cos x_1$

(c) $x_1' = (x_1 + x_2)^{1/3}, \quad x_2' = (x_1 - 1)^{1/2}$

(d) $x_1' = \tan x_2, \quad x_2' = \cot x_1$

## 8.2   CASE STUDY I: COMPETING SPECIES MODELS

### Single Species Models

In Chapter 1, we consider models of the growth of a single population. In the simplest case, we assume that the growth rate, $dx/dt$, is proportional to the present size of the population. This assumption leads to the exponential growth equation

$$\frac{dx}{dt} = kx.$$

A more sophisticated model arises if we assume that the factor of proportionality $k$ is not a constant but rather is a function, $k(x)$, of the present size of the population. One possibility is that the growth rate decreases linearly with the population size, which gives

$$k(x) = a - bx$$

for some positive constants $a$ and $b$. If the population becomes large enough, then the growth rate becomes negative, reflecting the lack of available resources. This assumption about the growth rate leads to the logistic equation

$$\frac{dx}{dt} = (a - bx)x.$$

The logistic equation has a stable fixed point at $x = a/b$. In contrast, the exponential growth equation implies that the population grows without bound. In this respect, the logistic equation can be considered a more realistic population model than the exponential growth equation.

## Two Species Models

Now suppose that there are two species, $x$ and $y$, that compete with each other for resources (habitat, for example). Suppose that each population grows at a rate that is proportional to its present size but that the factor of proportionality depends on the size of each population. The corresponding model has the form

$$x' = k_1(x, y)x,$$
$$y' = k_2(x, y)y.$$

One of the simplest assumptions is that $k_1$ and $k_2$ are linear in both $x$ and $y$, that is,

$$k_1(x, y) = a_1 - b_1x - c_1y,$$
$$k_2(x, y) = a_2 - b_2x - c_2y,$$

where all the parameters are nonnegative. The growth rates $k_1$ and $k_2$ of both populations can become negative as the population in one or both species becomes large enough. For simplicity, let us suppose that all the parameters equal 1 with the exception of $c_1$ and $b_2$, so that the model becomes

$$x' = (1 - x - c_1y)x,$$
$$y' = (1 - y - b_2x)y. \tag{8.4}$$

Let us check whether the assumptions about $k_1$ and $k_2$ lead to a model whose basic features are biologically plausible.

- If one population is extinct initially, then it remains extinct. The other population grows according to the logistic equation and eventually reaches an equilibrium value at 1.

- Neither population can go from a positive value to a negative value. Obviously, only nonnegative population values make sense.

- Suppose, for example, that the population of $y$ is at some value, say $y = 0.5$. At that instant, the growth rate for $y$ is

$$k_2(x, 0.5) = 1 - 0.5 - b_2x = 0.5 - b_2x.$$

If $x$ is sufficiently small, then $k_2(x, 0.5) > 0$, reflecting the idea that the $y$ population faces little competition and so continues to grow. If $x$ is sufficiently large, then $k_2(x, 0.5) < 0$, which implies that competitive pressures cause the $y$ population to decline. (A similar analysis applies to $k_1$.) Hence, $k_1$ and $k_2$ are consistent with the basic assumptions about the growth rates of real populations.

- Analogous reasoning shows that no solution of Eqs. (8.4) can increase indefinitely. If $x$ and $y$ are large enough, then $x'$ and $y'$ are both negative. Hence sufficiently large populations decrease in size.

Figure 8.2 shows the direction field and some representative solution curves for Eqs. (8.4), where we have set $c_1 = 0.5$ and $b_2 = 0.8$. The slope lines in the direction field are drawn by using the procedure described in Section 6.2. Only the first quadrant is relevant for a population model.

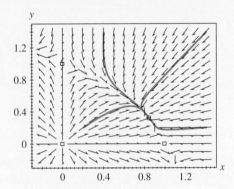

**FIGURE 8.2**    The direction field and some representative trajectories for Eqs. (8.4) for $c_1 = 0.5$ and $b_2 = 0.8$. The squares denote the location of fixed points.

The direction field reflects the basic features discussed above. First, notice that at any fixed value of the $y$ population with $0 < y < 1$, the slope lines point upward if $x$ is small (indicating that $y$ is growing) and downward if $x$ is large (indicating that $y$ is decreasing). Second, notice that the slope lines point toward the origin when $x$ and $y$ are both sufficiently large, indicating that neither population can grow indefinitely. The slope lines suggest that the trajectories starting at points along the coordinate axes remain confined there, implying that any population that is zero initially remains zero indefinitely. (An extinct population remains extinct.)

The origin corresponds to an initial condition that is a *fixed point*: If both populations are extinct, they remain extinct. Recall that **c** is a fixed point (also called an *equilibrium point*) of the autonomous system $\mathbf{x}' = \mathbf{f}(\mathbf{x})$ if $\mathbf{f}(\mathbf{c}) = \mathbf{0}$.

There are four fixed points for the system (8.4) when $c_1 = 0.5$ and $b_2 = 0.8$, each marked with a small box:

- $(0, 0)$, corresponding to two extinct populations;
- $(1, 0)$, corresponding to an extinct $y$;
- $(0, 1)$, corresponding to an extinct $x$; and
- $(x_0, y_0) = (5/6, 1/3)$.

The last fixed point corresponds to a *competitive coexistence*—both populations thrive at constant values.

In general, the fourth fixed point of Eqs. (8.4) is

$$(x_0, y_0) = \left( \frac{1-c_1}{1-c_1 b_2}, \frac{1-b_2}{1-c_1 b_2} \right).$$

Depending on the values of $c_1$ and $b_2$, which represent the effect of competition on the two species, the two populations modeled by Eqs. (8.4) may or may not have a competitive coexistence. The model makes sense only for nonnegative values of the populations, so both components of the fixed point must be positive for there to be a competitive coexistence.

The phase portrait in Figure 8.2 suggests that if neither population is extinct initially, then the populations eventually reach a competitive coexistence in which $x = 5/6$ and $y = 1/3$.

The phase portrait is a powerful tool for the graphical analysis of solutions. Although explicit solutions of Eqs. (8.4) cannot be found readily, the qualitative behavior of the model is easy to discern. In this example, the phase portrait suggests that if neither population is extinct at the start, then both populations eventually approach a constant value—neither species dies out.

The phase portraits in Fig. 8.3 show contrasting situations. Figure 8.3(a), corresponding to the case in which $c_1 = 0.5$ and $b_2 = 1.8$, suggests that most trajectories eventually approach the equilibrium at $(1, 0)$—the $y$ population dies out. Figure 8.3(b), corresponding to the case in which $c_1 = 1.5$ and $b_2 = 1.8$, suggests that in most cases, one or the other population eventually dies out, depending on the initial condition.

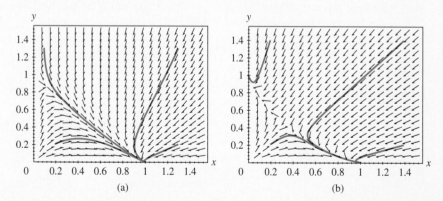

**FIGURE 8.3** (a) Phase portrait for Eqs. (8.4) with $c_1 = 0.5$, $b_2 = 1.8$; (b) with $c_1 = 1.5$, $b_2 = 1.8$.

## Stable and Unstable Fixed Points

The system (8.4) has as many as four fixed (equilibrium) points. Depending on the values of the coefficients, a given fixed point may be *asymptotically stable* in the sense of Definition 7.20, or it may be *unstable*, in the sense that trajectories starting from some nearby initial conditions eventually leave the neighborhood of the fixed point and do not return.

In Fig. 8.2, for instance, there is one stable fixed point (sink), located at (5/6, 1/3). Figure 8.2 depicts three unstable fixed points: $(0, 0)$, $(0, 1)$, and $(1, 0)$. Each of these fixed points is unstable because there exist initial conditions near each fixed point for which the solution trajectory does not return to the fixed point. For instance, suppose the initial populations are $(x_0, y_0) = (0.01, 0.01)$, near the fixed point at the origin. The direction field implies that both populations grow until they reach the competitive equilibrium at $(5/6, 1/3)$. Hence, the origin is an unstable fixed point.

In Fig. 8.3(b) there is a fixed point in the first quadrant (see Exercise 3), but it is not stable. Instead, $(1, 0)$ is attracting. In Fig. 8.3(b), $(0, 1)$ also is attracting.

The situations depicted in Figs. 8.2 and 8.3 raise several questions that we will address in subsequent sections.

- Can a fixed point disappear as parameters in the model assume different values? If so, what does the disappearance imply about the behavior of the underlying system?

- Can a fixed point change stability as the parameters assume different values? If so, is there a mathematical theory to explain why this happens?

- Does a given fixed point have "attracting directions" and "repelling directions"? If so, how can they be defined?

- Is there a way to determine analytically whether a given fixed point is stable?

## Nullclines

A more sophisticated graphical analysis of the competing species model can be undertaken by examining the regions of the phase diagram where the various populations are increasing or decreasing. Figure 8.4 shows the direction field for the model

$$x' = (1 - x - 0.5y)x,$$
$$y' = (1 - y - 0.8x)y,$$

(8.5)

which is Eq. (8.4) with $c_1 = 0.5$ and $b_2 = 0.8$.

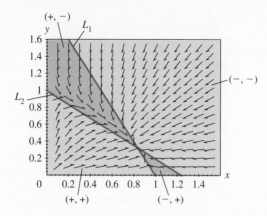

**FIGURE 8.4** The direction field and nullclines for Eqs. (8.5).

The indicated line segments are the *nullclines*. A nullcline is a set of points where one of the derivatives is 0. For example, the nullclines corresponding to $x' = 0$ are the vertical axis and the line $L_1$, defined by the equation $1 - x - 0.5y = 0$. Similarly, the nullclines corresponding to $y' = 0$ are the horizontal axis and the line $L_2$, defined by the equation $1 - y - 0.8x = 0$.

The nullclines divide the plane into regions. In each region, exactly one of the following possibilities holds:

1. $x' > 0$ and $y' > 0$, labeled $(+, +)$; both populations increase.
2. $x' > 0$ and $y' < 0$, labeled $(+, -)$; population $x$ increases while population $y$ decreases.
3. $x' < 0$ and $y' > 0$, labeled $(-, +)$; population $y$ increases while population $x$ decreases.
4. $x' < 0$ and $y' < 0$, labeled $(-, -)$; both populations decrease.

The line $L_1$ divides the first quadrant into two pieces. Notice that $x' > 0$ at points below $L_1$ and $x' < 0$ at points above $L_1$. Similarly, $L_2$ divides the first quadrant into two pieces. Here, $y' > 0$ at points below $L_2$ and $y' < 0$ at points above $L_2$. Notice that fixed points in Fig. 8.4 occur wherever a nullcline for $x'$ and a nullcline for $y'$ intersect. (See Exercise 4.)

If either population is sufficiently large, then the state of the system corresponds to a point in region $(-, -)$. The nullclines provide another way to verify that neither population can grow indefinitely if the growth rate is governed by the system (8.5).

## Exercises

**1.** Consider Fig. 8.3(a), which illustrates the direction field for the system

$$x' = (1 - x - 0.5y)x,$$
$$y' = (1 - y - 1.8x)y.$$

(a) Identify the nullclines where $x' = 0$.

(b) Identify the nullclines where $y' = 0$.

(c) Argue that the nullclines divide the first quadrant into three regions. Label each region as $(+, +)$, $(+, -)$, $(-, +)$, or $(-, -)$ according to the signs of $x'$ and $y'$. (See Fig. 8.4 for an example.)

(d) Argue that region $(+, -)$ is a *trapping region*. That is, explain why the trajectories starting at initial conditions in region $(+, -)$ are confined to region $(+, -)$ for all $t > 0$. Moreover, show that all trajectories starting from initial conditions in region $(+, +)$ eventually enter region $(+, -)$, where they remain forever thereafter.

(e) Explain why your analysis in (d) proves that the $y$ population eventually becomes extinct if the population growth is governed by this differential equation model.

**2.** Consider Fig. 8.3(b), which illustrates the direction field for the system

$$x' = (1 - x - 1.5y)x,$$
$$y' = (1 - y - 1.8x)y.$$

(a) Identify the nullclines where $x' = 0$.

(b) Identify the nullclines where $y' = 0$.

(c) Argue that the nullclines divide the first quadrant into four regions. Label each region as $(+, +)$, $(+, -)$, $(-, +)$, or $(-, -)$ according to the signs of $x'$ and $y'$. (See Fig. 8.4 for an example.)

(d) Argue that there are two trapping regions, $R_1 = (-, +)$ and $R_2 = (+, -)$. That is, explain why the trajectories starting at initial conditions in region $R_1$ are confined to region $R_1$ for all $t > 0$ and likewise for region $R_2$.

(e) Find the fixed points. Observe that, if the initial condition lies in the $(+, +)$ region, then trajectories eventually enter either region $R_1$ or region $R_2$, where it remains forever thereafter. Identify the one exceptional solution trajectory in that region. What happens to initial conditions in the $(-, -)$ region?

(f) Explain why your analysis in (e) shows that one of the populations most likely becomes extinct, depending on the initial condition, if the growth of the populations is governed by this differential equation model.

**3.** Figure 8.3(b) shows a case in which there is a competitive coexistence point, but the corresponding fixed point is not stable. Suppose the system initially is at the fixed point. How sensitive is such an ecological system to external disturbances? Discuss what such a model implies for an ecological system consisting of the two species.

**4.** Consider the system

$$x' = (1 - x - 0.5y)x,$$
$$y' = (1 - y - 0.8x)y.$$

(a) One nullcline corresponding to $x' = 0$ is given by $x = 1 - 0.5y$. When a trajectory crosses this nullcline, the derivative of its $x$ component is 0. Explain why this does not imply that $x(t) = $ constant for most initial conditions on this nullcline.

(b) Find solution trajectories that go through the nullclines, and give a biological interpretation of what it means when the trajectory intersects the nullcline.

(c) Explain why there must be a fixed point where a nullcline for $x'$ and a nullcline for $y'$ intersect.

**5.** Let

$$x' = (1 - x + ay)x,$$
$$y' = (1 - y - bx)y,$$

$$(8.6)$$

where $a$ and $b$ are positive constants.

(a) The system (8.6) models one of the following biological situations. Which one is it, and why?

- $x$ preys on $y$.
- $y$ preys on $x$.
- $x$ and $y$ prey on each other.

(b) Determine all the fixed (equilibrium) points.

(c) Let $a = 0.5$ and $b = 0.6$. Determine which of the fixed points correspond to a biologically plausible situation. Then determine the stability of each biologically plausible fixed point by examining whether trajectories starting from nearby initial conditions gradually get closer to the fixed point as $t$ becomes large.

(d) What happens to typical trajectories in (c)? How would you characterize the eventual outcome of the population model: competitive coexistence or extinction of one of the species?

(e) Repeat (c) for $a = 0.5$ and $b = 1.6$.

(f) Repeat (d) for $a = 0.5$ and $b = 1.6$.

**6.** Let

$$x' = (\lambda - ay)x,$$
$$y' = (bx - \rho)y. \tag{8.7}$$

(a) Assume that the parameters $\lambda$, $\rho$, $a$, and $b$ are all positive. The model (8.7), called the Lotka-Volterra model, describes the growth of a population $y$ that preys on $x$. (See Section 6.1.) Each population grows at a rate that is proportional to its present size, but the factors of proportionality are not constant; instead, they are given by the parenthesized terms. Explain the biological relevance of each of the terms $\lambda$, $ay$, $bx$, and $\rho$.

(b) Find all the fixed points.

(c) Let $\lambda = 3$, $a = 1$, $\rho = 0.2$, and $b = 0.05$. Sketch a phase portrait.

(d) Use an appropriate numerical method to compute and plot the trajectories corresponding to the initial conditions $(x_0, y_0) = (0.5, 3)$ and $(x_0, y_0) = (0.5, 0.5)$. Do the populations fluctuate periodically? Explain.

(e) Let $x_0 = 0.5$ and $y_0 = 3$. Plot $x(t)$ and $y(t)$ as functions of $t$. What does the phase lag between the graphs of $y$ and $x$ suggest about the growth of the two populations? Is this feature biologically plausible?

(f) Suppose that $x$ is an insect pest and that $y$ is a biological control. What would happen as the result of an application of pesticide, assuming that the pesticide kills most, but not all, of each population?

**7.** Consider the Lotka-Volterra model (8.7) discussed in Exercise 6.

(a) What happens to the prey if the predator becomes extinct?

(b) What happens to the predator if the prey becomes extinct?

(c) Modify the model (8.7) so that it yields a biologically plausible scenario when the predator becomes extinct.

## 8.3 CASE STUDY II: THE VAN DER POL OSCILLATOR

Chapter 5 discusses the linear second-order equation

$$x'' + ax' + bx = 0. \tag{8.8}$$

The solutions of Eq. (8.8) are linear combinations of exponential functions of the form $e^{\lambda t}$, where

$$\lambda = -\tfrac{1}{2}a \pm \tfrac{1}{2}\sqrt{a^2 - 4b}.$$

If $a > 0$ and $b > 0$, then $\lambda$ is negative (or has negative real part), and the solutions of Eq. (8.8) decay exponentially to zero, regardless of the initial conditions. In this case, we can regard the term $ax'$ as a linear frictional force that opposes the motion. If $a < 0$ and $b > 0$, then $\lambda$ is positive (or has a positive real part). Depending on the initial condition, the solution of Eq. (8.8) has a term that grows exponentially with time. When $a < 0$, we may regard the term $ax'$ as a kind of "feedback" or "negative resistance" term that enhances the motion.

The van der Pol oscillator is related in spirit to the linear pendulum equation (8.8). However, the frictional term is nonlinear. The van der Pol equation is given by

$$x'' + (\alpha + x^2)x' + x = 0, \tag{8.9}$$

where $\alpha$ is a constant. Equation (8.9) appears in a variety of physical situations. It models an electrical circuit that acts as a nonlinear resistor. The van der Pol equation also arises in models of electrical conduction in cardiac tissue and the deflection of certain metallic beams.

It is particularly interesting to compare the properties of the solutions of Eq. (8.9) to those of the linear second-order equation (8.8). As we will see, the qualitative behavior of a nonlinear differential equation can be very different from that of a related linear differential equation. The analysis of the van der Pol equation (8.9) consists of two cases, depending on the sign of $\alpha$.

**Case 1: $\alpha > 0$.**   In this case, $\alpha + x^2 > 0$ for every $x$. Therefore, the frictional term $(\alpha + x^2)x'$ always opposes the motion. In other words, the magnitude of the friction is always at least as large as $\alpha x'$, so the solutions decay to 0 in much the same way as those of the linear equation (8.8).

The solid curve in Fig. 8.5(a) is a time series plot of the solution of Eq. (8.9) when $\alpha = 1$. The initial conditions are $x(0) = 1$ and $x'(0) = 0$. For comparison, the dashed curve in Fig. 8.5(a) shows the solution of the linear oscillator equation

$$x'' + x' + x = 0, \tag{8.10}$$

whose damping term is never greater in absolute value than the damping term in Eq. (8.9). The same initial conditions are used. Both solutions appear to be pseudoperiodic, and both solutions appear to decay to 0 at comparable rates.

Figure 8.5(b) shows the phase portrait of the first-order system corresponding to Eq. (8.9) when $\alpha = 1$. All trajectories appear to spiral into the origin. Hence the origin is a spiral sink, just as it is for Eq. (8.10).

**Case 2: $\alpha < 0$.**   In this case the expression $\alpha + x^2$ is negative if $x^2$ is small enough. Thus, when $x$ is close enough to 0, the nonlinear term $(\alpha + x^2)x'$ can be regarded as a force that acts in the direction of the motion and increases the velocity. If $x$ is far enough

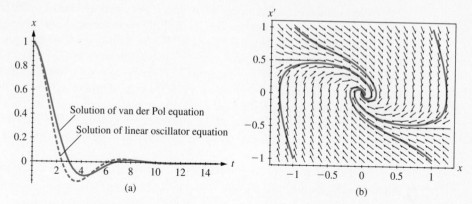

**FIGURE 8.5**   (a) Solid curve: a solution of the van der Pol equation (8.9) when $\alpha = 1$. Dashed curve: a solution of the linear oscillator equation (8.10) for the same initial conditions. (b) Phase portrait for the van der Pol equation.

away from 0, then $\alpha + x^2 > 0$, so the nonlinear term opposes the motion and slows it down.

Figure 8.6(a) shows the phase portrait of the first-order system corresponding to Eq. (8.9) when $\alpha = -1$. Two trajectories are shown: one starting from the initial condition $(-2, 2)$ and the other from $(0.1, 0.1)$. The two curves approach a common loop, called a *limit cycle*. A limit cycle is another kind of attractor. Trajectories starting

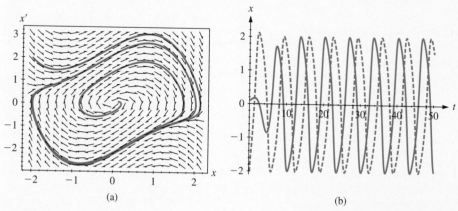

**FIGURE 8.6**   (a) Phase portrait of Eq. (8.9) when $\alpha = -1$. (b) Time series plot of two solutions.

from typical initial conditions form spirals that come closer and closer to the limit cycle as $t \to \infty$, but they never land exactly on the limit cycle for any finite value of $t$.

Limit cycles imply periodic behavior. Figure 8.6(b) shows a time series plot of the solutions corresponding to the same initial conditions as in Fig. 8.6(a). The solid curve corresponds to the initial condition $(x_0, x_0') = (0.1, 0.1)$, and the dashed curve corresponds to the initial condition $(x_0, x_0') = (-2, 2)$. There is an initial transient, but as the trajectories approach the limit cycle, the solutions approach a periodic function. In this example, the graph of each solution appears to be periodic for $t$ larger than about 15. The graphs of the two solutions are different in phase, but they have the same period.

Here are some of the most important differences in the behavior of the van der Pol equation (8.9) for the two values of $\alpha$ that we have considered.

- When $\alpha = 1$, the origin is asymptotically stable (it is a spiral sink). The origin is an attractor, and trajectories approach the origin as $t \to \infty$.

- When $\alpha = -1$, the origin is unstable; trajectories starting from initial conditions near the origin eventually move away, never to return. The limit cycle is an attractor, because most trajectories approach it as $t \to \infty$.

The solutions of nonlinear equations like the van der Pol oscillator often are very different in nature from those of linear differential equations. In particular, the van der Pol equation is an example of an autonomous, unforced differential equation whose solutions can oscillate. In contrast, the solutions of linear second-order equations always decay to 0 if there is any friction. (See also Exercise 1.)

## *Exercises*

**1.** The behavior of solutions of the van der Pol equation when $\alpha < 0$ is very different from that of the solutions of the linear equation $x'' + ax' + x = 0$ when $a < 0$. The van der Pol oscillator when $\alpha = -1$ is

$$x'' + \left(-1 + x^2\right)x' + x = 0. \qquad (8.11)$$

(a) Find the first-order system corresponding to Eq. (8.11).

(b) Consider the linear equation $x'' - x' + x = 0$. Find the corresponding system of first-order equations.

(c) Draw a phase portrait for each of the systems in (a) and (b). For each system, compute solu-

tion trajectories starting from the initial conditions $(0.1, 0.1)$ and $(1, 1)$ and add them to your plot.

(d) Compare and contrast the behavior between the two systems. Consider questions such as the following: Do solutions stay bounded? Do solutions oscillate periodically or quasiperiodically? Can solutions be regarded as having a transient part and a steady-state part?

(e) Does the equation in (b) correspond to a mass-spring system? Explain.

**2.** Let $x'' + (\alpha + x^2)x' + x = 0$. The numerical experiments outlined as follows suggest that the origin changes from a source to a sink as $\alpha$ goes from a negative value to a positive value. This qualitative change is called a *bifurcation*. A brief introduction to bifurcations is given in Sections 8.6 and 8.7.

(a) Let $\alpha = 0$. Determine the corresponding first-order system of equations and draw a direction field.

(b) Compute the solution trajectories starting from the initial conditions $(1, 0)$, $(0.1, 0)$, and $(0.05, 0)$. Try other initial conditions as you see fit.

(c) On the basis of your numerical experiments in (b), can you classify the origin as a sink, as a source, or is it hard to tell?

(d) Repeat (a)–(c) for $\alpha = -0.1$.

(e) Repeat (a)–(c) for $\alpha = +0.1$.

**3.** This exercise considers some of the geometric properties of limit cycles.

(a) Suppose that the closed curve in Fig. 8.7 is an attracting limit cycle. Draw trajectories that start from each of the indicated points and approach the limit cycle as $t$ becomes large. Show that you can draw the trajectories so that they do not cross each other or touch the limit cycle.

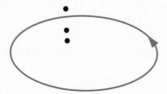

**FIGURE 8.7**   For Exercise 3.

_____

(b) Use a sketch that is similar in spirit to (a) to show that trajectories near a spiral sink can come very close to the fixed point for large values of $t$ but that the trajectories need not touch

each other or intersect the fixed point at any finite value of $t$.

(c) Show that the hypotheses of Theorem 8.1 apply to the van der Pol equation (8.9) for all values of the parameter $\alpha$.

(d) What can you conclude about the trajectories in the phase portraits in Figs. 8.5(b) and 8.6(a), provided that the numerical approximations of the solutions are sufficiently accurate?

**4.** The Fitzhugh-Nagumo equation is a model of the electrical excitation of a nerve cell membrane, given by

$$\begin{aligned} v' &= f(v) - w, \\ w' &= \alpha(v - w), \end{aligned} \tag{8.12}$$

where $\alpha$ is a constant and $f$ is a suitable function. For this exercise, let $\alpha = 0.01$ and

$$f(v) = v(0.25 - v)(v - 1).$$

The function $v(t)$ represents the electrical potential (voltage) across a nerve cell membrane, and $w(t)$ is the current through the membrane. Equation (8.12) models the following behavior, which is often seen in nature. After a sufficiently small electrical potential is applied to the nerve cell, then the potential decays to 0 rapidly. After a sufficiently large potential is applied, the potential remains large (or even increases), and current passes through the membrane for some interval of time before the potential decays to 0. We say that the nerve cell *fires* in response to a sufficiently large applied potential.

(a) Let the initial condition be $(v_0, w_0)$. Use a numerical solver to draw a time series plot of $v(t)$ for each of the following initial conditions: $(0.1, 0)$, $(0.2, 0)$, $(0.4, 0)$, and $(0.8, 0)$.

(b) For which of the four initial conditions in (a) might you say that the nerve cell fires? Explain your reasoning.

(c) How long does it take for the electrical potential to return to a small value for each of the initial conditions in (a)? Find a suitable "small" value and compare the interval of time needed for the potential to decay to this value for each initial condition.

(d) Suppose $w_0 = 0$, that is, there is no current through the nerve cell initially, and suppose $v_0 > 0$. The results in (a) suggest that if $v_0$ is sufficiently small, then the cell does not fire. If $v_0$ is larger than some threshold value, then the cell fires. Estimate this threshold value by experimenting with some appropri-

ately chosen initial conditions of the form $(v_0, 0)$.

5. Consider the Fitzhugh-Nagumo model in Exercise 4. (Use the same $\alpha$ and $f$.)

(a) Draw a phase portrait for $-0.2 \le v \le 1$ and $-0.02 \le w \le 0.16$.

(b) Draw the nullclines for $v$ and $w$.

(c) Discuss how the nullclines allow you to determine the threshold value of the initial voltage, discussed in Exercise 4(d), above which the nerve cell fires. (Assume that $w_0 = 0$.)

## 8.4  CASE STUDY III: THE LINEAR AND NONLINEAR PENDULUM

Section 5.1 discusses the derivation of the equation for a damped linear pendulum, given by

$$mx'' + \left(\frac{c}{\ell}\right)x' + \left(\frac{mg}{\ell}\right)x = 0, \tag{8.13}$$

where $x(t)$ is the angular position of the pendulum at time $t$, $m$ is the mass of the bob, $c$ is the coefficient of friction at the pivot, $\ell$ is the length of the pendulum, and $g$ is the acceleration due to gravity. The key assumption in the derivation—which leads to a linear differential equation—is that the displacement $x$ from the vertical (straight down) position is small. The acceleration on the bob due to gravity is proportional to $\sin x$; but if $x$ is small, then $\sin x \approx x$. For this reason, Eq. (8.13) is an accurate approximation of the motion of a real pendulum only if the displacement of the pendulum from the vertical is small.

In this section, we drop the assumption that $x$ is small and consider the nature of solutions of the nonlinear equation

$$mx'' + \left(\frac{c}{\ell}\right)x' + \left(\frac{mg}{\ell}\right)\sin x = 0.$$

As will be discussed, there are three qualitatively different types of solutions of the nonlinear pendulum; two are periodic and one is aperiodic. The period of the periodic solutions depends on the initial condition.

We also consider the case in which an external periodic forcing is applied to the pendulum. (Imagine a torque applied at the pivot.) For simplicity, we assume that $m = 1$, relabel the remaining constants, and write the linear equation as

$$x'' + ax' + bx = f(t) \tag{8.14}$$

and the nonlinear equation as

$$x'' + ax' + b\sin x = f(t). \tag{8.15}$$

Equations (8.14) and (8.15) describe a linear pendulum and nonlinear pendulum, respectively, with an external forcing. The solutions of Eq. (8.14) exhibit beating and resonance when $f$ is a linear combination of sine and cosine functions. In contrast, the solutions of Eq. (8.15) can exhibit erratic and unpredictable behavior for the same forcing function $f$. The remainder of this section shows examples of the differences between the solutions of Eqs. (8.14) and (8.15).

## The Undamped Case

We first consider the case in which there is no friction ($a = 0$) and no forcing. When $b = 1$, Eqs. (8.14) and (8.15) become

$$x'' + x = 0 \tag{8.16}$$

and

$$x'' + \sin x = 0, \tag{8.17}$$

respectively.

### Small-Amplitude Oscillations

Figures 8.8(a) and 8.8(b) show the phase portraits of Eqs. (8.16) and (8.17), respectively, when the amplitude of the displacement is small. Figure 8.8(c) shows time series plots of the solutions satisfying the initial conditions $x(0) = -1/2$ and $x'(0) = 0$. (The time series plots correspond to the outermost trajectory in Figs. 8.8(a) and 8.8(b).) In this case, $\sin x \approx x$, and the two phase portraits and time series plots are nearly identical. The solution of Eq. (8.16) satisfying these initial conditions is

$$x(t) = -\tfrac{1}{2}\cos t.$$

Although we do not know the analytical form of the solution of Eq. (8.17), the numerical solution suggests that it too is approximately $2\pi$-periodic with amplitude $1/2$.

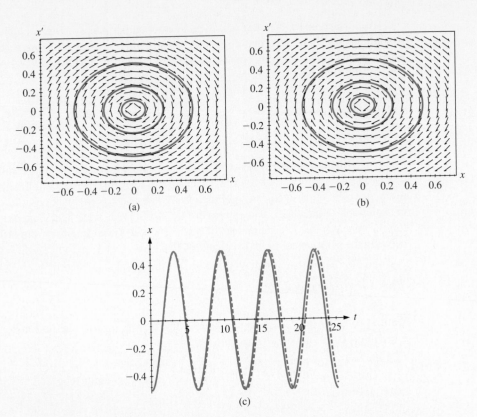

(a)          (b)

(c)

**FIGURE 8.8**   (a) The phase portrait for Eq. (8.16). (b) The phase portrait for Eq. (8.17). (c) Corresponding time series plots. The solution of Eq. (8.16) is shown as a solid curve, and the solution of Eq. (8.17) is shown as a dashed curve.

## Large-Amplitude Oscillations

The phase portraits of the linear and nonlinear pendulum are quite different when the amplitude of the motion is large. Figure 8.9(a) shows the phase portrait of the linear pendulum, Eq. (8.16), and Fig. 8.9(b) shows the phase portrait of the nonlinear pendulum, Eq. (8.17), for $|x| \leq 4\pi/3$.

The solutions of the linear pendulum exhibit simple harmonic motion. They are all $2\pi$-periodic, and the amplitude of the oscillations depends on the initial conditions. For each initial condition, the linear pendulum rotates anticlockwise for half a period and clockwise for half a period.

The situation for the nonlinear pendulum is considerably different. In particular, there are fixed points located at $x = \pm\pi$, $\pm 2\pi$, $\ldots$. Figure 8.9(b) shows three fixed

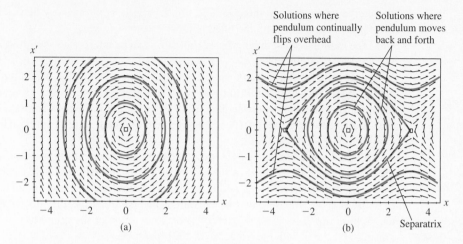

**FIGURE 8.9**   Phase portraits for (a) Eq. (8.16); (b) Eq. (8.17).

points, each marked with a small box. Since $x$ represents the angular position of the pendulum from the straight-down position, the fixed points at $x = \pm\pi$ correspond to a pendulum that is perfectly balanced in a straight-up position. (Because there are $2\pi$ radians in a circle, $x = -\pi$ and $x = +\pi$ refer to the same angular position.)

The existence of the fixed points at $x = \pm\pi$ is a significant difference between the nonlinear pendulum and the linear pendulum. The linear system (8.16) has only one fixed point, corresponding to the vertical (straight-down) position. In contrast, the nonlinear pendulum equation (8.17) admits the possibility that the pendulum can be balanced in a straight-up as well as a straight-down position.

Now suppose that the nonlinear pendulum starts from the straight-down position ($x = 0$) with some initial angular velocity, $x'(0) = v_0$. If $|v_0|$ is small, then the pendulum oscillates back and forth like the pendulum in a grandfather clock. The corresponding trajectories in the phase portrait in Fig. 8.9(b) form loops, indicating periodic motion in which the pendulum moves anticlockwise for half a period and clockwise for half a period.

If $|v_0|$ is large, then the trajectories in the phase portrait of the nonlinear pendulum no longer form loops. Instead, the solutions imply that the nonlinear pendulum always rotates in the same direction, repeatedly flipping over. (This behavior is consistent with one's physical intuition.) For instance, the topmost trajectory shown in Fig. 8.9(b) implies that $x' > 0$ always: The pendulum always rotates anticlockwise. The linear pendulum does not exhibit such solutions.

The phase portrait is separated into two regions: one where the trajectories form closed loops and another where they do not. The dividing line between the two regions is called a *separatrix*. In this example, the separatrix is a trajectory that connects the fixed point at $x = -\pi$ with the fixed point at $x = \pi$.

The motion corresponding to the separatrix can be interpreted as follows. Suppose we start the pendulum near $x = -\pi$ with a small initial velocity (say $x_0' > 0$). Physically, the initial condition corresponds to a situation in which the pendulum is placed nearly in the upright (equilibrium) position and given a small push. (If $x_0' > 0$, then the push is in the anticlockwise direction.) If the initial velocity is adjusted precisely, then the pendulum swings nearly 360 degrees and eventually comes to rest at the upright equilibrium position.

Clearly, such a balancing act would be difficult to perform in practice. However, the mathematical description of the motion is consistent with one's physical intuition: The separatrix is a single line, and if the initial position and velocity do not correspond to points that are precisely on the separatrix, then the true trajectory does not return to the equilibrium at $x = -\pi$. Exercise 4 explores some of the properties of the phase portrait associated with the existence of the separatrix.

## Forced, Undamped Motion

We next consider solutions of the undamped pendulum equations under the influence of a periodic forcing. As an example, we consider the equations

$$x'' + x = A \sin \tfrac{6}{5} t \tag{8.18}$$

and

$$x'' + \sin x = A \sin \tfrac{6}{5} t \tag{8.19}$$

for different values of the forcing amplitude $A$.

The solution of Eq. (8.18) satisfying the initial conditions $x(0) = x'(0) = 0$ is

$$x(t) = \frac{A \left( 30 \sin t - 25 \sin \tfrac{6}{5} t \right)}{11}. \tag{8.20}$$

Equation (8.20) is a beating solution. Its period is independent of the amplitude $A$ of the external forcing.

Figure 8.10 shows a plot of the solution (8.20) for two different values of $A$. Only the amplitude of the envelope changes in response to the increased amplitude of the forcing.

The solutions of the nonlinear equation (8.19) are quite different in character, depending on the amplitude of the forcing. Figure 8.11(a) shows a plot of the numerically computed solution of Eq. (8.19) for the initial conditions $x(0) = x'(0) = 0$ and $A = 1/5$. The solution is similar to the beating solution shown in Fig. 8.10. In contrast

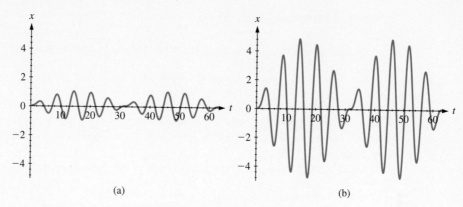

**FIGURE 8.10** The graph of Eq. (8.20) for (a) $A = 1/5$; (b) $A = 1$.

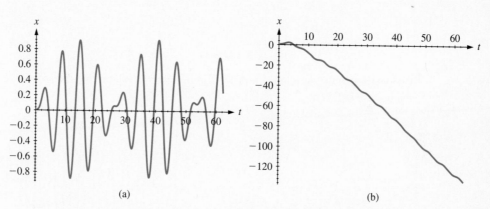

**FIGURE 8.11** Numerically computed solutions of Eq. (8.19) for (a) $A = 1/5$; (b) $A = 1$.

to the beating solution, however, the solution of the nonlinear equation is not exactly periodic.

When $A = 1$, the solution of the nonlinear equation is completely different from that of the linear equation. The nonlinear pendulum does not exhibit beating; rather, the pendulum rotates almost entirely in a clockwise direction.[1] During the time period plotted in Fig. 8.11(b), the pendulum rotates approximately 120 radians clockwise, corresponding to 19 full clockwise revolutions.

---

[1] We adopt the convention that negative angular velocity corresponds to clockwise motion. The pendulum makes one full clockwise rotation when $x$ goes from 0 to $-2\pi$.

## Forced, Damped Motion

The behavior of solutions of the forced, damped, linear pendulum equation,

$$x'' + cx' + \omega^2 x = f(t),$$

is summarized in Theorem 5.8 in Section 5.8. In particular, if $f$ is a linear combination of sine and cosine functions and has period $T$, then the steady-state solution $x_p$ is $T$-periodic. Moreover, the amplitude of the steady-state solution varies linearly with the amplitude of $f$.

Figure 8.12 shows a plot of the solution $x(t)$ of the forced, damped, linear equation

$$x'' + 0.2x' + x = A \sin t \tag{8.21}$$

for the initial conditions $x(0) = x'(0) = 0$. The steady-state solution is

$$x_{ss}(t) = -2.5 \cos t \qquad \text{when } A = 0.5$$

and

$$x_{ss}(t) = -12.5 \cos t \qquad \text{when } A = 2.5.$$

If we quintuple the amplitude $A$ of the forcing, then we quintuple the amplitude of the steady-state solution.

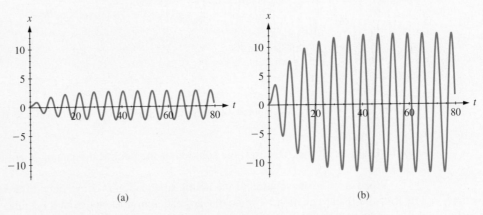

(a)                                    (b)

**FIGURE 8.12**   The solution $x(t)$ of Eq. (8.21) for (a) $A = 0.5$; (b) $A = 2.5$.

None of the conclusions of Theorem 5.8 holds for the nonlinear equation

$$x'' + 0.2x' + \sin x = A \sin t. \tag{8.22}$$

Figure 8.13 shows two numerically computed solutions of Eq. (8.22). The solution in Fig. 8.13(a) corresponds to $A = 0.5$. The steady-state solution is periodic, and the

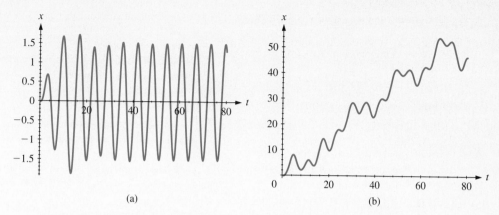

**FIGURE 8.13** The solution $x(t)$ of Eq. (8.22) for (a) $A = 0.5$; (b) $A = 2.5$.

amplitude is about 1.5. The solution in Fig. 8.13(b), corresponding to $A = 2.5$, no longer is periodic; instead, it is erratic and unpredictable. The nonlinear pendulum wobbles irregularly when $A = 2.5$. The angular position goes from 0 to approximately 50 from $t = 0$ to $t = 70$, which implies that the pendulum completes about eight anticlockwise revolutions.

Differential equations like Eq. (8.22) are the subject of considerable contemporary research. Numerical investigations show that the solutions can be quite complex, and many of the properties of the solutions still are not well understood. It is possible to carry out the numerical simulation for a very long time, and one continues to see erratic behavior. The steady-state behavior is said to be *chaotic*. Chaotic solutions are quite common in nonlinear differential equations. Section 8.8 discusses the challenges that chaos poses for predicting the future behavior of otherwise deterministic systems. The exercises ask you to explore some of the remarkable properties of the nonlinear pendulum equation in more detail.

## Exercises

**1.** Theorem 5.3 implies that the period of a solution of the linear undamped pendulum equation

$$x'' + x = 0 \qquad (8.23)$$

is independent of the initial condition. This exercise explores the period of solutions of the nonlinear pendulum equation

$$x'' + \sin x = 0. \qquad (8.24)$$

(a) Find the solution of Eq. (8.23) satisfying the initial conditions $x(0) = C$ and $x'(0) = 0$. What are the period and amplitude of the solution?

(b) Use a numerical solver to compute and graph an approximation of the solution of Eq. (8.24) satisfying the initial conditions $x(0) = 1$ and $x'(0) = 0$. Estimate the period and amplitude of the solution as accurately as you can from the graph.

(c) Repeat (b) for the initial conditions $x(0) = 2$ and $x'(0) = 0$.

(d) Repeat (b) for the initial conditions $x(0) = 3$ and $x'(0) = 0$.

(e) Compare and contrast your results in (b)–(d) with those in (a).

**2.** Let $x'' + 2 \sin x = 0$.

(a) Let $x(0) = x_0$ and $x'(0) = 0$. Use a numerical solver to estimate the period of the solution for $x_0 = 0.5, 1, 1.5, 2$, and $2.5$.

(b) Let $x_0 = 0.3$ and $x'(0) = 0$. Is the period of the solution greater than or less than the period of the solution that satisfies the initial conditions $x(0) = 0.5$ and $x'(0) = 0$?

**3.** Let $x'' + \sin x = 0$.

(a) Use a numerical solver to compute and graph an approximation of the solution satisfying the initial conditions $x(0) = 4$ and $x'(0) = 0$.

(b) Suppose that we regard the equation as a model of an undamped nonlinear pendulum. Can the motion of the pendulum be regarded as periodic? Explain.

(c) Use a numerical solver to compute and graph an approximation of the solution satisfying the initial conditions $x(0) = 4 - 2\pi$ and $x'(0) = 0$. Can the motion of the pendulum be regarded as the same as in (a)–(b)? Explain.

**4.** Let $x'' + \sin x = 0$.

(a) Let $x'(0) = 0$. Use a numerical solver to compute and graph an approximation of the solution satisfying the initial conditions $x(0) = 2.9$, $3.0$, and $3.1$. What is the approximate period of each solution?

(b) What do you think happens to the period as $x(0) \rightarrow \pi$? Experiment with some appropriate values of $x(0)$ as you form your conjecture.

(c) Describe the motion of the pendulum when $x(0) = 0$ and $x'(0) = 1.5, 2.0$, and $2.5$. Consider questions such as the following: Are the solutions periodic? If so, what is the approximate period of each solution? How is the motion of the pendulum qualitatively different in the three cases?

(d) Is it possible, at least in principle, to start the pendulum from the straight-down position with enough initial velocity so that it comes to rest eventually in the straight-up position? Explain, on the basis of your numerical experiments in (a)–(c).

**5.** Consider Fig. 8.13(a).

(a) Show how to devise an algorithm to predict the value of $x(t)$ for values of $t$ that are larger than those shown on the graph.

(b) Use your algorithm to estimate $x(100)$ and $x(120)$.

(c) Discuss the difficulty in predicting $x(100)$ on the basis of Fig. 8.13(b).

**6.** Let $x(0) = 0$ and $x'(0) = 3$.

(a) Use a numerical solver to approximate the solution of the nonlinear pendulum equation $x'' + 0.2x' + \sin x = 0$ that satisfies these initial conditions. Describe qualitatively the motion of the pendulum. (Does the pendulum flip over? Is the motion quasi-periodic?)

(b) Find the solution of the linear pendulum equation $x'' + 0.2x' + x = 0$ that satisfies these initial conditions. Compare and contrast the behavior of the linear pendulum with that in (a). Which equation, if either, appears to model the behavior of a real pendulum more realistically?

**7.** Let $x'' + 0.2x' + \sin x = 0$.

(a) Let $x'(0) = 0$ and use a numerical solver to approximate the solutions that satisfy $x(0) = 1$, 2, 3, and 4. Is the motion of the pendulum qualitatively different in one case compared to the others? Explain.

(b) Let $x(0) = 0$ and use a numerical solver to approximate the solutions that satisfy $x'(0) = 1$, 2, and 3. Is the motion of the pendulum qualitatively different from one case to another? Explain.

(c) Let $x(0) = 0$. Conduct a sequence of appropriate numerical experiments to determine approximately the range of values of $x'(0)$ for which the pendulum flips over exactly once.

(d) Suppose $x(0) = 0$. As $x'(0)$ gets larger, what happens to the motion of the pendulum? Is this behavior consistent with what you would expect from a real pendulum?

**8.** Theorem 5.8 asserts that if $f$ is a linear combination of sine and cosine functions with period $T$, then the steady-state solution of the linear pendulum equation

$$x'' + ax' + bx = f(t)$$

is $T$-periodic. This exercise asks you to explore this question using the nonlinear pendulum equation

$$x'' + 0.2x' + \sin x = f(t), \qquad (8.25)$$

where $f(t) = \cos \omega t$.

(a) Let $\omega = 1/2$. Use a numerical solver to integrate Eq. (8.25) from $t = 0$ to $t = 220$ with the initial conditions $x(0) = x'(0) = 0$. Plot $x$ only for $180 \le t \le 220$. (The interval $0 \le t \le 180$ is sufficient to allow transients to decay.)

(b) Estimate the period and amplitude of $x(t)$ from the graph in (a) as accurately as you can.

(c) Solve the corresponding linear equation $x'' + 0.2x' + x = \cos \omega t$ using the same value of $\omega$ as before. Compare the period and amplitude of $x$ with those in (b).

(d) Repeat (a)–(c) for $\omega = 1$.

(e) Repeat (a)–(c) for $\omega = 2$.

**9.** Let $x'' + 0.2x' + \sin x = 4 \cos \omega t$, where $x(0) = 0$ and $x'(0) = 0$.

(a) Use a numerical solver to approximate the solution of the initial value problem for $0 \le t \le 100$ with $\omega = 1/2$. Can the motion of the nonlinear pendulum be considered periodic? If so, what are the period and amplitude?

(b) Repeat (a) for $\omega = 1$ and $\omega = 2$.

**10.** Theorem 5.8 asserts that the steady-state solution of the linear pendulum equation

$$x'' + ax' + bx = f(t)$$

is independent of the initial condition. This exercise asks you to explore this question using the nonlinear pendulum equation

$$x'' + 0.2x' + \sin x = \cos \tfrac{1}{2}t. \qquad (8.26)$$

For each set of initial conditions in the following list, use a numerical solver to integrate and plot the solution of Eq. (8.26) for $0 \le t \le 280$. Then answer the following questions.

- What is the amplitude of the motion when $x$ approaches a steady state?

- The solution takes a certain amount of time before it appears to "settle down" to a steady state. How many complete revolutions does the pendulum execute during this interval?

- Contrast the behavior of Eq. (8.26) with the corresponding linear equation

$$x'' + 0.2x' + x = \cos \tfrac{1}{2}t$$

for the same initial conditions.

(a) $x(0) = 0, \quad x'(0) = 0$

(b) $x(0) = 0, \quad x'(0) = 0.1$

(c) $x(0) = 0.1, \quad x'(0) = 0.1$

## 8.5 LINEARIZATION

The examples discussed in Sections 8.2–8.4 show that nonlinear systems have fixed points that may be unstable or asymptotically stable, just like linear systems with constant coefficients. Chapter 7 contains a complete theory for linear systems of the form $\mathbf{x}' = \mathbf{A}\mathbf{x}$ that allows us to calculate solutions explicitly. These solutions show that the stability of an equilibrium can be decided by inspecting the eigenvalues of the matrix $\mathbf{A}$.

In contrast, no explicit solutions can be found for most nonlinear systems of differential equations. In this section, we show that it is sometimes possible to approximate a nonlinear differential equation by a linear differential equation for initial conditions near a given fixed point. This process is called the *linearization* of the differential equation.

The purpose of this section is to discuss the linearization of differential equations systematically. The basic idea is this: Under certain conditions, which are described in Theorems 8.4 and 8.7, we can replace a given nonlinear differential equation with a linear approximation such that the direction fields for the two equations are nearly identical near the fixed point. In particular, we can use the linear approximation to assess the stability of fixed points of nonlinear differential equations.

### The Linearization of First-Order Scalar Equations

Let $f(x)$ be a differentiable function. The *linearization* of $f$ at the point $x_f$ is

$$\ell(x) = f(x_f) + \frac{df}{dx}\bigg|_{x=x_f} (x - x_f). \tag{8.27}$$

The right-hand side of Eq. (8.27) is simply the first two terms of the Taylor expansion of $f$ around the point $x_f$. The graph of $\ell$ is a straight line that is tangent to the graph of $f$ at $x_f$. Linear approximations of functions are a useful computational tool in calculus and much of mathematics. (See Section 4.1.)

Differential equations of the form $x' = f(x)$ can be linearized at a fixed point $x_f$ by replacing $f$ with its linearization $\ell$ at $x_f$. Recall that $f(x_f) = 0$ if $x_f$ is a fixed point.

**Definition 8.2** *Let*

$$x' = f(x) \tag{8.28}$$

*be an autonomous differential equation, where $f$ is a differentiable function of $x$, and suppose that $x_f$ is a fixed point for Eq. (8.28). Let*

$$\ell(x) = \frac{df}{dx}\bigg|_{x=x_f} (x - x_f)$$

*be the linearization of f  at $x_f$. Then the equation*

$$x' = \frac{df}{dx}\Bigg|_{x=x_f} (x - x_f) \tag{8.29}$$

*is the* linearization of Eq. (8.28) at the fixed point $x_f$.

The linearization of a differential equation is useful only if $\ell(x)$ is not the zero function; $\ell$ is the zero function if and only if $df(x_f)/dx = 0$.

**Definition 8.3**   *Suppose that $x_f$ is a fixed point for the equation $x' = f(x)$. If $df(x_f)/dx \neq 0$, then $x_f$ is a* hyperbolic *fixed point. Otherwise, if $df(x_f)/dx = 0$, then $x_f$ is a* nonhyperbolic *fixed point.*

The key idea illustrated in the following examples is this:

*Near a hyperbolic fixed point, the direction fields and solutions of the linearization of a nonlinear differential equation are good approximations of those of the original equation.*

The approximations are sufficiently good that we can use the linearization to assess the stability of hyperbolic fixed points. Moreover, we can estimate how quickly solutions approach or move away from hyperbolic fixed points.

◆ *Example 1*   One form of the logistic equation is

$$p' = f(p) = p(1 - p). \tag{8.30}$$

Equation (8.30) has two fixed points: one at $p_f = 0$ and another at $p_f = 1$. By Eq. (8.27), the linearization of $f$ at 0 is

$$\ell_0(p) = (1 - 2p)|_{p=0}\, p = p,$$

and the linearization of $f$ at 1 is

$$\ell_1(p) = (1 - 2p)|_{p=1}\, (p - 1) = 1 - p.$$

Therefore, the linearization of Eq. (8.30) at $p_f = 0$ is

$$p' = p$$

and at $p_f = 1$ is

$$p' = 1 - p. \tag{8.31}$$

The linearization of Eq. (8.30) is represented in two ways in Fig. 8.14. Figure 8.14(a) shows a plot of $f(p)$ and its linearizations near $p_f = 0$ and $p_f = 1$. Although the graph of $f$ is a parabola, it is well approximated by straight lines in small neighborhoods of the points $p_f = 0$ and $p_f = 1$. This aspect of linearization is familiar from calculus.

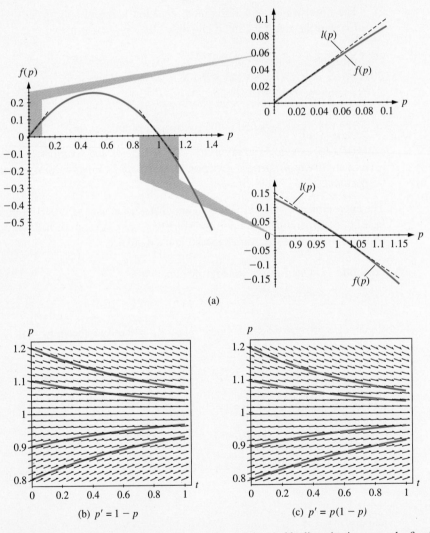

(a)

(b) $p' = 1 - p$                  (c) $p' = p(1 - p)$

**FIGURE 8.14**    (a) Phase diagram for the logistic equation and its linearizations near the fixed points. (b) Phase portrait for Eq. (8.31). (c) Phase portrait for Eq. (8.30).

Figure 8.14(b) shows the direction field and some representative solution curves corresponding to Eq. (8.31), which is the linearization of Eq. (8.30) at $p_f = 1$. Equation (8.31) is separable, and the general solution is

$$p(t) = 1 + Ce^{-t},$$

where $C$ is a constant that depends on the initial condition. Notice that $p(t) \to 1$ as $t \to \infty$. The fixed point at $p_f = 1$ is a sink for Eq. (8.31); solutions approach the fixed point exponentially quickly.

Figure 8.14(c) shows the direction field and some representative solutions corresponding to Eq. (8.30) in a vicinity of the fixed point $p_f = 1$. The direction field suggests that the solutions of the original, nonlinear equation (8.30) are close to those of the linear equation (8.31). In other words, we can conclude that the fixed point at $p_f = 1$ is also a sink for Eq. (8.30) and that solutions starting from initial conditions near the fixed point $p_f = 1$ approach it at a rate that is approximately proportional to $e^{-t}$.  ∎

It can be shown that the correspondence between the direction fields and solution curves illustrated in Example 1 holds for any hyperbolic fixed point. This correspondence is the idea behind the proof of the following theorem. (The proof is omitted here. See Section 1.3 for additional discussion and examples.)

**Theorem 8.4**  *Let $f$ be a differentiable function of $x$, and let $x' = f(x)$. Suppose that $x_f$ is a fixed point. If*

$$\left. \frac{df}{dx} \right|_{x=x_f} > 0,$$

*then $x_f$ is an unstable fixed point. If*

$$\left. \frac{df}{dx} \right|_{x=x_f} < 0,$$

*then $x_f$ is a stable fixed point.*

✦ *Example 2*   We can use Theorem 8.4 to determine the stability of the fixed points in Eq. (8.3), as follows. We have

$$\left. \frac{df}{dp} \right|_{p=0} = (1 - 2p)|_{p=0} = 1$$

and

$$\left. \frac{df}{dp} \right|_{p=1} = (1 - 2p)|_{p=1} = -1.$$

The slope of the tangent line to the graph of $f(p)$ is $+1$ at $p_f = 0$ and $-1$ at $p_f = 1$, as illustrated in Fig. 8.14. The fixed point at 0 is unstable, and the fixed point at 1 is stable.  ∎

◆ *Example 3*  Let

$$x' = f(x) = \sin 2x. \tag{8.32}$$

Any integer multiple of $\pi/2$ is a fixed point of Eq. (8.32). In this example, we consider the linearization of $f$ at $x_f = \pi$, which is

$$\ell(x) = f(\pi) + \left.\frac{df}{dx}\right|_{x=\pi} (x - \pi) = 2(x - \pi).$$

The linearization of Eq. (8.32) at $x_f = \pi$ is

$$x' = \ell(x) = 2(x - \pi). \tag{8.33}$$

Figure 8.15(a) shows the direction field and two representative solution curves for Eq. (8.33) near $x = \pi$, and Fig. 8.15(b) shows the direction field and solution curves for Eq. (8.32) in the same region. As in Example 1, the two pictures agree closely as long as we look in a sufficiently small neighborhood about $x_f = \pi$.

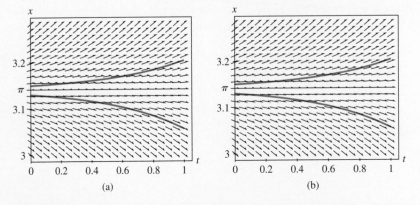

**FIGURE 8.15**  The direction field and two solution curves for (a) Eq. (8.33) and (b) Eq. (8.32) near the fixed point $x = \pi$.

The general solution of Eq. (8.33) is

$$x(t) = \pi + Ce^{2t}, \tag{8.34}$$

where $C$ is a constant that depends on the initial condition. Thus, $x_f = \pi$ is a source. The close correspondence between the two direction fields implies that $x_f = \pi$ is also a source for Eq. (8.32) and that solutions of Eq. (8.32) starting near the fixed point at $x_f = \pi$ move away from it at a rate that is proportional to $e^{2t}$. (See Exercise 3.)  ∎

Theorem 8.4 cannot be used to determine the stability of a fixed point at $x = x_f$ if $df(x_f)/dx = 0$. In such a case the corresponding linearization provides no information about whether solution curves leave or approach a vicinity of the fixed point. The next two examples illustrate some of the things that can happen.

✦ *Example 4* Let

$$x' = f(x) = x(1 - x)^2. \tag{8.35}$$

The fixed points are $x_f = 0$ and $x_f = 1$. Figure 8.16(a) shows the direction field near the fixed point $x_f = 1$. Notice that

$$\frac{df}{dx}\bigg|_{x=1} = (1 - x)^2 - 2x(1 - x)\bigg|_{x=1} = 0,$$

so the linearization of Eq. (8.35) at $x_f = 1$ is simply

$$x' = 0. \tag{8.36}$$

As Fig. 8.16(b) shows, the direction field of the linear equation (8.36) does *not* approximate the direction field of the nonlinear equation (8.35) near the fixed point at $x_f = 1$.

Solutions of Eq. (8.35) starting from initial conditions slightly below $x_f = 1$ approach the fixed point as $t \to \infty$. However, solutions starting from initial conditions slightly above $x_f = 1$ move away as $t$ increases. Thus, the fixed point at $x_f = 1$ is neither a sink nor a source; it is not hyperbolic, and it is not stable.

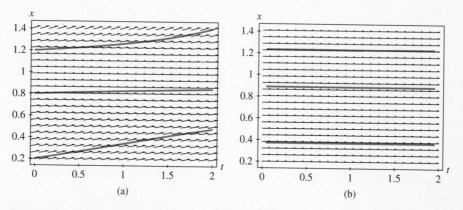

(a)  (b)

**FIGURE 8.16** Direction field for (a) Eq. 8.35; (b) Eq. (8.36).

♦ *Example 5* Let

$$x' = f(x) = x(1 - x)^3. \tag{8.37}$$

Equation (8.37) has the same fixed points as Eq. (8.35). Theorem 8.4 cannot be used to determine the stability of the fixed point at $x_f = 1$, because

$$\left.\frac{df}{dx}\right|_{x=1} = (1 - x)^3 - 3x(1 - x)^2\Big|_{x=1} = 0.$$

The corresponding linearized differential equation is $x' = 0$.

Figure 8.17 shows the phase diagram for Eq. (8.37). Notice that $x' = f(x) > 0$ if $0 < x < 1$, indicating that solutions increase with $t$ as long as they remain below 1. Also, $x' < 0$ if $x > 1$, indicating that solutions decrease with $t$ as long as they remain above 1. The equilibrium at $x_f = 1$ is not hyperbolic, but it is stable.

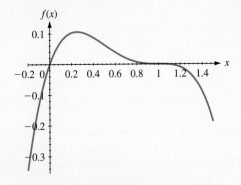

**FIGURE 8.17**    Phase diagram for Eq. (8.37).                                   ■

## The Linearization of Systems of Equations

The notion of linearization carries over to vector-valued functions and to first-order systems of differential equations.

**Definition 8.5** *Let* **f** *be a differentiable function from* $\mathbf{R}^n$ *to* $\mathbf{R}^n$. *The linearization of* **f** *at* $\mathbf{x}_f$ *is*

$$\ell(\mathbf{x}) = \mathbf{f}(\mathbf{x}_f) + \mathbf{D}_\mathbf{x}\mathbf{f}(\mathbf{x}_f)(\mathbf{x} - \mathbf{x}_f), \tag{8.38}$$

*where* $\mathbf{D}_\mathbf{x}\mathbf{f}(\mathbf{x}_f)$ *is the* $n \times n$ Jacobian matrix *of* **f** *evaluated at* $\mathbf{x}_f$, *given by*

$$\begin{pmatrix} \partial f_1/\partial x_1 & \partial f_1/\partial x_2 & \cdots & \partial f_1/\partial x_n \\ \partial f_2/\partial x_1 & \partial f_2/\partial x_2 & \cdots & \partial f_2/\partial x_n \\ & & \vdots & \\ \partial f_n/\partial x_1 & \partial f_n/\partial x_2 & \cdots & \partial f_n/\partial x_n \end{pmatrix}\Bigg|_{\mathbf{x}=\mathbf{x}_f}.$$

As in the scalar case, the linearization of differential equations is of particular interest near fixed points. Let $\mathbf{x}_f$ be a fixed point for the autonomous system $\mathbf{x}' = \mathbf{f}(\mathbf{x})$. Then $\mathbf{f}(\mathbf{x}_f) = \mathbf{0}$. The following definition is the analogue of Definition 8.2 for first-order systems of differential equations.

**Definition 8.6**   *Let*

$$\mathbf{x}' = \mathbf{f}(\mathbf{x}) \tag{8.39}$$

*be an autonomous first-order system of differential equations, where* $\mathbf{f}$ *is a differentiable function of* $\mathbf{x}$. *Suppose that* $\mathbf{x}_f$ *is a fixed point. The* linearization *of Eq. (8.39) about* $\mathbf{x}_f$ *is*

$$\mathbf{x}' = \mathbf{D_x}\mathbf{f}(\mathbf{x}_f)(\mathbf{x} - \mathbf{x}_f), \tag{8.40}$$

*where* $\mathbf{D_x}\mathbf{f}(\mathbf{x}_f)$ *is the Jacobian matrix of* $\mathbf{f}$ *evaluated at* $\mathbf{x}_f$.

The idea behind the linearization of an autonomous system of equations is the same as in the scalar case. We replace $\mathbf{f}(\mathbf{x})$ with its linearization in the vicinity of a fixed point of interest. The solutions of the linear equation closely resemble those of the original equation, provided that we restrict attention to a suitably small neighborhood of the fixed point and that the fixed point satisfies the criteria described in the following theorem. (The proof is beyond the scope of this book.)

**Theorem 8.7**   *Let* $\mathbf{x}' = \mathbf{f}(\mathbf{x})$ *be an autonomous system of differential equations, and suppose that* $\mathbf{x}_f$ *is a fixed point. Let* $\mathbf{A} = \mathbf{D_x}\mathbf{f}(\mathbf{x}_f)$ *be the Jacobian matrix of partial derivatives of* $\mathbf{f}$ *evaluated at* $\mathbf{x}_f$. *If none of the eigenvalues of* $\mathbf{A}$ *is 0 or is purely imaginary, then the stability of* $\mathbf{x}_f$ *may be determined as follows.*

- *If all of the eigenvalues of* $\mathbf{A}$ *are negative or have negative real part, then* $\mathbf{x}_f$ *is a sink.*
- *If all of the eigenvalues of* $\mathbf{A}$ *are positive or have positive real part, then* $\mathbf{x}_f$ *is a source.*
- *If at least one of the eigenvalues of* $\mathbf{A}$ *is positive or has positive real part, and if at least one of the eigenvalues of* $\mathbf{A}$ *is negative or has negative real part, then* $\mathbf{x}_f$ *is a saddle.*

**Definition 8.8**   *Let* $\mathbf{x}_f$ *be a fixed point of the autonomous differential equation* $\mathbf{x}' = \mathbf{f}(\mathbf{x})$, *and let* $\mathbf{A} = \mathbf{D_x}\mathbf{f}(\mathbf{x}_f)$ *be the Jacobian matrix of partial derivatives of* $\mathbf{f}$ *evaluated at* $\mathbf{x}_f$. *If the real part of each eigenvalue of* $\mathbf{A}$ *is nonzero, then* $\mathbf{x}_f$ *is a hyperbolic fixed point.*

In other words, a fixed point that satisfies the hypotheses of Theorem 8.7 is a hyperbolic fixed point. Theorem 8.7 says that the stability of a hyperbolic fixed point at $\mathbf{x}_f$ is determined by the eigenvalues of the Jacobian matrix of $\mathbf{f}$ evaluated at $\mathbf{x}_f$. The eigenvalues of the Jacobian matrix determine the stability of hyperbolic fixed points in nonlinear systems in a manner that is analogous to linear systems. (See Section 7.5.) However, if the fixed point is not hyperbolic (that is, it has one or more eigenvalues that are zero or have zero real part), then Theorem 8.7 cannot be used to determine the stability of the fixed point.

◆ *Example 6*    Figure 8.18(a) shows the phase portrait of the nonlinear system

$$\mathbf{x}' = \mathbf{f}(\mathbf{x}) = \begin{pmatrix} x_1(1 - x_1 - x_2/2) \\ x_2(1 - x_2 - 4x_1/5) \end{pmatrix}. \tag{8.41}$$

There is a fixed point at $\mathbf{x}_0 = (5/6, 1/3)$, and the numerically generated solutions in Fig. 8.18(a) suggest that it is stable. Theorem 8.7 can be used to prove that $\mathbf{x}_0$ is stable, as follows.

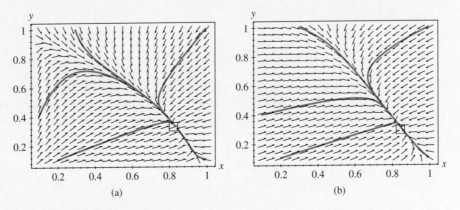

**FIGURE 8.18**    Direction field and some solution curves for (a) Eq. (8.41); (b) Eq. (8.42). The fixed point is marked with a small box.

The Jacobian matrix of partial derivatives evaluated at $\mathbf{x}_0 = (5/6, 1/3)$ is

$$\begin{aligned}
\mathbf{D_x f}(5/6, 1/3) &= \begin{pmatrix} \partial f_1/\partial x_1 & \partial f_1/\partial x_2 \\ \partial f_2/\partial x_1 & \partial f_2/\partial x_2 \end{pmatrix}\Bigg|_{\mathbf{x}=\mathbf{x}_0} \\
&= \begin{pmatrix} 1 - 2x_1 - x_2/2 & -x_1/2 \\ -4x_2/5 & 1 - 2x_2 - 4x_1/5 \end{pmatrix}\Bigg|_{(x_1,x_2)=(5/6,1/3)} \\
&= \begin{pmatrix} -5/6 & -5/12 \\ -4/15 & -1/3 \end{pmatrix},
\end{aligned}$$

and its eigenvalues are $-1$ and $-1/6$. Because both eigenvalues are negative, the fixed point at $(5/6, 1/3)$ is a sink (it is stable).

The linearization of Eq. (8.41) about $(5/6, 1/3)$ is

$$\mathbf{x}' = \mathbf{D_x f}(\mathbf{x}_0)(\mathbf{x} - \mathbf{x}_0) = \begin{pmatrix} -5/6 & -5/12 \\ -4/15 & -1/3 \end{pmatrix}\begin{pmatrix} x_1 - 5/6 \\ x_2 - 1/3 \end{pmatrix}. \tag{8.42}$$

Figure 8.18(a) shows the phase portrait corresponding to the nonlinear system (8.41). Figure 8.18(b) shows the corresponding linear system (8.42) in a region near the fixed point at $(5/6, 1/3)$. The phase portraits strongly resemble each other, but the similarities become less pronounced as one moves away from $(5/6, 1/3)$. Nevertheless, the linearization (8.42) is a good approximation of the original function $\mathbf{f}$ in the displayed region.
∎

There is an important theoretical result, called the Hartman-Grobman theorem, that relates nonlinear systems of equations and their linearizations. Although a precise statement of the theorem is beyond the scope of this book, its conclusions can be described in words as follows.

- Let $\mathbf{x}' = \mathbf{f}(\mathbf{x})$. If $\mathbf{x}_f$ is a hyperbolic fixed point, and if $\ell(\mathbf{x})$ is the linearization of $\mathbf{f}$ at $\mathbf{x}_f$, then there is a continuous change of coordinates that makes the direction field for the linear system $\mathbf{x}' = \ell(\mathbf{x})$ look exactly like the direction field for the nonlinear system $\mathbf{x}' = \mathbf{f}(\mathbf{x})$, at least in a neighborhood of $\mathbf{x}_f$. (The concept of a continuous change of coordinates can be described heuristically as follows. Suppose that the direction field of each equation were printed on a sheet of rubber. A continuous change of coordinates means that it is possible in principle to deform one direction field to look exactly like the other by stretching and shrinking portions of the rubber. The deformation can be done so that no portion of the rubber is torn or folded over onto itself.)

- Near a hyperbolic fixed point, solution trajectories approach or move away from the fixed point at a rate that is governed approximately by the solutions of the corresponding linearized system. That is, time series plots of each component of the solutions of the nonlinear system and those of the corresponding linear system are nearly the same for as long as the solutions remain near the fixed point. The rate of exponential growth or decay is determined by the eigenvalues of the Jacobian matrix evaluated at the fixed point.

One illustration of the Hartman-Grobman theorem is provided by Fig. 8.18, which shows the phase portraits for Eq. (8.41) and its linearization (8.42) near the fixed point $\mathbf{x}_0 = (5/6, 1/3)$. If each picture were printed on a sheet of rubber, then it would be possible to deform each one to look exactly like the other by stretching and shrinking portions of the rubber in a suitable way.

✦ *Example 7* The second-order equation

$$x'' + \sin x = 0 \tag{8.43}$$

models an undamped, nonlinear pendulum. In contrast to the corresponding linear equation $x'' + x = 0$, Eq. (8.43) has a fixed point at $x = \pi$. This fixed point corresponds to a situation in which the pendulum is perfectly balanced in a straight-up position. Physical intuition suggests that this fixed point is unstable, since a small push causes the pendulum to topple over and start swinging. Theorem 8.7 can be used to prove that the fixed point at $x = \pi$ is unstable, as follows.

The first-order system corresponding to Eq. (8.43) is

$$\mathbf{x}' = \mathbf{f}(\mathbf{x}) = \begin{pmatrix} x_1' \\ x_2' \end{pmatrix} = \begin{pmatrix} x_2 \\ -\sin x_1 \end{pmatrix}. \qquad (8.44)$$

The fixed point of the system (8.44) that corresponds to the straight-up position is $\mathbf{x}_f = (\pi, 0)$. The Jacobian matrix is

$$\mathbf{D}_\mathbf{x}\mathbf{f}(\pi, 0) = \begin{pmatrix} 0 & 1 \\ -\cos x_1 & 0 \end{pmatrix}\Bigg|_{\mathbf{x}=(\pi,0)} = \begin{pmatrix} 0 & 1 \\ 1 & 0 \end{pmatrix},$$

whose eigenvalues are $\lambda_1 = 1$ and $\lambda_2 = -1$. By Theorem 8.7, the fixed point is a saddle.

The linearization of Eq. (8.44) at $\mathbf{x}_f = (\pi, 0)$ is

$$\mathbf{x}' = \mathbf{D}_\mathbf{x}\mathbf{f}(\mathbf{x}_0)(\mathbf{x} - \mathbf{x}_f) = \begin{pmatrix} 0 & 1 \\ 1 & 0 \end{pmatrix}\begin{pmatrix} x_1 - \pi \\ x_2 \end{pmatrix} = \begin{pmatrix} x_2 \\ x_1 - \pi \end{pmatrix}. \qquad (8.45)$$

As was mentioned above, the eigenvalues of the Jacobian matrix $\mathbf{D}_\mathbf{x}\mathbf{f}(\mathbf{x}_f)$ are $\lambda_1 = 1$ and $\lambda_2 = -1$. The eigenvector corresponding to $\lambda_1$ is $\mathbf{v}_1 = (1, 1)$. If we take an initial condition in the direction of $\mathbf{v}_1$ away from the fixed point in the linear system (8.45), then the solution of Eq. (8.45) grows exponentially at a rate proportional to $e^{\lambda_1 t} = e^t$. The discussion above indicates that solutions of the nonlinear equation (8.44) also move away from the fixed point at approximately an exponential rate if the initial condition is in the direction of $\mathbf{v}_1$.

Figure 8.19 shows time series plots of the components $x_1(t)$ and $x_2(t)$ of the solutions of Eqs. (8.44) and (8.45) that satisfy the initial condition

$$\mathbf{x}_0 + 0.001\mathbf{v}_1 = (\pi + 0.001, 0.001).$$

This initial condition corresponds to a displacement of 0.001 from the fixed point $(\pi, 0)$ in the direction $\mathbf{v}_1$. It is straightforward to verify that the solution of the linear equation (8.45) is

$$\mathbf{x}(t) = \begin{pmatrix} \pi + 0.001e^t \\ 0.001e^t \end{pmatrix}.$$

Notice that the numerical approximation of the solution of the nonlinear equation (8.44) closely follows that of the linear equation (8.45) for $0 \le t < 7$ or so. The solutions diverge for $t > 7$ because they no longer lie in a neighborhood of the fixed point, and

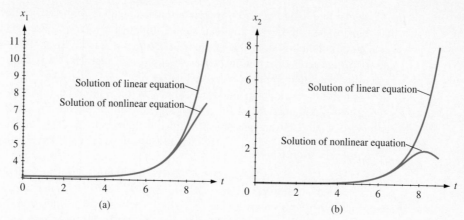

**FIGURE 8.19** Time series plots of each component of solutions of the linear system (8.45) and the nonlinear system (8.44) that satisfy the same initial condition.

the linear system (8.45) is no longer an accurate approximation of the nonlinear system (8.44).

Physically, this example implies that if we displace the pendulum by 0.001 rad from the vertical and give it an angular velocity of 0.001 rad/s, then the pendulum topples over. The angular velocity of the pendulum initially grows exponentially. ∎

Figure 8.20 shows the direction field for the linear system (8.45) near the fixed point at $(\pi, 0)$. The solid lines indicate the eigendirections. Points along the 45-degree line lie along the eigendirection corresponding to the eigenvalue 1, and points along the

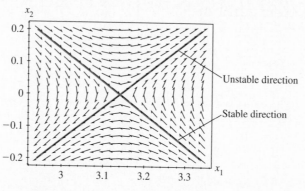

**FIGURE 8.20** The direction field for the linear system (8.45) near the fixed point $(\pi, 0)$.

−45-degree line lie along the eigendirection corresponding to the eigenvalue −1. We may think of the −45-degree line as a "stable" direction insofar as solution curves starting at initial conditions along this line lead back toward the fixed point. The +45-degree line is an "unstable" direction insofar as solutions move directly away from the origin from initial conditions on the 45-degree line.

A similar picture applies in the vicinity of the fixed point $(\pi, 0)$ for the nonlinear system (8.44). Although the stable and unstable directions are not straight lines, it can be shown that the stable and unstable directions are tangent to the eigenvectors $\mathbf{v}_2$ and $\mathbf{v}_1$, respectively, at $(\pi, 0)$. The stable and unstable directions are called *stable* and *unstable manifolds*.

Figure 8.21(a) shows the stable and unstable manifolds near the fixed point $(\pi, 0)$ for the nonlinear pendulum system (8.44). Figure 8.21(b) is a magnification of the region marked by the square in Fig. 8.21(a). When they are suitably magnified, the stable and unstable manifolds resemble straight lines. At this resolution, the manifolds are indistinguishable from the stable and unstable eigendirections in the corresponding linearization (compare Fig. 8.21(b) with the direction field in Fig. 8.20). There are many more interesting properties associated with stable and unstable manifolds that are important in understanding the qualitative behavior of many nonlinear systems, but the details are beyond the scope of this text. (See, for example, Exercise 9.)

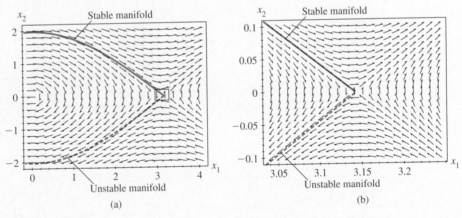

**FIGURE 8.21**    (a) Stable and unstable manifold near the fixed point $(x_1, x_2) = (\pi, 0)$ for Eq. (8.44). (b) A magnification of the region in (a) enclosed in the indicated box around the fixed point.

Theorem 8.7 allows us to study how the stability of a given fixed point changes with a parameter, as the next example shows.

◆ *Example 8*  The van der Pol oscillator, given by the system

$$x_1' = x_2,$$
$$x_2' = -x_1 - x_2 \left(\alpha + x_1^2\right),$$
(8.46)

has a fixed point at the origin. The linearization of Eq. (8.46) at $(0, 0)$ is

$$\mathbf{x}' = \mathbf{A}\mathbf{x} = \begin{pmatrix} 0 & 1 \\ -1 & -\alpha \end{pmatrix} \mathbf{x}.$$

The eigenvalues of $\mathbf{A}$ are

$$\lambda = \frac{-\alpha \pm \sqrt{\alpha^2 - 4}}{2}.$$

Let us see how the stability of the origin changes with $\alpha$.

- If $\alpha = -1$, then the eigenvalues of $\mathbf{A}$ are

$$\lambda = \frac{1 \pm i\sqrt{3}}{2}.$$

  The real part of each eigenvalue is positive, so the origin is a source. Because the corresponding linear system has complex conjugate eigenvalues, solutions near the origin oscillate with time. Therefore, the origin is a spiral source (see Fig. 8.6 in Section 8.3). In fact, the origin is a source whenever $\alpha < 0$.

- If $\alpha = 0$, then the real part of each eigenvalue of $\mathbf{A}$ is 0, and Theorem 8.7 cannot be used to assess the stability of the origin. (See also Exercise 2 in Section 8.3.)

- If $\alpha = 1$, then the eigenvalues of $\mathbf{A}$ are

$$\lambda = \frac{-1 \pm i\sqrt{3}}{2},$$

  so the origin is a spiral sink (see Fig. 8.5 in Section 8.3). In fact, the origin is a sink whenever $\alpha > 0$.                                                                          ∎

Example 8 shows that the nature of a given fixed point can change significantly as a parameter in the differential equation is varied. In particular, a source can change into a saddle or a sink, and conversely. Such qualitative changes in fixed points are called *bifurcations*, which are the subject of the next section.

## *Exercises*

**1.** Consider Eq. (8.37). This exercise considers the hyperbolic fixed point at $x_f = 0$ in more detail.

(a) Linearize Eq. (8.37) around the fixed point $x_f = 0$ and show that it is hyperbolic.

(b) Draw the direction field corresponding to Eq. (8.37) and the direction field correspond-

ing to the linear equation that you found in (a).

(c) Find the solutions of the linear equation in (a) that satisfy the initial conditions $x(0) = -0.1$ and $x(0) = 0.1$. (Plot them for $0 \le t \le 1$.) Add their graphs to the direction field.

(d) Using an appropriate numerical method, find approximate solutions for the initial conditions $x(0) = -0.1$ and $x(0) = 0.1$, and add them to the direction field for Eq. (8.37). Discuss how closely the direction fields and solution curves correspond.

**2.** Consider Eq. (8.37). This exercise considers in more detail the nonhyperbolic fixed point at $x_f = 1$.

(a) Draw the direction field corresponding to Eq. (8.37) near the fixed point at $x_f = 1$. Plot the solutions of Eq. (8.37) that satisfy the initial conditions $x(0) = 1.4$ and $x(0) = 0.6$ for $0 \le t \le 5$. Is the fixed point $x_f = 1$ stable or unstable?

(b) Show that $x_f = 1$ is not hyperbolic.

(c) Does the linearization give you any information about the behavior of solution curves near $x_f = 1$? Explain.

**3.** Show that $x = \pm\pi, \pm 2\pi, \ldots$ are unstable fixed points of Eq. (8.32) and that $x = \pm\pi/2, \pm 3\pi/2, \ldots$ are stable fixed points.

**4.** Let

$$x' = -\sin x. \qquad (8.47)$$

(a) Determine the linearization of Eq. (8.47) at $x = \pi$.

(b) Sketch a direction field for the linearized equation that you found in (a) and a direction field for Eq. (8.47) for some appropriate range in $x$ and $t$. Approximately how large can the range in $x$ be for the two direction fields to look similar?

(c) Solve the linear equation that you found in (a) for the initial condition $x(0) = 3$ and plot it.

(d) Determine either an analytical or a numerical approximation of the solution of Eq. (8.47) satisfying the initial condition $x(0) = 3$ and plot it.

(e) For how long is the relative difference between the solutions in (c) and (d) less than 10 percent?

**5.** For each equation in the following list:

- find all the fixed points;

- linearize the equation about each fixed point;

- determine the stability of each fixed point, where possible, using Theorem 8.4.

(a) $x' = \sin x$

(b) $y' = \cos y$

(c) $x' = (x - 1)(x - 2)$

(d) $y' = -(y - 1)(y - 2)$

**6.** Let $x' = x^3$.

(a) Show that $x = 0$ is a nonhyperbolic fixed point.

(b) Draw a phase diagram and describe qualitatively the solutions that start at initial conditions near $x = 0$. Do the solutions approach 0? Move away from 0? Explain.

(c) Find the solution satisfying the initial condition $x(0) = 1$ and plot it.

(d) The discussion in this section shows that solutions near a hyperbolic fixed point are well approximated by the solutions of the corresponding linearized differential equation. The latter are exponential functions. How is the situation different in this case, in which the fixed point is not hyperbolic?

**7.** Let

$$\mathbf{x}' = \mathbf{A}(\mathbf{x} - \mathbf{b}), \qquad (8.48)$$

where $\mathbf{A}$ is a constant $2 \times 2$ matrix and $\mathbf{b}$ is a constant vector.

(a) Show that the change of variable $\mathbf{y} = \mathbf{x} - \mathbf{b}$ makes Eq. (8.48) equivalent to the system

$$\mathbf{y}' = \mathbf{A}\mathbf{y},$$

which can be solved explicitly by using the methods described in Chapter 7.

(b) Use your result in (a) to find an explicit solution of Eq. (8.45) satisfying the initial condition $\mathbf{x}_0 = (a, b)$.

(c) Show that if $\mathbf{x}_0 = (\pi, 0) + c(1, 1)$, where $c$ is any nonzero constant, then solutions in (b) leave the fixed point at $(\pi, 0)$ at a rate that is proportional to $e^t$.

(d) Show that if $\mathbf{x}_0 = (\pi, 0) + c(-1, 1)$, where $c$ is any nonzero constant, then solutions in (b) approach the fixed point at $(\pi, 0)$ at a rate proportional to $e^{-t}$.

**8.** For each system of equations in the following list:

- find all the fixed points;

- linearize the system about each fixed point;

- determine the stability of each fixed point, where possible, using Theorem 8.7.

(a) $x' = x^2 - y$
    $y' = x^2 - y^2$

(b) $x' = x - y^2$
    $y' = x^2 - y^2$

(c) $x' = x - y^2$
    $y' = x^2 + y$

**9.** Consider Eq. (8.44).

(a) Use an appropriate numerical method to find an approximation of the solution satisfying the initial condition $\mathbf{x}(0) = (\pi - 0.1, 0.1)$.

(b) Repeat (a) for the initial condition $\mathbf{x}(0) = (\pi - 0.01, 0.01)$.

(c) Do your results in (a) and (b) suggest that you could balance a pendulum at the unstable equilibrium by giving it exactly the right initial velocity at some appropriate starting position different from $x_1 = \pi$? Explain.

**10.** Consider Eq. (8.41).

(a) Find all the fixed points and use Theorem 8.7 to determine their stability.

(b) Equation (8.41) is a model of two competing populations, $x$ and $y$. (See Section 8.2.) How can you interpret the meaning of each fixed point from a biological point of view?

**11.** Let

$$x_1' = (1 - x_1 - c_1 x_2)x_1,$$
$$x_2' = (1 - x_2 - b_2 x_1)x_2. \qquad (8.49)$$

Equation (8.49) is a simple model of two competing populations, where $x_1(t)$ and $x_2(t)$ are the population sizes of the two populations at time $t$. (See Section 8.2.) Only nonnegative values of $x_1$ and $x_2$ make sense from a biological point of view.

(a) Let $c_1 = 0.5$ and $b_2 = 1.8$. Find all the fixed points of Eq. (8.49) that are biologically plausible, and use Theorem 8.7 where possible to determine their stability. Classify each fixed point as a sink, source, or saddle. (Hint: Use Fig. 8.3(a) to check your answers.)

(b) Repeat (a) for $c_1 = 1.5$ and $b_2 = 1.8$. (See also Fig. 8.3(b).)

**12.** Figure 8.18 shows the direction field for the nonlinear equation (8.41) and its linearization near the fixed point $(5/6, 1/3)$. The neighborhood of the fixed point where Eq. (8.41) and its linearization have qualitatively similar solutions does not include the $x$ or $y$ axes. Why not? (Hint: Consider what happens to trajectories starting from initial conditions on one of the axes for both Eq. (8.41) and its linearization.)

**13.** The spruce budworm is a forest pest. (See Exercise 25 in Section 1.3.) Suppose the growth of the budworm population $p(t)$ is governed by the differential equation

$$\frac{dp}{dt} = ap - sp^2 - R(p), \qquad (8.50)$$

where $a$ and $s$ are constants and

$$R(p) = \frac{p^2}{1 + p^2}.$$

(a) Let $a = 0.5$ and $s = 0.1$. Find all the positive fixed points (use numerical approximations as necessary), linearize Eq. (8.50) around each fixed point, and determine the stability of each fixed point.

(b) Repeat (a) for $a = 0.6$ and $s = 0.1$.

**14.** Let $x' = \lambda x + x^2$.

(a) Determine the location of each fixed point as a function of the parameter $\lambda$.

(b) Determine the stability of each fixed point as a function of $\lambda$.

(c) Use your result in (a) to plot the location of each fixed point $x_{eq}$ as a function of $\lambda$. Such a plot is called a *bifurcation diagram*. The subject of bifurcation theory is introduced in Sections 8.6 and 8.7.

**15.** Repeat Exercise 14 for $x' = \lambda x + x^3$.

**16.** Repeat Exercise 14 for $x' = \lambda + x^2$.

**17.** Discuss how Theorem 8.4 may be regarded as a special case of Theorem 8.7. (Suggestion: What are the eigenvalues of a $1 \times 1$ matrix?)

**18.** The equation $x'' + 0.2x' + \sin x = 0$ may be regarded as a model of a damped, unforced, nonlinear pendulum.

(a) Write the equation as an equivalent first-order system, and find all the fixed points.

(b) Linearize the system in (a) about each of the fixed points that you find.

(c) Determine the stability of each fixed point.

(d) Interpret what your results in (a) and (c) imply about the behavior of a real pendulum. (For instance, can the pendulum oscillate about a

certain fixed point for appropriate initial conditions? Does the pendulum topple over in response to a disturbance?)

**19.** The Lotka-Volterra model, given by

$$\begin{aligned} x' &= x(\lambda - ky), \\ y' &= y(hx - \rho), \end{aligned} \tag{8.51}$$

where $\lambda$, $k$, $h$, and $\rho$ are positive constants, models a situation in which one species preys on another.

(a) Does $x$ prey on $y$, or does $y$ prey on $x$? Explain.

(b) Let $\lambda = 0.2$, $k = 0.2$, $\rho = 3$, and $h = 1$. Determine all the fixed points and their stability.

(c) Interpret the meaning of each fixed point from the point of view of a population biologist.

**20.** Consider the variation of the Lotka-Volterra model (8.51) given by

$$\begin{aligned} x' &= x(\lambda - x - ky), \\ y' &= y(hx - \rho), \end{aligned} \tag{8.52}$$

where $\lambda$, $k$, $h$, and $\rho$ are positive constants.

(a) Argue that system (8.52) is a more biologically realistic model than system (8.51). (Hint: Compare what the two equations imply if the predators become extinct.)

(b) Let $\lambda = 4$, $k = 2$, $\rho = 3$, and $h = 1$. Determine all the biologically relevant fixed points, and assess their stability.

(c) Repeat (b) for $\lambda = 2$, $k = 2$, $\rho = 3$, and $h = 1$.

(d) Briefly explain how the different values of $\lambda$ affect the qualitative nature of the two populations.

**21.** Consider the Lotka-Volterra model given by Eqs. (8.52).

(a) Show that the line $x = 0$ and the line $y = 0$ are both invariant, that is, solution curves starting from any initial condition on one of the axes remain on that axis for all $t$.

(b) Let $(x_0, y_0) = (0, \alpha)$, where $\alpha$ is a positive number. Show that the solution approaches the origin as $t$ becomes large. We say that the $y$ axis is the stable manifold of the fixed point at the origin.

(c) Let $(x_0, y_0) = (\beta, 0)$, where $\beta$ is a small positive number. Show that the solution approaches the fixed point at $(\lambda, 0)$. We say that the $x$ axis is part of the unstable manifold of the origin.

**22.** Theorem 8.7 cannot be used to determine the stability of a fixed point when the eigenvalues of the associated Jacobian matrix have zero real part. This exercise shows that such a fixed point may be stable or unstable, depending on higher-order terms in the Taylor expansion of **f**. Let

$$\mathbf{x}' = \mathbf{f}(\mathbf{x}) = \begin{pmatrix} -y - x(x^2 + y^2) \\ x - y(x^2 + y^2) \end{pmatrix}. \qquad (8.53)$$

(a) Evaluate the Jacobian matrix of **f** at the origin and show that the origin is not hyperbolic (that is, the eigenvalues of the Jacobian matrix are purely imaginary). Therefore, Theorem 8.7 cannot be used to determine the stability of the origin.

(b) We can convert Eq. (8.53) into polar coordinates by defining

$$\begin{pmatrix} x \\ y \end{pmatrix} = \begin{pmatrix} r \cos \phi \\ r \sin \phi \end{pmatrix}. \qquad (8.54)$$

Notice that $x' = r' \cos \phi - \phi' r \sin \phi$. An analogous formula holds for $y'$. Show that if you substitute the polar coordinate representation (8.54) into Eq. (8.53), you obtain the system

$$\begin{pmatrix} r' \\ \phi' \end{pmatrix} = \begin{pmatrix} -r^3 \\ 1 \end{pmatrix}. \qquad (8.55)$$

Argue that the origin is a stable fixed point for Eq. (8.55), either by solving the equation explicitly or by reasoning geometrically. Conclude that the origin is a sink for Eq. (8.53).

(c) Let

$$\mathbf{x}' = \mathbf{f}(\mathbf{x}) = \begin{pmatrix} -y + x(x^2 + y^2) \\ x + y(x^2 + y^2) \end{pmatrix}.$$

Show that the origin is not hyperbolic, and use the method outlined in (b) to show that the origin is not stable.

# 8.6   BIFURCATION THEORY I: FIRST-ORDER SCALAR EQUATIONS

The qualitative properties of the solutions of a given differential equation often depend on the parameters (constants) in the equation. As a parameter varies, for instance, a given fixed point may change from a sink into a source, or the number of fixed points may change. Such qualitative changes are called *bifurcations*.[2] Bifurcations can be classified and analyzed in a mathematically precise way.

Bifurcation theory is an important part of the analysis of nonlinear differential equations. Bifurcation theory is a large subject, and this section is intended only as a brief introduction. We first motivate the discussion with a simple example, then go on to introduce two types of bifurcations, the *saddle-node* and *transcritical* bifurcations, that occur frequently in practice.

---

[2]The word *bifurcate* comes from a Latin root that means *to split*.

## A Buckled Beam

To illustrate some of the questions that bifurcation theory addresses, we consider a very simple physical system, represented schematically in Fig. 8.22(a). Imagine a rod to which a load $\lambda$ is applied. The rod consists of two rigid pieces that are connected by pins that allow it to pivot either to the left or to the right. A torsional spring resists the compression of the two rods. One convenient way to describe the state of the system is to consider the angle $x$ that the rod makes with the vertical. A simplified model that describes this situation is the differential equation

$$x' = \lambda \sin x - x. \tag{8.56}$$

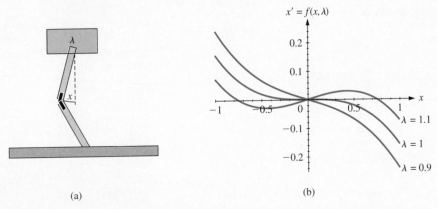

|       |       |
| :---: | :---: |
| (a)   | (b)   |

FIGURE 8.22   (a) A schematic diagram of a buckled beam. (b) Three phase diagrams for Eq. (8.56).

---

The load on the rod is $\lambda$. The position $x = 0$ is a fixed point, corresponding to the case in which the rod remains perfectly vertical. One question is whether the rod is likely to buckle if $\lambda$ increases past some threshold value. If the rod buckles, then it tends to one of two new equilibrium positions that depend on the load $\lambda$ and the torsion of the spring. (There are two equilibria because the rod is equally likely to buckle to the left or to the right.)

Explicit expressions for the fixed points of Eq. (8.56) cannot be found; therefore, it is not easy to find the critical load or an expression for a buckled state directly from the equation. However, we can graph the phase diagrams for various values of the load $\lambda$ and see whether new equilibrium solutions appear. Figure 8.22(b) shows the phase diagrams for $\lambda = 0.9$, $\lambda = 1$, and $\lambda = 1.1$. When $\lambda = 0.9$, there is only one fixed point, namely, $x = 0$, and it is stable. When $\lambda = 1.1$, the fixed point at 0 is unstable, and two stable fixed points appear, corresponding to cases in which the rod buckles to the left or to the right. (See Exercise 17.)

When $\lambda = 1$, there is only one fixed point at $x = 0$, but it is not hyperbolic. The value $\lambda = 1$ is a threshold value above which the rod is likely to buckle. The transition from one to three fixed points, accompanied by the change of stability of the fixed point at $x = 0$, is called a *bifurcation*. The *bifurcation point* is the point $(x_0, \lambda_0)$, where $\lambda_0$ is the threshold value and $x_0$ is the fixed point that is created or whose qualitative behavior changes. In this example, the bifurcation point is $(x_0, \lambda_0) = (0, 1)$. We often write first-order equations as

$$x' = f(x, \lambda)$$

to emphasize the dependence of the differential equation on $\lambda$.

Linear autonomous first-order equations have only one fixed point (see Exercise 1). In contrast, nonlinear autonomous first-order equations can have many fixed points, and the number of fixed points often depends on the parameter. For this reason, bifurcation theory is primarily a study of nonlinear differential equations.

## The Transcritical Bifurcation

Let

$$x' = f(x, \lambda) = \lambda x - x^2. \tag{8.57}$$

Equation (8.57) has two distinct fixed points if $\lambda \neq 0$: $x_f = 0$ and $x_f = \lambda$. There is a bifurcation at $\lambda = 0$, because the number of fixed points changes as the value of $\lambda$ crosses 0.

The qualitative nature of the fixed points also changes with $\lambda$. Figure 8.23(a) shows the phase diagram for Eq. (8.57) for three representative values of $\lambda$: $-0.5$, $0$, and $0.5$.

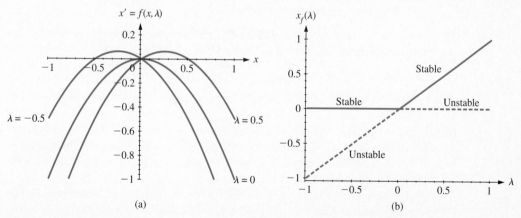

**FIGURE 8.23** (a) The phase diagram for Eq. (8.57). (b) The bifurcation diagram.

Notice the following aspects as $\lambda$ changes.

- The fixed point $x_f = 0$ is stable if $\lambda < 0$ and is unstable if $\lambda > 0$.
- The fixed point $x_f = \lambda$ is stable if $\lambda > 0$ and is unstable if $\lambda < 0$.

Figure 8.23(b) shows the location of the fixed points as a function of $\lambda$. The vertical axis is labeled $x_f(\lambda)$ to emphasize this dependence. The origin ($x_f(\lambda) = 0$) is a fixed point for every value of $\lambda$, as indicated by the horizontal line (part solid, part dashed) that extends across the plot. The location of the second fixed point depends on $\lambda$; it is simply $x_f(\lambda) = \lambda$, indicated by the diagonal line (also part solid, part dashed) that extends across the plot.

We indicate the stability of each fixed point by drawing the graph with a solid curve when the fixed point is stable and with a dashed curve when the fixed point is unstable. The resulting plot is called a *bifurcation diagram*. A bifurcation diagram summarizes the location and stability of one or more fixed points of a given differential equation. The bifurcation point in this example is $(x_0, \lambda_0) = (0, 0)$, because the fixed points coalesce at $x_0 = 0$ when $\lambda$ goes through the value $\lambda_0 = 0$.

This example illustrates a *transcritical bifurcation*. A transcritical bifurcation is characterized by an *exchange of stability*. The fixed points coalesce at the bifurcation point and separate again, but with the opposite stability. In this example, when $\lambda < \lambda_0 = 0$, the fixed point $x_f = 0$ is stable and the fixed point at $x_f = \lambda$ is unstable. The opposite is true when $\lambda > \lambda_0 = 0$.

## When Do Bifurcations Occur?

Generally speaking,

> *a bifurcation occurs whenever the number and/or the nature of the fixed points changes when a parameter reaches a threshold value.*

The discussion of the transcritical bifurcation, as well as the buckled beam modeled by Eq. (8.56), illustrate an important point about bifurcations:

> *Bifurcations can occur only when a fixed point becomes nonhyperbolic.*

To illustrate this idea more fully, consider the phase diagram in Fig. 8.23. If $\lambda = -0.5$, then $x_f = 0$ is a sink. If we change $\lambda$ slightly, then the graph of $f(x)$ shifts slightly, but the shift does not change the nature of $x_f = 0$; it remains a sink. Similarly, the other fixed point at $x_f = \lambda$ moves slightly when $\lambda$ changes, but it remains a source. In other words, sufficiently small changes to the differential equation do not change the qualitative nature of a hyperbolic fixed point.

On the other hand, when $\lambda = 0$, the fixed point at $x_f = 0$ is not hyperbolic. If $\lambda$ is made slightly negative, then the origin becomes a sink; but if $\lambda$ is made slightly positive, then the origin becomes a source. Moreover, there are two fixed points whenever $\lambda \neq 0$. In contrast to the hyperbolic case, *any* small change to the value of $\lambda$ changes the number and nature of the fixed points.

✦ *Example 1*  Let

$$x' = f(x, \lambda) = \tfrac{1}{2}(1 - \cos 2x) - (1 + \lambda)\sin x. \tag{8.58}$$

It is straightforward to verify that $x_f = 0$ is a fixed point of Eq. (8.58) for every value of $\lambda$. However, it is not obvious from direct inspection of Eq. (8.58) whether there are any additional fixed points or whether a bifurcation occurs as $\lambda$ changes. However, the situation can be analyzed graphically.

Figure 8.24(a) shows the phase diagram associated with Eq. (8.58) for some representative values of $\lambda$. For each fixed value of $\lambda$, the graph of $f(x, \lambda)$ resembles a parabola that goes through the origin. By drawing graphs for a sequence of values of $\lambda$, one can see that the graph of $f$ shifts to the right as $\lambda$ increases. (The graph also moves vertically as $\lambda$ changes.)

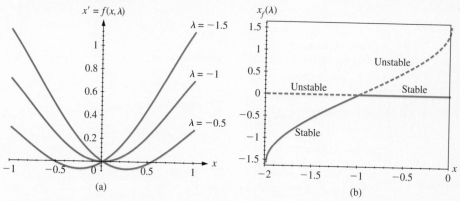

**FIGURE 8.24**  (a) Phase diagrams for Eq. (8.58) for some representative values of $\lambda$. (b) The associated bifurcation diagram.

When $\lambda = -3/2$, for instance, it is clear that $\partial f(0, \lambda)/\partial x > 0$, so the fixed point $x_f = 0$ is unstable. When $\lambda = -1$, the graph of $f$ appears to be tangent to the $\lambda$ axis at $x_f = 0$. When $\lambda = -1/2$, the graph of $f$ suggests that $\partial f(0, \lambda)/\partial x < 0$, so the fixed point $x_f = 0$ is stable.[3]

The phase diagram suggests that there are two fixed points in the interval $-1 \leq x \leq 1$, provided that $\lambda \neq -1$. Although there is no convenient formula for the second

---

[3]Theorem 8.4 can be applied to $f$ even though $f$ depends on $\lambda$ as well as $x$. However, the ordinary derivative $df/dx$ is replaced by the partial derivative $\partial f/\partial x$.

fixed point when $\lambda \neq -1$, its location can be found numerically. The phase diagram also suggests that the second fixed point is stable if $\lambda < -1$ and is unstable if $\lambda > -1$. (See Exercise 10.)

Figure 8.24(b) shows the associated bifurcation diagram. The fixed point $x_f = 0$ changes its stability as $\lambda$ goes through the value $-1$. Notice that the two fixed points coalesce at $\lambda = \lambda_0 = -1$. There is an exchange of stability at the bifurcation point. Equation (8.58) exhibits a transcritical bifurcation at $\lambda_0 = -1$. Notice that the bifurcation diagram in Fig. 8.24(b) is qualitatively similar to that in Fig. 8.23(b).    ∎

## The Saddle-Node Bifurcation

The *saddle-node bifurcation* refers to a situation in which a pair of fixed points, one stable and one unstable, is created or destroyed as a parameter is varied. The next example illustrates the basic idea.

✦ *Example 2*  Let

$$x' = f(x, \lambda) = -x^2 + \lambda. \tag{8.59}$$

If $\lambda > 0$, Eq. (8.59) has a stable fixed point at

$$x_f = \sqrt{\lambda} \tag{8.60}$$

and an unstable fixed point at

$$x_f = -\sqrt{\lambda}. \tag{8.61}$$

Equation (8.59) has two fixed points if $\lambda > 0$, one fixed point if $\lambda = 0$, and no fixed points if $\lambda < 0$. Figure 8.25(a) shows the phase diagram for Eq. (8.59) for three representative values of $\lambda$. As $\lambda$ varies, the graph of $f$ moves vertically. When $\lambda > 0$, the vertex of the parabola lies above the $x$ axis. When $\lambda = 0$, the vertex lies on the $x$ axis; and when $\lambda < 0$, the vertex lies below the $x$ axis.

The fixed points are located where the graph of $f$ crosses the $x$ axis, that is, at $x_f = \pm\sqrt{\lambda}$. Theorem 8.4 in Section 8.5 governs the stability of the fixed points. For each fixed value of $\lambda$, $\partial f/\partial x = -2x$. Therefore, the fixed point $x_f = \sqrt{\lambda}$ is stable, and the fixed point $x_f = -\sqrt{\lambda}$ is unstable.

Figure 8.25(b) shows the bifurcation diagram for Eq. (8.59). The solid curve is the graph of Eq. (8.60), and the dashed curve is the graph of Eq. (8.61). We say that the bifurcation diagram contains two *branches*. Each point on the solid branch corresponds to a stable fixed point, so we call it the *stable branch*. Similarly, each point on the dashed branch corresponds to an unstable fixed point, so it is the *unstable branch*. The bifurcation point is $(x_0, \lambda_0) = (0, 0)$.

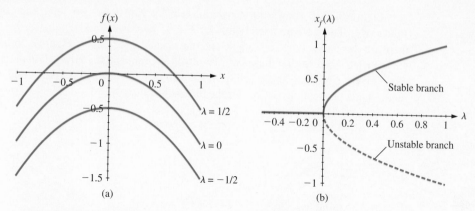

**FIGURE 8.25**   (a) Phase diagram for Eq. (8.59). (b) The corresponding bifurcation diagram.   ■

Figure 8.25 illustrates a saddle-node bifurcation. A saddle-node bifurcation occurs whenever either of the following circumstances holds as the parameter increases through the bifurcation point at $\lambda = \lambda_0$:

- a fixed point is created at $\lambda = \lambda_0$ that splits into two fixed points, one stable and one unstable, for $\lambda > \lambda_0$; or

- two fixed points, one stable and one unstable, coalesce into one fixed point at $\lambda = \lambda_0$, which then disappears for $\lambda > \lambda_0$.

Saddle-node bifurcations also occur in systems of equations. In a system of equations, the unstable fixed point is a saddle and the stable fixed point is a sink.[4] However, a detailed discussion of this topic is beyond our scope here.

✦ *Example 3*   The existence of a saddle-node bifurcation can be demonstrated graphically, even if there is no convenient expression for the fixed points. Let

$$x' = f(x, \lambda) = -e^x + x + \lambda. \tag{8.62}$$

A fixed point $x_f$ occurs wherever the right-hand side of Eq. (8.62) is 0. There is no convenient formula for $x_f$ as a function of $\lambda$, although the location of the fixed points can be determined numerically. However, for any fixed value of $\lambda$, it is possible to graph $f$.

---

[4]Sinks are also called *nodes*, which explains the origin of the term *saddle-node bifurcation*. The term is applied to scalar first-order equations as described above, even though scalar equations do not have saddle fixed points.

Figure 8.26(a) shows the phase diagram for Eq. (8.62) for three representative values of $\lambda$. There are no fixed points for $\lambda < 1$, there is one fixed point for $\lambda = 1$, and there are two fixed points, one stable and one unstable, for $\lambda > 1$. Figure 8.26(b) shows the associated bifurcation diagram. The phase diagram and the bifurcation diagram are qualitatively similar to those in Fig. 8.25. As in Example 2, Eq. (8.62) undergoes a saddle-node bifurcation as $\lambda$ passes the threshold value $\lambda_0 = 1$.

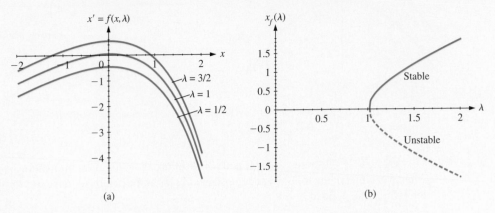

(a)                                                    (b)

**FIGURE 8.26**    (a) Phase diagram for Eq. (8.62); (b) the associated bifurcation diagram.    ■

✦ *Example 4*   Let

$$x' = f(x, \lambda) = -\sin x + \lambda, \tag{8.63}$$

where $-\pi \leq x \leq \pi$.

Figure 8.27(a) shows the phase diagram for Eq. (8.63) for three representative values of $\lambda$. If $\lambda > 1$, then there are no fixed points. If $\lambda = 1$, then there is one fixed point; and if $\lambda < 1$, then there is one stable and one unstable fixed point.

Figure 8.27(b) shows the associated bifurcation diagram. In this example, there are two fixed points, one stable and one unstable, that coalesce and disappear as $\lambda$ increases through the bifurcation point. In contrast to Examples 2 and 3, the bifurcation diagram opens to the left instead of to the right. Nevertheless, this scenario is consistent with a saddle-node bifurcation.    ■

It is possible to state precise analytical criteria under which transcritical and saddle-node bifurcations occur. However, the details are technical, and we omit the theorems. Often, the occurrence of transcritical and saddle-node bifurcations can be inferred from a careful examination of the phase diagrams that are associated with a given differential equation.

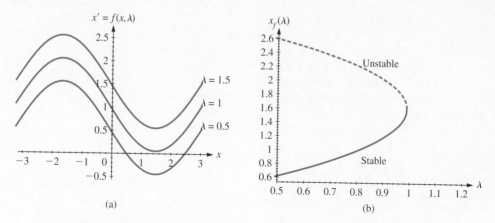

**FIGURE 8.27** (a) The phase diagram for Eq. (8.63) for three different values of $\lambda$. (b) The corresponding bifurcation diagram.

## A Physical Example of a Transcritical Bifurcation

Light is produced when an atom changes from a high-energy state to a low-energy state and emits a particle called a *photon*. In an ordinary light bulb, each atom on the filament emits photons whose frequency and phase are independent of those produced by every other atom on the filament. We say that the resulting radiation is *incoherent*. A laser is a device containing a cavity in which, under appropriate conditions, all the atoms emit photons at the same frequency and phase. The resulting radiation is *coherent* and has much higher intensity than an ordinary beam of light.

A simplified model of the emission of photons in a laser cavity can be derived in the following way. Let $n(t)$ be the number of photons in the laser cavity that are emitted with the same frequency and phase. The rate of change of the population of such photons is

$$\frac{dn}{dt} = \text{gain} - \text{loss}.$$

The loss represents the rate at which photons leave the laser cavity. The loss is proportional to the number of photons in the cavity, that is,

$$\text{loss} = kn$$

for some constant $k$.

The gain represents a process called *stimulated emission*, which results from the interaction of excited (high-energy) atoms and photons in the laser cavity. The gain is proportional to the number of excited atoms, $N(t)$, and the number of photons $n(t)$:

$$\text{gain} = GNn.$$

The constant $G$ is called the gain coefficient.

The number of high-energy atoms is determined by the energy that is supplied to the laser cavity. Assume that, without stimulated emission, the number of excited atoms stays constant at a level $N_0 = \lambda$. Each time a photon is produced, one atom is removed from the set of excited atoms. Therefore, $N(t) = \lambda - n(t)$, and

$$\frac{dn}{dt} = f(n, \lambda) = G(\lambda - n)n - kn = n(G\lambda - k - n). \qquad (8.64)$$

One equilibrium solution of Eq. (8.64) is $n(t) = 0$, which corresponds to a situation in which no lasing occurs. The stability of this equilibrium solution is determined by the sign of $G\lambda - k$. We may regard $\lambda$ as a parameter whose value depends on the amount of energy supplied to the laser cavity. If $G\lambda - k < 0$, then $n = 0$ corresponds to a stable fixed point of Eq. (8.64).

Equation (8.64) has a second fixed point, namely, $n_f = G\lambda - k$, that is distinct from the fixed point $n = 0$ if $G\lambda \neq k$. The second fixed point is unstable if $G\lambda - k < 0$ and is stable if $G\lambda - k > 0$.

The two fixed points coalesce when $G\lambda - k = 0$. There is an exchange of stability as the parameter $\lambda$ goes through the value $\lambda_0 = k/G$. Thus, Eq. (8.64) undergoes a transcritical bifurcation. (See Exercise 15.)

The physical interpretation of this bifurcation is rather interesting. If the amount of energy that is pumped into the laser cavity is sufficiently small, corresponding to a situation in which $\lambda < k/G$, then the fixed point $n = 0$ is stable. In this case, no photons in the laser cavity are produced with the same frequency and phase, so the device produces incoherent light (it acts like an ordinary light bulb).

If the amount of energy pumped into the laser cavity exceeds a certain threshold (that is, if $\lambda > k/G$), then the fixed point $n_f = G\lambda - k$ becomes stable. In practice, this means that if the device experiences an initial disturbance that causes some atoms to produce photons of the same frequency and phase (which is very likely in practice), then the solution approaches the fixed point $n_f = G\lambda - k$. In other words, the population of lasing photons reaches a constant value in response to small disturbances. Equation (8.64) is a model whose qualitative properties agree with laboratory observations: A laser cavity produces coherent light when it is "pumped" with enough energy.

## Exercises

**1.** Consider a linear, first-order, autonomous differential equation.

(a) What is the most general form of such an equation?

(b) Show that the equation in (a) has only one fixed point.

(c) How does the qualitative nature of the fixed point vary with the parameter? (That is, when does it change from a sink to a source?)

**2.** Let

$$x' = (x - 2)^2 - \lambda. \qquad (8.65)$$

(a) Let $\lambda = 1$ and plot a phase diagram for Eq. (8.65). Next draw phase diagrams for $\lambda = 0$ and $\lambda = -1$.

(b) Which fixed point is affected by the bifurcation? What is the bifurcation point?

(c) Determine all the fixed points $x$ of Eq. (8.65) as a function of $\lambda$ and classify the bifurcation as a transcritical or saddle-node bifurcation.

**3.** Repeat Exercise 2 for $x' = \lambda - x - e^{-x}$. Consider $\lambda = 0$, 1, and 2.

**4.** Let $x' = 1 + \lambda \cos \pi x$.

(a) Draw appropriate phase diagrams to show graphically that a bifurcation occurs at $(x_0, \lambda_0) = (1, 1)$.

(b) Classify the bifurcation in (a) as a transcritical or a saddle-node bifurcation.

**5.** Let $x' = x^3 + \lambda$. Is there a bifurcation as $\lambda$ changes? If so, is the bifurcation a saddle-node or transcritical bifurcation, or is it of a different nature? Explain.

**6.** Let $x' = -\lambda + x^2$. Determine the location of the fixed points as a function of $\lambda$ and plot the relation that you find. Label each branch in the bifurcation diagram as stable or unstable. Argue that a saddle-node bifurcation occurs as $\lambda$ crosses through 0.

**7.** Repeat Exercise 6 for the equations $x' = -\lambda$ $-x^2$ and $x' = \lambda + x^2$. The bifurcation diagram for a saddle-node bifurcation in a scalar first-order equation must be qualitatively similar to one of those in this exercise, the bifurcation diagram for Exercise 6, or the bifurcation diagram in Fig. 8.25.

**8.** Let

$$x' = f(x, \lambda) = \lambda + \cos x. \qquad (8.66)$$

(a) Draw appropriate phase diagrams for $\pi/2 \le x \le 3\pi/2$ to show graphically that a bifurcation occurs at $\lambda_0 = 1$. Classify the bifurcation as a transcritical or a saddle-node bifurcation.

(b) Plot the bifurcation diagram for Eq. (8.66) corresponding to the fixed points that you identified in (a). Label each branch as stable or unstable.

**9.** Consider Eq. (8.58).

(a) Show that $x = 0$ is a fixed point regardless of $\lambda$, and determine its stability as a function of $\lambda$. Explain why $\lambda = -1$ is a special case.

(b) Let $\lambda = -3/2$. Find the second fixed point in $-1 \le x \le 1$, and determine its stability using Theorem 8.4 in Section 8.5.

(c) Repeat (b) for $\lambda = -1/2$.

**10.** Let

$$x' = a\lambda + bx^2, \qquad (8.67)$$

where $a$, $b$, and $\lambda$ are constants and $ab \neq 0$.

(a) Let $\lambda > 0$. What criteria must $a$ and $b$ satisfy for Eq. (8.67) to have fixed points?

(b) Let $\lambda < 0$. What criteria must $a$ and $b$ satisfy for Eq. (8.67) to have fixed points?

(c) Suppose that $\lambda \neq 0$ and that $a$ and $b$ are such that Eq. (8.67) has fixed points. Show that Eq. (8.67) has one fixed point, $x_f^+ > 0$, and one fixed point, $x_f^- < 0$. Find an expression for each fixed point as a function of $\lambda$.

(d) Determine the conditions that $a$ and $b$ must satisfy for $x_f^-$ to be stable; for $x_f^+$ to be stable.

(e) If $\lambda \neq 0$, must it always be the case that one of the fixed points is stable and the other is unstable? Explain.

(f) Suppose $ab \neq 0$. What type of bifurcation does Eq. (8.67) undergo as $\lambda$ crosses 0?

**11.** Let $x' = x^2 + \lambda x$.

(a) Sketch a phase diagram for some representative values of $\lambda$. (Include the case $\lambda = 0$.)

(b) Determine the fixed points as a function of $\lambda$.

(c) Show that a bifurcation occurs. Identify the bifurcation point.

(d) Draw the corresponding bifurcation diagram. Label each branch of the bifurcation diagram as stable or unstable.

(e) Classify the bifurcation as a saddle-node or a transcritical bifurcation.

**12.** Let

$$x' = x(\lambda - 1 - x). \tag{8.68}$$

(a) Identify the fixed points in Eq. (8.68).

(b) Plot a phase diagram for appropriate values of $\lambda$, and determine the value of $\lambda$ at which a bifurcation occurs.

(c) Draw the bifurcation diagram and classify the bifurcation as a saddle-node or transcritical bifurcation. Label the branches of the bifurcation diagram as stable or unstable.

**13.** Repeat Exercise 12 for the equation

$$x' = -28 - 11x + 4\lambda + \lambda x - x^2.$$

**14.** Consider Eq. (8.64). Sketch a representative bifurcation diagram. (Plot the fixed points as functions of $\lambda$.) You may assume that $G$, $k$, and $\lambda$ are nonnegative constants.

**15.** Let

$$x' = \lambda x - x^3. \tag{8.69}$$

(a) Plot a phase diagram for Eq. (8.69) for $\lambda = 1$, $0$, and $-1$.

(b) Describe the bifurcation that occurs at $\lambda = 0$. How many fixed points are there for $\lambda > 0$ and $\lambda < 0$? What is the stability of each?

(c) Determine all the fixed points of Eq. (8.69), and plot their location as a function of $\lambda$ for $-1 \leq \lambda \leq 1$. The bifurcation that occurs at $\lambda = 0$ is called a *pitchfork bifurcation*. Why is the term appropriate?

**16.** Consider Eq. (8.56).

(a) Argue that there is only one fixed point in the interval $-1 \leq x \leq 1$ when $0 < \lambda < 1$ and that there are three fixed points in the same interval if $\lambda$ is slightly larger than 1.

(b) Show that the fixed point $x_f = 0$ is stable if $0 < \lambda < 1$ and is unstable if $\lambda > 1$.

(c) Let $\lambda = 1.1$. Find the location of the nonzero fixed points numerically and show that they are stable.

(d) Sketch a bifurcation diagram of Eq. (8.56) for $0.5 < \lambda < 1.5$ and show that the bifurcation is a pitchfork bifurcation. (See Exercise 15.)

# 8.7  BIFURCATION THEORY II: FIRST-ORDER SYSTEMS

Section 8.6 discusses examples of bifurcations that occur in first-order scalar equations. In this section, we show how bifurcations occur in certain systems of first-order differential equations. We also introduce a common bifurcation, the *Hopf* bifurcation, that occurs only in systems of two or more equations.

The discussion here is intended only as an introduction to a much larger subject. Many of the relevant theorems and the associated mathematical theory are beyond the scope of this textbook. However, the main ideas are similar in spirit to those in the previous section.

## A Transcritical Bifurcation

We consider first-order systems of the form

$$\mathbf{x}' = \mathbf{f}(\mathbf{x}, \lambda),$$

where $\lambda$ is a parameter. As in Section 8.6, we are interested in how the number and nature of the fixed points change as $\lambda$ varies.

Let

$$x' = x\left(1 - x - \tfrac{1}{2}y\right),$$
$$y' = y(1 - y - \lambda x). \tag{8.70}$$

The system (8.70) is a model of two competing species (see Section 8.2). Here, $\lambda$ is a constant that is related to the competitive advantage of species $x$ over species $y$. Larger values of $\lambda$ imply that increases in $x$ have more negative influence on the survival of $y$.

The system (8.70) has four fixed points. Two of them are of particular interest here:

- the fixed point at $(1, 0)$, corresponding to a situation in which species $y$ dies out, and
- the fixed point at

$$\left(\frac{1}{2 - \lambda}, \frac{2(1 - \lambda)}{2 - \lambda}\right),$$

corresponding to a competitive coexistence of the two species. We call this fixed point the *competitive coexistence point*.

Figure 8.28 shows the phase portraits for Eqs. (8.70) for $\lambda = 0.8$ and $\lambda = 1.1$. The phase portraits suggest that the stable (attracting) fixed point for $\lambda = 0.8$ is the competitive coexistence equilibrium and that the fixed point $(1, 0)$ is stable when $\lambda = 1.1$.

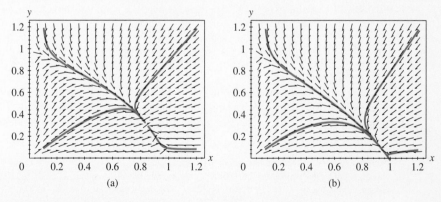

**FIGURE 8.28**   Phase portrait for Eqs. (8.70) for (a) $\lambda = 0.8$; (b) $\lambda = 1.1$.

The linearization of Eqs. (8.70) at $(1, 0)$ is

$$\mathbf{x}' = \begin{pmatrix} -1 & -1/2 \\ 0 & 1 - \lambda \end{pmatrix} \begin{pmatrix} x - 1 \\ y \end{pmatrix}. \tag{8.71}$$

The Jacobian matrix in Eq. (8.71) has two eigenvalues, $\mu_1 = -1$ with eigenvector $\mathbf{v}_1 = (1, 0)$ and $\mu_2 = 1 - \lambda$ with eigenvector $\mathbf{v}_2 = (1, 2\lambda - 4)$. Notice that the equilibrium at $(1, 0)$ is hyperbolic if $\lambda \neq 1$. Equation (8.71) suggests that the fixed point at $(1, 0)$ is unstable if $\lambda < 1$ (it is a saddle) and is stable if $\lambda > 1$ (it is a sink).

If $\lambda = 1$, then the equilibrium at $(1, 0)$ is not hyperbolic. In fact, a bifurcation occurs, as we now illustrate.

Figure 8.29 shows the location of the competitive coexistence point on the phase plane for various values of $\lambda$. As $\lambda$ approaches 1, the competitive coexistence point approaches the fixed point at $(1, 0)$. When $\lambda = 1$, the two fixed points coincide; and for values of $\lambda$ slightly larger than 1, the competitive coexistence point sits in the fourth quadrant.[5]

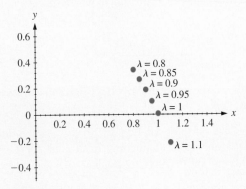

**FIGURE 8.29**    The location of the competitive coexistence point for various values of $\lambda$.

It is straightforward to check that the competitive coexistence point is stable if $\lambda < 1$ and is unstable if $1 < \lambda < 2$. Therefore, an exchange of stability occurs at $\lambda = 1$: The competitive coexistence point becomes unstable as $\lambda$ increases past 1, and the equilibrium at $(1, 0)$ becomes stable. This situation corresponds to a transcritical bifurcation.

---

[5]The competitive coexistence point is not biologically plausible once it moves outside of the first quadrant. However, from a strictly mathematical point of view, the fact that one component of the fixed point is negative poses no difficulty.

Figure 8.30 shows the bifurcation diagram for the two fixed points as a function of $\lambda$. (The vertical axis shows the $y$ coordinate of each fixed point.) The exchange of stability is apparent at $\lambda = 1$.

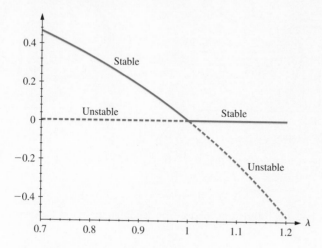

FIGURE 8.30    Bifurcation diagram for Eqs. (8.70).

## The Hopf Bifurcation

The previous discussion shows an example in which a fixed point becomes nonhyperbolic when one of the eigenvalues of the associated Jacobian matrix goes through 0. A fixed point also can become nonhyperbolic if the eigenvalues of the associated Jacobian matrix are complex conjugate pairs, and the real part of each eigenvalue becomes 0.

Let

$$\mathbf{x}' = \mathbf{f}(\mathbf{x}, \lambda) = \begin{pmatrix} y + \lambda x - x(x^2 + y^2) \\ -x + \lambda y - y(x^2 + y^2) \end{pmatrix}. \tag{8.72}$$

Figures 8.31(a) and 8.31(b) show the phase portraits for Eqs. (8.72) for $\lambda = -1$ and $\lambda = 1$, respectively. The two phase portraits are qualitatively different. The origin is a fixed point in both cases; it is stable for $\lambda = -1$ and is unstable for $\lambda = 1$. The phase portrait in Fig. 8.31 suggests that solution curves starting from initial conditions near the origin approach a limit cycle when $\lambda = 1$. In other words, the stable fixed point at the origin is replaced by a limit cycle (periodic orbit) as $\lambda$ increases.

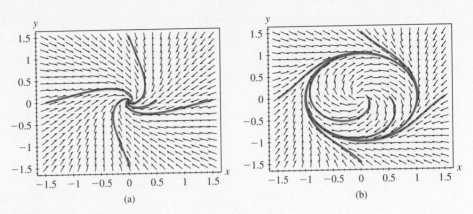

**Figure 8.31**   The phase portrait for Eqs. (8.72) for (a) $\lambda = -1$; (b) $\lambda = 1$.

Let us see how the bifurcation arises. The Jacobian matrix at the origin is

$$\mathbf{D_x f(0)} = \begin{pmatrix} \lambda & 1 \\ -1 & \lambda \end{pmatrix},$$

with complex conjugate eigenvalues $\lambda \pm i$. If $\lambda \neq 0$, then the origin is hyperbolic. By Theorem 8.7, the origin is a sink if $\lambda < 0$ and is a source if $\lambda > 0$. The origin is nonhyperbolic when $\lambda = 0$, which implies that $\lambda = 0$ is a possible bifurcation value. The nature of this bifurcation is different from the transcritical bifurcation in that a complex conjugate pair of eigenvalues crosses the imaginary axis in the complex plane as the parameter crosses a certain threshold.

Equations (8.72) can be written in polar coordinates as

$$
\begin{aligned}
r' &= -r\left(-\lambda + r^2\right), \\
\phi' &= -1.
\end{aligned}
\tag{8.73}
$$

(see Exercise 1). The polar representation (8.73) makes the following aspects of solution curves apparent:

- The only fixed point is the origin. (Although $\phi'$ is never 0, solutions have zero distance from the origin when $r = 0$.)
- Solution curves in the plane form spirals, because they rotate with constant speed.
- If $\lambda < 0$, then all solution curves approach the origin as $t \to \infty$.
- If $\lambda > 0$ and $r < \sqrt{\lambda}$, then the distance of solution curves from the origin increases with time.

- If $\lambda > 0$ and $r > \sqrt{\lambda}$, then the distance of solution curves from the origin decreases with time.

- If $\lambda > 0$ and $r = \sqrt{\lambda}$, then the distance of solution curves from the origin remains constant. In other words, solution curves starting from the initial conditions $(r_0, \phi_0) = (\sqrt{\lambda}, \phi_0)$ form circles. Such solutions are periodic.

The observations above imply that *if $\lambda > 0$, then Eqs. (8.72) and (8.73) have a stable limit cycle.* When $\lambda > 0$, the corresponding phase portrait must look qualitatively similar to that in Fig. 8.31(b).

Physical systems that are governed by equations like Eqs. (8.72) often exhibit oscillatory behavior after a parameter crosses a threshold value. An initial disturbance corresponding to an initial condition that is slightly away from the fixed point leads to solution curves that approach a limit cycle.

## *Exercises*

**1.** Show that Eq. (8.72) is equivalent to Eq. (8.73) under the change of coordinates

$$x(t) = r(t)\cos\phi(t),$$

$$y(t) = r(t)\sin\phi(t).$$

**2.** Draw the phase diagram for the equation $r' = -r(-\lambda + r^2)$. (See Eq. (8.73).) Use the phase diagram to prove each of the following statements about Eqs. (8.72) and (8.73).

- If $\lambda < 0$, then all solutions approach the origin as $t \to \infty$.

- If $\lambda > 0$ and $r < \sqrt{\lambda}$, then the distance of solution curves from the origin increases with time but eventually approaches a limiting value as $t \to \infty$.

- If $\lambda > 0$ and $r > \sqrt{\lambda}$, then the distance of solution curves from the origin decreases with time but eventually approaches a limiting value as $t \to \infty$.

- There exists a periodic solution (limit cycle) if $\lambda > 0$.

**3.** Let

$$r' = -r\left(a\lambda + br^2\right),$$

$$\phi' = c,$$

where $a$, $b$, $c$, and $\lambda$ are constants. (This system is expressed in polar coordinates.)

(a) Show that the sign of $c$ determines the direction in which solution curves spiral.

(b) Determine the relationship that $a$, $b$, and $\lambda$ must satisfy for there to be a limit cycle.

(c) Show that the sign of $b$ determines the stability of the limit cycle.

(d) Choose a set of values for $a$, $b$, $c$, and $\lambda$ for which there is a stable limit cycle, and plot a phase portrait.

(e) Choose a set of values for $a$, $b$, $c$, and $\lambda$ for which there is an unstable limit cycle, and plot a phase portrait.

**4.** Let

$$x' = x(1 - x - \lambda y),$$

$$y' = y\left(1 - y - \tfrac{1}{2}x\right).$$

(a) Determine each of the fixed points and discuss how their stability changes, if at all, as $\lambda$ increases from 0. Describe qualitatively the kinds of bifurcations that you observe.

(b) Interpret your findings in (a) assuming that the system models a pair of competing species. (See Section 8.2.)

**5.** A damped nonlinear pendulum with an applied torque can be modeled as

$$x_1' = x_2,$$
$$x_2' = \lambda - x_2 - \sin x_1,$$

where $\lambda$ depends on the size of the torque. Here, $x_1(t)$ denotes the angular position of the pendulum and $x_2(t)$ denotes the angular velocity at time $t$. We restrict attention to the case in which $0 \le x < 2\pi$.

(a) Let $\lambda = 0.5$. Find all the fixed points and determine their stability.

(b) Draw a phase portrait for the following values of $\lambda$: 0.5, 0.7, 0.9, and 1.1. Discuss the qualitative differences, if any, between the four pictures.

(c) Show that a possible bifurcation occurs at $\lambda = 1$ at one of the fixed points by showing that the fixed point is nonhyperbolic when $\lambda = 1$.

(d) Discuss how the nature of the fixed points changes as $\lambda$ increases through 1.

(e) Draw a phase portrait for $\lambda = 1.5$. Discuss what the phase portrait implies about the motion of the pendulum for various initial conditions.

## 8.8   CHAOS

### The Predictability Problem

We have seen that first-order systems of the form $\mathbf{x}' = \mathbf{f}(\mathbf{x})$ have unique solutions if the Jacobian matrix of $\mathbf{f}$ is continuous. (See the existence and uniqueness theorem 8.1 in Section 8.1.) Differential equation models of a wide variety of phenomena satisfy the hypotheses of the existence and uniqueness theorem. This property is what makes mathematical models so useful in science and engineering.

One of the basic goals of science is prediction. Electrical engineers often simulate the behavior of a particular collection of components before building the circuit. Aerospace engineers run elaborate computer simulations of an aircraft to test various design strategies. Meteorologists have developed sophisticated models of weather patterns to make better forecasts.

The existence and uniqueness theorem (Theorem 8.1) is a crucial theoretical tool, because it says that under rather general hypotheses, only one outcome is possible from a given initial condition. This property gives mathematical models their predictive power, assuming of course that the models accurately reflect the actual system.

However, even if we assume that a given nonlinear differential equation is a completely accurate model, it still may be impossible to determine the precise value of the solution a long time in advance. The reasons include the following:

- It is usually not possible to determine an explicit analytical solution, so the solution must be approximated by using a numerical method. Such methods have their own intrinsic errors, as discussed in Chapter 9.

- Even if the numerical method is accurate, the computed solution may exhibit no apparent pattern. Given an initial condition at $t = 0$, it may not be possible to predict the value of the solution at, say, $t = 1000$ without actually running a computer simulation from $t = 0$ to $t = 1000$.

- Solutions may be sensitive to initial conditions. As we will see, slight differences in the starting point, in the number of digits retained in computer calculations, and in the parameters in the problem may cause solutions to differ by a significant amount after a certain amount of time.

Mathematicians use the word *chaos* to describe solutions of differential and difference equations that are sensitive to small changes in initial conditions. Chaotic solutions tend to have erratic behavior. There is no universally accepted definition of chaos, and many of the mathematical details that are involved in formulating a definition are beyond the scope of this text. However, in this section, we illustrate the basic ideas and stress some of the important implications of chaos.

## The Lorenz Equations

Consider the following question: Will it rain in New York City next July 4? There are two kinds of scientifically valid answers:

1. We can consult historical records and compute the fraction of previous July 4's on which it has rained.
2. If July 4 is only a few days from now, we can feed a database of present conditions across North America into a weather prediction model and compute a forecast.

Method 2 probably provides the most accurate answer, but it can be applied only if July 4 lies in the near future. Method 1 gives only a statistical likelihood of rain, but it generally is the best answer that one can give for a long-term forecast.

Edward Lorenz is a renowned meteorologist who has been interested in the question of how long in advance the weather can be accurately predicted. In the early 1960s, Lorenz and others began to investigate simple models of fluid convection in hopes of gaining better insight into the behavior of the atmosphere and other factors that affect weather. *Convection* refers to the process by which warm fluids (or gases) tend to rise and cool fluids tend to sink. Convection is an important factor in determining the weather, because warm and cold plumes of air constantly churn in the atmosphere.

Lorenz used a digital computer to calculate numerical approximations of the solutions of the following system of differential equations, which form a crude model of a convection process:

$$x' = \sigma(y - x),$$
$$y' = \rho x - y - xz, \qquad (8.74)$$
$$z' = -\beta z + xy.$$

Here, $\sigma$, $\rho$, and $\beta$ are positive constants that depend on the physical characteristics of the fluid. Lorenz was particularly interested in the case $\sigma = 10$, $\rho = 28$, and $\beta = 8/3$.

Lorenz noticed that the solutions of Eqs. (8.74) oscillate irregularly. Figure 8.32(a) shows a plot of $x$ as a function of $t$. The initial condition is $(x_0, y_0, z_0) = (10.123456, 10, 10)$. Among other questions, Lorenz wanted to know whether there might be a pattern to the oscillations. (The other components have qualitatively similar behavior.) Lorenz programmed the computer to print out the values of $x$, $y$, and $z$ from time to time. In this way, Lorenz could stop the simulation, record the current values, and use them to restart the calculation at some later time.

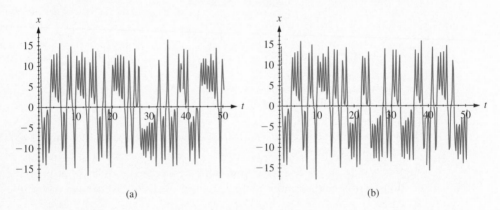

(a)                    (b)

**FIGURE 8.32**    Two numerically generated solutions of the Lorenz equations.

To double-check some of his work, Lorenz repeated some of the computer calculations. Immediately, a problem arose: He was unable to duplicate some of the previous calculations. After some investigation, Lorenz realized one source of difficulty. The computer printed out only the four most significant digits of the values of $x$, $y$, and $z$; but internally, the numbers were stored with greater accuracy. In other words, when Lorenz restarted a given calculation, his manually entered data were slightly different from the computed data from the earlier run.

Figure 8.32(b) shows an example of the phenomenon that Lorenz encountered. The curve shows the numerically computed solution starting from the initial condition $(x_0, y_0, z_0) = (10.12, 10, 10)$. Notice that the curve is nearly identical to that in

Fig. 8.32(a) for $0 \leq t \leq 15$ or so. However, for $t > 15$, noticeable differences appear. Although both curves are *qualitatively* similar, there are significant *quantitative* differences after a long enough time. In this example, the differences arise from a relative change in the initial condition of about 0.03 percent.

The Lorenz equations exhibit *sensitive dependence on initial conditions*. That is, a small change in the initial condition produces large changes in the solution after a while. Lorenz concluded that if his simple model of convection exhibits sensitive dependence on initial conditions, then it is likely that sophisticated weather models do, too.

One implication of sensitive dependence on initial conditions is that detailed long-term weather forecasts may be intrinsically impossible to make. For example, no one can have complete knowledge of the state of the atmosphere at any instant. The best that one can expect is a collection of measurements from various satellites and monitoring stations. These data are subject to uncertainties in measurement (the instruments are not completely accurate), and the measurements reflect only the state of the atmosphere at selected points. Often, weather forecasts deviate significantly from the actual weather after a couple of days. Meteorologists attribute these deviations in part to sensitive dependence on initial conditions.

The sensitivity to initial conditions in the Lorenz equations does not arise because of stochastic (random) terms in the equations themselves. In fact, it can be shown that the solutions of the Lorenz equations for any given initial value problem are unique. However, numerical evidence suggests that solution curves tend to separate quickly, even though they may start close together. The Lorenz equations are an example of a deterministic system that is not predictable in the long term.

Moreover, the sensitive dependence on initial conditions is a robust phenomenon in the Lorenz equations. It is possible to change the values of $\sigma$, $\rho$, and $\beta$ slightly and still see this sensitivity.

The advent of powerful personal computers has allowed scientists to perform extensive numerical studies of a wide variety of nonlinear equations. The nonlinear pendulum, discussed in Section 8.4, is another example of a simple physical model that exhibits chaos. Chaotic behavior has been found in a large number of relatively simple mathematical models, and the subject remains an active area of research.

## Chaos in Difference Equations

Differential equations pose several challenges in studying chaotic behavior. One challenge is in obtaining accurate numerical solutions. Another challenge is the question of how to visualize solution curves in three or more dimensions. For this reason, mathematicians often investigate simpler models, called *difference equations*, which are easier to simulate numerically and often are easier to analyze graphically. One model of considerable interest is the *logistic map*, which is also discussed in Section 0.3 and Section 9.3.

The logistic map is

$$x_{n+1} = f(x_n, \lambda) = \lambda x_n(1 - x_n), \tag{8.75}$$

where $\lambda$ is a constant. Equation (8.75) may be regarded as a simple model of a population with nonoverlapping generations. For instance, $x_n$ might be the population of a mosquito in year $n$. The population in year $n + 1$ depends on the size of the population in the previous year. (We restrict our attention to the case $0 \le x_n \le 1$, in which we measure the population in some normalized units.) The logistic map, like its differential equation counterpart, is a simple model of resource limitations on a population. In the case of Eq. (8.75), the value $x = 1$ is a limiting upper value on the size of the population. If the population in year $n$ is close to 1, then the population in year $n + 1$ is close to 0.

Suppose $\lambda = 0.9$ and the population in year 0 is $x_0 = 0.8$. (This is an example of an initial condition for Eq. (8.75).) We *iterate* the logistic map to obtain the values of the populations in future years. In this case, we have

$$x_0 = 0.8,$$

$$x_1 = 0.144,$$

$$x_2 = 0.1109\ldots,$$

$$x_3 = 0.0887\ldots,$$

$$x_4 = 0.0727\ldots,$$

$$x_5 = 0.06074\ldots.$$

and so on. The sequence $x_0, x_1, x_2, \ldots$ is called the *orbit* of $x_0$. In this example, if we continue the process long enough, we see that the population eventually dies out, because the orbit tends to 0.

Figure 8.33 shows a graphical representation of Eq. (8.75) along with the orbit of $x_0$. The parabola is the graph of $f$. The 45-degree line represents the relation $x_n = x_{n+1}$.

The orbit is drawn as a sequence of line segments in the following way. The initial condition is $x_0 = 0.8$. The image of $x_0$ is $x_1 = f(x_0) = 0.144$. Graphically, we move vertically along the line $x = 0.8$ until we hit the graph of $f$. We move left until we hit the 45-degree line, because the abscissa of the point of intersection on the 45-degree line is $x_1 = 0.144$. We move vertically along the line $x = 0.144$ until we hit the graph of $f$ again; this yields $x_2 = f(x_1) \approx 0.111$. We again move horizontally until we hit the 45-degree line, so that the abscissa of the point of intersection is $x_2 \approx 0.111$, and so on. The resulting plot is called a *staircase diagram*.

The staircase gradually descends toward the origin, implying that $x_n \to 0$ as $n \to \infty$. The origin is a stable fixed point when $\lambda = 0.9$. This fact can be demonstrated graphically, because for any initial condition such that $0 < x_0 < 1$, the corresponding orbit generates a staircase diagram that eventually leads toward 0.

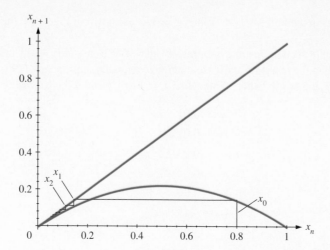

**FIGURE 8.33**   A plot of the logistic map, Eq. (8.75), for $\lambda = 0.9$ and the orbit starting from $x_0 = 0.8$.

Now consider the case in which $\lambda = 1.8$. Figure 8.34 shows a staircase diagram for the orbit starting at $x_0 = 0.8$. In contrast to the case shown in Fig. 8.33, the orbit no longer tends to 0 but rather approaches another point that is located at the intersection of the 45-degree line and the graph of $f$.

Such intersections correspond to fixed points of the logistic map. To see this, observe that if we start at a point of intersection, then we are already on the graph of $f$, that is, $f(x) = x$. The fixed points for Eq. (8.75) occur at $x = 0$ and $x = 1 - 1/\lambda$. If $0 < \lambda < 1$, then there is only one fixed point in the unit interval. When $\lambda > 1$, there are

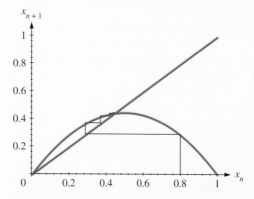

**FIGURE 8.34**   A plot of the logistic map, Eq. (8.75), for $\lambda = 1.8$ and the orbit starting from $x_0 = 0.8$.

two fixed points in the unit interval. When $\lambda = 1.8$, as shown in Fig. 8.34, the nonzero fixed point is $x = 4/9$.

It is possible to demonstrate graphically that $4/9$ is a stable fixed point. Let $0 < x_0 < 1$. Then the resulting staircase diagram eventually approaches $4/9$.

In contrast, the origin is not stable when $\lambda = 1.8$. Consider the following orbit:

$$x_0 = 0.01,$$
$$x_1 = 0.01782,$$
$$x_2 = 0.03150\ldots,$$
$$x_3 = 0.05492\ldots,$$
$$x_4 = 0.09342\ldots,$$
$$x_5 = 0.1524\ldots,$$
$$x_6 = 0.2325\ldots,$$
$$x_7 = 0.3212\ldots,$$
$$x_8 = 0.3925\ldots.$$

The corresponding staircase diagram consists of a sequence that heads up and to the right, away from 0 (and toward the fixed point at $4/9$).

Evidently, a bifurcation has occurred as the value of $\lambda$ increases from $\lambda = 0.9$ to $\lambda = 1.8$. The following theorem, which we state without proof, allows us to check the stability of the fixed point.

**Theorem 8.9**  *Let* $x_{n+1} = f(x_n)$, *and suppose that* $x_{eq}$ *is a fixed point, that is,*

$$x_{eq} = f\left(x_{eq}\right).$$

*If* $|f'(x_{eq})| < 1$, *then* $x_{eq}$ *is stable. If* $|f'(x_{eq})| > 1$, *then* $x_{eq}$ *is unstable.*

We say that a fixed point $x_{eq}$ for which $|f'(x_{eq})| \neq 1$ is *hyperbolic*. Theorem 8.9 is the analogue of Theorem 8.4 for maps.

◆ *Example 1*    We can use Theorem 8.9 to check the stability of the fixed point at 0 as follows. Since $f(x) = \lambda x(1 - x)$, we have

$$f'(0) = \lambda(1 - 2x)|_{x=0} = \lambda.$$

If $0 < \lambda < 1$, then the fixed point at 0 is stable. If $\lambda = 1$, then the origin is not hyperbolic; and if $\lambda > 1$, then the origin is unstable. Hence, a bifurcation occurs as $\lambda$ increases past 1.                                                                        ■

## The Period Doubling Sequence

Figure 8.35(a) shows the orbit of an initial condition near 0.8 for the logistic map when $\lambda = 3.3$. In contrast to the case in which $\lambda = 1.8$, the fixed point at $1 - 1/\lambda$ is no longer stable. Instead, the orbit approaches a *period-2* orbit. In other words, after a while, the orbit oscillates between two values, $p = 0.4794\ldots$ and $q = 0.8236\ldots$. It is straightforward to check that $f(p) = q$ and $f(q) = p$.

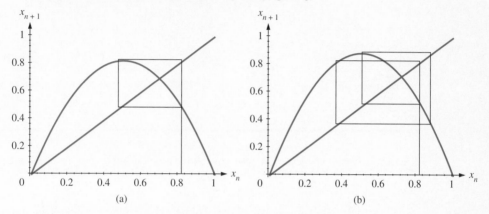

**FIGURE 8.35**   The logistic map and an associated periodic orbit for (a) $\lambda = 3.3$; (b) $\lambda = 3.54$.

Figure 8.35(b) shows the case in which $\lambda = 3.54$. Now the orbit approaches a *period-4* orbit. (The period-2 orbit is unstable.) That is, the orbit eventually oscillates among four values:

$$p = 0.3648\ldots,$$
$$q = 0.8203\ldots,$$
$$r = 0.5217\ldots,$$
$$s = 0.8833\ldots.$$

It is straightforward to verify that $f(p) = q$, $f(q) = r$, $f(r) = s$, and $f(s) = p$, completing the cycle. In fact, typical initial conditions approach this orbit after many iterations.

Figure 8.35 shows the effect of two *period-doubling* bifurcations that occur as $\lambda$ continues to increase. In other words, a stable period-1 orbit is replaced by a stable period-2 orbit and subsequently by a stable period-4 orbit.

In the late 1970s, Mitchell Feigenbaum investigated the behavior of the quadratic map and found that there is an infinite sequence of period-doubling bifurcations as $\lambda$ increases. In other words, there are values of $\lambda$ for which typical orbits approach cycles of period 16, 32, 64, 128, and so on. Table 8.1 shows the bifurcation values $\lambda_n$ at which a stable period $2^n$ orbit appears in the logistic map. As $\lambda$ approaches the value $\lambda_\infty = 3.5699\ldots$, the period of the orbit becomes infinite.

| $\lambda$ | Period |
|---|---|
| $\lambda_1 = 3$ | 2 |
| $\lambda_2 = 3.449\ldots$ | 4 |
| $\lambda_3 = 3.544\ldots$ | 8 |
| $\lambda_4 = 3.564\ldots$ | 16 |
| $\lambda_5 = 3.568\ldots$ | 32 |
| $\lambda_\infty = 3.5699\ldots$ | $\infty$ |

**TABLE 8.1**   Values of $\lambda$ leading to successive period doubling bifurcations in the logistic map.

Moreover, Feigenbaum has shown that period-doubling sequences are not unique to the logistic map. In fact, there is a large class of maps that exhibit period-doubling bifurcations.

## Chaos in the Logistic Map

So far, we have described a sequence of bifurcations that leads to periodic behavior for typical initial conditions. For instance, if $\lambda = 3.54$, then most initial conditions lead to a period-4 orbit. (There are exceptions, however. For instance, the orbit starting from $x_0 = 0$ stays at 0, and points that start exactly on the unstable period-1 and period-2 orbits remain on those orbits.)

One question concerns the behavior of typical orbits after the parameter $\lambda$ increases past $\lambda_\infty$. Although a complete answer is beyond the scope of this book, we can indicate a partial answer. In fact, chaotic behavior is seen for many values of $\lambda$ larger than $\lambda_\infty$.

Figure 8.36 shows a staircase diagram for the logistic map with $\lambda = 3.9$. The orbit wanders erratically all over the unit interval. There is no apparent pattern to the sequence of generated numbers.

Let us consider the following problem. The logistic map

$$x_{n+1} = \lambda x_n (1 - x_n)$$

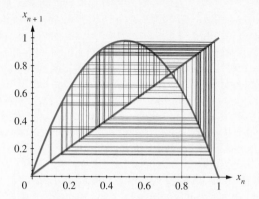

**FIGURE 8.36** A staircase diagram for the logistic map when $\lambda = 3.9$.

is an unambiguous rule for generating numbers, given an initial value $x_0$. Suppose $\lambda = 3.9$ and $x_0 = 0.8$. What is $x_{50}$?

This question is not easy to answer. The logistic map is easy to program on a computer. First, the authors wrote a program using a computer algebra system, requesting it to retain 10, 12, and 14 decimal digits of precision during each run. Next, the authors wrote a C-language program and did the calculation twice: once using single precision (corresponding to about 8 decimal digits of precision) and again using double precision (approximately 16 decimal digits). Table 8.2 lists the resulting answers. The answers clearly depend on the number of decimal digits of precision that are kept in the calculations.

Although the question of the value of $x_{50}$ is unambiguous from a mathematical point of view, the numerical answers that we have obtained are ambiguous. The nature

| Method | Computed Value of $x_{50}$ |
| --- | --- |
| 10 decimal digits | 0.896708 |
| 12 decimal digits | 0.103797 |
| 14 decimal digits | 0.372158 |
| C program, single precision | 0.231189 |
| C program, double precision | 0.368215 |

**TABLE 8.2** Estimates of $x_{50}$ for the logistic map with $\lambda = 3.9$ and $x_0 = 0.8$.

of the difficulty arises from the sensitive dependence on initial conditions in the logistic map for certain values of the parameter $\lambda$.

Imagine the following experiment, which we ask you to complete in the exercises (see Exercise 11). Let $x_0 = 0.8$ and calculate $x_1, x_2, \ldots, x_{20}$ in two ways using a software program that allows you to adjust the precision of numerical calculations. First, use 8 decimal digits of accuracy; call the resulting sequence $\hat{x}_1, \hat{x}_2, \ldots$. Second, use 10 decimal digits of accuracy; call the resulting sequence $\hat{y}_1, \hat{y}_2, \ldots$. After several steps, the computed values of $\hat{x}$ and $\hat{y}$ begin to differ in the last few decimal places. As the calculation proceeds, the differences affect more of the significant figures; eventually, the two sequences appear to be completely independent.

If a process is sensitive to small changes in initial conditions, then the numerical accuracy that is used to simulate it strongly influences the quantitative results that are obtained. In the preceding example, the differences between $\hat{x}_1$ and $\hat{y}_1$ initially are on the order of $10^{-8}$, owing to the difference in the precision. The errors continue to grow until they effectively swamp the calculation. The only way to determine the approximate value of $x_{50}$ is to use very high numerical accuracy.

## *Exercises*

**1.** A simple example of a chaotic process is

$$x_{n+1} = (10x_n) \bmod 1, \qquad (8.76)$$

where $x_0$ is an initial condition between 0 and 1. The modulus means to discard the integer part of the product $10x_n$. For example, if $x_0 = 0.123$, then $x_1 = 0.23$ and $x_2 = 0.3$.

(a) Suppose $x_0 = \pi/10$. What is the orbit that you obtain if you approximate $\pi$ as 3.1416?

(b) What is the orbit that you obtain if you approximate $\pi$ as 3.141592?

(c) How quickly do differences between the orbits in (a) and (b) grow as a function of $n$ for $n \le 5$?

(d) Suppose you need to know the value of $x_{10}$ to three significant digits. How accurately must you know the value of $\pi$?

(e) Suppose you need to know the value of $x_k$ to three significant digits. How accurately must you know the value of $\pi$?

**2.** Use Theorem 8.9 to assess the stability of the fixed point at $1 - 1/\lambda$ for the logistic map (8.75). For what range of values of $\lambda$ is the fixed point stable? Unstable? Not hyperbolic?

**3.** Let $x_{n+1} = ax_n$, where $a$ is a constant, and let $x_0 \ne 0$. Describe the orbit that results for each of the following values of $a$.

(a) $a = 1$            (b) $a = 2$

(c) $a = 1/2$        (d) $a = -1/2$

(e) $a = -1$

**4.** Let $x_{n+1} = ax_n$.

(a) Use Theorem 8.9 to determine the stability of the fixed point at 0 as a function of $a$.

(b) For which values of $a$ do points in typical orbits all have the same sign and approach 0 as $n \to \infty$?

(c) For which values of $a$ do points in typical orbits all have the same sign and approach infinity as $n \to \infty$?

(d) For which values of $a$ do points in typical orbits alternate sign and approach 0 as $n \to \infty$?

(e) For which values of $a$ do points in typical orbits alternate sign and approach infinity as $n \to \infty$?

**5.** Let $x_{n+1} = ax_n - x_n^2$, where $a$ is a positive constant.

(a) Find all the fixed points and use Theorem 8.9 to determine their stability as a function of $a$.

(b) Describe the behavior of typical orbits for $a = 0.9$, $a = 1$, and $a = 1.1$.

(c) Discuss the nature of the bifurcation that occurs as $a$ increases past 1.

**6.** Let $x_{n+1} = ax_n - x_n^2$, where $a$ is a negative constant.

(a) Find all the fixed points and use Theorem 8.9 to determine their stability as a function of $a$.

(b) Describe the behavior of typical orbits for $a = -0.9$, $a = -1$, and $a = -1.1$.

(c) Discuss the nature of the bifurcation that occurs as $a$ decreases past $-1$.

**7.** For each map in the following list, determine all the fixed points and use Theorem 8.9 to analyze their stability.

(a) $x_{n+1} = x_n(x_n - 1)(x_n - 2)$

(b) $x_{n+1} = x_n^2 + 1$

(c) $x_{n+1} = \sin x_n$

**8.** Generate staircase diagrams in an appropriate interval for each of the maps in the following list. Discuss the qualitative properties of typical orbits. Consider questions such as the following: Do the orbits all approach a periodic orbit or fixed point? Do the orbits wander erratically? Do orbits tend to $\infty$ or to $-\infty$?

(a) $x_{n+1} = x_n(x_n - 1)(x_n - 2)$

(b) $x_{n+1} = x_n^2 + 1$

(c) $x_{n+1} = \sin x_n$

**9.** Let $x_{n+1} = \lambda x_n(1 - x_n)$.

(a) Let $\lambda = 3.5$. Choose some different initial conditions and compute the resulting orbits. Do the initial conditions that you try eventually "settle down" to a predictable pattern? (You may need to compute several hundred points on the orbit before you can answer the question.) Discuss the pattern that you obtain.

(b) Repeat (a) for $\lambda = 3.55$.

(c) Repeat (a) for $\lambda = 3.565$.

(d) Repeat (a) for $\lambda = 3.569$.

**10.** Let $x_{n+1} = 3.9x_n(1 - x_n)$.

(a) Let $x_0 = 0.6$ and generate the next 15 iterates, that is, compute $x_1, x_2, \ldots, x_{15}$. Use a computer program, programmable calculator, or spreadsheet.

(b) Repeat (a) using the initial condition $y_0 = 0.61$.

(c) Let $\delta_n = |x_n - y_n|$. Plot $\ln \delta_n$ as a function of $n$. Do the points fall along a straight line, at least for sufficiently small values of $n$?

(d) What do your results in (c) imply about the rate of growth of $\delta$ for small values of $n$? (Is the growth rate linear, exponential, polynomial, or something else?)

**11.** Let $x_{n+1} = 3.9x_n(1 - x_n)$.

(a) Let $x_0 = 0.8$ and generate $x_1, x_2, \ldots, x_{30}$. Use a software package that allows you to control the number of significant digits in the calculation. Use 10 decimal digits here, if possible.

(b) Repeat the calculation in (a) using higher precision, say 12 decimal digits. Call the resulting orbit $y_1, y_2, \ldots, y_{30}$.

(c) Calculate $|x_n - y_n|$. This quantity allows you to estimate the error due to the differences in numerical precision. Plot $\ln |x_n - y_n|$ as a function of $n$. (Ignore those values of $n$ where the difference is 0.) Do the points tend to fall along a straight line?

(d) Use your results in (c) to interpret the rate of growth in the error, at least for moderate values of $n$. Is the growth exponential, at least for a time?

**12.** Let $x_{n+1} = 4.1x_n(1 - x_n)$.

(a) Show that there is an interval $I_1$ contained in the unit interval such that $x_1 > 1$ if $x_0 \in I_1$. What happens to the orbit if $x_0 \in I_1$?

(b) Show that there are two intervals $I_2^{(1)}$ and $I_2^{(2)}$, contained in the unit interval, such that if $x_0$ is in either $I_2^{(1)}$ or $I_2^{(2)}$, then $0 < x_1 < 1$ and $x_2 > 1$. What happens to the orbit starting from such an $x_0$?

(c) Show that there are four intervals $I_3^{(1)}, \ldots, I_3^{(4)}$ such that if $x_0$ is in one of them, then $x_1$ and $x_2$ lie between 0 and 1 but $x_3 > 1$.

(d) Can you determine a pattern here? How many intervals are there for which $x_4 > 1$ but $x_1, x_2$, and $x_3$ lie between 0 and 1?

**13.** Period-doubling bifurcations occur in differential equations as well as in difference equations like the logistic map. The Rössler equations are the system

$$x' = -y - z,$$
$$y' = x + ay,$$
$$z' = b + z(x - c),$$

where $a$, $b$, and $c$ are constants.

(a) Let $a = b = 0.2$ and let $c = 2.5$. Integrate the Rössler equations numerically from $0 \le t \le 200$ using the initial condition $(x_0, y_0, z_0) = (1, 1, 1)$. Eventually, $x$ "settles down" to a periodic function. Estimate the period.

(b) Repeat (a) for $c = 3.5$. What is the period now?

(c) Repeat (a) for $c = 4.2$. What is the period now?

(d) Let $c = 5$ and $a = b = 0.2$. Is $x$ periodic?

# Chapter 9

---

# NUMERICAL METHODS

---

In many cases, it is not possible to express the solution of an ordinary differential equation as the sum of a finite number of elementary functions (e.g., polynomials, exponentials or trigonometric functions of the independent variable). Many mathematical models, such as weather-forecasting models or simulations of air flow over an aircraft, consist of systems of thousands of differential equations. Even if it were possible to find an explicit solution, the enormous size and complexity of many systems would probably make meaningful interpretation of the resulting expressions very difficult.

From a theoretical perspective, the existence and uniqueness theorem (see Section 8.1) guarantees that many systems of differential equations do possess a unique solution, at least in principle. Because convenient analytical expressions for the solutions often do not exist, we usually resort to some sort of numerical approximation of the solution. The simplest example of such an approximation is Euler's method (see Sections 1.4 and 6.4).

The numerical solution of differential equations is an enormous subject. Much contemporary research is underway to improve existing algorithms and to develop algorithms that work effectively on parallel computers for applications like meteorology and aerospace engineering. In this brief chapter, we introduce some of the fundamental concepts and discuss some of the popular methods for finding numerical approximations of small systems of differential equations. Section 9.1 introduces some of the intrinsic errors that arise in virtually all numerical methods. Section 9.2 contains a brief introduction to difference equations, which is background material for Section 9.3. Section 9.3 discusses the stability of numerical methods; that is, whether a numerical method generates an approximation that is qualitatively and quantitatively similar to the true solution. Section 9.4 introduces two popular methods for solving differential equations: Heun's method and the fourth-order Runge-Kutta method.

## 9.1   ERRORS IN NUMERICAL METHODS

Every numerical method necessarily contains some error in the approximation of the solution of a given differential equation. Errors often are classified into three types:

1. the *local error*, which describes the error that occurs after one time step of the method;

2. the *global error*, which describes the cumulative effect of the local errors after many time steps; and

3. the *roundoff error*, which arises from the finite precision with which computers store and process numbers.

Let us see how these errors arise in the context of Euler's method for the initial value problem

$$x' = f(t, x), \qquad x(t_0) = x_0. \tag{9.1}$$

Suppose that $f$ satisfies the hypotheses of the existence and uniqueness theorem (Theorem 3.3). Euler's method is the rule

$$x_{n+1} = x_n + hf(t_n, x_n). \tag{9.2}$$

Euler's method simply approximates the solution at time $t_n + h$ by the tangent line through $(t_n, x_n)$. (See Section 1.4.) Here, $h$ is the step size. The graph of the true solution $x(t)$ is approximated by a sequence of points $(t_n, x_n)$ generated by using Eq. (9.2).

**Definition 9.1**   *Let the set* $\{(t_n, x_n) : n = 0, 1, \ldots, N\}$ *be an approximation to the solution* $x(t)$ *of the initial value problem (9.1). The* global error *at time* $t_n$ *is the difference between the true solution at* $t_n$ *and the approximation* $x_n$:

$$E_n = x(t_n) - x_n. \tag{9.3}$$

*The* absolute error *in the numerical approximation at* $t = t_n$ *is* $|E_n|$, *that is, the absolute value of the global error at* $t_n$.

*The* local error *is the difference between the true solution* $x(t_{n+1})$ *and the approximation after one time step, assuming that* $(t_n, x_n)$ *is a point on the graph of the true solution.*

Let $u(t)$ be the solution of the initial value problem

$$u' = f(t, u), \qquad u(t_n) = x_n.$$

The local error at the $n$th step of Euler's method is

$$E_n^{\text{loc}} = u(t_{n+1}) - [x_n + hf(t_n, x_n)]. \tag{9.4}$$

Figure 9.1 shows a schematic illustration of the global error $E_n$ and the local error $E_n^{loc}$. Notice that $E_n^{loc}$ error refers to the error that results when the numerical method is applied to advance the approximate solution from $t_n$ to $t_{n+1}$, whereas $E_n$ refers to the global error at $t_n$.

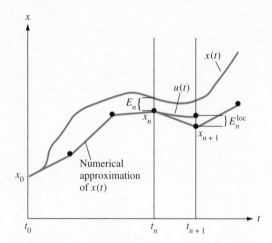

**FIGURE 9.1**   A schematic diagram of the local and global errors.

✦ *Example 1*   Let

$$x' = x, \qquad x(0) = 1. \tag{9.5}$$

The exact solution is $x(t) = e^t$. Table 9.1 shows the local and global errors for the Euler approximation to the solution of Eq. (9.5) for step sizes $h = 0.1$ and $h = 0.05$.

For each of the two values of $h$, the local error at the initial point and the global error after the first time step are identical. The local error at the second step is defined as the error in approximating the solution curve *assuming that the true solution curve goes through the point* $(t_1, x_1)$.

Although the local errors remain relatively small at each step, their effect is cumulative, and the global error grows with time. We observe the following:

- The local error grows as the value of the solution becomes larger.
- When the step size is halved from 0.1 to 0.05, the global error at $t = 1$ is also halved.
- When the step size is halved from 0.1 to 0.05, the local errors at $t = 0.1, 0.2, \ldots$, are reduced by approximately a factor of four.

| $t$ | $x(t)$ | $x_n$ | $E_n^{loc}$ | $E_n$ |
|---|---|---|---|---|
| | | $h = 0.1$ | | |
| 0 | 1 | 1 | 0.0052 | 0 |
| 0.10 | 1.105 | 1.100 | 0.0057 | 0.005 |
| 0.20 | 1.221 | 1.210 | 0.0063 | 0.011 |
| 0.30 | 1.350 | 1.331 | 0.0069 | 0.019 |
| 0.40 | 1.492 | 1.464 | 0.0076 | 0.028 |
| 0.50 | 1.649 | 1.611 | 0.0083 | 0.038 |
| 0.60 | 1.822 | 1.772 | 0.0092 | 0.051 |
| 0.70 | 2.014 | 1.949 | 0.0101 | 0.065 |
| 0.80 | 2.226 | 2.144 | 0.0111 | 0.082 |
| 0.90 | 2.460 | 2.358 | 0.0122 | 0.102 |
| 1.00 | 2.718 | 2.594 | 0.0134 | 0.125 |
| | | $h = 0.05$ | | |
| 0 | 1 | 1 | 0.0013 | 0 |
| 0.05 | 1.051 | 1.050 | 0.0013 | 0.001 |
| 0.10 | 1.105 | 1.103 | 0.0014 | 0.003 |
| 0.15 | 1.162 | 1.158 | 0.0015 | 0.004 |
| 0.20 | 1.221 | 1.216 | 0.0016 | 0.006 |
| 0.25 | 1.284 | 1.276 | 0.0016 | 0.008 |
| 0.30 | 1.350 | 1.340 | 0.0017 | 0.010 |
| 0.35 | 1.419 | 1.407 | 0.0018 | 0.012 |
| 0.40 | 1.492 | 1.477 | 0.0019 | 0.014 |
| 0.45 | 1.568 | 1.551 | 0.0020 | 0.017 |
| 0.50 | 1.649 | 1.629 | 0.0021 | 0.020 |
| 0.55 | 1.733 | 1.710 | 0.0022 | 0.023 |
| 0.60 | 1.822 | 1.796 | 0.0023 | 0.026 |
| 0.65 | 1.916 | 1.886 | 0.0024 | 0.030 |
| 0.70 | 2.014 | 1.980 | 0.0026 | 0.034 |
| 0.75 | 2.117 | 2.079 | 0.0027 | 0.038 |
| 0.80 | 2.226 | 2.183 | 0.0028 | 0.043 |
| 0.85 | 2.340 | 2.292 | 0.0030 | 0.048 |
| 0.90 | 2.460 | 2.407 | 0.0031 | 0.053 |
| 0.95 | 2.586 | 2.527 | 0.0033 | 0.059 |
| 1.00 | 2.718 | 2.653 | 0.0034 | 0.065 |

TABLE **9.1** The exact solution, the Euler approximation, the local error, and the global error for Eq. (9.5) for two values of the step size $h$.

An interesting question is whether the global error at time $t_N$ is the sum of all of the local errors $E_n^{\text{loc}}$ for $n < N$. The answer in general is "no" (but see Exercise 11). When $h = 0.1$, the sum of the local errors from $t = 0$ to $t = 1$ is

$$\sum_{n=0}^{9} E_n^{\text{loc}} \approx 0.0826.$$

When $h = 0.05$, the sum of the local errors from $t = 0$ to $t = 1$ is

$$\sum_{n=0}^{19} E_n^{\text{loc}} \approx 0.0420.$$

These values are comparable to, but not equal to, the global error at $t = 1$ for each value of $h$. As we will discuss in more detail, the local errors often allow one to estimate the global error in a numerical solution.   ∎

Of course, the local and global errors are not known exactly unless the solution is known exactly. In practice, the best that one can do is to try to estimate the errors, but accurate estimates can be difficult to obtain. Some rules of thumb are possible. Chapter 1 illustrates the following fact:

> *The global error in Euler's method decreases linearly with the step size.*

In other words, if the step size is cut in half, then the global error is cut in half. Example 1 also suggests the following observation, which we will discuss in more detail:

> *The local error in Euler's method decreases as the square of the step size.*

Thus, if the step size is cut in half, then the local error at each step is reduced by a factor of four, as illustrated in Example 1.

## Estimating Local and Global Errors

It is possible to determine an explicit relation between the local and global error in Euler's method. In this section, we discuss a theoretical analysis that gives some bounds on the size of the error as a function of the step size.

Let $x' = f(t, x)$, and assume that $x$ is at least twice differentiable. (It can be shown that this assumption is valid if $f$ is continuously differentiable.) Taylor's theorem implies that

$$x(t_0 + h) = x(t_0) + hx'(t_0) + \tfrac{1}{2}h^2 x''(\xi)$$

$$= x_0 + hf(t_0, x_0) + \tfrac{1}{2}h^2 x''(\xi), \tag{9.6}$$

where $\xi$ is a value between $t_0$ and $t_0 + h$. The local error in Euler's method is the error after one step of the method:

$$E^{\text{loc}} = \tfrac{1}{2}h^2 x''(\xi).$$

In other words, the local error is proportional to the square of the step size.

◆ *Example 2*    Let

$$x' = -x, \qquad x(0) = 1 \tag{9.7}$$

and determine the local error after one step of Euler's method for $h = 0.1, 0.01, 0.001$, and $0.0001$.

*Solution*    We have

$$x_1 = x_0 + hx_0 = 1 - h,$$

so the local error is

$$E^{\text{loc}} = e^{-h} - (1 - h).$$

Table 9.2 shows the numerical results. The table suggests that if $h$ is reduced by 10, then the local error is reduced by a factor of 100.

| $h$ | 0.1 | 0.01 | 0.001 | 0.0001 |
|---|---|---|---|---|
| $E^{\text{loc}}$ | $4.837 \times 10^{-3}$ | $4.983 \times 10^{-5}$ | $4.998 \times 10^{-7}$ | $4.999 \times 10^{-9}$ |

**TABLE 9.2**    Local error in Eq. (9.7) after one step of Euler's method for various step sizes $h$.   ■

The preceding discussion suggests that the local error in Euler's method is proportional to $h^2$ and the global error is proportional to $h$. A rigorous proof of this statement is possible. However, the following heuristic argument motivates the basic idea behind the proof.

If $h$ is made smaller, then more time steps are required to approximate the solution at some prespecified value of $t$. For instance, if $x(0) = x_0$ and we want to know $x(T)$, then the number of time steps needed is approximately $T/h$. If the local error is roughly constant for $0 \le t \le T$, then the global error $E$ at $t = T$ is proportional to $h$, because

$$\begin{aligned} E(T) &\propto \text{number of iterations} \times E^{\text{loc}} \\ &\propto (T/h) \times h^2 \\ &\propto h. \end{aligned}$$

(See Exercises 11–12.)

We say that Euler's method is a *first-order* method, because the global error is proportional to $h$. The numerical methods that we discuss later in the chapter are better in the sense that the global error is proportional to a larger power of $h$. For example, the global error in the *fourth-order* Runge-Kutta method is proportional to $h^4$ (see Section 9.4).

It is possible to estimate the local errors in each step of Euler's method more precisely. Let

$$x' = f(t, x), \tag{9.8}$$

where $x(t_0) = x_0$. Equation (9.6) shows that the local error is

$$E^{\text{loc}} = \tfrac{1}{2}h^2 x''(\xi)$$

for some value of $\xi$ between $t_0$ and $t_0 + h$. Although the value of $x''(\xi)$ is not known exactly, we can estimate it as follows.

Suppose that we want to find a numerical approximation of the solution of Eq. (9.8) in a box $S$ around the initial condition $(t_0, x_0)$. Figure 9.2 shows a schematic diagram of the situation. As long as the solution curve remains in $S$, we have

$$
\begin{aligned}
x''(t) &= \frac{d}{dt}x'(t) \\[2mm]
&= \frac{d}{dt}f(t, x(t)) && \text{by Eq. (9.8)} \\[2mm]
&= \frac{\partial f}{\partial t} + \frac{\partial f}{\partial x}x'(t) && \text{by the chain rule} \\[2mm]
&= \frac{\partial f}{\partial t} + f(t, x(t))\frac{\partial f}{\partial x} && \text{by Eq. (9.8).} \tag{9.9}
\end{aligned}
$$

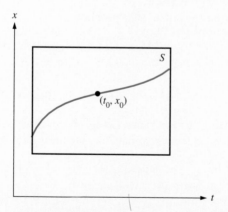

**FIGURE 9.2**   Schematic diagram of the solution of Eq. (9.8).

Therefore, $|x''(\xi)|$ is bounded on $S$ by the maximum absolute value on $S$ of each term on the right-hand side of Eq. (9.9). Let

$$M_1 = \max_{(t,x)\in S}\left|\frac{\partial f}{\partial t}\right|, \qquad M_2 = \max_{(t,x)\in S}|f(t,x)|, \qquad \text{and} \qquad M_3 = \max_{(t,x)\in S}\left|\frac{\partial f}{\partial x}\right|.$$

Then

$$\left|x''(\xi)\right| \leq M_1 + M_2 M_3.$$

Notice that $M_3$ is just the Lipschitz constant for $f$ on $S$, as is discussed in Chapter 3.

Although the local error in Euler's method is proportional to $h^2$, the proportionality constant is much larger for some differential equations than for others. As a result, the solution of some differential equations is difficult to approximate accurately, even with a relatively small step size.

◆ *Example 3*  Let

$$x' = -x, \qquad x(0) = 1,$$

and suppose we want to compute the solution for $0 \leq t \leq 1$. Let

$$S = \{(t,x) : 0 \leq t \leq 1,\ 0 \leq x \leq 1\}.$$

Then $M_1 = 0$ and $M_3 = 1$. The solution decreases with $t$ (why?), so $M_2 = 1$. Hence, the local error in Euler's method satisfies

$$E^{\text{loc}} \leq \tfrac{1}{2}h^2$$

on $S$. (In fact the inequality holds for all $t > 0$.)

A contrasting situation arises for the initial value problem

$$x' = 1 + x^2, \qquad x(0) = 0 \tag{9.10}$$

on the box

$$S = \{(t,x) : 0 \leq t \leq 1.5,\ 0 \leq x \leq \tan 1.5 \approx 14.10\}.$$

The true solution of Eq. (9.10) is $x(t) = \tan t$. Let us investigate the size of the local error when Euler's method is used to approximate the value of $x$ in $S$. We have

$$M_1 = 0,$$
$$M_2 = \max_{S}\left|1 + x^2\right| \approx 199.85,$$
$$M_3 = \max_{S}|2x| \approx 28.2.$$

Therefore, the local error satisfies

$$E^{\text{loc}} = \tfrac{1}{2}h^2 \left| x''(\xi) \right|$$
$$\leq \tfrac{1}{2}h^2(M_1 + M_2 M_3)$$
$$\approx 2818h^2$$

as long as the solution curve remains in $S$.

Of course, the preceding calculation is only an upper bound for the local error. However, the solution of Eq. (9.10) is always increasing and eventually reaches the top of the box. (In fact, there is a singularity at $t = \pi/2$.) The second derivative $x''(\xi)$, and hence the local error, increases as the solution reaches the top of the box, and the global error becomes rather large. The true solution is $x(1.5) = \tan 1.5 \approx 14.10$; Table 9.3 shows the approximation generated by Euler's method for the indicated step sizes $h$.

| $h$ | 0.1 | 0.01 | 0.001 | 0.0001 |
|---|---|---|---|---|
| $\hat{x}(1.5)$ | 4.638 | 10.60 | 13.59 | 14.04 |

TABLE 9.3   Numerical estimates of the solution of Eq. (9.10) at $t = 1.5$ using Euler's method for the indicated step sizes $h$.

Only for a step size of $10^{-4}$ is the global error in the solution at $t = 1.5$ less than 0.1. ∎

## The Effect of Roundoff Errors

The preceding examples have shown that the global error in Euler's method decreases at a rate proportional to $h$. However, as the next example shows, it is sometimes possible to observe a paradoxical situation in which the observed errors grow even when $h$ continues to decrease.

✦ *Example 4*   We can approximate the value of $e^{-1/10}$ by integrating the initial value problem

$$x' = -x, \qquad x(0) = 1 \tag{9.11}$$

from $t = 0$ to $t = 1/10$. (Why?) Figure 9.3 shows a Fortran program that implements Euler's method for Eq. (9.11). If we take 10 time steps, that is, if $n = 10$, then $h = 1/100$. The global error in the resulting solution is $4.55 \times 10^{-4}$. If we take 100 time steps

```
x = 1.0
read *, n
h = 0.1/n
do j=1,n
    x = x - h*x
enddo
print *, x, abs(x - exp(-0.1))
end
```

**FIGURE 9.3**    A Fortran program to integrate Eq. (9.11).

instead, then the step size is $h = 1/1000$, and the global error in the resulting solution is reduced to $4.55 \times 10^{-5}$. If we continue in this way, then we obtain the results in Table 9.4.

| $h$ | $n$ | $\hat{x}(0.1)$ | Error |
|---|---|---|---|
| $10^{-2}$ | $10$ | 0.904382 | $4.55 \times 10^{-4}$ |
| $10^{-3}$ | $10^2$ | 0.904792 | $4.55 \times 10^{-5}$ |
| $10^{-4}$ | $10^3$ | 0.904833 | $4.77 \times 10^{-6}$ |
| $10^{-5}$ | $10^4$ | 0.904838 | $3.58 \times 10^{-7}$ |
| $10^{-6}$ | $10^5$ | 0.904924 | $8.64 \times 10^{-5}$ |
| $10^{-7}$ | $10^6$ | 0.887430 | $1.740 \times 10^{-2}$ |

**TABLE 9.4**    Numerical estimates of the solution of Eq. (9.11) and the error at $t = 0.1$ using the indicated step sizes $h$.

The third column shows the numerical estimate of the solution of Eq. (9.11) at $t = 1/10$, denoted $\hat{x}(0.1)$. When $h$ is $10^{-4}$ or larger, the global error decreases linearly with $h$. However, if $h$ is made still smaller, then the global error starts to increase again. Instead of being the most accurate approximation, a step size of $10^{-7}$ produces the *least* accurate approximation!

Clearly, something has gone wrong. The culprit is roundoff error, and the difficulty involves the way in which most computers store and process floating-point numbers. The variables x and h in the program in Fig. 9.3 are single-precision variables; on a typical computer, single-precision variables are stored with approximately eight significant digits.

Initially, the value of x is stored as

$$x_0 = 1.00\,000\,00,$$

which is "exact." The first step of Euler's method produces

$$x_1 = x_0 + hx_0 = 1 + 10^{-7} \times 1.$$

Assuming eight significant figures, this result in the computer produces

$$x_1 = 1.00\,000\,01.$$

In principle, the second step of Euler's method produces

$$x_2 = x_1 + hx_1$$
$$= 1.0000001 + 10^{-7} \times 1.0000001$$
$$= 1.00\,000\,020\,000\,001.$$

However, the computed result is different. Because the computer retains only eight significant digits, all the digits to the right of the 2 are truncated, and the stored value of $x_2$ is

$$x_2 = 1.00\,000\,02.$$

This truncation may seem trivial. In fact, it would be if we were interested only in the results of the first few steps of Euler's method. However, if $h = 10^{-7}$, then *one million* steps of Euler's method are needed to approximate the solution at $t = 1/10$. After a while, the continued truncations affect the computed results in a significant way. Thus, roundoff error causes the global error to increase, rather than decrease, as the step size becomes sufficiently small.                                                                          ■

Roundoff error effectively limits the accuracy with which the solution of a differential equation can be computed numerically. It is not uncommon in practice for a numerical simulation to proceed for millions of time steps. In such situations, roundoff error can be a serious problem.

Much scientific computation is performed in *double* precision, which typically affords about 16 significant digits of precision. Double precision can mitigate the effects of roundoff error, but there is no way to avoid roundoff error entirely in present-day computers. In practice, there is always a tradeoff: A smaller step size reduces the global error in a mathematical sense but often increases the roundoff error in real computations.

## *Exercises*

**1.** Let $x' = \sin t$. Use Euler's method to calculate a numerical approximation of the solution at $t = 1$ that satisfies the initial condition $x(0) = 1$ for the step sizes $h = 1$, $h = 0.1$, and $h = 0.01$. Use the exact solution to determine the global error in the solution at $t = 1$. Does the global error decrease at a rate proportional to $h$?

**2.** Use an analysis similar to that in Example 3 to find an upper bound on the local error in Euler's method in a suitable box for the initial value problem in Exercise 1.

**3.** Let $x' = t \sin x$.

(a) Use Euler's method to calculate a numerical approximation of the solution at $t = 1$ that satisfies the initial condition $x(0) = 1$ for step sizes $h = 10^{-1}$, $10^{-2}$, $10^{-3}$, and $10^{-4}$.

(b) Determine the global error in the solution for each of the step sizes in (a). Does the global error decrease linearly with $h$?

(c) For $h = 10^{-1}$ and $h = 10^{-2}$, compute the local errors at each step of the method. What is the approximate ratio between the local errors at $t = 0.1$, $0.2$, $\ldots$, $1.0$? Do the ratios that you obtain suggest that the local error decreases at a rate proportional to $h^2$?

**4.** Using a suitable modification of the program in Fig. 9.3 (which can just as well be recast in C or some other language), rerun the calculation in Example 4 using double-precision arithmetic. Do step sizes of $10^{-5}$, $10^{-6}$, and $10^{-7}$ reduce the global error appreciably? Explain.

**5.** Suppose you have a calculator that can add, subtract, multiply, and divide but that cannot calculate any of the trigonometric functions with the touch of a key. It is possible to compute a quantity such as $\tan 0.5$ on such a device by applying Euler's method

to Eq. (9.10). Using single-precision arithmetic to approximate $\tan 0.5$ with Euler's method, determine a step size $h$ that minimizes the absolute error in the solution at $t = 0.5$.

**6.** One way to determine the precision with which floating-point numbers are represented on a given computer is to add successively smaller numbers to 1. Of course, in a mathematical sense, there is no nonzero number $x$ such that $1.0 + x = 1.0$. However, on a real computer with, say, eight significant digits of accuracy, the *numerical* result of the expression $1.0 + x$ is $1.0$ whenever $|x|$ is less than about $10^{-7}$.

For instance, let $x = 10^{-8}$. With eight significant digits, the sum $1.0 + x$ is performed as

$$1.00\,000\,00 + 0.00\,000\,001,$$

and the result is truncated after the seventh decimal place. Therefore, the computed result is $1.00\,000\,00$. The smallest positive number $x$ for which the *numerical* relation $1 + x > 1$ is called the *machine epsilon*. One way to compute the machine epsilon is to set $x_0 = 1$, then compute the sequence $x_1 = x_0/b$, $x_2 = x_1/b$, and so on, where $b$ is a suitable constant. The machine epsilon is the number $x_n$ such that $1 + x_n > 1$ and $1 + x_{n+1} = 1$ when the calculations are performed on the computer.

The number $b$ is the base in which numbers are stored internally in the computer. Many pocket calculators use base 10. Most digital computers (including virtually all personal computers) use base 2.

(a) Determine the machine epsilon on your pocket calculator.

(b) Determine the machine epsilon on your computer using single-precision arithmetic. (This computation requires some care. Some C compilers use double precision for all arithmetic calculations, even if the variables are declared to be single precision. Pascal compilers vary in their treatment of `real` declarations. Other languages assume that all floating-point variables are double precision. Some symbolic algebra programs use a base other than 2 and do floating-point arithmetic using specially coded routines instead of the computer's native floating-point instructions. Many personal computers retain extra precision in intermediate floating-point calculations, then round the results when they are stored into memory. Read your language manual carefully to determine how floating-point variables are represented.)

(c) Determine the machine epsilon on your computer using double-precision arithmetic.

**7.** One solution of the initial value problem $x' = \sqrt{x}$, $x(0) = 0$ is $x(t) = t^2/4$. Apply Euler's method for your choice of step size, and plot the numerically generated solution along with $x(t)$ for $0 \le t \le 1$. What happens? Can you suggest a possible cause of the different answers?

**8.** Let $x' = x^2$, where $x(0) = 0.9$.

(a) Determine the exact solution of the initial value problem and find $x(1)$.

(b) Use Euler's method to find a numerical approximation of $x(1)$ using step sizes $h = 10^{-1}, 10^{-2}, \ldots, 10^{-5}$. Determine the global error in the solution at $t = 1$ for each step size.

(c) Is there a range of values of $h$ for which the absolute error in the solution at $t = 1$ decreases linearly with $h$? If so, what is that range of values?

**9.** Let $x' = x^2 + 1$ and suppose that Euler's method is used to approximate the solution satisfying the initial condition $x(0) = 0$. How small must $h$ be so that the local error that occurs at $t = 1.5$ is less than 1? (See Example 3.)

**10.** Let $x' = x^2$.

(a) Use Euler's method to compute the solution satisfying the initial condition $x(0) = 1/2$ using

a step size of $1/10$, and find the global error at $t = 1$.

(b) Determine the local error that occurs at each time step in (a).

(c) Repeat your computations using a step size of $1/100$. Is the global error at $t = 1$ about 10 times smaller than that in (a)?

(d) Determine the local errors at $t = 0.1, 0.2, \ldots$, $1.0$ when $h = 1/100$. Are they also 10 times smaller than the local errors in (b)? Explain.

**11.** (a) Let $x' = 3t^2$. Use Euler's method with a step size of $h = 1/10$ to find a numerical approximation of the solution satisfying the initial value $x(0) = 0$. Compute the local error and the global error at each step of your calculation.

(b) Find the sum of all the local errors in (a) and show that it equals the global error at $t = 1$.

(c) Repeat (a) and (b) with $h = 1/100$.

**12.** Let $x' = -x$. Use Euler's method with a step size of $0.01$ to approximate the solution satisfying the initial condition $x(0) = 1$.

(a) Determine the sum of the local errors committed at each step of the calculation.

(b) Determine the global error in the solution at $t = 1$ and compare it to the absolute value of your answer in (a). Are they the same?

**13.** If the step size is negative, then Euler's method computes the solution of an initial value problem backward in time.

(a) Let $x' = t - x$ and use Euler's method to compute the solution satisfying the initial condition $x(0) = 1$ using a step size $h = +0.2$. (In other words, go forward in time from $t = 0$ to $t = 1$.)

(b) Denote your answer in (a) by $\hat{x}_5$. (The subscript refers to the numerical solution after five Euler steps.) Use Euler's method with a step size of $h = -0.2$ to compute the solution satisfying the initial condition $x(1) = \hat{x}_5$. (That is, go backward in time from $t = 1$ to $t = 0$.) Does your solution go back to where it started from? In other words, is the value of your numerically generated solution equal to 1 when $t = 0$?

(c) Explain why the numerical solution does not go back to where it started from. (Hint: Examine the analytical solutions that satisfy the initial conditions $x(0) = 1$ and $x(1) = \hat{x}_5$.)

## 9.2   A PRIMER ON LINEAR DIFFERENCE EQUATIONS

### Difference Equations and Numerical Methods

Every numerical method approximates the solution of a given differential equation, which is a continuous function $x(t)$, by a discrete set of points of the form $\{(t_n, x_n)\}$. The objective is to compute the points $(t_n, x_n)$ so that they lie very close to the graph of $x(t)$ over some interval of interest.

For example, let us apply Euler's method to the initial value problem

$$x' = -x, \qquad x(0) = 1. \tag{9.12}$$

Euler's method begins with the initial point $x_0 = 1$, corresponding to the initial condition. Subsequent points on the numerical solution are produced by the equation

$$x_{n+1} = x_n - hx_n = (1 - h)x_n, \tag{9.13}$$

where $h$ is the step size. Thus, if $h = 1/10$, then Euler's method produces the sequence $x_1 = 0.9$, $x_2 = 0.9^2 = 0.81$, and so on.

Equation (9.13) is an example of a *difference equation*. In the simplest case, a difference equation is simply a rule for generating $x_{n+1}$ given $x_n$. The sequence $x_0, x_1, x_2, \ldots$ is called the *orbit* of $x_0$ or the *iterates* of $x_0$. The point $x_0$ is the *initial condition* for the difference equation. We can write $x_n$ in terms of the initial condition $x_0$; when $h = 1/10$, we have $x_n = 0.9^n x_0$. This relation is a *solution* of Eq. (9.13) when $h = 0.1$, because we can write $x_n$ as an explicit function of $x_0$. (See also Exercise 1.)

Section 9.1 introduces the question of how accurately the orbit of the difference equation (9.13) approximates the true solution of Eq. (9.12). The global error in the numerical approximation is small provided that the step size $h$ is sufficiently small. However, that discussion raises as many questions as it answers. How small should the step size be to get an accurate numerical solution? What happens if the step size is too large?

No complete answer can be given to these questions that applies to an arbitrary differential equation. The discussion in Section 9.3 shows that a numerical solution may bear no resemblance at all to the true solution if the step size is not carefully chosen. To investigate this issue in more detail, we need to know some basic facts about difference equations, which are the subject of the rest of this section.

## A Simple One-Dimensional Map

One of the simplest difference equations is

$$x_{n+1} = ax_n, \tag{9.14}$$

where $a$ is a constant. We often refer to a difference equation as a *map*.[1] Equation (9.14) is a *one-dimensional* map because $x_n$ is a single number, and $x_{n+1}$ depends only on its immediate predecessor, $x_n$.

There is a close parallel between the behavior of the orbits of Eq. (9.14) and the solutions of the exponential growth equation $x' = ax$. For example, suppose $a = 2$. Then Eq. (9.14) is equivalent to

$$x_{n+1} = 2x_n. \tag{9.15}$$

The orbit of the initial condition $x_0 = 1$ is $1, 2, 4, 8, 16, 32, \ldots$. Equation (9.15) is simply a doubling map; the orbits grow exponentially.

In contrast, suppose $a = 1/2$. Then the orbit of the initial condition $x_0 = 1$ is $1, 1/2, 1/4, 1/8, 1/16, \ldots$; the orbits shrink exponentially to 0. If $a$ is negative, then the iterates alternate in sign. For instance, if $a = -1$, then the orbit of $x_0 = 1$ is $1, -1, 1, -1, \ldots$.

These examples above illustrate the following theorem, whose proof is straightforward (see Exercise 1).

**Theorem 9.2** *Let $x_{n+1} = ax_n$, where $a$ is a constant. Suppose $x_0 \neq 0$.*

- *If $|a| < 1$, then $\lim_{n\to\infty} |x_n| = 0$.*
- *If $|a| > 1$, then $\lim_{n\to\infty} |x_n| = \infty$.*
- *If $a > 0$, then $x_1, x_2, \ldots$ have the same sign as $x_0$.*
- *If $a < 0$, then the iterates of $x_0$ alternate in sign.*

## Linear Two-Dimensional Maps

A difference equation of the form $x_{n+1} = f(x_n)$ is said to be *linear* if $f$ is a linear function. For example, the equation $x_{n+1} = ax_n$ is linear because $f(x) = ax$ is a linear function.

If $\mathbf{x}$ is a 2-vector, and if $\mathbf{f}$ maps 2-vectors to 2-vectors, then the difference equation

$$\mathbf{x}_{n+1} = \mathbf{f}(\mathbf{x}_n) \tag{9.16}$$

---

[1] In contrast, a differential equation is often called a *flow* because the solution curves can be pictured as flowing along the slope lines of the direction field.

is a *two-dimensional* map. If $\mathbf{f}$ is linear, then Eq. (9.16) can be written in the form

$$\mathbf{x}_{n+1} = \mathbf{A}\mathbf{x}_n, \tag{9.17}$$

where $\mathbf{A}$ is a matrix. The simplest example of such a map is when $\mathbf{A}$ is a constant diagonal matrix; then Eq. (9.17) is equivalent to the system

$$\mathbf{x}_{n+1} = \begin{pmatrix} x_{n+1}^{(1)} \\ x_{n+1}^{(2)} \end{pmatrix} = \begin{pmatrix} a_{11}x_n^{(1)} \\ a_{22}x_n^{(2)} \end{pmatrix}, \tag{9.18}$$

where $a_{11}$ and $a_{22}$ are the diagonal elements of $\mathbf{A}$.

Theorem 9.2 can be used to analyze the orbits of Eq. (9.18). There are three basic cases, assuming that $\mathbf{x}_0 \neq \mathbf{0}$:

- If $|a_{11}| < 1$ and $|a_{22}| < 1$, then $\lim_{n\to\infty} \|\mathbf{x}_n\| = 0$; that is, the orbits all tend to the origin. We say that the origin is a *sink*.

- If $|a_{11}| > 1$ and $|a_{22}| > 1$, then $\lim_{n\to\infty} \|\mathbf{x}_n\| = \infty$. We say that the origin is a *source*.

- If $|a_{11}| > 1$ and $|a_{22}| < 1$, and if $x_0^{(1)} \neq 0$, then the orbits approach the horizontal axis as they move toward infinity; that is,

$$\lim_{n\to\infty} \left\|x_n^{(1)}\right\| = \infty \qquad \text{and} \qquad \lim_{n\to\infty} \left\|x_n^{(2)}\right\| = 0.$$

Figure 9.4 shows a schematic representation of what two different orbits might look like in the case where $a_{11} > 1$ and $0 < a_{22} < 1$. An analogous statement can be made

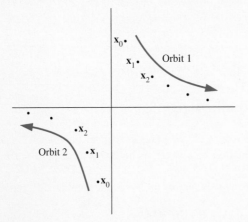

**FIGURE 9.4**    A schematic diagram of two different orbits near a saddle fixed point at the origin.

when $|a_{11}| < 1$ and $|a_{22}| > 1$. We say that the origin is a *saddle fixed point* or simply a *saddle*.

Of course, there are other special cases in which $a_{11} = 1$ or $a_{22} = 1$. Notice that if $a_{11}$ or $a_{22}$ is negative, then the corresponding component in the orbit of $\mathbf{x}_0$ alternates in sign, assuming that the corresponding component of $\mathbf{x}_0$ is nonzero.

As in the case of linear systems of differential equations with constant coefficients, the eigenvalues and eigenvectors of $\mathbf{A}$ are crucial to analyzing the orbits of Eq. (9.17). Suppose that $\mathbf{v}$ is an eigenvector of $\mathbf{A}$ corresponding to the eigenvalue $\lambda$; for simplicity, assume that $\lambda$ is real. Also suppose that $\mathbf{x}_0 = c\mathbf{v}$. Then

$$\mathbf{x}_1 = \mathbf{A}\mathbf{x}_0 = c(\mathbf{A}\mathbf{v}) = c\lambda\mathbf{v}.$$

(Why?) More generally,

$$\mathbf{x}_n = c\lambda^n\mathbf{v}.$$

Thus, if the initial condition lies on an eigenvector, then the orbit remains on the eigenvector. The orbit tends toward the origin if $|\lambda| < 1$ and tends toward infinity if $|\lambda| > 1$.

Now suppose that $\mathbf{A}$ is a $2 \times 2$ matrix with real, distinct eigenvalues $\lambda_1$ and $\lambda_2$ and corresponding eigenvectors $\mathbf{v}_1$ and $\mathbf{v}_2$. The following theorem allows us to express any 2-vector $\mathbf{x}$ as a unique linear combination of $\mathbf{v}_1$ and $\mathbf{v}_2$.

**Theorem 9.3**  *Let $\mathbf{A}$ be a $2 \times 2$ matrix with real, distinct eigenvalues $\lambda_1$ and $\lambda_2$ and corresponding eigenvectors $\mathbf{v}_1 = (v_{11}, v_{21})$ and $\mathbf{v}_2 = (v_{12}, v_{22})$. Given any 2-vector $\mathbf{x} = (x_1, x_2)$, there exist unique constants $c_1$ and $c_2$ such that*

$$\mathbf{x} = c_1\mathbf{v}_1 + c_2\mathbf{v}_2.$$

**Proof**   The vectors $\mathbf{v}_1$ and $\mathbf{v}_2$ are linearly independent (see Theorem 7.14 in Section 7.4). The relation $\mathbf{x} = c_1\mathbf{v}_1 + c_2\mathbf{v}_2$ is equivalent to the $2 \times 2$ system

$$c_1v_{11} + c_2v_{12} = x_1,$$

$$c_1v_{21} + c_2v_{22} = x_2.$$

This system has a unique solution, because its determinant is nonzero (see Exercise 12 in Section. 7.1).    ∎

Theorem 9.3 allows us to analyze the orbits of any two-dimensional linear difference equation of the form $\mathbf{x}_{n+1} = \mathbf{A}\mathbf{x}_n$, provided that $\mathbf{A}$ has real, distinct eigenvalues. Given $\mathbf{x}_0$, we can find unique constants $c_1$ and $c_2$ such that $\mathbf{x}_0 = c_1\mathbf{v}_1 + c_2\mathbf{v}_2$. Since matrix multiplication is a linear operation, we have

$$
\begin{aligned}
\mathbf{x}_1 &= \mathbf{A}\mathbf{x}_0 \\
&= \mathbf{A}(c_1\mathbf{v}_1 + c_2\mathbf{v}_2) \\
&= c_1\mathbf{A}\mathbf{v}_1 + c_2\mathbf{A}\mathbf{v}_2 \\
&= c_1\lambda_1\mathbf{v}_1 + c_2\lambda_2\mathbf{v}_2
\end{aligned}
$$

because $\mathbf{v}_1$ and $\mathbf{v}_2$ are eigenvectors. More generally,

$$
\mathbf{x}_n = \mathbf{A}^n\mathbf{x}_0 = c_1\lambda_1^n\mathbf{v}_1 + c_2\lambda_2^n\mathbf{v}_2. \tag{9.19}
$$

The preceding analysis can be used to prove the following theorem.

**Theorem 9.4** *Let $\mathbf{x}_{n+1} = \mathbf{A}\mathbf{x}_n$ be a two-dimensional difference equation in which $\mathbf{A}$ is a constant $2 \times 2$ matrix with real, distinct eigenvalues $\lambda_1$ and $\lambda_2$.*

- *If $|\lambda_1| < 1$ and $|\lambda_2| < 1$, then the origin is a sink.*
- *If $|\lambda_1| > 1$ and $|\lambda_2| > 1$, then the origin is a source.*
- *If $|\lambda_1| < 1$ and $|\lambda_2| > 1$, or if $|\lambda_1| > 1$ and $|\lambda_2| < 1$, then the origin is a saddle.*

◆ *Example 1*    Let

$$
\mathbf{x}_{n+1} = \mathbf{A}\mathbf{x}_n = \begin{pmatrix} 17/12 & -1/4 \\ 1/4 & 7/12 \end{pmatrix} \mathbf{x}_n. \tag{9.20}
$$

The dots in Fig. 9.5 represent the first few iterates of $\mathbf{x}_0 = (3/2, 4)$ under Eq. (9.20).

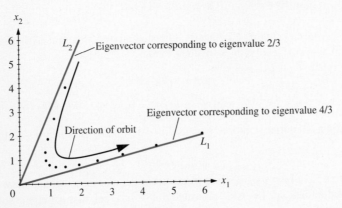

**FIGURE 9.5**    An orbit of Eq. (9.20).

The orbit moves in along the line $L_2$ for a time, then moves away along the line $L_1$. We can understand the motion of the orbit as follows.

One eigenvalue of $\mathbf{A}$ is $\lambda_1 = 4/3$ with corresponding eigenvector $\mathbf{v}_1 = (3, 1)$, and the other eigenvalue is $\lambda_2 = 2/3$ with corresponding eigenvector $\mathbf{v}_2 = (1, 3)$. If

$$\mathbf{x}_0 = c_1\mathbf{v}_1 + c_2\mathbf{v}_2,$$

then

$$\begin{aligned}\mathbf{x}_1 &= \mathbf{A}(c_1\mathbf{v}_1 + c_2\mathbf{v}_2)\\ &= c_1\mathbf{A}\mathbf{v}_1 + c_2\mathbf{A}\mathbf{v}_2\\ &= \tfrac{4}{3}c_1\mathbf{v}_1 + \tfrac{2}{3}c_2\mathbf{v}_2,\end{aligned}$$

and more generally,

$$\mathbf{x}_n = \left(\tfrac{4}{3}\right)^n c_1\mathbf{v}_1 + \left(\tfrac{2}{3}\right)^n c_2\mathbf{v}_2.$$

As $n$ increases, the component of $\mathbf{x}_n$ along the eigenvector $\mathbf{v}_2$ shrinks to zero, because $(2/3)^n$ tends to 0. Conversely, the component along $\mathbf{v}_1$ continues to grow, because $(4/3)^n$ becomes large. Therefore, we can think of $\mathbf{v}_1$ as the expanding direction and $\mathbf{v}_2$ as the contracting direction.

The line $L_2$ contains the eigenvector $\mathbf{v}_2$. Notice that the orbit initially approaches the origin along $L_2$. Soon, however, the term $(4/3)^n$ dominates, and the orbit begins to move away from the origin along the direction of $\mathbf{v}_1$, indicated by the line $L_1$. The analysis confirms that the origin is a saddle fixed point. ∎

The exercises explore additional properties of two-dimensional linear difference equations that are useful in the analysis of numerical methods for solving ordinary differential equations. See also Section 8.8 for an introduction to nonlinear difference equations.

## Exercises

**1.** Let $x_{n+1} = ax_n$, where $a$ is a constant, and suppose $x_0 \neq 0$.

(a) Find an expression for $x_n$ in terms of the initial condition $x_0$.

(b) Show that if $|a| < 1$, then $\lim_{n\to\infty} |x_n| = 0$.

(c) Show that if $|a| > 1$, then $\lim_{n\to\infty} |x_n| = \infty$.

(d) Show that if $a < 0$, then the orbit of $x_0$ alternates in sign.

**2.** There are many parallels between the difference equation

$$x_{n+1} = ax_n \qquad (9.21)$$

and the exponential growth equation

$$x' = kx. \qquad (9.22)$$

(a) For what values of $k$ is $x = 0$ a sink for Eq. (9.22)?

(b) For what values of $a$ is $x = 0$ a sink for Eq. (9.21)?

(c) Suppose $x_0 \neq 0$. For what values of $a$ is the orbit of $x_0$ a constant?

(d) Suppose $x_0 \neq 0$. For what values of $k$ are solutions of Eq. (9.22) constant?

**3.** Determine whether the origin is a source, a sink, or a saddle for each of the difference equations in the following list.

(a) $x_{n+1} = 5x_n - 9y_n$
    $y_{n+1} = (3x_n - 5y_n)/2$

(b) $x_{n+1} = (5x_n + y_n)/2$
    $y_{n+1} = (x_n + 5y_n)/2$

(c) $x_{n+1} = \frac{3}{2}y_n$
    $y_{n+1} = 6x_n$

**4.** Let

$$\mathbf{x}_{n+1} = \mathbf{Ax}_n = \begin{pmatrix} 1 & c \\ c & 1 \end{pmatrix}\mathbf{x}_n.$$

For which values of $c$, if any, is the origin a sink? A source? A saddle? Explain.

**5.** Repeat Exercise 4 for the case in which

$$\mathbf{A} = \begin{pmatrix} 2 & 1+c \\ 1+c & 1 \end{pmatrix}.$$

**6.** Let

$$\mathbf{x}_{n+1} = \mathbf{Ax}_n = \begin{pmatrix} 0 & -\frac{1}{2} \\ -\frac{1}{2} & 0 \end{pmatrix}\mathbf{x}_n. \qquad (9.23)$$

(a) Show that $(-1, 1)$ is an eigenvector of $\mathbf{A}$. What is the corresponding eigenvalue?

(b) Let $\mathbf{x}_0 = (-1, 1)$. Compute the first five iterates of $\mathbf{x}_0$ and plot them.

(c) Show that $(1, 1)$ is another eigenvector of $\mathbf{A}$. What is the corresponding eigenvalue?

(d) Let $\mathbf{x}_0 = (1, 1)$. Compute the first five iterates of $\mathbf{x}_0$ and plot them.

(e) What is the qualitative difference between the orbits of the initial condition in (b) and that in (d)?

(f) On the basis of your results here, what can be said in general about the qualitative behavior of orbits of equations like (9.23) when the matrix $\mathbf{A}$ has a negative eigenvalue?

**7.** Prove Theorem 9.4.

**8.** This exercise asks you to prove that if $\| \cdot \|$ denotes the Euclidean norm, and if $c$ is any constant, then

$$\|c\mathbf{x}\| = |c| \, \|\mathbf{x}\| \qquad (9.24)$$

for any vector $\mathbf{x}$. (This result is used in Exercise 9.)

(a) Show that Eq. (9.24) holds whenever $c$ is a real constant and $\mathbf{x}$ is a real vector.

(b) Show that Eq. (9.24) holds whenever $c$ is a complex constant and $\mathbf{x}$ is a complex vector. (Note: If $c = a + bi$, then $|c| = \sqrt{a^2 + b^2}$ and $\|\mathbf{x}\| = \sqrt{|x_1|^2 + |x_2|^2}$ for a 2-vector $\mathbf{x}$.)

**9.** Let

$$\mathbf{x}_{n+1} = \mathbf{Ax}_n = \frac{1}{2}\begin{pmatrix} 1 & -1 \\ 1 & 1 \end{pmatrix}\mathbf{x}_n.$$

(a) Let $\mathbf{x}_0 = (1, 1)$. Compute and plot the first 10 iterates of $\mathbf{x}_0$.

(b) On the basis of your results in (a), does the origin appear to be a sink, a source, or a saddle?

(c) Find the eigenvalues of $\mathbf{A}$.

(d) More generally, suppose $\mathbf{x}_{n+1} = \mathbf{Bx}_n$, where $\mathbf{B}$ is a $2 \times 2$ matrix with complex conjugate eigenvalues, and let $\mathbf{v}$ be an eigenvector corresponding to one of the eigenvalues $\lambda$. If $\mathbf{x}_0 = c\mathbf{v}$, find an expression for $\|\mathbf{x}_n\|$. (Hint: Use the result of Exercise 8.)

(e) If $\mathbf{B}$ is a $2 \times 2$ matrix with complex conjugate eigenvalues, under what conditions is the origin for the map $\mathbf{x}_{n+1} = \mathbf{Bx}_n$ a sink? Justify your answer using your results in (a)–(d).

(f) What condition must the eigenvalues of $\mathbf{B}$ satisfy for the origin to be a source? Justify your answer.

# 9.3   STABILITY OF NUMERICAL ALGORITHMS

Every numerical method approximates the solution of the initial value problem

$$x' = f(t, x), \qquad x(t_0) = x_0 \tag{9.25}$$

by a suitable difference equation. A *single-step* numerical method is a difference equation of the form

$$
\begin{aligned}
t_{n+1} &= t_n + h, \\
x_{n+1} &= x_n + g(t_n, x_n, h),
\end{aligned}
\tag{9.26}
$$

where $h$ is the step size and $g$ is a function that depends on $f$ and the numerical method itself. The term *single step* refers to the fact that $x_{n+1}$ depends only on $x_n$ and not on other previous values. (A *multistep* method is one in which $x_{n+1}$ depends on $x_n, x_{n-1}$, etc.) We say that Eq. (9.26) is a *fixed-step* method if $h$ is a constant. A *variable-step* method is one in which $h$ varies depending on $t_n$ and $x_n$.[2] An important question is whether the sequence $\{(t_0, x_0), (t_1, x_1), \ldots\}$ lies close to the graph of $x(t)$, the "true" (but usually unknown) solution of Eq. (9.25). Ideally, the approximation should be as accurate as possible. As is illustrated in Section 9.1, there is a tradeoff between the accuracy provided by a small step size and the additional error that results from numerical roundoff in a computer.

There is an additional source of problems that are intrinsic to numerical methods. In particular, if the step size $h$ is badly chosen, then the orbit of Eq. (9.26) may bear little or no resemblance to the true solution of Eq. (9.25). The *stability* of a numerical method refers to the question of whether the method generates a faithful approximation to the true solution for a given value of $h$.

## Numerical Instabilities in Euler's Method

The stability problem is most easily illustrated with Euler's method. Let

$$x' = -kx, \qquad x(0) = x_0, \tag{9.27}$$

where $k$ is a positive constant. When it is applied to Eq. (9.27), Euler's method is the difference equation

$$t_{n+1} = t_n + h, \tag{9.28}$$

$$x_{n+1} = x_n - hkx_n, \tag{9.29}$$

---

[2]The best numerical methods are variable-step methods, because they adjust $h$ to try to minimize the local error at each step. However, the analysis of these methods is beyond the scope of this text.

where $t_0 = 0$ and $x_0 = x(0)$. Equations (9.28) and (9.29) are decoupled, and it is straightforward to show that

$$t_n = nh, \tag{9.30}$$

$$x_n = (1 - hk)^n x_0. \tag{9.31}$$

The true solution of Eq. (9.27) is $x(t) = x_0 e^{-kt}$, which decays exponentially to 0 as $t \to \infty$. At a minimum, then, we must have $x_n \to 0$ as $n \to \infty$ for the orbit of Eq. (9.29) to approximate $x(t)$; that is, 0 must be a sink for Eq. (9.29). This condition requires that $|1 - hk| < 1$. (Why?)

Notice that if $h$ is too large, that is, if $|1 - hk| > 1$, then Euler's method "blows up" in that the orbit of Eq. (9.29) tends to infinity as $n$ becomes large. In this case, the numerical approximation generated by Euler's method bears no resemblance to the true solution of Eq. (9.27). We say that Euler's method is *numerically unstable* if $|1 - hk| > 1$.

Another important question is whether the numerical approximation is *qualitatively* similar to the true solution. For example, suppose $1/k < h < 2/k$. Then $-1 < 1 - hk < 0$. In this case, the iterates of Eq. (9.29) alternate in sign. In effect, the Euler solution oscillates, even though it converges to 0 as $n$ becomes large. This behavior is not consistent with the true solution of Eq. (9.27), whose sign does not change.

If $0 < h < 1/k$, then all iterates of Eq. (9.29) have the same sign as $x_0$, and $x_n \to 0$ as $n \to \infty$. In this respect, the difference equation (9.29) faithfully reproduces the behavior of the true solution of Eq. (9.27). Therefore, for Euler's method to produce a qualitatively and quantitatively accurate numerical approximation, we must have $0 < h < 1/k$.

◆ *Example 1*    Let

$$x' = -x, \qquad x(0) = 10. \tag{9.32}$$

Figure 9.6(a) shows a graph of the true solution of Eq. (9.32), given by $x(t) = 10e^{-t}$.

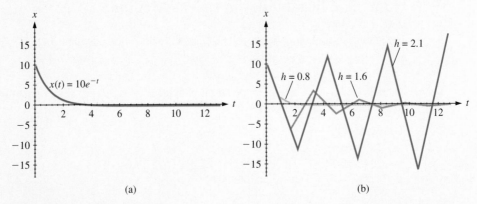

(a)                                                                              (b)

**FIGURE 9.6**    (a) The true solution of Eq. (9.32), $x(0) = 10$. (b) Three numerical approximations using Euler's method.

Figure 9.6(b) shows the graphs of three numerical approximations to $x(t)$ using Euler's method. If $h = 0.8$, then the numerical solution is quantitatively and qualitatively accurate. If $h = 1.6$, then the approximation alternates in sign; the oscillatory solution is qualitatively inconsistent with the true solution. If $h = 2.1$, then the numerical solution does not resemble the true solution.

Equation (9.32) is just Eq. (9.27) with $k = 1$ and $x_0 = 10$. The preceding analysis implies that we must have $0 < h < 1$ for Euler's method to produce an accurate approximation of the true solution. The results shown in Fig. 9.6 are consistent with this prediction.

∎

We say that Euler's method is *numerically unstable* if $h$ is too large. The numerical instability produces an approximation whose quantitative and qualitative behavior does not correspond to the true solution of the original differential equation. As the preceding discussion shows, the value of the step size $h$ at which a numerical instability occurs depends on the differential equation that is being integrated.

## Numerical Instabilities in Nonlinear Equations

The discussion in Section 9.2 can be used to assess the stability of Euler's method when it is applied to autonomous linear differential equations. In this section, we show that numerical methods can exhibit other kinds of instabilities when they are applied to nonlinear differential equations. The following example assumes some acquaintance with the logistic map discussed in Sections 8.8 and 0.3.

✦ *Example 2*   Euler's approximation to the logistic equation provides a nice example of a numerical instability in a nonlinear differential equation. Euler's method leads to a nonlinear difference equation, as we now show. The logistic equation, given by

$$p' = p(a - bp), \tag{9.33}$$

where $a$ and $b$ are two positive constants, is a simple model of population growth in the presence of resource constraints (see Chapter 1). If $p_0 > 0$, then $p(t) \to a/b$ as $t \to \infty$. (The nonzero equilibrium solution, $p = a/b$, is called the *carrying capacity* of the population.)

Euler's method applied to Eq. (9.33) leads to the difference equation

$$p_{n+1} = p_n + hp_n(a - bp_n)$$
$$= p_n(1 + ha - hbp_n). \tag{9.34}$$

(Henceforth, we omit the equation $t_{n+1} = t_n + h$ when we discuss fixed-step methods.) Under the change of variable

$$p_n = \left(\frac{1 + ha}{hb}\right) x_n, \tag{9.35}$$

Eq. (9.34) is equivalent to the *logistic map*

$$x_{n+1} = (1 + ha)x_n(1 - x_n). \tag{9.36}$$

The constant $1 + ha$ plays the role of the parameter $\lambda$ as discussed in Section 8.8.

Equation (9.36) has two fixed points, $x = 0$ and $x = ha/(1 + ha)$. These fixed points correspond to the equilibrium solutions of the logistic equation (9.33) under the change of variable in Eq. (9.35). In particular, the equilibrium at $x = ha/(1 + ha)$ is stable if $0 < h < 2/a$ (see Exercise 10). Hence, if $0 < h < 2/a$, then the orbits of the logistic map arising from Euler's method approach a fixed point that corresponds to the nonzero equilibrium solution of the logistic equation (9.33). However, the qualitative behavior is not the same unless $h$ satisfies other restrictions, as discussed in Exercise 10.

Figure 9.7(a) shows the numerical approximations generated by Euler's method when it is applied to the initial value problem

$$p' = p(2 - p), \qquad p(0) = 1. \tag{9.37}$$

(Here, $a = 2$ and $b = 1$.) The line segments connect successive iterates of Eq. (9.36) for the indicated choices of the step size $h$. For $h = 0.2$, the numerically generated solution accurately reflects the true solution of the corresponding initial value problem. The numerical solution generated by $h = 0.9$ oscillates in a manner that is not consistent with the true solution, although the numerical solution does converge to the carrying capacity (equilibrium solution). If $h = 1.05$, then the numerical solution does not agree with the true solution. If $h$ is chosen larger, say $h = 1.5$, then the behavior is chaotic, as illustrated in Fig. 9.7(b).

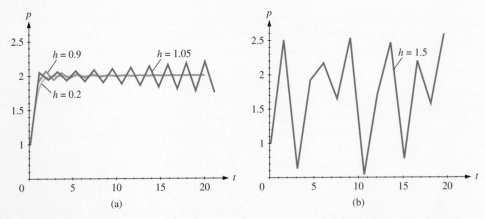

**FIGURE 9.7**   (a) Three Euler approximations to the logistic equation (9.33) with $a = 2$ and $b = 1$ using step sizes $h = 0.2$, 0.9, and 1.05. (b) Euler approximation using $h = 1.5$.   ∎

At first glance, it might appear that step sizes such as $h = 1.05$ are unusually large in comparison to the values that we have seen up until now in various examples. It is important to keep in mind that the the notion of a "large" step size depends heavily on the nature of the solution to be approximated. For instance, in a population model like Eq. (9.37), the independent variable might be measured in days as we follow the evolution of the population through a single growing season. In this case, $h = 1$ corresponds to a step size of one day—not an unreasonably large value if the growing season is several months long.

Nevertheless, the analysis above shows that values of $h$ that are much larger than 1 produce numerical solutions that do not mimic the true solution. In Example 2, a step size of $h = 0.2$ produces an acceptable numerical solution, but it implies that the growth of the population must be followed at intervals of only a few hours ($0.2$ day $= 4.8$ hours).

## Stiff Differential Equations

In many cases, a small step size must be used to integrate the equation, even though we are interested only in the solution over much longer time scales. Such equations are said to be *stiff*. This situation occurs primarily in systems of differential equations, as the following linear system illustrates.

✦ *Example 3*   Let

$$\mathbf{x}' = \mathbf{A}\mathbf{x}, \tag{9.38}$$

where

$$\mathbf{A} = \begin{pmatrix} -0.1 & 0 \\ 0 & -100 \end{pmatrix}.$$

The system (9.38) corresponds to the uncoupled pair of equations

$$x' = -0.1x, \tag{9.39}$$

$$y' = -100y, \tag{9.40}$$

whose solutions tend exponentially to 0 as $t \to \infty$.

Euler's method approximates the solution of Eq. (9.38) using the difference equation

$$\mathbf{x}_{n+1} = \mathbf{x}_n + h\mathbf{A}\mathbf{x}_n. \tag{9.41}$$

Because we integrate Eqs. (9.39) and (9.40) as a pair, the step size $h$ must be small enough for the method to be numerically stable for both equations.

We can rewrite Eq. (9.41) as

$$\mathbf{x}_{n+1} = (\mathbf{I} + h\mathbf{A})\mathbf{x}_n = \begin{pmatrix} 1 - \frac{1}{10}h & 0 \\ 0 & 1 - 100h \end{pmatrix} \mathbf{x}_n. \tag{9.42}$$

The origin is a sink for Eq. (9.38). Therefore, the step size $h$ must be chosen so that the orbit of Eq. (9.42) tends to $\mathbf{0}$ monotonically for the numerical solution to be qualitatively similar to the true solution. This is true only if $0 < 1 - \frac{1}{10}h < 1$ and $0 < 1 - 100h < 1$. The latter condition implies that $h$ must be less than 0.01.

The true solution is

$$\begin{pmatrix} x(t) \\ y(t) \end{pmatrix} = \begin{pmatrix} x_0 e^{-t/10} \\ y_0 e^{-100t} \end{pmatrix}.$$

The $y$ component decays to zero very rapidly. For instance, if $(x_0, y_0) = (1, 1)$, then $y(t)$ is negligible in comparison to $x(t)$ if $t$ is larger than about 0.1. Nevertheless, to have a numerical solution that is even qualitatively similar to the true solution, the step size must be restricted to $h < 0.01$.

Figure 9.8 shows a plot of the solution of Eq. (9.40) satisfying the initial condition $y_0 = 1$, together with the Euler approximation with a step size $h = 0.02$ (which is twice as large as the maximum allowed value that has just been identified). Although the value of $y$ is negligible for $t$ greater than about 0.05, Euler's method is numerically unstable for this value of $h$ and does not converge to the true solution.

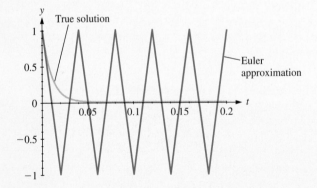

**FIGURE 9.8**   The solution of Eq. (9.40), corresponding to one of the coordinates of Eq. (9.38), and its numerical approximation using Euler's method with $h = 0.02$.  ■

Equation (9.38) is an example of a *stiff* differential equation. Although there is no rigorous definition, stiffness refers to a situation in which the stability of the numerical solution is governed by a component of the true solution that decays rapidly relative to the other components. In this example, the value of $y(t)$ rapidly becomes negligible in comparison to that of $x(t)$. Nevertheless, the stability of the numerical method is determined by the need to follow $y(t)$ accurately, even though it tends almost immediately to 0.

✦ *Example 4*   Let us determine the largest step size $h$ for which Euler's method is numerically stable for the equation

$$\mathbf{x}' = \mathbf{A}\mathbf{x} = \begin{pmatrix} -299.8 & 99.9 \\ -599.4 & 199.7 \end{pmatrix} \mathbf{x}. \qquad (9.43)$$

The eigenvalues of $\mathbf{A}$ are $\lambda_1 = -100$ and $\lambda_2 = -1/10$ with corresponding eigenvectors $\mathbf{v}_1 = (1, 2)$ and $\mathbf{v}_2 = (1, 3)$. Therefore, the general solution of Eq. (9.43) is

$$\mathbf{x}(t) = c_1 e^{-100t} \mathbf{v}_1 + c_2 e^{-t/10} \mathbf{v}_2.$$

All initial conditions tend ultimately to the origin, which is a sink.

Euler's method approximates Eq. (9.43) by the difference equation

$$\mathbf{x}_{n+1} = (\mathbf{I} + h\mathbf{A})\mathbf{x}_n = \begin{pmatrix} 1 - 299.8h & 99.9h \\ -599.4h & 1 + 199.7h \end{pmatrix} \mathbf{x}_n. \qquad (9.44)$$

A straightforward calculation shows that the eigenvalues of the matrix $\mathbf{I} + h\mathbf{A}$ are $\lambda_1 = 1 - 100h$ and $\lambda_2 = 1 - \frac{1}{10}h$, with corresponding eigenvectors $\mathbf{v}_1 = (1, 2)$ and $\mathbf{v}_2 = (1, 3)$, respectively. Thus, the orbits of Eq. (9.44) are qualitatively similar to the true solutions of Eq. (9.43) provided that $0 < 1 - 100h < 1$ and $0 < 1 - \frac{1}{10}h < 1$. Therefore, we must have $0 < h < 0.01$ for the orbits of Eq. (9.44) to be qualitatively and quantitatively similar to the solutions of Eq. (9.43). ∎

As in Example 3, Eq. (9.43) is stiff. (See Exercise 4.) The component of the solution along $\mathbf{v}_1$ decays very rapidly in comparison to the component along $\mathbf{v}_2$, yet the component along $\mathbf{v}_1$ dictates the maximum step size that can be used. In general, the maximum value of the step size for which Euler's method is numerically stable depends on the eigenvalue of the matrix $\mathbf{A}$ that is largest in absolute value.

When no explicit solution is available, it is not always easy to determine ahead of time whether a given system of differential equations is stiff. Often, trial and error are the best way to determine a suitable step size for a given problem. However, as Examples 3 and 4 illustrate, it is possible to give a criterion for stiffness for linear systems of equations with constant coefficients, as follows.

*Let $\mathbf{A}$ be a constant matrix. The equation $\mathbf{x}' = \mathbf{A}\mathbf{x}$ is stiff if the ratio of the largest eigenvalue to the smallest eigenvalue of $\mathbf{A}$ is large.*

There is no precise criterion to determine what is meant by "large." The details depend on the numerical method, the time interval that needs to be followed, and the minimum practicable time step. The exercises explore this idea in more detail.

Stiff differential equations arise when there are two or more disparate time scales in the solution—a situation that occurs frequently in practice. For instance, computer models of ozone depletion in the upper atmosphere consist of very stiff systems of differential equations. Many of the relevant chemical reactions occur over small fractions of a second, but the transport of ozone-destroying chemicals through the earth's atmosphere occurs over intervals of many years. Numerical methods that can deal effectively with stiff systems of differential equations have been essential in predicting the evolution of the "ozone hole" over the Southern Hemisphere.

There are two main challenges in solving stiff differential equations:

- The step size must be small to retain numerical stability, but as the step size gets smaller, the equations require more time to solve on a computer.

- Supercomputers alone cannot overcome the first problem, because if the step size is too small, then numerical roundoff errors become significant, decreasing the accuracy of the solution.

Special numerical methods have been developed to handle stiff systems of equations. One strategy is to devise an integration scheme that is stable for a wide range of possible step sizes. A detailed discussion of such numerical methods is beyond the scope of this text. However, it is worth noting that numerical methods for solving stiff differential equations are an active area of research.

## *Exercises*

**1.** Let $x' = -x$.

(a) Draw a direction field in the box

$$0 \le t \le 10, \quad -5 \le x \le 5.$$

(b) Let $x(0) = 2$ and integrate from $t = 0$ to $t = 10$ using Euler's method for the following step sizes: $h = 0.5$, $h = 1.5$, and $h = 2.5$. (Note: Just use six time steps for the case $h = 1.5$.)

(c) Sketch your results in (b) on the direction field that you generated in (a). Use a geometric argument to explain why the numerical solutions either approach the origin monotonically, approach the origin and alternate sign, or fail to

approach the origin at all as $n \to \infty$. Which outcomes do you see for the step sizes in (b)?

**2.** Determine the maximum step size for which Euler's method produces numerical solutions that are qualitatively similar to the true solution of the logistic equation

$$p' = p(a - bp)$$

for the following values of $a$ and $b$:

(a) $a = 1, b = 1$

(b) $a = 1/100, b = 1$

(c) $a = 100, b = 1$

**3.** This exercise discusses the logistic map that arises from the logistic equation.

   (a) Show that the change of variable in Eq. (9.35) makes Eq. (9.34) equivalent to Eq. (9.36).

   (b) Show that if $0 \leq x_0 \leq 1$, then the orbit of $x_0$ under the logistic map (9.36) remains in the unit interval provided that $0 \leq 1 + ha \leq 4$.

**4.** (a) Find the exact solution of Eq. (9.43) that satisfies the initial condition $(x_0, y_0) = (1, 1)$.

   (b) Show that the origin is a sink for Eq. (9.44) if $0 < h < 0.02$ (see Example 4). Why is it necessary to make the restriction $0 < h < 0.01$ as stated at the end of Example 4?

**5.** Let
$$\mathbf{x}' = \mathbf{A}\mathbf{x} = \begin{pmatrix} -10 & 1 \\ 0 & 0.1 \end{pmatrix}\mathbf{x}.$$

   (a) Find the general solution and describe the qualitative behavior of typical solution trajectories in the phase plane.

   (b) Determine the set of step sizes $h$ for which Euler's method is numerically stable.

   (c) Determine the set of step sizes $h$ for which the Euler solution is qualitatively similar to the true solution.

   (d) Would you characterize the system as stiff? Explain.

**6.** Repeat Exercise 5 for the equation
$$\mathbf{x}' = \begin{pmatrix} -7 & 2 \\ -2 & -2 \end{pmatrix}\mathbf{x}.$$

**7.** Repeat Exercise 5 for the equation
$$\mathbf{x}' = \begin{pmatrix} -61 & 59 \\ 59 & -61 \end{pmatrix}\mathbf{x}.$$

**8.** Repeat Exercise 5 for the equation
$$\mathbf{x}' = \begin{pmatrix} -45 & -5 \\ -5 & -45 \end{pmatrix}\mathbf{x}.$$

**9.** This exercise considers the question of stiffness in the case in which one of the eigenvalues of the matrix is 0. Let
$$\mathbf{x}' = \begin{pmatrix} -5 & 5 \\ 5 & -5 \end{pmatrix}\mathbf{x}.$$

   (a) Find the general solution.

   (b) Determine the largest step size $h$ for which Euler's method is numerically stable.

   (c) Would you describe the equation as stiff? (Does the step size have to be very small in comparison to the length of time it takes for the exponentially decaying solution to go to 0?)

**10.** Example 2 shows that Euler's method applied to the logistic equation (9.33) is equivalent to the logistic map (9.36).

   (a) Show that the point $x_f = ha/(1 + ha)$ is a fixed point for Eq. (9.36), and show that it corresponds to the maximum sustainable population for the logistic differential equation (9.33).

   (b) Let $f(x) = (1 + ha)x(1 - x)$. Find $f'(x_f)$.

   (c) Show that if $ha > 1$, then there are initial conditions $x_0$ such that $x_f < x_0 < 1$ but $x_1 < x_f$, where $x_1$ is the image of $x_0$ under the logistic map (9.36).

   (d) Show that if $ha < 1$, then the orbit of every initial condition in $(0, 1)$ goes monotonically to $x_f$. In other words, if $0 < x_0 < x_f$, then $x_0 < x_1 < x_2 < \cdots$, but $x_n \to x_f$ as $n \to \infty$; likewise, if $x_f < x_0 < 1$, then $x_0 > x_1 > x_2 > \cdots$, but $x_n \to x_f$ as $n \to \infty$.

   (e) Show that if $ha > 2$, then $x_f$ is no longer a stable fixed point.

   (f) What do your results in (c)–(e) say about the qualitative behavior of the Euler solution? For which values of the step size $h$ is the Euler solution qualitatively similar to the true solution of Eq. (9.33)?

**11.** Let

$$p' = ap - bp^3, \qquad (9.45)$$

where $a$ and $b$ are positive constants.

(a) Identify the equilibrium solutions of Eq. (9.45) and assess their stability.

(b) Identify the difference equation that arises from applying Euler's method to Eq. (9.45). Find the fixed points and determine their stability. What is the correspondence between the fixed points found here and the equilibrium solutions in (a)?

(c) Use an analysis similar to that in Example 2 to determine the set of step sizes $h$ for which Euler's method is numerically stable (that is, the values of $h$ for which the orbits of the difference equation in (b) converge to the fixed points).

(d) For which step sizes $h$ are the orbits of Euler's method qualitatively similar to the solutions of Eq. (9.45)? (In other words, how large can $h$ be so that the numerical solutions do not oscillate?)

**12.** Let

$$v' = -g + kv^2. \qquad (9.46)$$

(Equation (9.46) is a model for the air resistance encountered by a falling body; see Chapter 2.) Here, $g$ and $k$ are positive constants.

(a) Determine the difference equation that is obtained when Euler's method is applied to Eq. (9.46) with a fixed step size $h$.

(b) Identify the equilibrium solution of Eq. (9.46) that corresponds to terminal velocity. What is the corresponding fixed point in the equation that you identified in (a)?

(c) Determine the set of step sizes $h$ for which Euler's method is numerically stable.

(d) Determine the set of step sizes $h$ for which Euler's method yields numerical solutions that are qualitatively similar to the true solutions.

**13.** Let

$$x'' + 9x = 0. \qquad (9.47)$$

(a) Is the origin a source, sink, center, or saddle for Eq. (9.47)? Justify your answer.

(b) Identify the difference equation that arises when Euler's method is applied to Eq. (9.47) using a fixed step size $h$. (You can write the difference equation as $\mathbf{x}_{n+1} = \mathbf{B}\mathbf{x}_n$ for an appropriate matrix $\mathbf{B}$. What is $\mathbf{B}$?)

(c) Find the eigenvalues of $\mathbf{B}$.

(d) Show that the orbits of the difference equation are not qualitatively similar to the true solutions of Eq. (9.47) for any step size $h$. Is the origin a source, sink, or saddle for the difference equation in (b)?

## 9.4   BETTER NUMERICAL METHODS

One of the difficulties of Euler's method is that small step sizes are required to obtain an accurate approximation of the solution of a given differential equation (see, for example, Section 6.4). The global error decreases at a rate that is proportional to $h$. Hence, if $h$ is cut in half, then the global error is cut in half.

The theory of numerical quadrature suggests that there are more accurate ways of integrating differential equations.[3] That is, the global error in the solution may decrease at a rate that is proportional to a power of $h$. Consider a numerical method for which

---

[3] Numerical quadrature refers to the numerical approximation of integrals.

the global error decreases at a rate proportional to $h^2$. If the step size is cut in half, then the global error is reduced by a factor of four. If the global error decreases at a rate proportional to $h^4$, then a reduction in the step size by a factor of two cuts the global error by a factor of 16. This section investigates some popular numerical methods whose global errors decrease as a power of $h$.

## Integral Equations

Euler's method is a simple geometrical rule for approximating a solution of a given initial value problem by moving along a succession of tangent lines. To improve on Euler's method, we need to look at the solution of the initial value problem from a different point of view.

Let

$$x' = f(t, x) \tag{9.48}$$

with $x(t_0) = x_0$. If we integrate both sides of Eq. (9.48), we obtain

$$\int_{t_0}^{t} x'(s) \, ds = \int_{t_0}^{t} f(s, x(s)) \, ds.$$

The fundamental theorem of calculus implies that the *differential* equation (9.48) is equivalent to the *integral* equation

$$x(t) = x_0 + \int_{t_0}^{t} f(s, x(s)) \, ds. \tag{9.49}$$

We have not "solved" the differential equation (9.48) because the unknown function $x(t)$ appears on both sides of the integral equation (9.49). However, the reformulation of the original differential equation in the form of Eq. (9.49) expresses the solution in terms of an integral over a time interval. We can use the theory of numerical quadrature to estimate the integral on the right-hand side of Eq. (9.49) and hence determine a numerical approximation to the solution $x(t)$ of the differential equation (9.48). The next example illustrates this idea for a very simple differential equation.

◆ *Example 1*   Let $x' = f(t)$, where $x(0) = 0$. The derivative depends only on $t$. The equivalent integral equation is

$$x(t) = \int_{0}^{t} f(s) \, ds. \tag{9.50}$$

Even if the integral is difficult or impossible to express as a sum of elementary functions, it can be approximated by Riemann sums.

The first step of Euler's method produces the approximation $x_1 = hf(0)$. In effect, Euler's method approximates the solution of Eq. (9.50) by rectangles (Riemann sums) as illustrated in Fig. 9.9. In this example, the value of $x(t)$ can be regarded as the area between the $t$ axis and the graph of $f$.

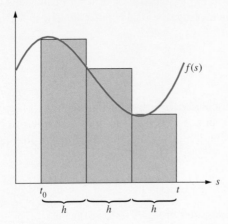

**FIGURE 9.9**   Schematic diagram of the approximation of the integral in Eq. (9.50) using Riemann sums.                                                                   ∎

The situation depicted in Example 1 remains qualitatively similar when $x' = f(t, x)$, that is, when $x'$ depends both on $t$ and $x$. Since $x$ is a function of $t$, the function $f$ ultimately is a function only of $t$. Let the initial condition be $x(t_0) = x_0$. One step of Euler's method produces the approximation

$$\hat{x}(t_0 + h) = x_0 + hf(t_0, x_0). \tag{9.51}$$

The equivalent integral formulation of the solution at $t = t_0 + h$ is

$$x(t_0 + h) = x_0 + \int_{x_0}^{t_0+h} f(s, x(s))\, ds.$$

In this case, Euler's method approximates the integral by a rectangle of width $h$ and height $f(t_0, x_0)$, just as in Fig. 9.9.

## Heun's Method

Although a sequence of rectangles can be used to find a numerical estimate of the value of a definite integral, other procedures can be used. For instance, the *trapezoidal rule* approximates a definite integral by the average of the values of $f$ at the left- and right-hand sides of each subinterval. Figure 9.10 shows a schematic depiction of the trapezoidal rule for approximating a definite integral between $t_0$ and $t$.

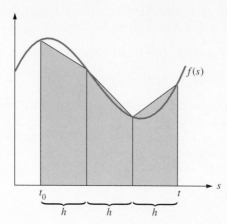

**FIGURE 9.10** Approximations of the integral $\int_{t_0}^{t} f(s)\,ds$ by trapezoids.

The trapezoidal rule approximates the integral from $t = t_0$ to $t = t_0 + h$ as

$$\hat{x}(t_0 + h) = x_0 + \tfrac{1}{2}h[f(t_0, x_0) + f(t_0 + h, x(t_0 + h))]. \tag{9.52}$$

The term $f(t_0 + h, x(t_0 + h))$ is simply the value of the function $f$ at the right-hand side of the subinterval $[t_0, t_0 + h]$. One difficulty, of course, is that we do not know $x(t_0 + h)$, because it is the solution of the differential equation that we are trying to approximate.

However, it is possible to *estimate* $x(t_0 + h)$: Just use Euler's method. That is, we use Euler's method to estimate the height of the trapezoid at $t = t_0 + h$, but we use the trapezoidal rule to approximate the definite integral from $t_0$ to $t_0 + h$. The resulting numerical procedure is called *Heun's method*[4] or the *improved Euler method*.

Let

$$x' = f(t, x), \qquad x(t_0) = x_0 \tag{9.53}$$

and fix a step size $h$. Heun's method approximates the solution of the initial value problem (9.53) with the difference equation

$$x_{n+1} = x_n + \tfrac{1}{2}h\left[f(t_n, x_n) + f\left(t_{n+1}, x_{n+1}^{E}\right)\right], \tag{9.54}$$

---

[4] *Heun* rhymes with *coin*.

where

$$x_{n+1}^E = x_n + hf(t_n, x_n).$$

Heun's method is sometimes called a *predictor-corrector* method. Euler's method "predicts" the value of the solution at $t = t_n + h$, and Eq. (9.54) "corrects" it.

✦ *Example 2*    Let us apply Heun's method to the initial value problem

$$x' = x, \qquad x(0) = 1 \tag{9.55}$$

using a step size of $h = 0.1$. The following calculations show explicitly how to apply Heun's method to obtain an estimate of $x(0.2)$.

Observe first that $t_0 = 0$ and $x_0 = 1$, corresponding to the initial conditions. In the following computations, $x_1$ is an estimate of $x(0.1)$ and $x_2$ is an estimate of $x(0.2)$. We have

$$\begin{aligned}
x_1^E &= x_0 + hf(t_0, x_0) \\
&= 1 + 0.1 \qquad \text{because } f(t, x) = x \text{ and } h = 0.1 \\
&= 1.1,
\end{aligned}$$

so that

$$\begin{aligned}
x_1 &= x_0 + \tfrac{1}{2}h\left[f(t_0, x_0) + f\left(t_1, x_1^E\right)\right] \\
&= 1 + 0.05[1 + 1.1] \\
&= 1.1050.
\end{aligned}$$

For the second step, we first compute

$$\begin{aligned}
x_2^E &= x_1 + hf(t_1, x_1) \\
&= 1.1050 + (0.1)(1.1050) \\
&= 1.2155,
\end{aligned}$$

so that

$$\begin{aligned}
x_2 &= x_1 + \tfrac{1}{2}h\left[f(t_1, x_1) + f\left(t_2, x_2^E\right)\right] \\
&= 1.1050 + 0.05[1.1050 + 1.2155] \\
&= 1.2210,
\end{aligned}$$

where we have rounded to four decimal places.

Notice that the true solution is $x(0.2) = e^{0.2} \approx 1.2214$. The absolute error in the numerical method at $t = 0.2$ is approximately $4 \times 10^{-4}$, which corresponds to a relative error of approximately 0.03 percent.                                                          ■

A basic question is how the error in the numerical solution decreases as the step size is made smaller. It can be shown that the global error in Heun's method is proportional to $h^2$. In other words, if the step size is cut in half, then the global error in the solution is reduced by a factor of four. There is an additional computational cost: The function $f$ must be evaluated twice during each time step instead of once. However, the additional accuracy generally more than compensates for the extra computation.

✦ *Example 3*   It is straightforward to program Heun's method on a calculator or personal computer so that the effect of smaller step sizes can be assessed on Eq. (9.55). Table 9.5 shows the numerical approximations of $x(1)$ using Euler's method and Heun's method, along with the global errors, for the indicated step sizes $h$. It is easy to see from the table that if $h$ is reduced by a factor of 10, then the global error in Euler's method is reduced by a factor of 10 but the global error in Heun's method is reduced by a factor of 100. When $h = 0.001$, the numerical solution using Heun's method is accurate to seven decimal digits, whereas the numerical solution using Euler's method is accurate to only three digits. In other words, when $h = 0.001$, the numerical approximation generated by Heun's method for this initial value problem is 10,000 times more accurate than the one generated by Euler's method for the same time step.

| $h$ | Euler's Method | Euler Error | Heun's Method | Heun Error |
|---|---|---|---|---|
| 0.1 | 2.59374246 | $1.24539368 \times 10^{-1}$ | 2.714080847 | $4.20098 \times 10^{-3}$ |
| 0.01 | 2.704813829 | $1.3468 \times 10^{-2}$ | 2.718236860 | $4.4968 \times 10^{-5}$ |
| 0.001 | 2.716923932 | $1.3579 \times 10^{-3}$ | 2.718281382 | $4.46 \times 10^{-7}$ |

TABLE 9.5   Approximations to the solution of Eq. (9.55) at $t = 1$ and the global error using Euler's method and Heun's method.   ∎

✦ *Example 4*   Let

$$x' = 1 - tx, \qquad x(0) = 1. \tag{9.56}$$

In contrast to Example 2, the exact solution of Eq. (9.56) cannot be expressed in terms of elementary functions. Nevertheless, we can infer the approximate decrease in the global error as a function of $h$ by comparing the differences in the numerical solutions computed for various fixed values of $h$.

Let $\hat{x}_h(1)$ denote the numerical approximation to the solution at $t = 1$ using a step size of $h$. Define

$$r_h = \hat{x}_{h/10}(1) - \hat{x}_h(1).$$

Here, $r_h$ is the difference between numerical approximations as the step size $h$ is reduced by 10.

Table 9.6 shows the numerical approximation of the solution of Eq. (9.56) using Heun's method for the indicated value of $h$. The third line shows the corresponding value of $r_h$. Notice that $r_h$ decreases approximately by a factor of 100 when $h$ is cut by 10. This numerical result suggests that the global error in the solution decreases at a rate that is approximately proportional to $h^2$.

| $h$ | 1 | 0.1 | 0.01 | 0.001 | 0.0001 |
|---|---|---|---|---|---|
| $\hat{x}_h(1)$ | 1 | 1.329 | 1.3312 | 1.33130900 | 1.331309092 |
| $r_h$ | — | 0.3299 | $1.342 \times 10^{-3}$ | $1.190 \times 10^{-5}$ | $9.2 \times 10^{-8}$ |

**TABLE 9.6**   The numerical solution of Eq. (9.56) using Heun's method and the differences in the approximations as a function of $h$.    ∎

## Taylor Series Methods

When discussing the global error in Euler's method, we resorted to an argument based on Taylor's theorem to show that the local error in Euler's method is proportional to $h^2$. *Taylor series methods* have theoretical interest because they can be used to construct numerical methods for which the local and global errors decrease at a rate that is proportional to a prespecified integer power of $h$. Taylor series methods are based on the idea that if a series expansion of the solution $x(t)$ of a given differential equation is available, then we can estimate the error in a numerical procedure that is based on approximating $x$ with a finite number of terms in the expansion.

Suppose that $x$ is differentiable for a large number of times. The *$k$th-order Taylor series expansion of $x$ about $t_0$* is

$$x(t_0 + h) = x(t_0) + hx'(t_0) + \frac{h^2}{2}x''(t_0) + \frac{h^3}{3!}x'''(t_0) + \cdots + \frac{h^k}{k!}x^{(k)}(t_0) + R(t_0, h),$$

(9.57)

where $R(t_0, h)$ is a remainder term of the form

$$R(t_0, h) = \frac{h^{k+1}}{(k+1)!}x^{(k)}(\xi)$$

and $\xi$ is a number between $t_0$ and $t_0 + h$. The remainder term can be regarded as the local error in the approximation of $x$ by the first $k$ terms in the Taylor expansion. Notice that the local error is proportional to $h^{k+1}$.

If $x$ is the solution of the initial value problem

$$x' = f(t, x), \qquad x(t_0) = x_0,$$

then the derivatives in Eq. (9.57) can be expressed by using the chain rule as follows:

$$x'(t_0) = f(t_0, x_0),$$

$$x''(t_0) = \frac{\partial f}{\partial t} + \frac{\partial f}{\partial x} \times f,$$

$$x'''(t_0) = \frac{\partial^2 f}{\partial t^2} + 2\frac{\partial^2 f}{\partial t \partial x} \times f + \frac{\partial^2 f}{\partial x^2} \times f^2 + \frac{\partial f}{\partial x}\left(\frac{\partial f}{\partial t} + \frac{\partial f}{\partial x} \times f\right),$$

$$\vdots$$

where $f$ and its partial derivatives are evaluated at $(t_0, x_0)$.

Equation (9.57) can be applied at any point along the solution curve. The formula is the same, except that $t_0$ is replaced with $t_n$, where $t_n = t_0 + hn$. Numerical methods based on Eq. (9.57) are called *kth-order Taylor series methods*.

From a theoretical perspective, the global error in Taylor series methods is relatively easy to analyze. However, from a practical point of view, Taylor series methods are undesirable because the derivatives of $f$ are cumbersome to evaluate. The next examples show how Taylor series methods can be applied in simple cases.

✦ *Example 5* Let

$$x' = tx, \qquad x(1) = 2. \tag{9.58}$$

The exact solution of Eq. (9.58) can be found by using the separation of variables; it is $x(t) = 2e^{(t^2-1)/2}$.

The first step in the fourth-order Taylor series method for Eq. (9.58) is

$$x(1 + h) = x(1) + hx'(1) + \tfrac{1}{2}h^2 x''(1) + \tfrac{1}{6}h^3 x'''(1) + \tfrac{1}{24}h^4 x^{(4)}(1).$$

The derivatives of $x$ can be determined from Eq. (9.58) and the chain rule. We have

$$\begin{aligned}
x''(1) &= \frac{d}{dt}(tx)\Big|_{t=1}\\
&= x + tx'\big|_{t=1}\\
&= x + t^2 x\big|_{t=1}\\
&= 4,
\end{aligned}$$

and

$$x'''(1) = \frac{d}{dt}x + t^2 x \Big|_{t=1}$$

$$= x' + 2tx + t^2 x' \big|_{t=1}$$

$$= 3\,tx + t^3 x \big|_{t=1}$$

$$= 8,$$

and

$$x^{(4)}(1) = \frac{d}{dt}(3tx) + t^3 x \Big|_{t=1}$$

$$= 3x + 3tx' + 3t^2 x + t^3 x' \big|_{t=1}$$

$$= 3x + 6t^2 x + t^4 x \big|_{t=1}$$

$$= 20.$$

Therefore, the approximation to the solution at $t = 1 + h$ is

$$\hat{x}(1 + h) = 2 + 2h + 2h^2 + \tfrac{4}{3}h^3 + \tfrac{5}{6}h^4.$$

We continue in a similar manner at $t = 1 + h$, $t = 1 + 2h, \ldots$, until we arrive at $t = 2$. Table 9.7 shows the exact and the numerical approximation to the solution of Eq. (9.58) at $t = 1 + h$ as a function of $h$. As predicted by the theory, when $h$ is reduced by 10, the error in the solution is reduced approximately a factor of $10^5$.

| $h$ | $x(1 + h)$ | $\hat{x}(1 + h)$ | Approximate Error |
|---|---|---|---|
| 1 | 8.9633781406761296454 | 8.1666666666666666667 | $7.967 \times 10^{-1}$ |
| 0.1 | 2.2214212207114104646 | 2.2214166666666666666 | $4.554 \times 10^{-6}$ |
| 0.01 | 2.0202013417102120356 | 2.0202013416666666666 | $4.355 \times 10^{-11}$ |
| 0.001 | 2.0020020013341671002 | 2.0020020013341666666 | $4.336 \times 10^{-16}$ |

TABLE 9.7   The true solution $x(1 + h)$ and its approximation $\hat{x}(1 + h)$ for Eq. (9.58) by using a fourth-order Taylor series method.                                        ■

Notice that a truncation of Taylor's formula (9.57) is *exact* if the $(k+1)$st and higher derivatives of $x$ are zero. In particular, the remainder term $R(t_0, h)$ in Eq. (9.57) is identically zero if $x$ is a polynomial of degree $k$.

An interesting question concerns the class of differential equations that can be solved precisely (in the absence of roundoff error) by using Taylor series methods of a specified order. Euler's method is simply a Taylor series method of order 1, and the class of differential equations that can be solved exactly is

$$x' = \text{constant}.$$

✦ *Example 6*  This example discusses the class of autonomous differential equations that can be solved exactly by using a second-order Taylor series method. To be an exact representation of the solution, all derivatives of third- and higher-order must be zero if we carry out the expansion (9.57) to $k = 2$. Therefore, we must have

$$x'' = \text{constant}.$$

Because the differential equation is autonomous, $x' = f(x)$, and the chain rule implies that

$$x'' = \left. \frac{\partial f}{\partial x} \times f \right|_{(t,x)} = \text{constant}. \tag{9.59}$$

For simplicity, take the constant as 1. The condition (9.59) yields a differential equation for $f$ of the form

$$f(x) \times \frac{df}{dx} = 1$$

that can be solved by using the separation of variables. We have

$$\int f(x) \, df = \int dx,$$

which implies that

$$\tfrac{1}{2}[f(x)]^2 = x + C$$

or

$$f(x) = \sqrt{2x}$$

if we require $f(0) = 0$.

Therefore, a second-order Taylor series method provides an exact solution of differential equations of the form

$$x' = \sqrt{2x}.$$

Let $x(0) = x_0$ be a positive initial condition. (Why is the requirement $x_0 > 0$ important?) The solution is

$$x(t) = \left( \frac{t}{\sqrt{2}} + \sqrt{x_0} \right)^2.$$

The graph of $x$ is a parabola. Notice that $x''' = x^{(4)} = \cdots = 0$, as expected. ∎

It is straightforward to generalize Example 9.4 to higher-order Taylor series algorithms. In general, a Taylor series method of order $k$ gives exact solutions of those differential equations whose solutions are polynomials of degree $k$ or less.

## Runge-Kutta Methods

Although Taylor series methods permit the construction of accurate numerical methods, they are cumbersome because many derivatives of $f$ are required. One popular class of numerical methods are the *Runge-Kutta* methods, which provide a good compromise between ease of use and numerical accuracy.

As is discussed above, the initial value problem

$$x' = f(t, x), \qquad x(t_0) = x_0 \tag{9.60}$$

is equivalent to the integral equation

$$x(t) = x_0 + \int_{t_0}^{t} f(s, x(s)) \, ds. \tag{9.61}$$

Heun's method approximates the integral by trapezoids, producing a global error that is proportional to $h^2$.

An alternative quadrature method is *Simpson's rule*, which approximates the integral by a parabola containing the points of the graph at the endpoints and the midpoint of each subinterval. Simpson's rule approximates the integral

$$\int_{t}^{t+h} f(s) \, ds$$

by the formula

$$\tfrac{1}{6} h \big( f(t) + 4f \left( t + \tfrac{1}{2} h \right) + f(t + h) \big). \tag{9.62}$$

It can be shown that the error in the approximation (9.62) is proportional to $h^5$.

The *fourth-order Runge-Kutta method* is a generalization of Simpson's rule to the integral equation (9.61). To apply Simpson's rule, we need to know

$$f\left(t_0 + \tfrac{1}{2}h, x\left(t_0 + \tfrac{1}{2}h\right)\right),$$

which gives the height of the function at the midpoint of the interval $[t_0, t_0 + h]$, and we need to know

$$f(t_0 + h, x(t_0 + h)),$$

which gives the height of the function at the right endpoint.

In general, we do not have a formula for $x(t)$, and we do not know the value of $x$ except at the left endpoint $t_0$. However, as in the case of Heun's method, we can estimate $x(t_0 + h/2)$ and $x(t_0 + h)$ using Euler's method. The estimates can be used to construct the numerical procedure.

The fourth-order Runge-Kutta method advances the solution from $t = t_n$ to $t = t_n + h$ as follows. Let

$$f_1 = f(t_n, x_n), \tag{9.63}$$

$$f_2 = f\left(t_n + h/2, x_n + \tfrac{1}{2}hf_1\right), \tag{9.64}$$

$$f_3 = f\left(t_n + h/2, x_n + \tfrac{1}{2}hf_2\right), \tag{9.65}$$

$$f_4 = f(t_n + h, x_n + hf_3). \tag{9.66}$$

The value of $x_{n+1} = x(t_n + h)$ is estimated to be

$$x_{n+1} = x_n + \tfrac{1}{6}h(f_1 + 2f_2 + 2f_3 + f_4). \tag{9.67}$$

Equation (9.63) gives the height of the function $f$ at the left endpoint of the interval $[t_n, t_n + h]$. Equation (9.64) is an estimate of $f$ at the midpoint of the interval $[t_n, t_n + h]$; the quantity $x_n + hf_1/2$ is simply Euler's method applied to Eq. (9.60) with a step size of $h/2$. Equation (9.65) is a second estimate of $f$ at the midpoint of the interval $[t_n, t_n + h]$. Here, $x$ is estimated with Euler's method, using the result in Eq. (9.64). Finally, Eq. (9.66) is an estimate of $f$ at the right endpoint that uses Euler's method for $x(t_n + h)$. The formula (9.67) is Simpson's rule.

It can be shown that the local error in Eqs. (9.63)–(9.67) is proportional to $h^5$ and that the global error is proportional to $h^4$. For example, if the step size is cut in half, then the global error is reduced by a factor of 16.

The Runge-Kutta formulas are a good compromise between solution accuracy and computational complexity. Four evaluations of $f$ are required at each time step (compared to two evaluations for Heun's method and one evaluation for Euler's method). However, the higher accuracy usually is worth the extra computational effort.

Figure 9.11 shows a complete Fortran program to implement the fourth-order Runge-Kutta method to the exponential growth equation $x' = f(t, x) = x$. The input consists of the initial time $t_0$, the initial value $x(t_0)$, the number of steps $n$, and the step size $h$. The program applies the Runge-Kutta formulas (9.63)–(9.67) for $n$ time steps. The program can be applied in principle to any first-order scalar equation by making appropriate changes to the function $f$.

```
external f                  subroutine rk(f,t,x,n,h)
read *, t,x,n,h             external f
call rk(f,t,x,n,h)          do j=1,n
print *, t,x                    f1 = f(t,x)
end                            f2 = f(t+h/2, x+h*f1/2)
                               f3 = f(t+h/2, x+h*f2/2)
function f(t,x)                f4 = f(t+h, x+h*f3)
f = x                          x = x + (h/6)*(f1+2*f2+2*f3+f4)
end                            t = t+h
                            enddo
                            end
```

**FIGURE 9.11**　Fortran program to integrate the differential equation $x' = x$ using the fourth-order Runge-Kutta method.

Figure 9.11 shows only the simplest possible implementation of the Runge-Kutta method. In practice, it is best to use a professionally written library subroutine. A good library routine attempts to estimate the size of the local error, adjusts the step size as necessary to satisfy a user-specified local error tolerance, and tries to determine whether the equations are stiff.

✦ *Example 7*　Let

$$x' = x, \qquad x(0) = 1. \tag{9.68}$$

The solution is $x(t) = e^t$. Table 9.8 shows the global error in the solution at $t = 1$ using a double-precision version of the program illustrated in Fig. 9.11.

The numerical results support the claim that the global error decreases at a rate proportional to $h^4$. When $h = 0.001$, the computed result is accurate to 14 significant digits. In particular, the global error at $t = 1$ in the Runge-Kutta method for this problem is *20 million times* less for the same step size than Heun's method. (See Table 9.5.)

| $h$ | Approximate Global Error |
| --- | --- |
| 0.1 | $2.08 \times 10^{-6}$ |
| 0.01 | $2.25 \times 10^{-10}$ |
| 0.001 | $2.04 \times 10^{-14}$ |

TABLE 9.8   The global error in the solution of Eq. (9.68) at $t = 1$ using the fourth-order Runge-Kutta method for the indicated step sizes $h$. ∎

## Exercises

**1.** For each of the initial value problems in the following list,

- approximate the solution at $t = 1$ using Heun's method with $h = 0.1$;

- approximate the solution at $t = 1$ using Heun's method with $h = 0.01$;

- approximate the solution at $t = 1$ using the fourth-order Runge-Kutta method with $h = 0.1$;

- approximate the solution at $t = 1$ using the fourth-order Runge-Kutta method with $h = 0.01$;

- use the exact solution to compute the absolute (global) error in the solution at $t = 1$. What is the ratio of the absolute errors between the two methods for each step size?

(a) $x' = 2x$, $x(0) = 3$

(b) $x' = \sin x$, $x(0) = 1$

(c) $x' = x^2$, $x(0) = 1/10$

(d) $x' = e^{-x}$, $x(0) = 1$.

**2.** Let $x' = x^2$ and $x(0) = 1$.

(a) Use Heun's method with step sizes of $h = 0.1$, 0.01, and 0.001 to estimate $x(0.9)$.

(b) Calculate the exact solution and determine the global error in each of the approximations that you found in (a). If $h$ is reduced by a factor of 10, by how much does the absolute error at $t = 0.9$ decrease?

(c) Use the fourth-order Runge-Kutta method with double-precision arithmetic and step sizes of $h = 0.1, 0.01$, and 0.001 to estimate $x(0.9)$.

(d) Repeat your analysis in (b) with the results computed in (c).

**3.** Repeat Exercise 2 for the initial value problem $x' = x^2 + 1$, $x(0) = 0$. Estimate $x(1.5)$. Compare your results with Example 3 in Section 9.1.

**4.** Let $x' = 1 - tx$, where $x(0) = 1$.

(a) Use the Runge-Kutta method and double-precision arithmetic with step sizes of $h = 1$, $0.1, 0.01$, and 0.001 to estimate $x(1)$. Call your estimates $\hat{x}_1(1)$, $\hat{x}_{0.1}(1)$, $\hat{x}_{0.01}(1)$, and $\hat{x}_{0.001}(1)$, respectively.

(b) Compute the differences

$$|\hat{x}_1(1) - \hat{x}_{0.1}(1)|,$$
$$|\hat{x}_{0.1}(1) - \hat{x}_{0.01}(1)|,$$
$$|\hat{x}_{0.01}(1) - \hat{x}_{0.001}(1)|$$

between the different approximations, and construct a table similar to Table 9.6.

(c) Do your results in (b) support the claim that the global error is proportional to $h^4$? Explain.

**5.** Find an example of an autonomous differential equation $x' = f(x)$ for which a third-order Taylor series method approximates the solution exactly (in the absence of roundoff error).

**6.** Let $x' = t - x$, where $x(0) = 1$.

(a) Use a fourth-order Runge-Kutta algorithm and double-precision arithmetic to estimate $x(1)$ with the following step sizes: $h = 0.1$, $h = 0.01$, $h = 0.001$, and $h = 0.0001$.

(b) Find the exact solution and determine the global error in the solution for each of the step sizes in (a).

(c) For which step size in (a), if any, is the solution essentially accurate to the machine precision?

**7.** Show how to apply a fourth-order Taylor series method to the initial value problem

$$x' = x^2, \quad x(0) = 1.$$

Compare your answer to the fourth-order Taylor series expansion of the exact solution about $t_0 = 0$.

**8.** Let

$$x' = t \sin x, \quad x(0) = 1.$$

(a) Find the exact solution.

(b) Conduct appropriate numerical experiments to find the largest step size $h$ for which Heun's method produces a global error at $t = 1$ of less than $10^{-3}$.

(c) Repeat (b) for the fourth-order Runge-Kutta method.

**9.** Let $x' = -kx$, where $k$ is a positive constant. For a fixed value of $k$, find the largest value of the step size $h$ for which Heun's method is numerically stable.

# Chapter 10

---

# THE LAPLACE TRANSFORM

---

The goal of this chapter is to show how the Laplace transform, introduced in Chapter 4, can be used to find analytic representations of the solutions of certain kinds of differential equations. The method is an alternative to the method of undetermined coefficients, described in Section 5.6. In addition, the Laplace transform is the analytical method of choice for linear differential equations with constant coefficients that have a discontinuous forcing term, as we explain in Section 10.4.

## 10.1  DEFINITION AND FUNDAMENTAL PROPERTIES

The Laplace transform is a linear operator that is of particular interest in electrical engineering and control theory. In this section, we define the Laplace transform and explore its basic properties. (See also Section 4.2.)

### Functions of Exponential Order

The domain of the Laplace transform operator is the set of functions of *exponential order*. Roughly speaking, a function is of exponential order if it can be bounded by a suitable exponential function.

**Definition 10.1**  *Let $f$ be a function whose domain includes the nonnegative real numbers. We say that $f$ is of* exponential order *if there exist positive constants $M$ and $k$ such that*

$$|f(t)| \leq Me^{kt} \tag{10.1}$$

*whenever $t \geq 0$.*

Definition 10.1 implies that the graph of any function of exponential order never rises above the graph of a suitable exponential function.

✦ *Example 1*   The function $f(t) = t^2 + 1$ is of exponential order, because the inequality (10.1) can be satisfied for every $t \geq 0$ by choosing $M = 1$ and $k = 2$; that is, $|t^2 + 1| \leq e^{2t}$ for every $t \geq 0$. Figure 10.1 shows the graphs of $t^2 + 1$ and $e^{2t}$.

There is nothing special about these choices of $M$ and $k$. They are simply two constants that place the graph of $Me^{kt}$ above the graph of $f$. We could just as well have chosen $M = 3$ and $k = 1$, because $|t^2 + 1| \leq 3e^t$ for every $t \geq 0$. Definition 10.1 does not require us to find the *smallest* numbers $M$ and $k$ that satisfy Eq. (10.1). We need only *one* suitable choice of $M$ and $k$.

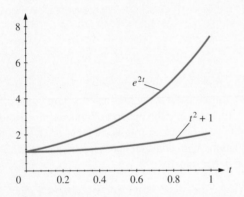

**FIGURE 10.1**   The function $t^2 + 1$ is of exponential order because it can be bounded by the function $e^{2t}$ for all $t \geq 0$.

✦ *Example 2*   The cosine function is of exponential order, because

$$|\cos t| \leq 1 \leq e^t$$

for every $t \geq 0$. Thus, Definition 10.1 is satisfied with $M = 1$ and $k = 1$.  ■

✦ *Example 3*   Let $p(t) = 2t^2 + 3t + 1$. L'Hôpital's rule implies that

$$\lim_{t \to \infty} \frac{e^t}{p(t)} = \infty.$$

This fact can be used to show that $p$ is of exponential order, as follows.

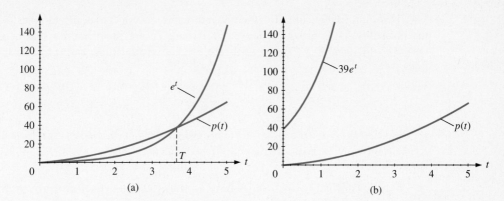

**FIGURE 10.2**   (a) The graph of $p(t) = 2t^2 + 3t + 1$ and the graph of $e^t$. (b) The graph of $p(t)$ and the graph of $39e^t$.

Because the preceding limit is infinite, there must be a time $T$ past which $e^t > p(t)$. Figure 10.2(a) shows the graphs of $p(t)$ and $e^t$. The two curves cross at $T \approx 3.66$. Thus, if $t > T$, then $e^t > p(t)$.

A suitable value of $M$ can be found by estimating

$$\max_{0 \le t \le T} |p(t)|,$$

which occurs when $t = T$. Here, $p(T) \approx 38.7$. One possible choice for $M$ is $M = 39$, where we round up the value of $p(T)$ just to be safe. Figure 10.2(b) shows the graph of $39e^t$. Clearly, the graph of $39e^t$ lies above the graph of $p$ for all $t \ge 0$. Therefore,

$$|p(t)| \le 39e^t$$

whenever $t \ge 0$, which proves that $p$ is of exponential order.   ■

◆ *Example 4*   This example generalizes the previous one. Let $a_0, a_1, \ldots, a_n$ be constants, and let

$$p(t) = a_n t^n + a_{n-1} t^{n-1} + \cdots + a_1 t + a_0.$$

L'Hôpital's rule implies that

$$\lim_{t \to \infty} \left| \frac{e^t}{p(t)} \right| = \infty.$$

(See Exercises 9 and 10 in Section 2.1.)

This limit implies that if $t$ is sufficiently large, then $e^t \geq |p(t)|$. In other words, the inequality holds provided that $t \geq T$, where $T$ is a finite number that depends on the coefficients of $p$. Polynomials are continuous functions, and it can be shown that

$$M = \max_{0 \leq t \leq T} |p(t)|$$

is guaranteed to exist, to be finite, and to be positive. Therefore, $|p(t)| \leq Me^t$ whenever $t \geq 0$. Hence, polynomials are functions of exponential order.                              ∎

Functions of exponential order have an important analytical property, which is summarized in the following theorem.

**Theorem 10.2**   *If $f(t)$ is a continuous function of exponential order, then*

$$\int_0^\infty e^{-st} f(t)\, dt \tag{10.2}$$

*exists and is finite for every sufficiently large positive number $s$.*

**Proof**   If $f$ is of exponential order, then $|f(t)| \leq Me^{kt}$ for suitable constants $M$ and $k$. In addition,

$$\left| e^{-st} f(t) \right| = e^{-st} |f(t)|$$

$$\leq e^{-st} Me^{kt}$$

$$= Me^{(k-s)t}. \tag{10.3}$$

Hence,

$$\left| \int_0^\infty e^{-st} f(t)\, dt \right| \leq \int_0^\infty \left| e^{-st} f(t) \right| dt$$

$$\leq \int_0^\infty Me^{(k-s)t}\, dt \tag{10.4}$$

by virtue of Eq. (10.3). The integral (10.4) exists and is finite whenever $s > k$. Hence, the integral (10.2) exists and is finite for $s > k$. This proves the theorem.                    ∎

## The Definition of the Laplace Transform

The Laplace transform is a linear operator (see Chapter 4) that is defined as follows.

**Definition 10.3** *Let $f$ be a function of exponential order. The Laplace transform of $f$, denoted by $\mathcal{L}(f)$ or $\mathcal{L}(f(t))$, is the function $F(s)$ given by*

$$F(s) = \mathcal{L}(f(t)) = \int_0^\infty e^{-st} f(t)\, dt. \tag{10.5}$$

Equation (10.5) has several important features:

- The Laplace transform takes a function $f$ of the variable $t$ and replaces it with a new function, $F$, of the variable $s$.[1]

- Equation (10.5) involves an improper integral. By Theorem 10.2, the integral exists and is finite, provided that $f$ is a continuous function of exponential order and $s$ is a sufficiently large positive number. Hence, the Laplace transform exists for functions of exponential order.

- Definition 10.1 does not require $f$ to be a continuous function. If $f$ is of exponential order and has only a finite number of discontinuities, then the integral (10.5) exists. We say that $f$ is *piecewise continuous* if $f$ is bounded and is discontinuous only at a finite number of points in any bounded interval in its domain. Hence, the Laplace transform exists for piecewise continuous functions of exponential order.

The following examples show how to compute the Laplace transform of some simple functions.

✦ *Example 5*   Let $f$ be the constant function 1, that is, $f(t) = 1$ for all $t$. Clearly, $f$ is continuous and of exponential order. The Laplace transform of $f$ is

$$\mathcal{L}(f(t)) = \mathcal{L}(1)$$
$$= \int_0^\infty e^{-st}\, dt$$
$$= \lim_{t\to\infty} -\frac{e^{-st}}{s} + \frac{1}{s}$$
$$= \frac{1}{s} \tag{10.6}$$

---

[1] Strictly speaking, $s$ should be regarded as a complex number in Definition 10.3. In other words, $\mathcal{L}$ maps the real function $f$ to the complex function $F$. For simplicity, we regard $s$ as a real number.

whenever $s > 0$. In this example, the Laplace operator maps the constant function 1 (whose domain is all real numbers) to the function $F(s) = 1/s$ (whose domain is all numbers $s > 0$). ∎

✦ *Example 6*   Let $f(t) = e^{kt}$, where $k$ is any fixed constant. The Laplace transform of $f$ is

$$\mathcal{L}(f(t)) = \mathcal{L}\left(e^{kt}\right)$$

$$= \int_0^\infty e^{-st} e^{kt}\, dt$$

$$= \int_0^\infty e^{(k-s)t}\, dt$$

$$= \frac{1}{k-s} e^{(k-s)t} \bigg|_{t=0}^{t=\infty}$$

$$= \frac{1}{s-k} \tag{10.7}$$

whenever $s > k$. (The improper integral diverges if $s \leq k$.) Thus, the Laplace operator maps the exponential function $f(t) = e^{kt}$ to the function $F(s) = 1/(s-k)$. The improper integrals shown above are defined if $s > k$. ∎

Table 10.1 summarizes the Laplace transforms of the functions that we have considered. There are two additional properties of the Laplace transform that make it a useful tool in the solution of linear differential equations. These properties are summarized in the next two theorems.

|      | $f(t)$ | $\mathcal{L}(f)$ | **Domain of $F(s)$** |
|------|--------|------------------|----------------------|
| (1)  | 1      | $1/s$            | $s > 0$              |
| (2)  | $e^{kt}$ | $\dfrac{1}{s-k}$ | $s > k$            |

**TABLE 10.1**   Brief table of Laplace transforms.

**Theorem 10.4**   *The Laplace transform is a linear operator.*

Theorem 10.4 is proved in Chapter 4.

**Theorem 10.5**  *Let $f$ and $g$ be continuous functions of exponential order. Then the following properties hold.*

- *If $\mathcal{L}(f)$ exists, then it is unique.*
- *If $f$ and $g$ are continuous functions such that $f(t) = g(t)$ for all $t \geq 0$, then $\mathcal{L}(f) = \mathcal{L}(g)$.*
- *If $F$ and $G$ are the Laplace transforms of the continuous functions $f$ and $g$, respectively, and if $F = G$, then $f(t) = g(t)$ for all $t \geq 0$.*
- *The inverse Laplace transform operator, denoted $\mathcal{L}^{-1}$, is linear.*

The third property of Theorem 10.5 is especially important in the context of differential equations. It says that *if two continuous functions of exponential order have the same Laplace transform, then the functions are identical for $t \geq 0$.* The continuity assumption is essential; it is possible for two different discontinuous functions to have the same Laplace transform, as is illustrated in Exercise 15.

The next two examples show how the analytical properties described in Theorems 10.4 and 10.5 can be exploited to determine the Laplace transform of a broad class of functions using the entries in Table 10.1. The next section describes the table lookup method in more detail.

✦ *Example 7*  We can compute $\mathcal{L}(3e^{4t} + 2)$ using the linearity of the Laplace operator as follows:

$$\mathcal{L}\left(3e^{4t} + 2\right) = \mathcal{L}\left(3e^{4t}\right) + \mathcal{L}(2)$$
$$= 3\mathcal{L}\left(e^{4t}\right) + 2\mathcal{L}(1)$$
$$= \frac{3}{s-4} + \frac{2}{s},$$

using entries (1) and (2) in Table 10.1.                                        ■

✦ *Example 8*  This example illustrates how to compute an inverse Laplace transform. The computations are analogous to those in the previous example but in reverse order. Let

$$\mathcal{L}^{-1}\left(\frac{2}{s} + \frac{3}{s+2}\right).$$

Theorem 10.5 says that the inverse Laplace operator is linear. We can exploit this property as follows:

$$\mathcal{L}^{-1}\left(\frac{2}{s} + \frac{3}{s+2}\right) = \mathcal{L}^{-1}\left(\frac{2}{s}\right) + \mathcal{L}^{-1}\left(\frac{3}{s+2}\right)$$
$$= 2\mathcal{L}^{-1}\left(\frac{1}{s}\right) + 3\mathcal{L}^{-1}\left(\frac{1}{s+2}\right)$$
$$= 2 + 3e^{-2t}.$$

Entry (2) in Table 10.1 matches the term $1/(s + 2)$ with $k = -2$. The third property of Theorem 10.5 is also important here, because it implies that $2 + 3e^{-2t}$ is the *only* continuous function whose Laplace transform is $2/s + 3/(s + 2)$. ∎

✦ *Example 9*   The entries in Table 10.1 can be used to compute

$$\mathcal{L}^{-1}\left(\frac{3}{s^2 + 3s}\right)$$

as follows. Although neither of the entries in the table matches the given expression, an equivalent expression can be derived by using partial fraction decompositions. In this case,

$$\frac{3}{s^2 + 3s} = \frac{1}{s} - \frac{1}{s + 3},$$

so

$$\mathcal{L}^{-1}\left(\frac{3}{s^2 + 3s}\right) = \mathcal{L}^{-1}\left(\frac{1}{s}\right) - \mathcal{L}^{-1}\left(\frac{1}{s + 3}\right) = 1 - e^{-3t}.$$

Partial fraction decompositions are very useful in working with Laplace transforms. Appendix C contains a brief review of partial fraction decompositions. ∎

## Exercises

**1.** Show that each function in the following list is of exponential order by finding a value of $M$ and a value of $k$ that satisfy Eq. (10.1) for $t \geq 0$.

(a) $f(t) = t$

(b) $f(t) = t^2$

(c) $f(t) = e^{3t} - 5e^{-t}$

**2.** Determine whether each function in the following list is of exponential order. If it is of exponential order, then exhibit a value of $M$ and a value of $k$ that satisfy Eq. (10.1) for the function. Otherwise, explain why it is not possible to find such constants.

(a) $f(t) = 2t^2$

(b) $f(t) = 2^{2t}$

(c) $f(t) = 2^{2^t}$

(d) $f(t) = (2^2)^t$

(e) $g(t) = \ln t$

(f) $g(t) = \ln(t + 1)$

(g) $h(t) = \cos 2t$

(h) $h(t) = \cos(2^t)$

(i) $q(t) = t^{1000} - 10^4 t^2$

(j) $q(t) = 10^{10^{10}} - 10t^2$

**3.** Use Table 10.1 and the linearity of the Laplace transform to compute the Laplace transform of each function in the following list.

(a) $f(t) = e^{3t}$          (b) $f(t) = 2e^{-t} + 1$

(c) $f(t) = 2^t + 6$          (d) $f(t) = 1/3^t$

**4.** Use the definition of the Laplace transform, Eq. (10.5), to find the Laplace transform of each function in Exercise 1. Use the linearity of the Laplace transform where convenient.

**5.** The function $f$ is *bounded* if there is a nonnegative number $M$ such that $|f(t)| \leq M$ for every $t$ in the domain of $f$.

(a) Prove or disprove: A bounded function is of exponential order.

(b) Is a function of exponential order necessarily bounded? Explain.

**6.** Suppose that $f(t)$ and $g(t)$ are functions of exponential order.

(a) Prove or disprove: $f(t) + g(t)$ is of exponential order.

(b) Prove or disprove: $f(t)g(t)$ is of exponential order.

**7.** Suppose that, for some positive constant $k$,

$$\lim_{t \to \infty} e^{-kt} f(t) = C,$$

where $C$ is a nonzero constant. Is $f$ a function of exponential order? Explain.

**8.** The hyperbolic cosine and the hyperbolic sine are defined as

$$\cosh x = \frac{e^x + e^{-x}}{2}$$

and

$$\sinh x = \frac{e^x - e^{-x}}{2},$$

respectively. Use Table 10.1 and the linearity of the Laplace transform to compute $\mathcal{L}(\cosh kx)$ and $\mathcal{L}(\sinh kx)$, where $k$ is any constant.

**9.** Does the Laplace transform of $1/t$ exist? Does the Laplace transform of $1/(t+1)$ exist? Explain.

**10.** Use Table 10.1 and a partial fraction decomposition to determine the inverse Laplace transform of each function in the following list.

(a) $\dfrac{1}{s^2 + 5s + 6}$

(b) $\dfrac{1}{s(s+1)}$

(c) $\dfrac{1}{s(s+3)}$

(d) $\dfrac{1}{s^2 - 4}$

(e) $\dfrac{1}{s^2 + 7s + 12}$

(f) $\dfrac{1}{s^2 + 6s + 8}$

(g) $\dfrac{1}{(s+1)(s+2)(s+3)}$

(h) $\dfrac{s+4}{(s+1)(s+2)(s+3)}$

**11.** Euler's formula, derived in Chapter 5, says that

$$e^{it} = \cos t + i \sin t \qquad (10.8)$$

for any real number $t$. (Here, $i = \sqrt{-1}$.) Use Table 10.1, the linearity of the Laplace transform, and Eq. (10.8) to determine $\mathcal{L}(\cos t)$ and $\mathcal{L}(\sin t)$. You can treat $i$ just as you would a real constant when using the properties of linearity.

**12.** Is it true that $\mathcal{L}(fg) = \mathcal{L}(f)\mathcal{L}(g)$? Explain.

**13.** Let

$$f(t) = \begin{cases} (\sin t)/t, & t > 0, \\ 1, & t = 0. \end{cases}$$

Is $f$ of exponential order? Explain.

**14.** Show that

$$\mathcal{L}(t^n) = \frac{n!}{s^{n+1}} \qquad (10.9)$$

as follows.

(a) Find $\mathcal{L}(t)$ using the definition of the Laplace transform, Eq. (10.5).

(b) Use Eq. (10.5) to find $\mathcal{L}(t^2)$. (See also Exercise 4.)

(c) Establish Eq. (10.9) by induction as follows. Suppose that $\mathcal{L}(t^n) = n!/s^{n+1}$ for some $n$. (Parts (a) and (b) show that Eq. (10.9) holds for $n = 1$ and $n = 2$.) Then $\mathcal{L}\left(t^{n+1}\right)$, which is defined by

$$\mathcal{L}\left(t^{n+1}\right) = \int_0^\infty t^{n+1} e^{-st}\, dt,$$

can be integrated by parts to obtain an expression involving $\mathcal{L}(t^n)$.

**15.** Theorem 10.5 says that if $f$ and $g$ are continuous and if $\mathcal{L}(f) = F = \mathcal{L}(g) = G$, then $f(t) = g(t)$ for all $t \geq 0$. This exercise shows that the conclusion is false if $f$ and $g$ are not continuous.

(a) Let $f$ be the function defined by

$$f = \begin{cases} t, & t \neq 3, \\ 0, & t = 3. \end{cases}$$

Use Eq. (10.5) to find $\mathcal{L}(f)$.

(b) What is $\mathcal{L}(t)$? How does it compare to your answer in (a)? Explain.

(c) Let $f$ be the function defined by

$$f = \begin{cases} t, & 0 \leq t \leq 3 \text{ or } t \geq 4, \\ 0, & \text{if } 3 < t < 4. \end{cases}$$

Use Eq. (10.5) to determine $\mathcal{L}(f)$, and compare it to your answers in (a) and (b).

(d) Let $f$ be the function defined by

$$f = \begin{cases} t, & t \leq 3, \\ 0, & t > 3. \end{cases}$$

Use Eq. (10.5) to determine $\mathcal{L}(f)$, and compare it to your answers in (a)–(c).

(e) How might Theorem 10.5 be changed to include a statement about discontinuous functions? On the basis of your results in (a)–(d), what kinds of discontinuities might be allowed?

## 10.2 THE TRANSFORM TABLE LOOKUP METHOD

This section discusses how the Laplace transform can be used to compute explicit solutions of first- and second-order ordinary differential equations with constant coefficients, that is, differential equations of the form

$$x' + ax = g(t) \qquad \text{or} \qquad x'' + ax' + bx = g(t),$$

where $a$ and $b$ are constants and $g$ is a function of exponential order. Examples of functions $g$ for which the Laplace transform method is applicable include

- the sine and cosine functions,
- polynomials,
- exponential functions,
- Heaviside functions, and
- a sum or product of any of these.

Table 10.2 lists the Laplace transforms of these commonly encountered functions. The solutions are derived by using a table lookup procedure that is described later in this section.

| | $f(t)$ | $\mathcal{L}(f) = F(s)$ | Domain of $F(s)$ |
|---|---|---|---|
| (1) | $1$ | $\dfrac{1}{s}$ | $s > 0$ |
| (2) | $e^{kt}$ | $\dfrac{1}{s - k}$ | $s > k$ |
| (3) | $f'(t)$ | $sF(s) - f(0)$ | same as $F(s)$ |
| (4) | $f''(t)$ | $s^2 F(s) - sf(0) - f'(0)$ | same as $F(s)$ |
| (5) | $\sin \omega t$ | $\dfrac{\omega}{s^2 + \omega^2}$ | $s > 0$ |
| (6) | $\cos \omega t$ | $\dfrac{s}{s^2 + \omega^2}$ | $s > 0$ |
| (7) | $t^n, \quad n = 0, 1, 2, \ldots$ | $\dfrac{n!}{s^{n+1}}$ | $s > 0$ |
| (8) | $t \sin \omega t$ | $\dfrac{2\omega s}{(s^2 + \omega^2)^2}$ | $s > 0$ |
| (9) | $t \cos \omega t$ | $\dfrac{s^2 - \omega^2}{(s^2 + \omega^2)^2}$ | $s > 0$ |
| (10) | $\dfrac{1}{\omega} \sin \omega t - t \cos \omega t$ | $\dfrac{2\omega^2}{(s^2 + \omega^2)^2}$ | $s > 0$ |
| (11) | $(-t)^n f(t)$ | $\dfrac{d^n}{ds^n} F(s)$ | same as $F(s)$ |
| (12) | $e^{kt} f(t)$ | $F(s - k)$ | depends on $F(s)$ |
| (13) | $f(t - a)H(t - a)$ | $e^{-sa} F(s)$ | same as $F(s)$ |
| (14) | $\delta(t - t_0), \quad t_0 \geq 0$ | $e^{-st_0}$ | |

TABLE 10.2 Table of Laplace transforms.

## A Derivative Formula

There is a simple relationship between the Laplace transform of a differentiable function $f(t)$ of exponential order and its derivative $f'(t)$. Let $f(t)$ be a continuously differentiable function of exponential order, and let $\mathcal{L}(f)$ be its corresponding Laplace transform.

Then $\mathcal{L}(f')$ also exists, as the following integration by parts shows:

$$\mathcal{L}\big(f'(t)\big) = \int_0^\infty e^{-st} f'(t)\,dt$$

$$= e^{-st} f(t)\big|_{t=0}^\infty + \int_0^\infty s e^{-st} f(t)\,dt$$

$$= -f(0) + s \int_0^\infty f(t) e^{-st}\,dt$$

$$= -f(0) + s\mathcal{L}(f(t)). \tag{10.10}$$

The derivation of Eq. (10.10) requires that $\lim_{t\to\infty} f(t)e^{-st} = 0$, a fact that is straightforward to establish when $f$ is of exponential order (see Exercise 7). Equation (10.10) corresponds to entry (3) in Table 10.2.

   This discussion suggests the following theorem, whose proof is left as an exercise. (See Exercise 7.)

**Theorem 10.6**  *Let $f$ be a continuously differentiable function of exponential order. Then $\mathcal{L}(f')$ is defined, and*

$$\mathcal{L}(f') = s\mathcal{L}(f(t)) - f(0).$$

## The Basic Solution Procedure

We can use Eq. (10.10) to find solutions of differential equations of the form

$$\frac{dx}{dt} + ax(t) = g(t), \tag{10.11}$$

where $a$ is a constant, $g$ is a function of exponential order, and the initial condition is specified as $x(0) = x_0$. The linearity and uniqueness properties of the Laplace transform (Theorems 10.4 and 10.5) in Section 10.1 are used repeatedly in deriving a solution of Eq. (10.11).

   We begin by finding the Laplace transform of each side of Eq. (10.11) as follows:

$$\mathcal{L}\big(x' + ax\big) = \mathcal{L}\big(x'\big) + a\mathcal{L}(x)$$

$$= [s\mathcal{L}(x) - x(0)] + a\mathcal{L}(x)$$

$$= \mathcal{L}(g). \tag{10.12}$$

The first equality follows from the linearity of the Laplace transform. The second equality follows from Theorem 10.6 (entry (3) in Table 10.2). The third equality holds because the Laplace transform of a continuous function is unique (see Theorem 10.5).

Since the Laplace transform maps a function of $t$ to a function of $s$, we let $\mathcal{L}(g(t)) = G(s)$ and $\mathcal{L}(x(t)) = X(s)$. Then Eq. (10.12) is equivalent to

$$X(s) = \frac{G(s) + x(0)}{s + a}. \qquad (10.13)$$

We use Table 10.2, possibly with some algebraic manipulations such as partial fraction decompositions, to find the inverse Laplace transform of $X(s)$, which yields the solution $x(t)$. In this way,

> the Laplace transform turns the solution procedure into an algebraic problem wherein we solve for the unknown function $X(s)$.

The Laplace transform solution method for Eq. (10.11) consists of three steps:

1. Determine the Laplace transform of both sides of the differential equation.
2. Substitute the initial condition $x(0) = x_0$, and solve for $X(s)$.
3. Determine $\mathcal{L}^{-1}(X(s))$, which yields the solution $x(t)$.

The following example illustrates the procedure.

◆ *Example 1*   The initial value problem

$$x' + 5x = -1, \qquad x(0) = 2 \qquad (10.14)$$

can be solved by using the methods described in Section 1.5 and Section 5.6. It can also be solved by using the Laplace transform, as follows.

**Step 1.** Calculate the Laplace transform of both sides of Eq. (10.14). The left-hand side of Eq. (10.14) yields

$$\mathcal{L}\left(x' + 5x\right) = \mathcal{L}\left(x'\right) + 5\mathcal{L}(x)$$
$$= [sX(s) - x(0)] + 5X(s),$$

where we have used entry (3) in Table 10.2 and the linearity of the transform. The right-hand side yields

$$\mathcal{L}(-1) = -\frac{1}{s},$$

by entry (1). Therefore, the Laplace transform of Eq. (10.14) is

$$sX(s) - x(0) + 5X(s) = -\frac{1}{s}. \qquad (10.15)$$

**Step 2.** We substitute for the initial condition $x(0) = 2$ into Eq. (10.15) and solve for $X(s)$, which yields

$$X(s) = \frac{-1}{s(s+5)} + \frac{2}{s+5}. \tag{10.16}$$

**Step 3.** We apply the inverse Laplace transform to $X(s)$ to recover $x(t)$. The first expression on the right-hand side of Eq. (10.16) can be split by using a partial fraction decomposition. We have

$$x(t) = \mathcal{L}^{-1}\left(\frac{-1}{s(s+5)} + \frac{2}{s+5}\right)$$

$$= \mathcal{L}^{-1}\left(\frac{-\frac{1}{5}}{s} + \frac{\frac{1}{5}}{s+5}\right) + \mathcal{L}^{-1}\left(\frac{2}{s+5}\right)$$

$$= -\frac{1}{5}\mathcal{L}^{-1}\left(\frac{1}{s}\right) + \frac{1}{5}\mathcal{L}^{-1}\left(\frac{1}{s+5}\right) + 2\mathcal{L}^{-1}\left(\frac{1}{s+5}\right)$$

$$= -\frac{1}{5} + \frac{1}{5}e^{-5t} + 2e^{-5t}$$

$$= -\frac{1}{5} + \frac{11}{5}e^{-5t}.$$

The linearity of the inverse Laplace transform is used repeatedly to split the operator $\mathcal{L}^{-1}$ across each term in a sum and to factor out constants. Entries (1) and (2) in Table 10.2 are used to determine the inverse transforms after the partial fraction decompositions are computed. ∎

The solution procedure proceeds in the same manner for second-order differential equations, as the next example shows.

✦ *Example 2*   Let

$$x'' + 4x = \cos t, \qquad x(0) = 1, \quad x'(0) = 2. \tag{10.17}$$

**Step 1.** Find the Laplace transform of both sides of Eq. (10.17) using the transform table lookup method. The left-hand side is transformed by using entry (4) in Table 10.2:

$$\mathcal{L}(x'' + 4x) = \mathcal{L}(x'') + 4\mathcal{L}(x) \qquad \text{by linearity}$$
$$= s^2 X(s) - sx(0) - x'(0) + 4X(s) \qquad \text{using entry (4).}$$

Entry (6) yields the Laplace transform of the right-hand side:

$$\mathcal{L}(\cos t) = \frac{s}{s^2 + 1}.$$

Thus, the Laplace transform of Eq. (10.17) is

$$s^2 X(s) - sx(0) - x'(0) + 4X(s) = \frac{s}{s^2 + 1}. \qquad (10.18)$$

**Step 2.** We substitute the initial conditions into Eq. (10.18) and solve for $X(s)$ to find

$$X(s) = \frac{s + 2}{s^2 + 4} + \frac{s}{(s^2 + 1)(s^2 + 4)}.$$

**Step 3.** We invert the transform to recover the solution $x(t)$. Because the inverse Laplace transform is linear, we have

$$x(t) = \mathcal{L}^{-1}(X(s)) = \mathcal{L}^{-1}\left(\frac{s + 2}{s^2 + 4}\right) + \mathcal{L}^{-1}\left(\frac{s}{(s^2 + 1)(s^2 + 4)}\right).$$

We proceed term by term to break up each expression into one that matches an expression in the middle column of Table 10.2. The first term leads to

$$\mathcal{L}^{-1}\left(\frac{s + 2}{s^2 + 4}\right) = \mathcal{L}^{-1}\left(\frac{s}{s^2 + 4}\right) + \mathcal{L}^{-1}\left(\frac{2}{s^2 + 4}\right)$$

$$= \cos 2t + \sin 2t, \qquad (10.19)$$

where we use entries (5) and (6) in Table 10.2 and the linearity of the inverse transform.

A partial fraction decomposition is used to find the inverse transform of the second term. Here,

$$\frac{s}{(s^2 + 1)(s^2 + 4)} = \frac{\frac{1}{3}s}{s^2 + 1} - \frac{\frac{1}{3}s}{s^2 + 4},$$

so

$$\mathcal{L}^{-1}\left(\frac{s}{(s^2 + 1)(s^2 + 4)}\right) = \mathcal{L}^{-1}\left(\frac{\frac{1}{3}s}{s^2 + 1}\right) - \mathcal{L}^{-1}\left(\frac{\frac{1}{3}s}{s^2 + 4}\right)$$

$$= \tfrac{1}{3}\cos t - \tfrac{1}{3}\cos 2t \qquad (10.20)$$

by entry (6).

Finally, we add (10.19) and (10.20) together to obtain the solution of Eq. (10.17),

$$x(t) = \tfrac{1}{3}\cos t + \tfrac{2}{3}\cos 2t + \sin 2t. \qquad \blacksquare$$

✦ *Example 3*   Let

$$q'' + \omega^2 q = A \sin \alpha t, \qquad q(0) = 0, \quad q'(0) = 0, \qquad (10.21)$$

where $\alpha \neq \omega$. Equation (10.21) models an LC circuit with a periodic forcing and negligible resistance. (Notice that Eq. (10.21) is just the equation for simple harmonic motion if $A = 0$.) The Laplace transformation can be used to derive the solution of Eq. (10.21) as follows.

**Step 1.** The Laplace transform of the left-hand side of Eq. (10.21) is found in a manner similar to that in the previous example. Entry (5) in Table 10.2 implies that

$$\mathcal{L}(A \sin \alpha t) = \frac{A\alpha}{s^2 + \alpha^2}.$$

Therefore, the Laplace transform of Eq. (10.21) is

$$s^2 Q(s) - sq(0) - q'(0) + \omega^2 Q(s) = \frac{A\alpha}{s^2 + \alpha^2},$$

where $q(0)$ and $q'(0)$ are the initial charge and current in the circuit, respectively.

**Step 2.** The initial conditions $q(0) = 0$ and $q'(0) = 0$ lead to

$$Q(s) = \frac{A\alpha}{\left(s^2 + \alpha^2\right)\left(s^2 + \omega^2\right)}.$$

**Step 3.** The inverse transform can be found by partial fractions. The assumption $\alpha \neq \omega$ implies that all the roots of the denominator are simple. Therefore, the partial fraction decomposition has the form

$$\frac{A\alpha}{\left(s^2 + \alpha^2\right)\left(s^2 + \omega^2\right)} = \frac{Bs + C}{s^2 + \alpha^2} + \frac{Ds + E}{s^2 + \omega^2}$$

for appropriate constants $B$, $C$, $D$, and $E$. A straightforward (but somewhat lengthy) calculation yields $B = D = 0$, $C = -A\alpha/(\alpha^2 - \omega^2)$, and $E = A\alpha/(\alpha^2 - \omega^2)$.
   Therefore,

$$Q(s) = \left(\frac{A\alpha}{\alpha^2 - \omega^2}\right)\left(\frac{1}{s^2 + \omega^2}\right) - \left(\frac{A\alpha}{\alpha^2 - \omega^2}\right)\left(\frac{1}{s^2 + \alpha^2}\right).$$

Applying entry (5) in Table 10.2 to each expression in the preceding equation yields the solution,

$$q(t) = \left(\frac{A}{\alpha^2 - \omega^2}\right)\left(\frac{\alpha}{\omega} \sin \omega t - \sin \alpha t\right),$$

which exhibits beating because $\alpha \neq \omega$.                                            ■

✦ *Example 4*   Let

$$q'' + \omega^2 q = A \sin \omega t, \qquad q(0) = q'(0) = 0. \qquad (10.22)$$

Equation (10.22) is identical to Eq. (10.21) except that $\alpha = \omega$, that is, the period of the forcing is identical to the period of the simple harmonic oscillator. We can compute the solution using the Laplace transform method as follows.

**Step 1.** The Laplace transform of Eq. (10.22) can be computed in a manner similar to that in Example 3; we have

$$s^2 Q(s) - sq(0) - q'(0) + \omega^2 Q(s) = \frac{A\omega}{s^2 + \omega^2}.$$

**Step 2.** The initial conditions $q(0) = 0$ and $q'(0) = 0$ lead to

$$Q(s) = \frac{A\omega}{\left(s^2 + \omega^2\right)^2}.$$

**Step 3.** To invert the transform, we exploit the linearity of the inverse Laplace operator to obtain

$$q(t) = \mathcal{L}^{-1}\left(\frac{A\omega}{\left(s^2 + \omega^2\right)^2}\right)$$

$$= \left(\frac{A}{2\omega}\right)\mathcal{L}^{-1}\left(\frac{2\omega^2}{\left(s^2 + \omega^2\right)^2}\right)$$

$$= \left(\frac{A}{2\omega}\right)\left(\frac{1}{\omega}\sin \omega t - t\cos \omega t\right),$$

by entry (10) of Table 10.2.

Notice that the solution exhibits resonance.                                ∎

## Exercises

**1.** Find the solution of each of the following initial value problems using the Laplace transform. In each case, $k$ is an arbitrary constant.

(a) $x' = kx, x(0) = 10$

(b) $x' = kx - 100, x(0) = 1000$

(c) $10x' = -x + 100, x(0) = 3$

**2.** Find the solution of each of the following initial value problems using the Laplace transform.

(a) $x'' + 2x = \cos t, x(0) = x'(0) = 1$

(b) $x'' + 4x = 5e^{-t}, x(0) = 1, x'(0) = 0$

**3.** Find the solution of each of the following initial value problems using the Laplace transform.

(a) $x'' + x = \cos t, x(0) = x'(0) = 1$

(b) $x'' + 9x = 9t, x(0) = 0, x'(0) = -1$

**4.** Find the solution of each of the following initial value problems using the Laplace transform.

(a) $x'' + 5x = 4\sin 2t, x(0) = 0, x'(0) = -1$

(b) $x'' + 4x = 4\sin 2t, x(0) = 0, x'(0) = -1$

**5.** Find the solution of each of the following initial value problems using the Laplace transform.

(a) $x'' + 4x' + 3x = 4e^{-2t}$, $x(0) = 1$, $x'(0) = 2$

(b) $x'' + 5x' + 6x = \cos t$, $x(0) = 0$, $x'(0) = 0$

**6.** Find the solution of each of the following initial value problems using the Laplace transform.

(a) $x'' + 3x' + 2x = 5\cos t$, $x(0) = 1$, $x'(0) = -1$

(b) $x'' + \frac{5}{2}x' + x = t$, $x(0) = 0$, $x'(0) = 1$

**7.** Let $f$ be a continuously differentiable function of exponential order.

(a) Show that

$$\lim_{t \to \infty} f(t)e^{-st} = 0$$

whenever $s$ is a suitably large positive number. (This relation is used in the derivation of Eq. (10.10).)

(b) Suppose that $\mathcal{L}(f)$ is defined whenever $s > k$. Show that $\mathcal{L}(f')$ is defined for $s > k$.

**8.** Let $f(t)$ be a twice differentiable function, and suppose that $f'$ and $f''$ are of exponential order. Show that $\mathcal{L}(f''(t)) = s^2 F(s) - sf(0) - f'(0)$, where $F(s) = \mathcal{L}(f(t))$. (Hint: Use integration by parts twice.)

**9.** Let $f^{(n)} = d^n f/dt^n$, and suppose that $f$, $f'$, $f''$, ..., $f^{(n-1)}$ are of exponential order. Show by induction that

$$\mathcal{L}\left(f^{(n)}\right) = s^n \mathcal{L}(f) - s^{n-1} f(0)$$
$$- s^{n-2} f'(0) - \cdots$$
$$- sf^{(n-2)}(0) - f^{(n-1)}(0).$$

(Hint: Theorem 10.2 and Exercise 8 show that the statement is true for $n = 1$ and $n = 2$.)

## 10.3   SHIFTS, DERIVATIVES, AND SYSTEMS

This section describes two formulas, a shift formula and a derivative formula, that are often useful in practice. These formulas allow you to obtain the Laplace transforms of certain functions in terms of other Laplace transforms. This section also shows how the Laplace transform can be applied to systems of linear differential equations with constant coefficients.

### Shifts in the $s$ Domain

Let $f(t)$ be a function of exponential order whose Laplace transform is $F(s)$. The shift formula in the $s$ domain is simply

$$\mathcal{L}\left(e^{kt} f(t)\right) = F(s - k). \tag{10.23}$$

Equation (10.23), which corresponds to entry (12) in Table 10.2, can be derived directly from the definition of the Laplace transform (see Exercise 11). Equation (10.23) is called a *shift* because it gives a relation between the change of variable $u = s - k$ in the $s$ domain and its effect on the corresponding function in the $t$ domain. The shift formula is used for underdamped homogeneous equations and whenever a forcing function, such as a polynomial or sine or cosine function, is multiplied by $e^{kt}$.

✦ *Example 1*   The Laplace transform of $g(t) = e^{-t} \cos 2t$ can be computed by using the shift formula as follows. We have

$$F(s) = \mathcal{L}(\cos 2t) = \frac{s}{s^2 + 4}$$

by entry (6) of Table 10.2.  The shift formula (10.23) can be applied with $k = -1$ to obtain

$$\mathcal{L}(g) = F(s - (-1)) = F(s + 1) = \frac{s + 1}{(s + 1)^2 + 4}.$$  ∎

The shift formula often can be used to find the inverse Laplace transform in cases in which the denominator of an expression in the $s$ domain cannot be factored over the set of real numbers.  When the denominator is quadratic, it is possible to complete the square and apply the shift formula to invert the transform, as the next examples show.

✦ *Example 2*   We can compute

$$\mathcal{L}^{-1}\left(\frac{s}{s^2 - 2s + 5}\right) \tag{10.24}$$

as follows.  The denominator cannot be factored over the real numbers, but we can complete the square.  That is,

$$s^2 - 2s + 5 = (s - 1)^2 + 4.$$

Therefore,

$$\mathcal{L}^{-1}\left(\frac{s}{s^2 - 2s + 5}\right) = \mathcal{L}^{-1}\left(\frac{s}{(s - 1)^2 + 4}\right)$$

$$= \mathcal{L}^{-1}\left(\frac{(s - 1) + 1}{(s - 1)^2 + 4}\right)$$

$$= \mathcal{L}^{-1}\left(\frac{s - 1}{(s - 1)^2 + 4}\right) + \mathcal{L}^{-1}\left(\frac{1}{(s - 1)^2 + 4}\right). \tag{10.25}$$

Let us examine the first term in Eq. (10.25) in more detail.  First observe that

$$\mathcal{L}^{-1}\left(\frac{s}{s^2 + 4}\right) = \cos 2t,$$

by entry (6) in Table 10.2. In Eq. (10.25), however, $s$ is replaced by $s - 1$. Thus, the shift formula (12) can be applied with $k = 1$, and we have

$$\mathcal{L}^{-1}\left(\frac{s-1}{(s-1)^2+4}\right) = e^t \cos 2t.$$

To invert the second term in Eq. (10.25), we first observe that

$$\mathcal{L}^{-1}\left(\frac{1}{s^2+4}\right) = \mathcal{L}^{-1}\left(\frac{\frac{1}{2}\cdot 2}{s^2+4}\right)$$

$$= \tfrac{1}{2}\sin 2t.$$

However, $s$ is replaced by $s - 1$ in Eq. (10.25), so we again apply the shift formula (12) with $k = 1$ to obtain

$$\mathcal{L}^{-1}\left(\frac{1}{(s-1)^2+4}\right) = \frac{e^t \sin 2t}{2}.$$

Therefore,

$$\mathcal{L}^{-1}\left(\frac{s}{s^2-2s+5}\right) = e^t \cos 2t + \frac{e^t \sin 2t}{2}. \qquad \blacksquare$$

◆ *Example 3*   The solution of the initial value problem

$$x'' + 4x' + 13x = 0, \qquad x(0) = 1, \quad x'(0) = 2 \qquad (10.26)$$

can be obtained by using the usual three-step method, as follows.

**Step 1.** The Laplace transform of Eq. (10.26) is

$$s^2 X(s) - sx(0) - x'(0) + 4sX(s) - 4x(0) + 13X(s) = 0.$$

**Step 2.** We substitute the initial conditions $x(0) = 1$ and $x'(0) = 2$ and solve for $X(s)$ to obtain

$$X(s) = \frac{s+6}{s^2+4s+13}.$$

**Step 3.** The solution $x(t)$ can be recovered by computing $\mathcal{L}^{-1}(X)$. However, the denominator cannot be factored as $(s+a)(s+b)$ for real numbers $a$ and $b$. Instead, we can complete the square; that is,

$$s^2 + 4s + 13 = (s+2)^2 + 9,$$

so that

$$\mathcal{L}^{-1}\left(\frac{s+6}{s^2+4s+13}\right) = \mathcal{L}^{-1}\left(\frac{s+2}{(s+2)^2+9}\right) + \mathcal{L}^{-1}\left(\frac{4}{(s+2)^2+9}\right). \qquad (10.27)$$

Entry (6) shows that

$$\mathcal{L}^{-1}\left(\frac{s}{s^2+9}\right)=\cos 3t,$$

and entry (5) shows that

$$\mathcal{L}^{-1}\left(\frac{4}{s^2+9}\right)=\mathcal{L}^{-1}\left(\frac{\frac{4}{3}\cdot 3}{s^2+9}\right)$$

$$=\tfrac{4}{3}\sin 3t.$$

However, $s$ is replaced by $s+2$ in Eq. (10.27). Therefore, we apply the shift formula (12) with $k=-2$ to each term on the right-hand side of Eq. (10.27). The computations above imply that

$$\mathcal{L}^{-1}\left(\frac{s+6}{(s+2)^2+9}\right)=e^{-2t}\cos 3t+\tfrac{4}{3}e^{-2t}\sin 3t.\qquad\blacksquare$$

The shift formula can be combined with partial fraction decompositions to find explicit solutions of linear, constant coefficient differential equations whose forcing functions are the product of an exponential and a sine, cosine, or polynomial.

◆ *Example 4*   The solution of the initial value problem

$$x''+x=10e^{-t}\cos 2t,\qquad x(0)=x'(0)=0\qquad\qquad(10.28)$$

can be computed by using the Laplace transform method as follows. The shift formula applied to the right-hand side of Eq. (10.28) yields

$$\mathcal{L}\left(10e^{-t}\cos 2t\right)=10\left(\frac{s+1}{(s+1)^2+4}\right).$$

We apply the Laplace transform to the left-hand side of Eq. (10.28), substitute the initial conditions, and solve for $X(s)$ to obtain

$$X(s)=\frac{10(s+1)}{\left((s+1)^2+4\right)\left(s^2+1\right)}.$$

The denominator of the right-hand side is a product of irreducible quadratic factors, so the partial fraction decomposition has the form

$$\frac{As+B}{(s+1)^2+4}+\frac{Cs+D}{s^2+1},$$

which leads to

$$X(s) = \frac{s+3}{s^2+1} - \frac{(s+1)+4}{(s+1)^2+4}.$$

Therefore,

$$x(t) = \cos t + 3\sin t - e^{-t}\cos 2t - 2e^{-t}\sin 2t,$$

by entries (5) and (6) of Table 10.2, together with the shift formula (12).  ∎

## The Derivative Formula

The derivative formula, entry (11) in Table 10.2, often is useful when the denominator has complex roots that are not simple, as the following example shows.

✦ *Example 5*   Find

$$\mathcal{L}^{-1}\left(\frac{-2bs}{\left(s^2+b^2\right)^2}\right). \tag{10.29}$$

*Solution*   Observe that

$$\frac{-2bs}{\left(s^2+b^2\right)^2} = \frac{d}{ds}\left(\frac{b}{s^2+b^2}\right).$$

Now

$$\mathcal{L}^{-1}\left(\frac{b}{s^2+b^2}\right) = \sin bt,$$

so the derivative formula implies that

$$\mathcal{L}^{-1}\left(\frac{-2bs}{\left(s^2+b^2\right)^2}\right) = -t\sin bt,$$

which is equivalent to entry (8) in Table 10.2.  ∎

## Linear Systems of Equations

The Laplace transform can be used to solve linear systems of differential equations with constant coefficients. The method is particularly useful for nonhomogeneous systems, as the computations discussed in Section 7.8 often are quite tedious. The same three-step solution procedure is used, as the next example illustrates.

✦ *Example 6*   Let

$$x_1' = x_1 + x_2,$$
$$x_2' = -x_1 + x_2,$$

(10.30)

where $x_1(0) = x_2(0) = 1$. The Laplace transform method is applied as follows.

**Step 1.** The Laplace transform of Eq. (10.30) is

$$sX_1(s) - x_1(0) = X_1(s) + X_2(s),$$
$$sX_2(s) - x_2(0) = -X_1(s) + X_2(s).$$

**Step 2.** We substitute the initial conditions to find

$$sX_1(s) - 1 = X_1(s) + X_2(s),$$
$$sX_2(s) - 1 = -X_1(s) + X_2(s),$$

which is just a linear system of equations for $X_1(s)$ and $X_2(s)$. Some straightforward algebra shows that

$$X_1(s) = \frac{s}{(s-1)^2 + 1},$$

$$X_2(s) = \frac{s-2}{(s-1)^2 + 1}.$$

**Step 3.** We can use the shift formula to invert the transforms and recover the solution. We have

$$\mathcal{L}^{-1}(X_1(s)) = \mathcal{L}^{-1}\left(\frac{(s-1) + 1}{(s-1)^2 + 1}\right) = e^t \cos t + e^t \sin t,$$

using entries (5), (6), and (12) in Table 10.2. Similarly,

$$\mathcal{L}^{-1}(X_2(s)) = \mathcal{L}^{-1}\left(\frac{(s-1) - 1}{(s-1)^2 + 1}\right) = e^t \cos t - e^t \sin t.$$

Therefore, the solution is

$$x_1(t) = e^t(\cos t + \sin t),$$
$$x_2(t) = e^t(\cos t - \sin t).$$

∎

## *Exercises*

**1.** Find the inverse Laplace transform of each function in the following list.

(a) $\dfrac{s+4}{(s-1)^2}$        (b) $\dfrac{s^2+4}{(s-1)^3}$

(c) $\dfrac{1}{s^2+4s+5}$        (d) $\dfrac{1}{s^2-6s+13}$

**2.** Find the inverse Laplace transform of each function in the following list.

(a) $\dfrac{1}{s(s-1)^2}$        (b) $\dfrac{1}{s^2(s-1)^2}$

(c) $\dfrac{s+3}{(s+1)^2}$        (d) $\dfrac{64s^2}{(s^2+4)^3}$

**3.** Find the solution of each of the initial value problems in the following list.

(a) $x''+4x=10e^{-t}\sin t,\ x(0)=1,\ x'(0)=1$

(b) $x''+x=10te^{-t},\ x(0)=1,\ x'(0)=2$

**4.** Find the solution of each of the initial value problems in the following list.

(a) $x''+x=10e^{-t}\sin t,\ x(0)=1,\ x'(0)=2$

(b) $x''+x=8e^{-2t}\sin t,\ x(0)=1,\ x'(0)=0$

**5.** Find the solution of each of the initial value problems in the following list.

(a) $x''+2x'+2x=0,\ x(0)=1,\ x'(0)=-2$

(b) $x''+4x'+8x=0,\ x(0)=2,\ x'(0)=4$

(c) $x''-6x'+10x=0,\ x(0)=1,\ x'(0)=1$

**6.** Find the solution of each of the initial value problems in the following list.

(a) $x''+x'+x=\sin t,\ x(0)=1,\ x'(0)=1$

(b) $x''+2x'+5x=3e^{-t}\cos t,\ x(0)=1,\ x'(0)=1$

**7.** Use the Laplace transform to find the solution of the system

$$x_1'=2x_1+3x_2,$$

$$x_2'=3x_1+2x_2$$

satisfying the initial conditions $x_1(0)=1$ and $x_2(0)=3$.

**8.** Use the Laplace transform to find the solution of the system

$$x_1'=-3x_1+x_2,$$

$$x_2'=x_1-3x_2$$

satisfying the initial conditions $x_1(0)=1$ and $x_2(0)=2$.

**9.** Use the Laplace transform to find the solution of the system

$$x_1'=18x_1-10x_2,$$

$$x_2'=29x_1-16x_2$$

satisfying the initial conditions $x_1(0)=10$ and $x_2(0)=5$.

**10.** Use the Laplace transform to find the solution of the system

$$x_1'=2x_1-2x_2+e^{-t},$$

$$x_2'=6x_1-5x_2+e^{-2t}$$

satisfying the initial conditions $x_1(0)=1$ and $x_2(0)=-1$.

**11.** Establish entry (12) in Table 10.2 by using the definition of the Laplace transform, Eq. (10.5).

**12.** Let $F(s)=\int_0^\infty e^{-st}f(t)\,dt$ for some function $f(t)$ of exponential order. Assume that the integral converges for all values of $s$ that are larger than some number $k$. It can be shown that

$$\frac{dF(s)}{ds} = \int_0^\infty \frac{d}{ds}\left[e^{-st} f(t)\right] dt$$

for all $s > k$.

(a) Use this result to show that

$$\mathcal{L}(tf(t)) = -\frac{dF(s)}{ds}.$$

This relation corresponds to entry (11) in Table 10.2 with $n = 1$.

(b) Use your result in (a) to establish entry (8) in Table 10.2.

(c) Use the result in (a) and induction to establish entry (11) in Table 10.2.

**13.** Derive entry (9) in Table 10.2 using entry (6) and the derivative formula, entry (11).

**14.** Establish entry (10) in Table 10.2 as follows. Use the identity

$$\frac{1}{\left(s^2 + \omega^2\right)^2} = \frac{1}{2\omega^3} \frac{\omega}{s^2 + \omega^2} - \frac{1}{2\omega^2} \frac{s^2 - \omega^2}{\left(s^2 + \omega^2\right)^2}$$

and entries (5) and (9) to find

$$\mathcal{L}^{-1}\left(\frac{1}{\left(s^2 + \omega^2\right)^2}\right).$$

Then derive the general formula.

**15.** This exercise and Exercises 16, 17, 18, and 20 discuss the basis of a qualitative theory of the Laplace transform of linear constant coefficient differential equations.
  Let

$$x'' + ax' + bx = 0. \qquad (10.31)$$

(a) Show that

$$\mathcal{L}(x) = X(s) = \frac{G(s)}{H(s)},$$

where $G(s)$ is a polynomial that depends on the initial conditions and $H(s)$ is the characteristic equation corresponding to Eq. (10.31).

(b) Suppose that Eq. (10.31) is a model of an over-damped linear spring. Show that the roots of $H$ are real, negative, and distinct.

(c) Find the form of the inverse Laplace transform, but do not evaluate the coefficients in the partial fraction decomposition. What can you say about the qualitative behavior of the solution $x(t)$ as $t \to \infty$?

**16.** Let

$$x'' + ax' + bx = 0. \qquad (10.32)$$

(a) Use the result of Exercise 15(a) to show that if Eq. (10.32) is a model of an underdamped linear spring, then the roots of $H$ are complex conjugates with negative real parts.

(b) Find the form of the inverse Laplace transform, but do not evaluate the coefficients in the partial fraction decomposition. What can you say about the qualitative behavior of the solution $x(t)$ as $t \to \infty$?

**17.** Let $x'' - x = 0$.

(a) Show that $\mathcal{L}(x)$ can be written in the form $X(s) = G(s)/H(s)$ for appropriate functions $G$ and $H$.

(b) Show that if $G(1) = 0$, then $x(t) = ke^{-t}$ for some constant $k$.

(c) Show that if $G(-1) = 0$, then $x(t) = me^{t}$ for some constant $m$.

(d) Show that if the roots of $G(s)$ are not 1 or $-1$, then $x(t)$ is a linear combination of $e^t$ and $e^{-t}$.

**18.** Prove the following result. Let

$$X(s) = \frac{As + B}{s^2 + as + b},$$

where $A$, $B$, $a$, and $b$ are real constants.

- If the roots of the denominator are real and distinct, then $x(t) = \mathcal{L}^{-1}(X)$ is a linear combination of exponential functions. If the roots are positive and distinct, then $x(t)$ tends to infinity as $t$ becomes large. If the roots are negative and distinct, then $x(t)$ tends to 0 as $t$ becomes large.

- If the roots of the denominator are complex conjugate, then $x(t)$ oscillates with time. If the real part of the roots is positive, then $x$ has an exponentially increasing amplitude; and if the real part of the roots is negative, then $x$ has an exponentially decaying amplitude.

**19.** For each Laplace transform in the following list, determine the qualitative behavior of $x(t)$ without explicitly inverting the transform. Consider questions such as the following: Does $x(t)$ go to infinity as $t$ becomes large, or does $x(t)$ tend to 0? Does $x(t)$ oscillate?

(a) $X(s) = \dfrac{s+1}{s^2 + 4s + 1}$

(b) $X(s) = \dfrac{1}{s^2 + 6s + 5}$

(c) $X(s) = \dfrac{1}{s^2 - 4s + 3}$

(d) $X(s) = \dfrac{s}{s^2 + s + 1}$

**20.** Let $x'' + ax' + bx = f(t)$.

(a) Assume that the Laplace transform of $f(t)$ can be written in the form $F(s) = M(s)/N(s)$. Show that

$$X(s) = \frac{G(s)}{H(s)} + \frac{M(s)}{N(s)H(s)}.$$

(b) Assume that $a > 0$ and $b > 0$. Discuss how the type of roots of the equation $N(s) = 0$ determines the steady-state behavior of the solution $x(t)$.

## 10.4  DISCONTINUOUS AND RAMPED FORCING

As is mentioned at the beginning of this chapter, the Laplace transform is the analytical tool of choice for solving linear, constant coefficient differential equations that contain a discontinuous forcing term. This section discusses how such equations arise and the qualitative properties of their solutions.

### Jump Discontinuities

Consider the response of an electrical circuit when a switch is thrown and an electrical potential suddenly is applied. For instance, no potential might be applied to a circuit for $0 \le t \le 1$. However, at $t = 1$, a switch is thrown, connecting the circuit to a battery that supplies a constant potential of 1 V.

Figure 10.3 shows the graph of the electrical potential as a function of $t$ when a switch is thrown at $t = 1$. The dashed vertical line is intended only to guide the eye—the potential is discontinuous at $t = 1$. (For our purposes, it does not matter whether we take $f(1) = 0$ or $f(1) = 1$. The important point is that $|f(t_1) - f(t_2)| = 1$ whenever $t_1 < 1$ and $t_2 > 1$.) We say that the function $f$ has a *jump discontinuity* at $t = 1$.

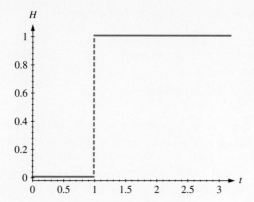

**FIGURE 10.3**   The graph of $H(t - 1)$.

---

**Definition 10.7**   *The function $f(t)$ has a* jump discontinuity *at $t = a$ if there are finite values A and B such that*

1. $\lim\limits_{t \nearrow a} f(t)$ *exists and equals A,*

2. $\lim\limits_{t \searrow a} f(t)$ *exists and equals B, and*

3. $A \neq B.$

*The limit (1) is taken as $t$ increases up to $a$ and is called the* left-hand limit *at $a$. The limit (2) is taken as $t$ decreases down to $a$ and is called the* right-hand limit *at $a$.*

Definition 10.7 says that if the left- and right-hand limits of $f$ at $a$ are different and finite, then $f$ has a jump discontinuity. The value of $f(a)$ is not important—only the left- and right-hand limits at $a$. In fact, $f(a)$ need not be defined at all.

✦ *Example 1*   The function whose graph is depicted in Fig. 10.3 has a jump discontinuity at $t = 1$, because the left-hand limit is 0 and the right-hand limit is 1.   ∎

✦ *Example 2*   The function $f(t) = 1/t$ does not have a jump discontinuity at $t = 0$, because the left- and right-hand limits of $f$ are not finite at $t = 0$.   ∎

## The Heaviside Function

Functions with jump discontinuities can often be regarded as linear combinations of a particular discontinuous function called the *Heaviside function.*

**Definition 10.8**   *The Heaviside function, denoted $H(t)$, is defined as*

$$H(t) = \begin{cases} 0, & t \le 0, \\ 1, & t > 0. \end{cases}$$

*The Heaviside function is also called the* step function.

    The Heaviside function has a jump discontinuity at $t = 0$, because the left-hand limit is 0 and the right-hand limit is 1. The location of the discontinuity can be altered by adding a shift to the argument of $H$. For example, Fig. 10.3 shows the graph of $H(t - 1)$. If $t < 1$, then $t - 1 < 0$ and $H(t) = 0$; and if $t > 1$, then $t - 1 > 0$ and $H(t) = 1$. The amplitude of the discontinuity can be altered by multiplying $H$ by an appropriate constant or other function.

    Definition 10.8 specifies that $H(0) = 0$, but the actual value is not important. For our purposes, we could just as well take $H(0) = 1$. The important point is that virtually any pulse can be constructed with a suitable combination of shifts of Heaviside functions.

◆ *Example 3*   Figure 10.4(a) shows a rectangular pulse between $t = 2$ and $t = 6$ of height 3. This function would model an electrical potential if we switched on a connection to a 3-volt battery for four seconds starting at $t = 2$ s. (The dashed vertical lines are intended only to guide the eye and emphasize that the function is discontinuous at $t = 2$ and $t = 6$.) We denote the function in Fig. 10.4(a)

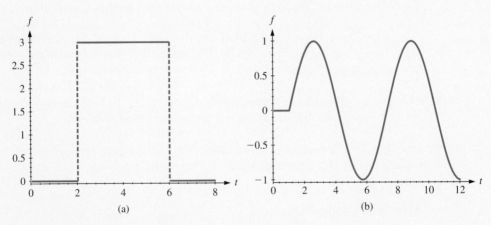

**Figure 10.4**   (a) A rectangular pulse. (b) The graph of $H(t - 1)\sin(t - 1)$.

as

$$f(t) = 3(H(t - 2) - H(t - 6)).$$

The first term "switches on" at $t = 2$, but the second one remains 0 until $t = 6$. Thus, $f(t) = 3$ if $2 < t < 6$. When $t > 6$, both Heaviside functions are 1, so $f(t) = 0$.   ■

✦ *Example 4*   Figure 10.4(b) shows the graph of the function $f(t) = H(t - 1) \sin(t - 1)$. Notice that $H(t - 1) = 0$ if $t \le 1$, so $f(t) = 0$ if $t \le 1$. When $t > 1$, $f(t) = \sin(t - 1)$, and its graph is the same as the usual graph of the sine function, except that it is translated by one unit along the $t$ axis.   ■

The Heaviside function is bounded, so it is of exponential order. Its Laplace transform is

$$\mathcal{L}(H(t - a)) = \int_0^\infty H(t - a) e^{-st} \, dt$$

$$= \int_a^\infty e^{-st} \, dt$$

$$= -\frac{e^{-st}}{s} \Big|_a^\infty$$

$$= \frac{e^{-as}}{s}$$

whenever $s > 0$. Notice that $\mathcal{L}(H)$ does not depend on the value of $H(0)$.

In a similar way, we can determine the Laplace transform of any function of the form $g(t - a) = H(t - a) f(t - a)$ provided that $f$ is of exponential order. If we let $u = t - a$, then

$$\mathcal{L}(g(t - a)) = \int_0^\infty f(t - a) H(t - a) e^{-st} \, dt$$

$$= \int_a^\infty f(t - a) e^{-s(t-a) - sa} \, dt$$

$$= e^{-sa} \int_0^\infty f(u) e^{-su} \, du$$

$$= e^{-sa} \mathcal{L}(f). \tag{10.33}$$

Equation (10.33) is is called the *time shift formula* and corresponds to entry (13) in Table 10.2.

The next two examples illustrate how to use the shift formula (10.33) to compute the Laplace transform of some simple pulses.

◆ *Example 5*    The Laplace transform of the rectangular pulse

$$f(t) = 3(H(t-2) - H(t-6)),$$

whose graph is shown in Fig. 10.4(a), can be computed as follows:

$$\begin{aligned} \mathcal{L}(f) &= \mathcal{L}[3(H(t-2) - H(t-6))] \\ &= 3\mathcal{L}(H(t-2)) - 3\mathcal{L}(H(t-6)) \\ &= e^{-2s}\left(\frac{3}{s}\right) - e^{-6s}\left(\frac{3}{s}\right). \end{aligned}$$

■

◆ *Example 6*    To calculate the Laplace transform of the single sine pulse shown in Fig. 10.5, we represent the pulse as a product of a sine function of period 1 with a sum of Heaviside functions. The pulse "turns on" at $t = 1$ and "turns off" at $t = 2$, as in Example 5. The amplitude of the pulse is 1, so an appropriate formula is

$$g(t) = (\sin 2\pi t)(H(t-1) - H(t-2)).$$

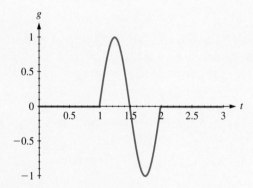

**FIGURE 10.5**    A single sine pulse.

The shift formula (10.33) requires that the time $t$ be measured from a common origin. In this example, however, the sine function is 1-periodic, so the identity

$$\sin 2\pi t = \sin 2\pi (t-1) = \sin 2\pi (t-2)$$

can be used. We find

$$
\begin{aligned}
\mathcal{L}(g) &= \mathcal{L}\big[(\sin 2\pi t)(H(t-1) - H(t-2))\big] \\
&= \mathcal{L}\big[(\sin 2\pi t)H(t-1)\big] - \mathcal{L}\big[(\sin 2\pi t)H(t-2)\big] \\
&= \mathcal{L}\big[(\sin 2\pi(t-1))H(t-1)\big] - \mathcal{L}\big[(\sin 2\pi(t-2))H(t-2)\big] \\
&= e^{-s}\left(\frac{2\pi}{s^2 + 4\pi^2}\right) - e^{-2s}\left(\frac{2\pi}{s^2 + 4\pi^2}\right).
\end{aligned}
$$                                                                                    ∎

✦ *Example 7*   Consider the single sine pulse in Fig. 10.6. It is shifted by 1 unit from the origin, its amplitude is 3, and it exists for an interval of length $2\pi$. Since the pulse completes one full cycle, the graph shows a portion of a $2\pi$-periodic sine function, namely, $3\sin(t-1)$. Because the graph is zero for $t < 1$ and $t > 1 + 2\pi$, the sine function is multiplied by a term of the form $H(t-1) - H(t-1-2\pi)$. Therefore, the function shown in the graph can be expressed as

$$
f(t) = 3\big[H(t-1) - H(t-(1+2\pi))\big]\sin(t-1).
$$

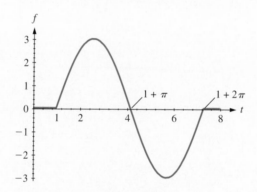

**FIGURE 10.6**   A single sine pulse shifted by 1 unit from the origin.                     ∎

✦ *Example 8*   Consider the initial value problem

$$
x'' + 4x = f(t), \qquad x(0) = x'(0) = 0, \tag{10.34}
$$

where $f(t)$ is the forcing term discussed in Example 7. Since $f(t) = 0$ for $0 \le t \le 1$, the solution of Eq. (10.34) must also be zero for $0 \le t \le 1$, because it corresponds to an unforced, undamped oscillator at equilibrium during this time interval. (This kind of observation is useful for checking the final answer.)

The Laplace transform of $f(t)$ can be found by applying entry (13) in Table 10.2. We have

$$\mathcal{L}\left[3(H(t-1) - H(t-(1+2\pi)))\sin(t-1)\right]$$
$$= 3\mathcal{L}\left[H(t-1)\sin(t-1)\right] - 3\mathcal{L}\left[H(t-(1+2\pi))\sin(t-(1+2\pi))\right]$$
$$= \frac{3e^{-s}}{s^2+1} - \frac{3e^{-(1+2\pi)s}}{s^2+1}.$$

We apply the Laplace transform to the left-hand side of Eq. (10.34), substitute the initial conditions, and solve for $X(s)$ to obtain

$$X(s) = \frac{3e^{-s} - 3e^{-(1+2\pi)s}}{(s^2+1)(s^2+4)}.$$

A partial fraction decomposition gives

$$\frac{3}{(s^2+1)(s^2+4)} = \frac{1}{s^2+1} - \frac{1}{s^2+4}.$$

Therefore,

$$X(s) = \left(\frac{e^{-s}}{s^2+1} - \frac{e^{-s}}{s^2+4}\right) - \left(\frac{e^{-(1+2\pi)s}}{s^2+1} - \frac{e^{-(1+2\pi)s}}{s^2+4}\right).$$

The shift formula (13), entry (5), and the periodicity of the sine function imply that

$$\mathcal{L}^{-1}\left(\frac{e^{-s}}{s^2+1} - \frac{e^{-s}}{s^2+4}\right) = H(t-1)\sin(t-1) - \frac{H(t-1)\sin 2(t-1)}{2}$$

and

$$\mathcal{L}^{-1}\left(\frac{e^{-(1+2\pi)s}}{s^2+1} - \frac{e^{-(1+2\pi)s}}{s^2+4}\right)$$
$$= H(t-1-2\pi)\sin(t-1-2\pi) - \frac{H(t-1-2\pi)\sin 2(t-1-2\pi)}{2}.$$

Therefore,

$$x(t) = H(t-1)\left[\sin(t-1) - \tfrac{1}{2}\sin 2(t-1)\right]$$
$$- H(t-1-2\pi)\left[\sin(t-1-2\pi) - \tfrac{1}{2}\sin 2(t-1-2\pi)\right]$$
$$= \left[H(t-1) - H(t-1-2\pi)\right]\left[\sin(t-1) - \tfrac{1}{2}\sin 2(t-1)\right]. \quad (10.35)$$

The nature of the solution is easier to understand if it is broken up into the cases $0 \le t \le 1$, $1 < t < 1 + 2\pi$, and $t \ge 1 + 2\pi$. Equation (10.35) is equivalent to

$$x(t) = \begin{cases} 0, & 0 \le t \le 1, \\ \sin(t-1) - \frac{1}{2}\sin(2t-2), & 1 < t < 1 + 2\pi, \\ 0, & t \ge 1 + 2\pi. \end{cases}$$

Therefore, the solution is nonzero only for the duration of the pulse.                 ∎

## Discontinuous Forcing and Continuous Solutions

Theorem 5.9 implies that the solutions of the second-order differential equation

$$x'' + ax' + bx = f(t) \tag{10.36}$$

are defined and twice differentiable on the entire real line if $f$ is continuous everywhere and $a$ and $b$ are constants. However, in this section, we consider the possibility that $f$ has jump discontinuities. Theorem 5.9 does not apply to such forcing functions. A fundamental question is whether the solutions exist under such circumstances and, if so, whether the solutions themselves have discontinuities.

Suppose that $f$ has a single jump discontinuity at $t = c$. Then $f$ is continuous on the interval $I_L$ to the left of $c$ and on the interval $I_R$ to the right of $c$. Theorem 5.9 implies that the solutions of Eq. (10.36) are twice differentiable on each subinterval $I_L$ and $I_R$.

The only difficulty concerns the nature of the solution at $t = c$. Equation (10.36) is equivalent to

$$x''(t) = f(t) - ax'(t) - bx(t).$$

We would like $x(t)$ and all its derivatives to be continuous at $t = c$. However, because $f$ is discontinuous at $t = c$, $x''(t)$ must be discontinuous at $t = c$, even if $x(t)$ and $x'(t)$ are continuous at $t = c$. Therefore, the best that we can do is to find a solution $x(t)$ that has one continuous derivative on any interval containing $t = c$.

This idea can be understood in everyday terms. Suppose you drive down a highway at a constant velocity of 90 km/h (25 m/s). After 30 seconds, you floor the accelerator to pass a slower vehicle. Suppose the resulting acceleration is 1 m/s$^2$. The velocity of the car increases continuously from 90 km/h to a higher value that depends on the length of time that the car accelerates. The position of the car also changes continuously.

Let us idealize the situation and assume that you depress the accelerator instantaneously and that the car responds instantaneously. In this case, the acceleration of the car jumps from 0 to 1 m/s$^2$ at $t = 30$ s. Thus, the differential equation governing the position of the car is

$$x'' = H(t - 30), \tag{10.37}$$

which is Eq. (10.36) with $a = b = 0$. Figure 10.7 shows the graph of the acceleration function described by Eq. (10.37); the dashed vertical line indicates the discontinuity at $t = 30$.

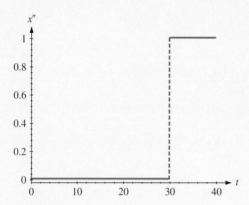

**FIGURE 10.7**    A graph of Eq. (10.37).

We expect both the position $x(t)$ and the velocity $x'(t)$ to be continuous at $t = 30$ s. The solution of Eq. (10.37) can be determined readily by using basic calculus. When $0 \leq t < 30$ s, the acceleration $x_L''(t) = 0$. If we take $x_L(0) = 0$, then the position $x_L$ before the accelerator is depressed is

$$x_L(t) = 25t \text{ m.}$$

After the accelerator is depressed, $x_R''(t) = 1$ m/s$^2$. In addition, $x_L(30) = x_R(30)$ and $x_L'(30) = x_R'(30)$. Thus, the position $x_R$ after the accelerator is depressed is

$$x_R(t) = 25t + \tfrac{1}{2}(t - 30)^2 \text{ m.}$$

Therefore, the car's position $x(t)$ in meters at time $t$ is

$$x(t) = \begin{cases} 25t, & 0 \leq t \leq 30, \\ 25t + \tfrac{1}{2}(t - 30)^2, & t > 30. \end{cases} \tag{10.38}$$

The corresponding velocity in meters per second is

$$x'(t) = \begin{cases} 25, & 0 \leq t \leq 30, \\ 25 + (t - 30), & t > 30. \end{cases} \tag{10.39}$$

Figure 10.8 shows the graphs of $x$ and $x'$. It is obvious that both $x$ and $x'$ are continuous at $t = 30$ s. The effect of the Heaviside function is apparent in Fig. 10.8(b), insofar as the slope of $x'$ changes discontinuously at $t = 30$ s. The acceleration $x''$ is continuous

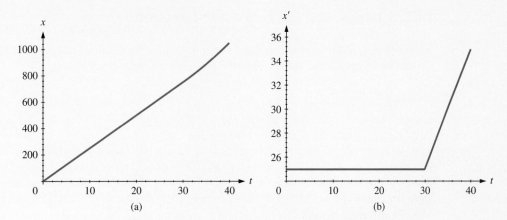

**FIGURE 10.8**   (a) The graph of $x$, Eq. (10.38). (b) The graph of $x'$, Eq. (10.39).

on the subintervals $0 \leq t < 30$ and $t > 30$, but it has a jump discontinuity at $t = 30$, corresponding to the instant at which the accelerator is depressed. Equation (10.37) has a unique solution that is twice differentiable on each of the subintervals $t < 30$ and $t > 30$.

If you take your foot off the accelerator at $t = 40$ s, so that the net acceleration of the car drops to 0, the position and velocity still change continuously. In fact, the position and velocity of the car change continuously, regardless of the number of jump discontinuities in the acceleration. The only requirement is that any jump discontinuities in the acceleration be isolated, that is, each jump discontinuity must be separated by an interval.

This example of the automobile acceleration is intended to motivate a more general result, which we summarize in Definition 10.9 and Theorem 10.10.

**Definition 10.9**   *Let $I$ be a finite interval, that is, $I = (a, b)$ for some real numbers $a$ and $b$. The function $f(t)$ is piecewise continuous on $I$ if $f$ is defined, bounded, and continuous except possibly for a finite number of jump discontinuities in $I$.*

**Theorem 10.10**   *Let*

$$x'' + a(t)x' + b(t)x = f(t), \qquad\qquad (10.40)$$

*where $a$ and $b$ are continuous functions. Suppose that $f$ is piecewise continuous and that the initial conditions are specified at a value of $t$ at which $f$ is continuous. Equation (10.40) has a unique solution that is defined on $I$ if the solution $x$ and its derivative $x'$ are chosen to be continuous at every point of $I$. In this case, $x''$ is piecewise continuous on $I$.*

## LRC Circuits and the Heaviside Function

We conclude our discussion with some examples of how the solutions of models of LRC circuits behave when the forcing function has jump discontinuities.

✦ *Example 9* Let

$$x'' + 4x = H(t-1)\cos(t-1) \tag{10.41}$$

with the initial conditions $x(0) = 1$, $x'(0) = 2$. Figure 10.9 shows a graph of the forcing function on the right-hand side of Eq. (10.41). (The dashed vertical line is intended only as a visual indicator of the jump discontinuity at $t = 1$.) Such a function might arise if an LC circuit is connected to a signal generator by means of a switch that is thrown part of the way through a cycle.

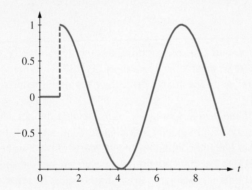

**FIGURE 10.9** A graph of $H(t-1)\cos(t-1)$.

The solution can be derived by using the methods discussed previously. Our emphasis here is on the qualitative features of the solution, which is

$$x(t) = \cos 2t + \sin 2t + H(t-1)\left[\tfrac{1}{3}\cos(t-1) - \tfrac{1}{3}\cos 2(t-1)\right]. \tag{10.42}$$

Equation (10.41) represents an undamped oscillator. Therefore, the solution (10.42) is a linear combination of the homogeneous solution (a $\pi$-periodic function) and a particular solution that depends on the forcing (a $2\pi$-periodic function). Notice, however, that the particular solution does not "turn on" until $t = 1$, corresponding to the imposed forcing that begins at $t = 1$.

Figure 10.10(a) shows a plot of the solution (10.42). The graph suggests that the solution is continuous and differentiable throughout the illustrated interval; in particular,

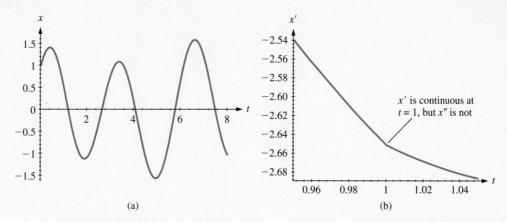

(a)

(b)

**FIGURE 10.10**   (a) A graph of the solution (10.42) of Eq. (10.41). (b) A graph of the derivative, Eq. (10.44), near $t = 1$.

---

$x$ appears to be differentiable at $t = 1$. We can demonstrate that $x$ is in fact differentiable at $t = 1$ by rewriting Eq. (10.42) as

$$x(t) = \begin{cases} \cos 2t + \sin 2t, & 0 \le t \le 1, \\ \cos 2t + \sin 2t + \frac{1}{3}[\cos(t-1) - \cos 2(t-1)], & t > 1. \end{cases} \quad (10.43)$$

Clearly, $x(t)$ is continuous everywhere except possibly at $t = 1$. We can investigate the continuity of $x$ at $t = 1$ by taking left- and right-hand limits. We have

$$\lim_{t \nearrow 1} x(t) = \lim_{t \nearrow 1} (\cos 2t + \sin 2t) = \cos 2 + \sin 2$$

and

$$\lim_{t \searrow 1} x(t) = \lim_{t \searrow 1} \left( \cos 2t + \sin 2t + \tfrac{1}{3}[\cos(t-1) - \cos 2(t-1)] \right) = \cos 2 + \sin 2.$$

Therefore,

$$\lim_{t \nearrow 1} x(t) = \lim_{t \searrow 1} x(t) = \cos 2 + \sin 2.$$

Because the left- and right-hand limits of Eq. (10.43) are equal at $t = 1$, $x$ is continuous at $t = 1$.

The velocity $x'(t)$ is

$$x'(t) = \begin{cases} -2\sin 2t + 2\cos 2t, & 0 \le t \le 1, \\ -2\sin 2t + 2\cos 2t + \frac{1}{3}[-\sin(t-1) + 2\sin 2(t-1)], & t > 1. \end{cases}$$

$$(10.44)$$

Here,

$$\lim_{t \nearrow 1} x'(t) = -2\sin 2 + 2\cos 2,$$

and

$$\lim_{t \searrow 1} x'(t) = -2\sin 2 + 2\cos 2.$$

Thus, the left- and right-hand limits of $x'$ are the same at $t = 1$, and $x'$ is continuous at $t = 1$.

Figure 10.10(b) shows a plot of Eq. (10.44) in a small interval around $t = 1$. The graph has a "corner" at $t = 1$, and it is not possible to draw a unique tangent line to the curve at $t = 1$. Notice that the acceleration is given by

$$x''(t) = H(t - 1)\cos(t - 1) - 4x(t),$$

which is discontinuous at $t = 1$.                                                               ∎

✦ *Example 10*    Consider the LRC circuit discussed in Example 2 of Section 5.8, where the periodic forcing is replaced by a jump discontinuity. Let

$$q'' + 100q' + 10{,}000q = 100H(t - 2). \tag{10.45}$$

Equation (10.45) models a situation in which the circuit is connected at $t = 2$ to a battery that supplies an electrical potential of 100 V. Suppose that $q(0) = q'(0) = 0$. The solution can be found readily with the aid of a computer algebra system; it is given by

$$q(t) = \frac{1}{100}H(t - 2)\left(1 - \frac{1}{\sqrt{3}}e^{-50(t-2)}\sin 50\sqrt{3}(t - 2)\right.$$

$$\left. - e^{-50(t-2)}\cos 50\sqrt{3}(t - 2)\right), \tag{10.46}$$

where $q(t)$ represents the amount of charge in the circuit at time $t$.

Let us analyze the behavior of $q$. Clearly, $q(t) = 0$ for $t < 2$. We check whether $q$ is continuous at $t = 2$ by taking the right-hand limit:

$$\lim_{t \searrow 2} q(t) = \tfrac{1}{100}\left[1 - \left(e^0/\sqrt{3}\right)\sin 0 - e^0\cos 0\right]$$

$$= \tfrac{1}{100}(1 - 0 - 1)$$

$$= 0.$$

Thus, $q$ is continuous at $t = 2$.

The solution (10.46) contains two transient terms. The term $e^{-50(t-2)}$ rapidly tends to 0 as $t$ increases from 2. As $t$ becomes large, $q$ approaches the steady-state solution, which is constant. Thus, the charge in the circuit described by Eq. (10.45) exhibits a momentary transient but rapidly approaches the constant value $1/100$.   ∎

✦ *Example 11*   As a last example, we consider a ramped forcing. Let

$$x'' + x = f(t), \qquad x(0) = x'(0) = 0, \tag{10.47}$$

where $f$ is the function whose graph is depicted in Fig. 10.11(a). The function $f$ models a situation in which an LRC circuit is connected to a potential that increases linearly from 0 to 3 volts over a 10-second interval. It is straightforward to derive an analytical representation of $f$ if we consider the case $0 \leq t \leq 10$ and the case $t > 10$ separately. Now

$$f(t) = \begin{cases} \frac{3}{10}t, & 0 \leq t \leq 10, \\ 3, & t > 10, \end{cases}$$

so

$$f(t) = \tfrac{3}{10}t\left[1 - H(t - 10)\right] + 3H(t - 10)$$
$$= \tfrac{3}{10}t + H(t - 10)\left[3 - \tfrac{3}{10}t\right]. \tag{10.48}$$

Then the solution of Eq. (10.47) is

$$x(t) = \tfrac{3}{10}(t - \sin t) - \tfrac{3}{10}H(t - 10)\left[t - 10 - \sin(t - 10)\right],$$

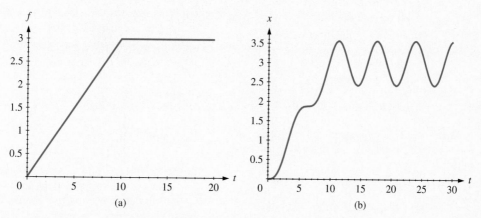

FIGURE 10.11   (a) The forcing function for Eq. (10.47). (b) The solution $x$.

which we can rewrite as

$$x(t) = \begin{cases} \frac{3}{10}(t - \sin t), & 0 \le t \le 10, \\ 3 + \frac{3}{10}\left[\sin(t - 10) - \sin t\right], & t > 10. \end{cases} \qquad (10.49)$$

Figure 10.11(b) shows a graph of Eq. (10.49). The graph suggests that $x(t)$ is continuous; in fact, $x$, $x'$, and $x''$ are all continuous functions, because the forcing $f$ is continuous.

The behavior of the solution changes when the forcing is turned off. During the ramping interval $(0 \le t \le 10)$, the solution is a linear combination of a $2\pi$-periodic function (the homogeneous solution) and a linearly increasing function (the particular solution). When the forcing is turned off, the solution reverts to a $2\pi$-periodic function. There is no transient because there is no damping. ∎

## *Exercises*

**1.** Find an analytical representation of each function in Fig. 10.12.

**2.** For each function in the following list:

- Express the function as a linear combination of Heaviside functions.

- Determine whether $f$ has any jump discontinuities, and if so, state where they occur.

(a) $f(t) = \begin{cases} \cos 5t, & 0 \le t < 10, \\ 1, & t \ge 10 \end{cases}$

(b) $f(t) = \begin{cases} 3t, & 0 \le t < 5, \\ \sin(t - 5), & t \le 5 \end{cases}$

(c) $f(t) = \begin{cases} 2, & 0 \le t < 3, \\ -2, & 3 < t \le 10, \\ \sin(t - 10), & t > 10 \end{cases}$

**3.** Figure 10.13 shows a graph of a ramp function followed by a $2\pi$-periodic sine function of unit amplitude.

(a) Explain why the expression

$$[1 - H(t - 10)]t + H(t - 10)\sin t$$

is *not* a correct analytical representation of the function depicted in Fig. 10.13.

(b) Determine a correct analytical representation of the function.

**4.** Use the Laplace transform method to derive Eq. (10.46) in Example 10.

**5.** This exercise considers the analysis of the differential equation (10.47) in Example 11.

(a) Derive the solution using the method of Laplace transforms. Assume that $x(0) = x'(0) = 0$.

(b) Verify that $x(t)$ is continuous at $t = 10$ by taking left- and right-hand limits.

(c) Verify that $x'(t)$ is continuous at $t = 10$ by taking left- and right-hand limits.

(d) Verify that $x''(t)$ is continuous at $t = 10$ by taking left- and right-hand limits.

**6.** Let

$$f(t) = \begin{cases} 0, & 0 \le t < \pi, \\ \sin 2t, & \pi \le t \le 2\pi, \\ 0, & t > 2\pi. \end{cases}$$

(a) Let $x'' + 4x = f(t)$. Use the method of Laplace transforms to find the solution of the initial value problem $x(0) = x'(0) = 0$.

(b) Discuss the qualitative nature of the solution for

- $0 \le t \le \pi$,
- $\pi \le t \le 2\pi$,
- $t \ge 2\pi$.

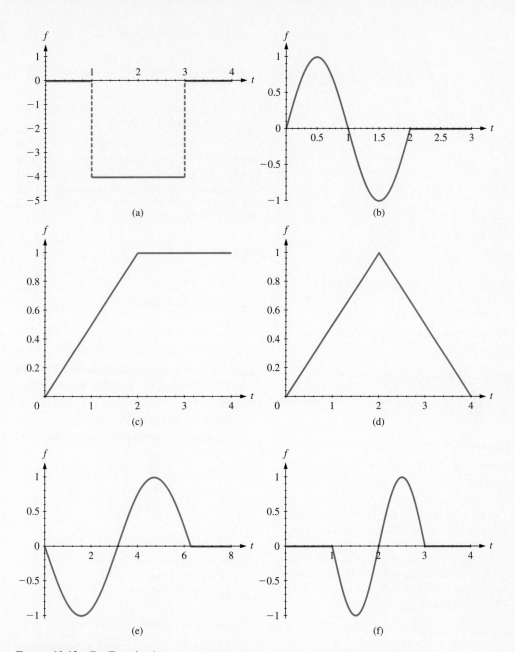

**FIGURE 10.12**  For Exercise 1.

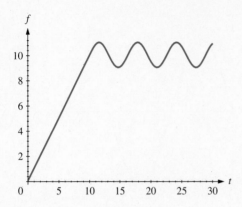

**FIGURE 10.13**    For Exercise 3.

_____

Consider questions such as the following: Is the solution constant? Does it oscillate? If so, what are the period and amplitude?

(c) Determine whether $x$, $x'$, and $x''$ are continuous at $\pi$ and $2\pi$.

**7.** Repeat Exercise 6 for the differential equation $x'' + x = f(t)$. (Let $f$ be the function given in Exercise 6.) Let $x(0) = x'(0) = 0$.

**8.** For each function $f(t)$ depicted in Fig. 10.12,

- Identify any jump discontinuities in the function.
- Find the solution of the initial value problem $x'' + x = f(t)$, $x(0) = 1$, $x'(0) = 1$.
- Discuss the qualitative behavior of the solution $x(t)$ after the forcing $f(t)$ is "turned off." Consider questions such as the following: Is the solution periodic? If so, what is the period? What is the amplitude of the solution? Does the solution tend to infinity or to 0 as $t \to \infty$, or does the solution remain bounded?

**9.** Let

$$x'' + 9x = \begin{cases} 3, & 0 \le t < 2\pi/3, \\ 0, & t \ge 2\pi/3. \end{cases}$$

(a) Solve the initial value problem $x(0) = x'(0) = 0$.

(b) Identify the homogeneous solution and the particular solution.

(c) Plot $x$, $x'$, and $x''$, and determine whether each is continuous at $t = 2\pi/3$.

**10.** Let $x'' + 4x = f(t)$, and assume that $x(0) = x'(0) = 0$. Give an example of a function $f(t)$ for which the solution $x(t)$ is nonzero for at least one point in the interval $\pi < t < 3\pi$ and for which $x(t) = 0$ for every $t$ outside this interval.

**11.** Let

$$x' + x = H(t - 1). \tag{10.50}$$

Notice that Eq. (10.50) is a first-order differential equation.

(a) Find a function $x_1$ that solves Eq. (10.50) for $0 \le t < 1$.

(b) Find a function $x_2$ that solves Eq. (10.50) for $t > 1$.

(c) Find a function $x_1$ that satisfies the initial condition $x_1(0) = 0$, and choose $x_2$ so that $x_1(1) = x_2(1)$.

(d) Let

$$x(t) = \begin{cases} x_1(t), & 0 \le t < 1, \\ x_2(t), & t > 1. \end{cases}$$

Is $x'(t)$ continuous at $t = 1$? Explain.

(e) Write a statement similar to Theorem 10.10 that describes the properties of the solutions of the differential equation

$$x' + a(t)x = f(t),$$

assuming that $a$ is a continuous function of $t$ and $f$ is piecewise continuous.

## 10.5   IMPULSIVE FORCING

This section discusses forcing functions whose action lasts only for a very short interval of time. We focus on a particular kind of impulsive force, called a *delta function*, that has many applications in physics. The Laplace transform method is very useful for solving differential equations with impulsive forcing.

Newton's second law (see Section 5.1) implies that

$$m\frac{dv}{dt} = f(t), \tag{10.51}$$

where $f$ is the force applied to the object. If the forcing $f$ acts only over an interval $[t_0, t_1]$, then $f$ changes the momentum of the object by an amount that is simply the integral of Eq. (10.51):

$$m(v(t_1) - v(t_0)) = \int_{t_0}^{t_1} f(t)\, dt. \tag{10.52}$$

Figure 10.14 shows a representative graph of such a forcing $f$.

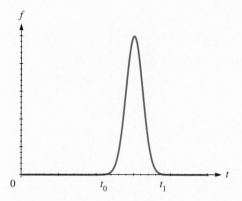

**FIGURE 10.14**   A representative graph of a force that acts over a finite interval.

In many cases, we are interested in forces that act only over a short time interval. For instance, imagine a block of mass $m$ that is hit with a hammer; the duration of the impact is very short. From a mathematical perspective, we are often more interested in the effects of the impact on the motion (that is, the change in the momentum of the object) than in the functional form of the impact itself.

Let $f$ be a force that changes the momentum of a 1-kg mass by one unit (1 kg m/s). Equations (10.51) and (10.52) imply that one such force is a force of 1 N that acts for 1 s. Another example is a force of 10 N that acts for 0.1 s. More generally, any force of $1/h$ N that acts for $h$ s changes the momentum of a 1-kg mass by 1 kg m/s.

Let

$$f_h(t - t_0) = \begin{cases} 1/h, & \text{if } -h/2 \le t - t_0 \le h/2, \\ 0, & \text{otherwise.} \end{cases} \tag{10.53}$$

Figure 10.15 shows the graph of $f_h$ for $h = 1$, $h = 1/2$, and $h = 1/4$. Each $f_h$ is a rectangular pulse of height $1/h$ and width $h$, centered on the vertical line through $t - t_0 = 0$. Each rectangle has unit area, that is,

$$\int_{-h/2}^{h/2} f_h(u)\, du = 1, \tag{10.54}$$

where $u = t - t_0$. If $f_h(t - t_0)$ represents the force in newtons that is applied to a 1-kg mass, then the momentum of the mass is changed by 1 kg m/s over the time interval of length $h$ over which $f_h$ is nonzero.

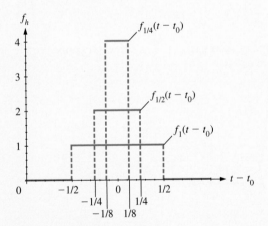

**FIGURE 10.15**    The graph of $f_h(t - t_0)$, defined by Eq. (10.53), for $h = 1$, $h = 1/2$, and $h = 1/4$.

The idea of an *impulsive* force is to consider the limit of the integral in Eq. (10.54) as $h \to 0$, that is, as the duration of the forcing goes to zero. As $h$ becomes smaller, the impact time decreases, but the value of the integral, and thus the total change in momentum, remains the same. This idea suggests the following definition.

**Definition 10.11**   *Let $h$ be any positive real number. The relation*

$$\int_{-h/2}^{h/2} \delta(u)\, du = 1$$

*defines the unit impulse forcing $\delta$, which is called the* delta function. *The notation $\delta(t - t_0)$ refers to an impulsive force at $t = t_0$.*

We can think of the delta function as the limit of $f_h$ as $h \to 0$. It is not obvious that the limit as $h \to 0$ of the integral in Eq. (10.54) is a well-defined quantity, because $\lim_{h \to 0} f_h(0) = \infty$. The delta function cannot be an ordinary function, because no ordinary function has an integral whose value is a nonzero constant that is independent of the width of the interval of integration. Despite these difficulties,[2] it can be shown that the delta function is a mathematically sensible way to model the effect of an impulsive force.

Although we cannot think of the delta function as having a specific nonzero value like an ordinary function, we can consider the effect of the delta function on other functions. The basic idea can be motivated as follows.

Let $f_h$ be the function defined in Eq. (10.53), and let $g$ be any continuous function. For any positive $h$,

$$\int_{-\infty}^{\infty} g(t) f_h(t - t_0)\, dt = \int_{-h/2}^{h/2} \frac{g(u + t_0)}{h}\, du,$$

where we make the change of variable $u = t - t_0$. The intermediate value theorem for integrals implies that there exists a value $\tau \in [-h/2, h/2]$ such that

$$\int_{-h/2}^{h/2} g(u + t_0)\, du = h g(\tau + t_0),$$

which means that

$$\int_{-h/2}^{h/2} g(u + t_0) f_h(u)\, du = g(\tau + t_0).$$

However, $\tau \to 0$ as $h \to 0$. This argument suggests that we can define

$$\int_{-\infty}^{\infty} g(t) \delta(t - t_0)\, dt = g(t_0) \tag{10.55}$$

for any continuous function $g$.

---

[2]Objects like the delta function are called *distributions* or *generalized functions*, for which there is a rigorous mathematical theory. However, the details are beyond the scope of this text.

Equation (10.55) implies that the Laplace transform of the delta function is

$$\mathcal{L}(\delta(t - t_0)) = e^{-st_0} \tag{10.56}$$

for any positive constant $t_0$. (See Exercise 1.) Equation (10.56), which corresponds to entry (14) in Table 10.2, lets us use the Laplace transform method to solve linear, constant-coefficient differential equations with impulsive forcing.

✦ *Example 1*    Consider a linear pendulum that is hit with a hammer. One representative model of this situation is

$$x'' + x = \delta(t - 10), \tag{10.57}$$

where the impulse is applied at $t = 10$. Let the initial conditions be $x(0) = 1$ and $x'(0) = 0$.

We can derive the solution of this initial value problem in the usual way. The Laplace transform of Eq. (10.57) is

$$s^2 X(s) - sx(0) - x'(0) + X(s) = e^{-10s},$$

which leads to

$$X(s) = \frac{s}{s^2 + 1} + \frac{e^{-10s}}{s^2 + 1}$$

after the initial conditions have been substituted. Entry (6) and the shift formula (12) in Table 10.2 imply that

$$x(t) = \cos t + H(t - 10) \sin (t - 10), \tag{10.58}$$

where $H$ is the Heaviside function.

Figure 10.16(a) shows a plot of $x(t)$. The plot suggests that $x(t)$ is continuous; in particular, it is continuous at $t = 10$, because

$$\lim_{t \nearrow 10} x(t) = \lim_{t \searrow 10} x(t) = \cos 10.$$

The angular velocity of the pendulum is

$$x'(t) = \begin{cases} -\sin t, & 0 \leq t \leq 10, \\ -\sin t + \cos (t - 10), & t > 10. \end{cases}$$

Notice that

$$\lim_{t \nearrow 10} x'(t) = -\sin 10 \qquad \text{but} \qquad \lim_{t \searrow 10} x'(t) = 1 - \sin 10;$$

therefore, $x'$ has a jump discontinuity at $t = 10$, indicated by the dashed vertical line in

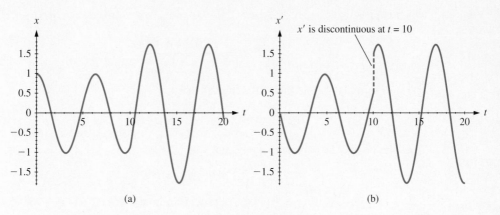

**FIGURE 10.16** (a) A plot of $x$, Eq. (10.58). (b) A plot of $x'$.

Fig. 10.16(b). The jump discontinuity reflects the instantaneous change in momentum that is imparted to the pendulum by the impulse at $t = 10$. ∎

Example 1 illustrates a basic property:

*Consider a second-order, constant-coefficient, linear differential equation with an impulsive forcing at $t = t_0$. It is possible to find a solution $x(t)$ that is continuous at every $t$ and whose derivative $x'(t)$ is continuous except for a jump discontinuity at $t = t_0$.*

✦ *Example 2*   Consider an exponentially growing population, some of whose members are harvested over a short interval. (Imagine a colony of bacteria from which a number of individuals are removed at some fixed instant.) An equation like

$$p' = p - k$$

is not a good model of this situation, because it represents the case in which individuals are removed from the population at a sustained, constant rate ($k$ units per unit time). A better model is

$$p' = p - k\delta(t - t_0), \tag{10.59}$$

which reflects the removal of $k$ units at one particular instant (or over a very short time interval).

A solution of Eq. (10.59) can be derived by using the Laplace transform. The Laplace transform of Eq. (10.59) is

$$sP(s) - p(0) = P(s) - ke^{-st_0},$$

which implies that

$$P(s) = \frac{p(0) - ke^{-st_0}}{s-1}.$$

The shift formula (12) in Table 10.2 implies that

$$p(t) = p(0)e^t - ke^{t-t_0}H(t-t_0). \tag{10.60}$$

The solution may be expressed equivalently as

$$p(t) = \begin{cases} p(0)e^t, & 0 \le t \le t_0, \\ (p(0)e^{t_0} - k)e^{t-t_0}, & t > t_0. \end{cases}$$

Figure 10.17 shows a plot of Eq. (10.60), where $p(0) = 1$, $k = 5$, and $t_0 = 2$. Each curve in the plot is the plot of an exponential function. However, the removal of five units of population at time $t_0 = 2$ corresponds to a jump discontinuity of the solution at $t = 2$. Exercise 4 discusses some further consequences of this harvesting model.

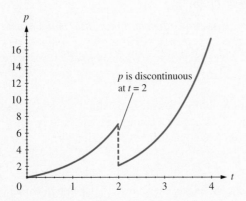

**FIGURE 10.17**   A plot of Eq. (10.60).                                             ∎

Example 2 illustrates another basic property:

*Consider a first-order, constant-coefficient, linear differential equation with an impulsive forcing at $t = t_0$. It is possible to find a solution $x(t)$ that is continuous except for a jump discontinuity at $t = t_0$.*

✦ *Example 3*   An LRC circuit that is subjected to an external potential can be modeled by the linear second-order equation

$$Lq'' + Rq' + q/C = f(t).$$

(See Section 5.1.) Consider the initial value problem

$$Lq'' + Rq' + q/C = \delta(t), \qquad q(0) = q'(0) = 0, \tag{10.61}$$

which represents a circuit that is subjected to an impulsive potential at $t = 0$. The Laplace transform of Eq. (10.61) is

$$L\big(s^2 Q(s) - sq(0) - q'(0)\big) + R(sQ(s) - q(0)) + Q(s)/C = 1,$$

and the initial conditions imply that

$$Q(s) = \frac{1}{Ls^2 + Rs + 1/C}.$$

The corresponding solution, $q(t)$, is called the *impulse response* of the circuit. The function $Q(s)$ is called the *transfer function* of the circuit. In other words, the transfer function is the Laplace transform of the impulse response, and the impulse response refers to the solution of the initial value problem (10.61). The notions of impulse response and transfer function are used frequently in engineering applications that consider the behavior of linear systems.   ■

## Exercises

**1.** Use Eq. (10.55) to derive Eq. (10.56).

**2.** Solve the initial value problem

$$x' = \delta(t - t_0), \qquad x(0) = 0.$$

What does your solution suggest about the relationship between the delta function and the Heaviside function?

**3.** What is the physical significance of the delta function in each of the following contexts?

(a) A linear spring modeled by $mx'' + cx' + kx = \delta(t)$.

(b) An LRC circuit modeled by $Lq'' + Rq' + 1/Cq = \delta(t)$.

(c) A population modeled by $p' = kp + \delta(t - 1)$.

(d) A falling ball modeled by $v' = -g + \delta(t - 1)$.

**4.** Suppose that the growth of an exponentially growing population is modeled by

$$p' = p - k\delta(t - t_0),$$

where $k\delta(t - t_0)$ reflects the harvesting of $k$ units of the population at time $t_0$.

(a) Suppose that, in some appropriate units, $p(0) = 1$. What is the maximum size of the harvest that can be obtained at $t_0 = 1$?

(b) Suppose that $p' = p$ in the absence of harvesting. However, we want to be able to harvest 1 unit of the population at equal intervals of time. (That is, at times $t = t_h, t = 2t_h, t = 3t_h, \ldots$, we remove 1 unit of the population.) What is the smallest value of $t_h$ that permits such a harvest without driving the population to extinction?

(c) Suppose that $p' = p$ in the absence of harvesting. However, at time $t = 1, t = 2, \ldots$, we remove $k$ units of the population. How large can $k$ be without driving the population to extinction?

**5.** Let $x'' + 9x = \delta(t - 3)$ with the initial conditions $x(0) = 1$ and $x'(0) = 2$.

(a) Solve the initial value problem.

(b) Is $x(t)$ continuous at $t = 3$? Justify your answer.

(c) Is $x'(t)$ continuous at $t = 3$? Justify your answer.

**6.** Repeat Exercise 5 for the initial value problem $x'' + 2x' + 5x = 5\delta(t - 3)$, $x(0) = 3$, $x'(0) = 1$.

**7.** Repeat Exercise 5 for the initial value problem $x'' + 4x = \cos 2t - 4\delta(t - 5)$, $x(0) = 2$, $x'(0) =$

$-5$, but examine the continuity of $x$ and $x'$ at $t = 5$ instead of at $t = 3$.

**8.** Find the impulse response of each of the following LRC circuits.

(a) $L = 1$ H, $R = 300$ Ω, $C = 10$ $\mu$F

(b) $L = 1$ H, $R = 10$ Ω, $C = 100$ $\mu$F

(c) $L = 4$ H, $R = 300$ Ω, $C = 100$ $\mu$F

**9.** Let $x'' + 4x = f(t)$, where $x(0) = 0$ and $x'(0) = 1$.

(a) Suppose $f(t) = \delta(t - \pi)$. Find the solution of the initial value problem, and plot it for $0 \le t \le 2\pi$. Comment on the behavior of the solution before and after the impulse.

(b) Suppose $f(t) = \delta(t - \pi) + \delta(t - 2\pi)$. (This corresponds to impulsive forces applied at $t = \pi$ and at $t = 2\pi$.) Solve the initial value problem, and plot the solution for $0 \le t \le 3\pi$. Comment on the behavior of the solution before and after each impulse.

(c) Suppose $f(t) = \delta(t - \pi) + \delta(t - 2\pi) + \delta(t - 3\pi)$. Find the solution, and plot it for $0 \le t \le 4\pi$. What is the behavior before and after each impulse?

(d) What is the pattern that arises here? Suppose we apply a delta function impulse every $\pi$ units of time. What happens to the solution? Is the situation akin to resonance? Explain.

# *Appendix A*

# MORE ON THE METHOD OF UNDETERMINED COEFFICIENTS

The method of undetermined coefficients, as outlined in Section 5.6, can be extended to a wide class of forcing functions. The method is applicable to nonhomogeneous, linear equations with constant coefficients whose forcing term is

- a sine or cosine function;
- a polynomial;
- an exponential function;
- a sum or product of these.

## A.1  THE ALGORITHM

The method of undetermined coefficients can be cast as an algorithm as follows.

**Step 0.** Write the differential equation in the form

$$ax'' + bx' + cx = f(t). \tag{A.1}$$

Let $L(x) = ax'' + bx' + cx$, and write $f(t)$ as the sum

$$f(t) = f_1(t) + f_2(t) + \cdots + f_k(t),$$

| Form of $f_i$ | Guess for $x_p^{(i)}$ |
|---|---|
| $p_n(t) = a_n t^n + a_{n-1} t^{n-1} + \cdots + a_1 t + a_0$ | $A_n t^n + A_{n-1} t^{n-1} + \cdots + A_1 t + A_0$ |
| $a \cos \omega t + b \sin \omega t$ | $A \cos \omega t + B \sin \omega t$ |
| $a e^{kt}$ | $A e^{kt}$ |
| $e^{kt} p_n(t)$ | $e^{kt} \left( A_n t^n + A_{n-1} t^{n-1} + \cdots + A_1 t + A_0 \right)$ |
| $a e^{kt} (b \cos \omega t + c \sin \omega t) p_n(t)$ | $e^{kt} \cos \omega t \left( A_n t^n + A_{n-1} t^{n-1} + \cdots + A_1 t + A_0 \right) + e^{kt} \sin \omega t \left( B_n t^n + B_{n-1} t^{n-1} + \cdots + B_1 t + B_0 \right)$ |

TABLE A.1   Table of guesses for the particular solution in the method of undetermined coefficients.

where each $f_i$ has the form of one of the functions in the left-hand column of Table A.1.

**Step 1.** Determine the homogeneous solution, that is, the solution $x_h$ of the equation $L(x) = 0$. This solution contains $j$ arbitrary constants if the order of the differential equation is $j$.

**Step 2.** For each $f_i$ identified in Step 0:

**Step A.** Guess a particular solution $x_p^{(i)}$ of the form indicated by the corresponding entry in the right-hand column of Table A.1.

**Step B.** If any term in the current guess is also a term in the homogeneous solution $x_h$, then replace $x_p^{(i)}$ with $t x_p^{(i)}$ (that is, multiply the current guess by $t$). Repeat this step with the new guess.

**Step C.** Substitute the guess into the differential equation; that is, compute $L\left( x_p^{(i)} \right)$.

**Step D.** Determine the values of the coefficients in $x_p^{(i)}$ (i.e., the coefficients in capital letters in the corresponding entry of Table A.1) so that $L\left( x_p^{(i)} \right) = f_i(t)$ for all $t$.

**Step 3.** Let

$$x_p(t) = x_p^{(1)}(t) + x_p^{(2)}(t) + \cdots + x_p^{(k)}(t),$$

and form the general solution,

$$x(t) = x_h(t) + x_p(t).$$

**Step 4.** If initial conditions are specified, then use them to determine the values of the arbitrary constants in the general solution. The resulting function is the solution of the initial value problem.

## A.2  EXAMPLES

The following examples illustrate the general algorithm presented in Section A.1.

✦ *Example 1*   We find the solution of the first-order equation

$$x' + x = t^2 + 4t + 6 + \cos t \tag{A.2}$$

satisfying the initial condition $x(0) = 3$ as follows.

**Step 0.** We write Eq. (A.2) in the form

$$L(x) = f_1(t) + f_2(t),$$

where $L(x) = x' + x$, $f_1(t) = t^2 + 4t + 6$, and $f_2(t) = \cos t$.

**Step 1.** The homogeneous solution is the solution of the equation

$$x' + x = 0,$$

or $x_h(t) = Ce^{-t}$. You may recognize that the homogeneous equation is just an exponential growth equation, or you can derive the solution by the separation of variables.

**Step 2.** We work first with $f_1$.
  **Step A.** Since $f_1$ is a polynomial of degree 2, we guess a polynomial of degree 2, that is,

$$x_p^{(1)} = A_2 t^2 + A_1 t + A_0.$$

**Step B.** No term in $x_p^{(1)}$ is contained in $x_h$, so the guess requires no modification.
**Step C.** We compute

$$L\left(x_p^{(1)}\right) = 2A_2 t + A_1 + A_2 t^2 + A_1 t + A_0.$$

**Step D.** We must find values of $A_2$, $A_1$, and $A_0$ such that the relation

$$L\left(x_p^{(1)}\right) = t^2 + 4t + 6$$

is satisfied identically. In this case, we have

$$A_2 t^2 + (2A_2 + A_1)t + (A_1 + A_0) = t^2 + 4t + 6.$$

We equate coefficients of like terms to derive the linear system

$$A_2 = 1,$$
$$2A_2 + A_1 = 4,$$
$$A_1 + A_0 = 6,$$

which implies that $A_2 = 1$, $A_1 = 2$, and $A_0 = 4$. Therefore,

$$x_p^{(1)} = t^2 + 2t + 4.$$

Next we consider $f_2(t) = \cos t$.

**Step A.** Since $f_2$ is a $2\pi$-periodic function, we try a guess of the form

$$x_p^{(2)} = A \cos t + B \sin t,$$

which is also $2\pi$-periodic.

**Step B.** No term in the guess is contained in $x_h$, so the guess requires no modification.

**Step C.** We compute

$$L\left(x_p^{(2)}\right) = -A \sin t + B \cos t + A \cos t + B \sin t.$$

**Step D.** We must find values of $A$ and $B$ such that the relation

$$L\left(x_p^{(2)}\right) = \cos t$$

is satisfied identically. In this case, we have

$$(-A + B) \sin t + (A + B) \cos t = \cos t.$$

We equate coefficients of like terms to derive the linear system

$$-A + B = 0,$$
$$A + B = 1,$$

which implies that $A = B = 1/2$. Therefore,

$$x_p^{(2)} = \tfrac{1}{2} \cos t + \tfrac{1}{2} \sin t.$$

**Step 3.** We form a particular solution,

$$x_p(t) = x_p^{(1)}(t) + x_p^{(2)}(t) = t^2 + 2t + 4 + \tfrac{1}{2} \cos t + \tfrac{1}{2} \sin t,$$

and find that the general solution is

$$x(t) = Ce^{-t} + t^2 + 2t + 4 + \tfrac{1}{2} \cos t + \tfrac{1}{2} \sin t.$$

Notice that the general solution contains one arbitrary constant $C$ because Eq. (A.2) is first order.

**Step 4.** We substitute the initial condition at $t = 0$ to find

$$x(0) = C + 4 + \tfrac{1}{2},$$

which implies that $C = -3/2$ when $x(0) = 3$. Therefore, the solution of the initial value problem is

$$x(t) = -\tfrac{3}{2}e^{-t} + t^2 + 2t + 4 + \tfrac{1}{2}\cos t + \tfrac{1}{2}\sin t.$$

The result of the calculations can readily be checked by substituting the solution into Eq. (A.2). ∎

✦ *Example 2*   We can find the solution of the equation

$$x'' + x = 5e^{-t}\cos t \qquad (A.3)$$

satisfying the initial conditions $x(0) = -1$ and $x'(0) = 3$ as follows.

**Step 0.** We have $L(x) = x'' + x$ and $f(t) = 5e^{-t}\cos t$.

**Step 1.** The homogeneous solution is

$$x_h(t) = c_1 \cos t + c_2 \sin t,$$

which has two arbitrary constants because Eq. (A.3) is second order.

**Step 2.** There is only one term in $f(t)$ to worry about.

  **Step A.** We use the last entry in Table A.1 to guess a solution of the form

$$x_p(t) = e^{-t}(A \cos t + B \sin t).$$

Notice that the exponential term in the guess for $x_p$ has the same decay constant (here $-1$) as the forcing $f$, and the oscillatory term in $x_p$ is $2\pi$-periodic, just like the corresponding term in $f$.

  **Step B.** No term in $x_p$ is contained in $x_h$, because each sine and cosine term in $x_p$ is multiplied by an exponential. Therefore, the guess requires no modification.

  **Step C.** We substitute the guess into Eq. (A.3) to obtain

$$e^{-t}(A \cos t + B \sin t + 2A \sin t - 2B \cos t) = 5e^{-t}\cos t$$

after some algebra.

  **Step D.** We rewrite the result of Step C as

$$e^{-t}(A - 2B) \cos t + e^{-t}(2A + B) \sin t = 5e^{-t}\cos t$$

and equate coefficients of like terms to derive the linear system

$$A - 2B = 5,$$
$$2A + B = 0,$$

which implies that $A = 1$ and $B = -2$.

**Step 3.** We let

$$x_p(t) = e^{-t}(\cos t - 2\sin t),$$

and the general solution is

$$x(t) = c_1 \cos t + c_2 \sin t + e^{-t}(\cos t - 2\sin t).$$

**Step 4.** The initial conditions yield

$$x(0) = -1 = c_1 + 1,$$
$$x'(0) = c_2 - 3 = 3,$$

which implies that the solution of the initial value problem is

$$x(t) = -2\cos t + 6\sin t + e^{-t}(\cos t - 2\sin t). \qquad \blacksquare$$

◆ *Example 3*   Let

$$x' + x = e^{-t}, \qquad x(0) = 2.$$

The method of undetermined coefficients is applied as follows.

**Step 0.** We have $L(x) = x' + x$ and $f(t) = e^{-t}$.

**Step 1.** The solution of the homogeneous equation $x' + x = 0$ is $x_h(t) = Ce^{-t}$.

**Step 2.** There is only one forcing term to worry about.
    **Step A.** According to Table A.1, we guess a particular solution of the form $x_p(t) = Ae^{-t}$.
    **Step B.** The initial guess is the same as the homogeneous solution. It is easy to check that $L\left(Ae^{-t}\right) = 0$. Therefore, we modify the guess by multiplying it by $t$; that is, we let $x_p(t) = Ate^{-t}$. The modified guess is not the same as the homogeneous solution.
    **Step C.** We compute

$$L(x_p) = L\left(Ate^{-t}\right) = Ae^{-t}.$$

    **Step D.** The relation

$$L(x_p) = e^{-t}$$

is satisfied identically if $A = 1$.

**Step 3.** A particular solution is $x_p(t) = te^{-t}$, and the general solution is

$$x(t) = Ce^{-t} + te^{-t}.$$

**Step 4.** The initial condition implies that $C = 2$. Therefore, the solution of the initial value problem is

$$x(t) = 2e^{-t} + te^{-t}.$$

■

✦ *Example 4*   We can determine the solution of the equation

$$x'' + 2x' + x = 12(t^2 + t)e^{-t} \tag{A.4}$$

satisfying the initial conditions $x(0) = 1$ and $x'(0) = 2$ as follows.

**Step 0.** We have $L(x) = x'' + 2x' + x$ and $f(t) = 12(t^2 + t)e^{-t}$.

**Step 1.** The homogeneous solution is

$$x_h(t) = c_1 e^{-t} + c_2 t e^{-t}.$$

(See Section 5.5.)

**Step 2.** We have only one forcing term to consider.

**Step A.** The forcing term is the product of a quadratic polynomial with an exponential term. Table A.1 suggests a guess of the form

$$x_p(t) = \left(A_2 t^2 + A_1 t + A_0\right) e^{-t}.$$

**Step B.** The guess in Step A contains the terms $A_1 t e^{-t}$ and $A_0 e^{-t}$, which are also in the homogeneous solution. Therefore, we replace the original guess with

$$x_p(t) = t \left(A_2 t^2 + A_1 t + A_0\right) e^{-t}.$$

However, this revised guess contains the term $A_0 t e^{-t}$, which is also in the homogeneous solution. Therefore, we modify the guess by multiplying it by another factor of $t$ to obtain

$$x_p(t) = t^2 \left(A_2 t^2 + A_1 t + A_0\right) e^{-t},$$

which has no terms that are contained in the homogeneous solution.

**Step C.** We compute $L(x_p)$, which, after some routine algebra, gives

$$L(x_p) = \left(12A_2 t^2 + 6A_1 t + 2A_0\right) e^{-t}.$$

**Step D.** We solve the relation

$$\left(12A_2 t^2 + 6A_1 t + 2A_0\right) e^{-t} = 12(t^2 + t)e^{-t}$$

by equating like terms. We find $A_0 = 0$, $A_1 = 2$, and $A_2 = 1$.

**Step 3.** A particular solution is

$$x_p(t) = t^2(t^2 + 2t)e^{-t} = (t^4 + 2t^3)e^{-t}.$$

**Step 4.** The general solution is

$$x(t) = c_1 e^{-t} + c_2 t e^{-t} + (t^4 + 2t^3)e^{-t}.$$

**Step 5.** We have

$$x(0) = 1 = c_1,$$
$$x'(0) = 2 = -c_1 + c_2,$$

so $c_1 = 1$ and $c_2 = 3$. Therefore, the solution of the initial value problem is

$$x(t) = (t^4 + 2t^3 + 3t + 1)e^{-t}. \qquad \blacksquare$$

## Exercises

**1.** Use the method of undetermined coefficients to find a solution of each initial value problem in the following list.

(a) $x' + x = e^t, x(0) = 1$

(b) $x'' + x = e^t, x(0) = 1, x'(0) = 1$

**2.** Use the method of undetermined coefficients to find a solution of each initial value problem in the following list.

(a) $x'' + x = \cos t + 3\sin 2t, x(0) = 1, x'(0) = -1$

(b) $x'' + 4x = \cos t + 3\sin 2t, x(0) = 1, x'(0) = -1$

**3.** Use the method of undetermined coefficients to find a solution of each initial value problem in the following list.

(a) $x'' + x' = 2t\cos t, x(0) = 1, x'(0) = -1$

(b) $x'' + x' = t^2 + 1, x(0) = 1, x'(0) = 0$

**4.** Use the method of undetermined coefficients to find a solution of each initial value problem in the

following list.

(a) $x'' + 3x' + 2x = e^{-t}, x(0) = 2, x'(0) = 3$

(b) $x'' + 3x' + 2x = (t + 1)e^{-2t}, x(0) = 1, x'(0) = 0$

**5.** For each equation in the following list, determine a suitable guess for a particular solution, but do not evaluate the coefficients.

(a) $x'' + 4x' + 3x = e^{-t} + \cos t$

(b) $x'' + 4x' + 3x = e^t - te^{-t}$

(c) $x'' + 2x' + 5x = \cos t$

(d) $x'' + 2x' + 5x = e^{-t}(\cos 2t + \cos t)$

(e) $x'' + 4x' + 13x = e^{-t}\sin 3t$

(f) $x'' + 4x' + 13x = (t + 1)e^{-2t}\sin 3t$

**6.** Let $L(x) = ax'' + bx' + cx$, where $a$, $b$, and $c$ are nonzero constants. Consider the differential equation

$$L(x) = p_n(t), \qquad \text{(A.5)}$$

where

$$p_n(t) = a_n t^n + a_{n-1} t^{n-1} + \cdots + a_1 t + a_0.$$

Prove that the method of undetermined coefficients always leads to a solution of Eq. (A.5). Suggestion: Let $x_p(t) = A_n t^n + A_{n-1} t^{n-1} + \cdots + A_1 t + A_0$, as suggested by Table A.1. Equate coefficients of like terms. Show that the resulting system of linear equations for the unknowns $A_0, A_1, \ldots, A_n$ is triangular; that is, there is an equation for $A_0$ that is trivial to solve and whose solution can be used to determine $A_1, A_2, \ldots$, in turn.

**7.** Let $L(x) = ax'' + bx'$, where $a$ and $b$ are nonzero constants, and consider the differential equation

$$L(x) = p_n(t), \qquad (A.6)$$

where $p_n$ is as in Exercise 6.

(a) Explain why a guess of the form

$$x_p(t) = A_n t^n + A_{n-1} t^{n-1} + \cdots + A_1 t + A_0,$$

as suggested by Table A.1, does *not* lead to a particular solution of Eq. (A.6).

(b) Show that a guess of the form

$$x_p(t) = t \left( A_n t^n + \cdots + A_1 t + A_0 \right)$$

always leads to a solution of Eq. (A.6).

(c) Explain why it is not necessary to include a constant term in the guess in (b).

# *Appendix B*

# THE VARIATION OF PARAMETERS

The method of *variation of parameters* can be used to find solutions of linear second-order equations of the form

$$x'' + ax' + bx = f, \tag{B.1}$$

where $f$ is an arbitrary continuous function. Unlike the method of undetermined coefficients, there is no restriction on $f$ except that it be continuous on an interval containing the initial conditions. In principle, the coefficients $a$ and $b$ need not be constant, but the method is most practical when $a$ and $b$ are constant coefficient functions.

Suppose that the homogeneous solution $x_h$ of Eq. (B.1) is known. It can be written as the linear combination

$$x_h(t) = c_1 x_1(t) + c_2 x_2(t), \tag{B.2}$$

where $x_1$ and $x_2$ are two linearly independent solutions of the homogeneous equation $x'' + ax' + bx = 0$. (The homogeneous solution is straightforward to find when $a$ and $b$ are constant; see Sections 5.3–5.5.)

The idea behind the variation of parameters is to find a particular solution of Eq. (B.1) of the form

$$x_p(t) = u_1(t)x_1(t) + u_2(t)x_2(t). \tag{B.3}$$

In other words, we replace the constants $c_1$ and $c_2$ in Eq. (B.2) with functions $u_1(t)$ and $u_2(t)$ that must be determined; hence the name *variation of parameters*. (The method is also called the *variation of constants*.)

The particular solution $x_p$ contains two unknown functions, $u_1$ and $u_2$, but the differential equation (B.1) provides only one constraint. There might be infinitely many choices of $u_1$ and $u_2$ that lead to a particular solution of Eq. (B.1) unless some additional condition is imposed on $u_1$ and $u_2$. In fact, it is possible to find an additional constraint that allows $u_1$ and $u_2$ to be computed relatively easily, at least in principle.

The product rule implies that

$$x'_p = u'_1 x_1 + u'_2 x_2 + u_1 x'_1 + u_2 x'_2.$$

The "trick" that makes the variation of parameters workable is to require that

$$u'_1(t)x_1(t) + u'_2(t)x_2(t) = 0 \tag{B.4}$$

for every $t$. (That is, $u'_1 x_1 + u'_2 x_2$ is the zero function.) With this additional constraint, we have

$$x'_p = u_1 x'_1 + u_2 x'_2,$$

and consequently,

$$x''_p = u'_1 x'_1 + u'_2 x'_2 + u_1 x''_1 + u_2 x''_2.$$

We substitute these expressions into Eq. (B.1) and collect terms to obtain

$$x''_p + ax'_p + bx_p = u'_1 x'_1 + u'_2 x'_2 + u_1 \left[x''_1 + ax'_1 + bx_1\right] + u_2 \left[x''_2 + ax'_2 + bx_2\right]. \tag{B.5}$$

Notice that each of the terms in square brackets is 0, because $x_1$ and $x_2$ are homogeneous solutions of Eq. (B.1). Equations (B.1) and (B.5) imply that

$$u'_1 x'_1 + u'_2 x'_2 = f. \tag{B.6}$$

Equations (B.4) and (B.6) form a linear system of equations for $u'_1$ and $u'_2$. A routine calculation (see Exercise 5) shows that

$$u'_1 = -\frac{x_2 f}{W(x_1, x_2)}$$

and

$$u'_2 = \frac{x_1 f}{W(x_1, x_2)},$$

where $W(x_1, x_2)$ is the Wronskian determinant of $x_1$ and $x_2$,

$$W(x_1, x_2) = \det \begin{pmatrix} x_1 & x_2 \\ x'_1 & x'_2 \end{pmatrix} = x_1 x'_2 - x_2 x'_1.$$

Because $x_1$ and $x_2$ are linearly independent solutions of the same linear second-order equation, Abel's theorem (Theorem 5.12 in Section 5.9) implies that their Wronskian is never 0. We integrate to obtain

$$u_1(t) = -\int_{t_0}^{t} \frac{x_2(s) f(s)}{W(x_1(s), x_2(s))} \, ds \tag{B.7}$$

and

$$u_2(t) = \int_{t_0}^{t} \frac{x_1(s) f(s)}{W(x_1(s), x_2(s))} \, ds, \tag{B.8}$$

assuming that the initial conditions are specified at $t_0$. Since $x_1$ and $x_2$ are known, the expressions for $u_1$ and $u_2$ determine the particular solution, Eq. (B.3). The general solution has the form

$$x(t) = x_p(t) + c_1 x_1(t) + c_2 x_2(t).$$

The constants $c_1$ and $c_2$ can be chosen to satisfy any initial conditions (see Exercise 8). We summarize the discussion in the following theorem.

**Theorem B.1** *Let $a$, $b$, and $f$ be continuous functions of $t$. Let $x_1$ and $x_2$ be two linearly independent homogeneous solutions of the equation*

$$x'' + ax' + bx = f.$$

*Then a particular solution is*

$$x_p(t) = u_1(t)x_1(t) + u_2(t)x_2(t),$$

*where $u_1$ and $u_2$ are given by Eqs. (B.7) and (B.8), respectively, when the initial conditions are specified at $t_0$. The general solution is*

$$x(t) = x_p(t) + c_1 x_1(t) + c_2 x_2(t),$$

*and the constants $c_1$ and $c_2$ can be chosen to satisfy any initial conditions of the form $x(t_0) = x_0$, $x'(t_0) = v_0$.*

The meaning of Theorem B.1 can be summarized as follows:

*The solution of any linear second-order ordinary differential equation with continuous coefficients and a continuous forcing function can be expressed as an integral.*

The following examples illustrate the procedure.

✦ *Example 1*   Let

$$x'' + x = f(t) = \sec t, \qquad x(0) = 1, \qquad x'(0) = -1. \tag{B.9}$$

The homogeneous solution of Eq. (B.9) is

$$x_h(t) = c_1 x_1(t) + c_2 x_2(t) = c_1 \cos t + c_2 \sin t.$$

We assume that a particular solution has the form

$$x_p(t) = u_1(t)x_1(t) + u_2(t)x_2(t) = u_1(t) \cos t + u_2(t) \sin t.$$

Notice that $W(x_1(t), x_2(t)) = 1$. Equations (B.7) and (B.8) imply that

$$u_1'(t) = -\sin t \sec t = -\tan t$$

and

$$u_2'(t) = \cos t \sec t = 1.$$

Therefore,

$$u_1(t) = \int_0^t -\tan s \, ds = \ln |\cos t|$$

and

$$u_2(t) = t,$$

so a particular solution is

$$x_p(t) = (\cos t)(\ln |\cos t|) + t \sin t.$$

The general solution is

$$x(t) = c_1 \cos t + c_2 \sin t + (\cos t)(\ln |\cos t|) + t \sin t. \qquad \text{(B.10)}$$

The initial conditions imply that $c_1 = 1$ and $c_2 = -1$, so the solution of the initial value problem is

$$x(t) = \cos t - \sin t + (\cos t)(\ln \cos t) + t \sin t.$$

Notice that the interval of existence is $-\pi/2 < t < \pi/2$, because the term $\ln \cos t$ is not defined at $t = \pm \pi/2$. (Why can the absolute value bars be dropped from Eq. (B.10) for the given initial conditions?) ∎

✦ *Example 2*  Let

$$x'' + x = \frac{1}{t+1}, \qquad x(0) = x_0, \qquad x'(0) = v_0. \qquad \text{(B.11)}$$

As in Example 1, the homogeneous solution is

$$x_h(t) = c_1 x_1(t) + c_2 x_2(t) = c_1 \cos t + c_2 \sin t,$$

and we assume that a particular solution has the form

$$x_p(t) = u_1(t)x_1(t) + u_2(t)x_2(t) = u_1(t) \cos t + u_2(t) \sin t.$$

Equations (B.7) and (B.8) imply that

$$u'_1(t) = -\frac{\sin t}{t+1} \qquad \text{and} \qquad u'_2(t) = \frac{\cos t}{t+1},$$

so

$$u_1(t) = -\int_0^t \frac{\sin s}{s+1}\, ds \qquad \text{and} \qquad u_2(t) = \int_0^t \frac{\cos s}{s+1}\, ds.$$

These integrals cannot be evaluated in closed form. However, they can be approximated by using Simpson's rule or another quadrature method. The solution of the initial value problem is

$$x(t) = x_0 \cos t + v_0 \sin t - (\cos t)\int_0^t \frac{\sin s}{s+1}\, ds + (\sin t)\int_0^t \frac{\cos s}{s+1}\, ds.$$

(See Exercise 7.) Chapter 9 discusses other numerical methods for approximating the solution of Eq. (B.11). ∎

## Exercises

**1.** Use the method of variation of parameters to find a solution of each of the initial value problems in the following list. Also determine the interval of existence of the solution.

(a) $x'' + 2x' + x = e^{-2t}$, $x(0) = 1$, $x'(0) = -1$

(b) $x'' + 2x' + x = e^{-t}$, $x(0) = 1$, $x'(0) = 1$

**2.** Let $x'' + x = \cos t$. Use the method of variation of parameters to find a solution satisfying the initial conditions $x(0) = 0$ and $x'(0) = 1$.

**3.** Use the method of variation of parameters to find a solution of each of the initial value problems in the following list. Also determine the interval of existence of the solution.

(a) $x'' + 4x = \tan t$, $x(0) = 1$, $x'(0) = 2$

(b) $x'' + 4x = \tan 2t$, $x(0) = 1$, $x'(0) = 2$

**4.** Use the method of variation of parameters to find a solution of each of the initial value problems in the following list.

(a) $x'' + x = \cos^2 t$, $x(0) = 2$, $x'(0) = 3$

(b) $x'' - x = e^t$, $x(0) = 1$, $x'(0) = 1$

**5.** Find the solution of the linear system given by Eqs. (B.4) and (B.6).

**6.** The functions $u_1$ and $u_2$, given by Eqs. (B.7) and (B.8), are determined only up to a constant. (That is, the indefinite integrals lead to formulas for $u_1$ and $u_2$ that contain an arbitrary constant.) Explain why the arbitrary constants can be taken as 0.

**7.** The general solution of Eq. (B.11) is

$$x(t) = c_1 \cos t + c_2 \sin t$$
$$- (\cos t)\int_0^t \frac{\sin s}{s+1}\, ds$$
$$+ (\sin t)\int_0^t \frac{\cos s}{s+1}\, ds.$$

Show that $c_1 = x_0$ and $c_2 = v_0$.

**8.** Let $x'' + ax' + bx = f$, and suppose that $x_1$ and $x_2$ are two linearly independent homogeneous solutions. Let a particular solution be computed by using the method of variation of parameters. The general solution has the form

$$x(t) = c_1 x_1(t) + c_2 x_2(t)$$
$$+ u_1(t) x_1(t) + u_2(t) x_2(t)$$

for appropriate functions $u_1$ and $u_2$. Show that there is a unique choice of the constants $c_1$ and $c_2$ that satisfies any initial conditions of the form $x(t_0) = x_0$ and $x'(t_0) = v_0$.

**9.** Let $x'' + x = (\sin t)/t$. Notice that the amplitude of the forcing tends to 0 as $t \to \infty$. Suppose $x(0) = x'(0) = 0$.

(a) Using the variation of parameters, show that a particular solution has the form

$$x_p(t) = u_1(t) \cos t + u_2(t) \sin t,$$

where

$$u_1(t) = -\int_0^t \frac{\sin^2 s}{s}\, ds$$

and

$$u_2(t) = \int_0^t \frac{\sin s \cos s}{s}\, ds.$$

(b) The *cosine integral* is the function defined by

$$\mathrm{Ci}\, t = \gamma + \ln t + \int_0^t \frac{\cos s - 1}{s}\, ds,$$

where $\gamma = 0.57721\ldots$, and the *sine integral* is the function defined by

$$\mathrm{Si}\, t = \int_0^t \frac{\sin s}{s}\, ds.$$

Show that

$$u_1(t) = \tfrac{1}{2}(\mathrm{Ci}\, 2t - \gamma - \ln 2t)$$

and that

$$u_2(t) = \tfrac{1}{2}\mathrm{Si}\, 2t.$$

(c) It can be shown that the expressions for $u_1$ and $u_2$ given in (b) lead to the following limits:

$$\lim_{t \to 0} u_1(t) = 0,$$

$$\lim_{t \to 0} u_2(t) = 0,$$

$$\lim_{t \to 0} u_1'(t) = 0,$$

$$\lim_{t \to 0} u_2'(t) = 1.$$

Use these limits to find the solution that satisfies the initial conditions $x(0) = x'(0) = 0$.

(d) It can be shown that

$$\lim_{t \to \infty} \mathrm{Ci}\, t = 0$$

and

$$\lim_{t \to \infty} \mathrm{Si}\, t = \pi/2.$$

What do these limits imply about the qualitative behavior of the solution in (c) as $t \to \infty$? (In particular, what happens to the amplitude of the solution as $t \to \infty$?) Is the behavior akin to resonance? Explain.

(e) What is the solution of the equation $x'' + x = \sin t$ satisfying the initial conditions $x(0) = 0$ and $x'(0) = 0$? Contrast the qualitative behavior of this solution to that in (d).

**10.** This exercise discusses the relationship between the method of variation of parameters for second-order equations and the solution of linear, first-order, inhomogeneous systems of two equations with constant coefficients. (See Section 7.8.)

Let

$$x'' + ax' + bx = f(t). \qquad (B.12)$$

(a) Write Eq. (B.12) as an equivalent system of the form

$$\mathbf{x}' = \mathbf{A}\mathbf{x} + \mathbf{F}(t) \qquad (B.13)$$

for an appropriate constant matrix $\mathbf{A}$ and function $\mathbf{F}$.

(b) Assume that the homogeneous solution of Eq. (B.12) is known; write it as

$$x_h(t) = c_1 x_1(t) + c_2 x_2(t),$$

where $x_1$ and $x_2$ are two linearly independent solutions of the homogeneous equation. What is a fundamental matrix $\mathbf{\Psi}(t)$ associated with the homogeneous solution of Eq. (B.13)?

(c) Assume that a particular solution of Eq. (B.13) has the form $\mathbf{x}_p = \mathbf{\Psi}(t)\mathbf{u}(t)$ and insert it into Eq. (B.13). What are the components of $\mathbf{u}(t)$? How are they related to Eqs. (B.4) and (B.6)?

**11.** This exercise discusses a variation of parameters method for first-order equations and its relationship to the method of integrating factors. Let

$$x'(t) + kx(t) = f(t).$$

(a) Find the homogeneous solution $x_h(t)$.

(b) Show that it is possible to find a particular solution by assuming that the particular solution has the form $x_p(t) = u(t)x_h(t)$.

(c) How does the solution procedure in (b) compare to the method of integrating factors discussed in Section 2.3?

# *Appendix C*

# PARTIAL FRACTION DECOMPOSITION

The objective of the partial fraction decomposition is to break up an expression of the form

$$\frac{N(s)}{P(s)},$$

where $N$ and $P$ are polynomial functions of $s$, into an equivalent expression of the form

$$\frac{N_1(s)}{P_1(s)} + \cdots + \frac{N_k(s)}{P_k(s)},$$

where each $P_i$ is a factor of $P$. There are two basic cases, depending on whether $P$ has distinct simple roots or multiple roots. We treat these cases in turn.

## C.1   THE BASIC CASE: SIMPLE ROOTS

In the simplest case, each of the factors of $P$ has one of the following forms:

1. $P_i(s) = s - r$, where $r$ is a constant, or
2. $P_i(s) = s^2 + b^2$, where $b$ is a constant, or
3. $P_i(s) = (s - a)^2 + b^2$, where $a$ and $b$ are constants.

Factors of these three types are *irreducible*, because they cannot be factored further over the set of real numbers when $a$, $b$, and $r$ are real numbers. If $P$ is a polynomial with real coefficients and distinct, simple roots, then the fundamental theorem of algebra implies that each of the factors of $P$ is of one of these three types.

The partial fraction decomposition is computed in three steps.

**Step 1.** Factor $P$. If each of the factors is distinct and is of one of the three types listed above, then continue to Step 2. Otherwise, $P$ has a root with multiplicity greater than one, and it is necessary to use the procedure described later in this section.

**Step 2.** Let

$$\frac{N(s)}{P(s)} = \frac{N_1(s)}{P_1(s)} + \cdots + \frac{N_k(s)}{P_k(s)}, \tag{C.1}$$

where each $P_i$ is a factor of $P$ and each $N_i$ is a polynomial in $s$ whose degree is one less than the degree of $P_i$.

**Step 3.** Determine the coefficients of each $N_i$ such that Eq. (C.1) is an identity.

◆ *Example 1*   Let

$$\frac{N(s)}{P(s)} = \frac{s+2}{s^3+s}.$$

Here, $P(s) = s(s^2+1)$, so all of the roots of $P$ are simple. (What are they?) The partial fraction decomposition is

$$\frac{s+2}{s(s^2+1)} = \frac{A}{s} + \frac{Bs+C}{s^2+1}, \tag{C.2}$$

because the degree of $P_1(s) = s$ is 1 and the degree of $P_2(s) = s^2+1$ is 2. Therefore, $N_1(s) = A$ (degree 0) and $N_2(s) = Bs+C$ (degree 1).

Next, we determine the values of $A$, $B$, and $C$ to make Eq. (C.2) an identity. We first find common denominators, obtaining

$$\frac{s+2}{s(s^2+1)} = \frac{A(s^2+1)}{s(s^2+1)} + \frac{s(Bs+C)}{s(s^2+1)}$$

$$= \frac{(A+B)s^2 + Cs + A}{s(s^2+1)}.$$

This relation is an identity provided that

$$A + B = 0,$$
$$C = 1,$$
$$A = 2,$$

that is, if $A = 2$, $B = -2$, and $C = 1$. Thus, Eq. (C.2) is equivalent to

$$\frac{s+2}{s(s^2+1)} = \frac{2}{s} + \frac{-2s+1}{s^2+1}. \qquad\blacksquare$$

✦ *Example 2*   Find the partial fraction decomposition of

$$\frac{N(s)}{P(s)} = \frac{s+5}{s^3 + 2s^2 + 5s}.$$

***Solution***   We have

$$P(s) = s(s^2 + 2s + 5) = s\left[(s+1)^2 + 4\right],$$

so the partial fraction decomposition has the form

$$\frac{s+5}{s\left[(s+1)^2 + 4\right]} = \frac{A}{s} + \frac{Bs + C}{(s+1)^2 + 4}.$$

As in the preceding example, we find common denominators and compare coefficients of like terms:

$$\frac{s+5}{s\left[(s+1)^2 + 4\right]} = \frac{A\left[(s+1)^2 + 4\right]}{s\left[(s+1)^2 + 4\right]} + \frac{s(Bs + C)}{s\left[(s+1)^2 + 4\right]}$$

$$= \frac{(A+B)s^2 + (2A+C)s + 5A}{s\left[(s+1)^2 + 4\right]},$$

which is an identity provided that

$$A + B = 0,$$
$$2A + C = 1,$$
$$5A = 5,$$

that is, if $A = 1$, $B = -1$, and $C = -1$. Therefore,

$$\frac{s+5}{s^3 + 2s^2 + 5s} = \frac{1}{s} - \frac{s+1}{(s+1)^2 + 4}. \qquad \blacksquare$$

## C.2   THE CASE OF REPEATED ROOTS

If $P$ has a factor of the form $(s - r)^m$, where $m > 1$, then we say that $r$ is a root of $P$ of *multiplicity* $m$. (In other words, $r$ is a repeated root.) In general, if $N(s)$ is a polynomial of degree $m - 1$ or less, then

$$\frac{N(s)}{(s - r)^m} = \frac{A_1}{s - r} + \frac{A_2}{(s - r)^2} + \cdots + \frac{A_m}{(s - r)^m}$$

is the corresponding partial fraction decomposition. The constants $A_1, \ldots, A_m$ are determined by finding common denominators in the manner illustrated previously.

✦ *Example 1*   The partial fraction decomposition of the expression

$$\frac{s+1}{(s-3)^2}$$

has the form

$$\frac{s+1}{(s-3)^2} = \frac{A_1}{s-3} + \frac{A_2}{(s-3)^2}.$$

We find common denominators and compare coefficients of like terms to obtain

$$\frac{s+1}{(s-3)^2} = \frac{A_1(s-3)}{(s-3)^2} + \frac{A_2}{(s-3)^2}$$

$$= \frac{A_1 s + (A_2 - 3A_1)}{(s-3)^2},$$

which is an identity provided that $A_1 = 1$ and $A_2 = 4$. Therefore,

$$\frac{s+1}{(s-3)^2} = \frac{1}{s-3} + \frac{4}{(s-3)^2}.$$     ∎

The previous discussion is not comprehensive in that we have not considered the case in which $P$ contains a factor of the form $\left[(s-a)^2 + b^2\right]^m$ where $m > 1$. The only common case in practice is $m = 2$; see entry (10) in Table 10.2. The corresponding inverse Laplace transform can be obtained by means other than partial fraction decomposition; see Exercise 14 in Section 10.3.

## *Exercises*

**1.** Compute the partial fraction decomposition of each expression in the following list.

(a) $\dfrac{4s}{s^2 + 6s + 5}$     (b) $\dfrac{2}{s^2 + 5s + 6}$

**2.** Compute the partial fraction decomposition of each expression in the following list.

(a) $\dfrac{s+1}{s^3 + 4s}$     (b) $\dfrac{3}{s^3 + 9s}$

**3.** Compute the partial fraction decomposition of each expression in the following list.

(a) $\dfrac{9}{s^3 - 6s^2 + 9s}$     (b) $\dfrac{s+2}{(s+4)^3}$

**4.** Compute the partial fraction decomposition of each expression in the following list.

(a) $\dfrac{1}{s^2(s+1)}$     (b) $\dfrac{1}{s^2(s+1)^2}$

# *Appendix D*

# THE FUNDAMENTAL THEOREM OF CALCULUS

An *antiderivative* of a function $f(t)$ is a function $F(t)$ such that $dF/dt = f(t)$. We usually write

$$F(t) = \int f(t)\, dt$$

or

$$F(t) = \int f(t)\, dt + C$$

to emphasize that the antiderivatives of an integrable function are identical up to a constant. The fundamental theorem of calculus is a statement about definite integrals.

**Theorem D.1  *The Fundamental Theorem of Calculus*** Let $f$ be a continuous function, and let $F$ be an antiderivative of $f$. Then

$$\int_a^b f(t)\, dt = F(b) - F(a). \tag{D.1}$$

✦ *Example 1*  Let $f(t) = t^2$. An antiderivative of $f$ is

$$F(t) = \tfrac{1}{3}t^3 + C,$$

where $C$ is any constant. Theorem D.1 implies that

$$\int_a^b f(t)\,dt = \left(\tfrac{1}{3}b^3 + C\right) - \left(\tfrac{1}{3}a^3 + C\right)$$

$$= \tfrac{1}{3}b^3 - \tfrac{1}{3}a^3.$$

Because the constant $C$ cancels out of the calculation, we can omit the constant of integration for the purpose of computing the definite integral.    ∎

We can interpret Eq. (D.1) as the area between the graph of $f$ and the $t$ axis between $t = a$ and $t = b$. If $a < b$, then the regions where the graph of $f$ lies above the $t$ axis add a positive contribution to the area, and the regions where the graph of $f$ lies below the $t$ axis add a negative contribution. (The reverse holds when $a > b$.)

In the study of differential equations, we frequently are interested in the case where the definite integral has a variable upper limit.

✦ *Example 2*   Let $f(t) = t^2$. Then

$$\int_0^t f(s)\,ds = \tfrac{1}{3}s^3 \big|_0^t = \tfrac{1}{3}t^3. \tag{D.2}$$

We can interpret Eq. (D.2) as giving the area between the graph of $f$ and the horizontal axis from $s = 0$ to $s = t$, as illustrated in Fig. D.1.

Here is another interpretation of Eq. (D.2). Imagine that we move to the right along the $s$ axis, starting at $s = 0$, at the rate of one unit per second. After $t$ seconds, the area swept out by the graph of $f$ is given by Eq. (D.2).

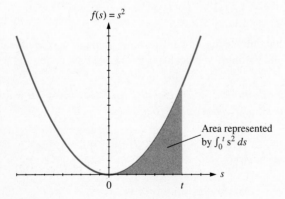

**FIGURE D.1**   A graphical illustration of Eq. (D.2).    ∎

The quantity $s$ in Eq. (D.2) is a dummy variable of integration. There is no difference in meaning between the expressions

$$\int_a^b t^2 \, dt \quad \text{and} \quad \int_a^b s^2 \, ds.$$

However, if we have a definite integral whose upper limit is $t$, then we use $s$ in the integrand to avoid confusion.

Let $f(t) = t^2$. Then Eq. (D.2) implies that

$$\frac{d}{dt} \int_0^t f(s) \, ds = \frac{d}{dt} \left( \frac{t^3}{3} \right) = t^2 = f(t).$$

In other words, when we differentiate the integral with respect to $t$, we get $f$ back again. This example illustrates an equivalent formulation of Theorem D.1, which emphasizes that integration is the inverse of differentiation.

**Theorem D.2**    *Let $f$ be a continuous function, and let $a$ be a constant. Then*

$$\frac{d}{dt} \int_a^t f(s) \, ds = f(t).$$

✦ *Example 3*    Let $x$ be a continuously differentiable function. (In other words, both $x(t)$ and $x'(t)$ are continuous functions.) Theorem D.1 implies that

$$\int_{t_0}^t x'(s) \, ds = x(t) - x(t_0).$$

Because $t_0$ is a constant, $x(t_0)$ is also a constant. We can think of $x(t_0)$ as the initial condition for a differential equation. Notice also that

$$\frac{d}{dt} \int_{t_0}^t x'(s) \, ds = x'(t). \qquad \blacksquare$$

## Exercises

**1.** Evaluate each of the integrals in the following list.

(a) $\displaystyle\int_0^t \cos s \, ds$      (b) $\displaystyle\int_0^t \sin s \, ds$

**2.** Evaluate each of the integrals in the following list.

(a) $\displaystyle\int_0^t t e^{-st} \, ds$      (b) $\displaystyle\int_0^t e^{-st} \cos s \, ds$

**3.** Evaluate each of the integrals in the following list. Assume that $f$ is a function that is continuous on the entire real line.

(a) $\displaystyle\int_0^t \frac{d}{ds}\big(f(s)\big)\,ds$ 　　(b) $\displaystyle\frac{d}{dt}\int_0^t f(s)\,ds$

**4.** Evaluate each of the integrals in the following list. Assume that $x$ is a function that is continuously differentiable on the entire real line.

(a) $\displaystyle\int x'(t)\,dt$ 　　(b) $\displaystyle\int_{t_0}^t x'(s)\,ds$

**5.** Evaluate each expression in the following list.

(a) $\displaystyle\lim_{t\to 0}\frac{d}{dt}\int_0^t \frac{\sin s}{s}\,ds$

(b) $\displaystyle\lim_{t\to 0}\frac{d}{dt}\int_0^t \frac{\cos s - 1}{s}\,ds$

# Answers and Hints to

# Selected Exercises

## Chapter 0

### 0.1

1. About $35,000.

### 0.2

1. (a) It overestimates the maximum height. (b) $mv' = -mg + kv$, where $k$ is a constant of proportionality.

### 0.3

1. Hint: If you use a calculator, make sure that it is set to radian mode.

2. (a) The fixed points are $x = 0$ and $x = 2/3$; both are unstable. (b) The fixed point is $x \approx 0.377$; unstable. (c) The fixed points are $x = 0$ and $x \approx 0.736$; both are unstable.

## Chapter 1

### 1.1

1. (a) Independent $t$, dependent $p$, first order. (b) Independent $t$, dependent $p$, second order. (c) Independent $t$, dependent $p$, second order. (d) Independent is not stated, but it is some variable like $t$; dependent $x$; first order.

3. (a) $k = \ln 1.03$; (b) $k = \frac{1}{12} \ln 1.03$.

5. (c) Approximately 54.1 million.

7. $\omega = \pm 2$.

11. (a) $x(t) = -e^{1.4t}$; (b) $x(t) = e^{1.4(t+1)}$; (c) $x(t) = -e^{1.4(t-1)}$.

## 1.2

1. Partial answer: Equation (i) matches field (c).

3. (a) If $x(0) = 1$, then $x \to 2$ as $t \to \infty$. If $x(0) = 0$, then $x$ remains 0 as $t \to \infty$.

5. (a) Solutions approach a straight line. (b) $p' = A = t - 2(At + B)$ if and only if $1 - 2A = 0$ and $-2B = A$, which implies $A = 1/2$, $B = -1/4$.

7. (a) The qualitative behavior is similar to $p' = p$, except that solutions grow more rapidly at any fixed value of $p$.

8. (a) The qualitative behavior of the solutions is similar to that of $p' = \sin p$, except that solutions "level off" more quickly because $p'$ is twice as large for any fixed value of $p$.

11. (a) Not realistic because it implies that the population becomes infinite if $p(0) > 2$. (b) Realistic insofar as it models a threshold situation wherein the population becomes extinct if it becomes small enough and recovers to an equilibrium value otherwise. A sufficiently large initial population also declines to a constant (finite) value.

13. Both solutions are increasing functions if $x_p(0) = x_q(0) = 1$. Therefore, $x^p > x^q$ if $p > q$, so the comparison theorem implies $x_p(t) > x_q(t)$ for $t > 0$ if $p > q$.

15. (a) $x_2(t) > x_1(t)$. (c) No.

17. Hint: Are both sides of the differential equation well defined when $x = \pi/2$?

## 1.3

1. No.

3. (a) Autonomous; (b) nonautonomous.

5. (a) Autonomous; (b) nonautonomous; (c) autonomous.

8. (a) $x = 1$, unstable and $x = -1$, stable. (b) $x = 0$, unstable. (c) $x = k\pi$, $k$ any integer, is unstable; $x = k\pi/2$, $k$ any odd integer, is stable.

10. (a) Autonomous.

12. (a) 1.

14. (a) $x' = (x - 1)(2 - x)$. (b) $x' = (1 - x)(2 - x)$.

16. (a) True; (b) false; (c) true; (d) false; (e) false; (f) false.

18. (a) If $f$ is a continuous function on the interval $[a, b]$, then $f$ assumes all values between $f(a)$ and $f(b)$.

20. Hint: If solutions oscillate, then they increase for a time and decrease for a time.

22. Hint: First observe that every solution must be always increasing or always decreasing.

23. (a) $p'' = df(p)/dp \times f(p)$; (b) no; (c) yes.

## 1.4

1. (a) If $h = 0.5$, then $\hat{x}(2) = 2.75$; if $h = 0.25$ then $\hat{x}(2) \approx 3.188$; if $h = 0.125$ then $\hat{x}(2) \approx 3.422$.

3. (a) Smaller; (b) larger.

5. Hint: Consider the initial value problem $x' = x$, $x(0) = 1$.

## 1.5

1. $x(t) = x_0 e^{kt}$, for all values of $x_0$.

3. We must exclude cases in which $x(t) = 0$ for some $t$. The equilibrium solution must be found by inspection; it is the zero function.

5. (a) $x(t) = \left(\frac{2}{3}t + 1\right)^{3/2}$, defined for all $t \geq -3/2$. As $t \to \infty$, $x(t) \to \infty$. (b) $x(t) = \left(\frac{3}{4}t + 1\right)^{4/3}$, defined for all $t$. As $t \to \infty$, $x(t) \to \infty$.

7. (a) Equilibrium at $x = -1$; the solutions of the initial value problems are $x(t) = -1 + 2e^t$ and $x(t) = -1$, respectively. (b) No equilibria. The solutions are $x(t) = \tan(t + \pi/4)$ and $x(t) = \tan(t - \pi/4)$.

9. (a) $\ln|\sec x + \tan x| = t + C$; (e) $\pi/2$ in each case.

12. (a) Hint: Integrate by partial fractions. (d) The limits are 0 and 2, respectively.

## Chapter 2

### 2.1

1. (a) $C = 1$, $k = \frac{1}{2}\ln 2$. (b) $C = 1$, $k = \frac{3}{2}\ln 3$. (c) $C = 5$, $k = \ln 10$. (d) $C = 1$, $k = \frac{1}{4}\ln\frac{7}{8}$.

3. (a) 3.88 years; (b) 3.86 years; (c) 3.85 years (approximately).

5. (a) $T_d = (\ln 2)/r$; (b) yes.

7. (a) Plutonium 239, 34.8 yr; americium 243, 10.6 yr; oxygen 14, 0.102 s; (b) the required time is 3 times the half life.

9. (a)–(d) Each limit is infinite. (e) The results are the same as for the case $k = 1$.

11. (a) An equivalent ratio is $(K \sin \omega t)/e^t$, but the numerator does not approach a limit. (b) 0.

13. (b) $225,000.

15. (a) About 243,000 yr; (b) yes.

17. (a) $B_0(1 + r/4)^4$; (b) $B_0(1 + r/12)^{12}$.

19. (a) $20,085.54; (b) $4321.94; (c) 4.6; (d) 3.50; (e) 8.27; (f) $v(30)/p(30) = e^{30r}/(1 + i)^{30}$.

## 2.2

1. (b) $1297.44.

3. (b) Hint: The rate at which the principal changes depends on the rate at which interest accrues and the rate at which payments are made. (c) Yes; unstable.

5. (b) $119.08; (c) yes; (d) no.

7. (a) $1206.05; (b) principal $148,982.25; interest $13,454.89.

9. (b) $599.02/month in the first year and $663.31/month in the second year.

11. Start with cold water.

13. (a) 51.9 min, 103.8 min, and 155.8 min, respectively. (b) We have $T(t) = (T_0 - R)e^{-kt} + R$, so the temperature difference between the object and its surroundings is $(T_0 - R)e^{-kt}$. The decay constant $k$ can be computed directly from measurements of how the temperature *difference* changes with time.

15. Hint: Use the separation of variables.

16. (a) $-27.97$ m/s; (b) no; (c) 26.41 m/s; (d) 71.18 m and 5.01 s, respectively.

18. (a) A larger value of $k$ means greater air resistance at a given velocity; (b) more quickly; (c) decreases; (d) decreases; (e) more slowly; (f) greater speed.

19. (a) $v(t)$ does not depend on the initial height. (b) 10 m/s. (c) No. The equation must be modified. Hint: Think about the direction in which air resistance acts when the object rises.

22. (c) No. He loses less than 17 lb. (e) Yes. An equilibrium solution means that caloric intake exactly balances the energy requirements needed to maintain a constant body weight.

24. (a) Individuals/time. (b) If $k - m > 0$, then the population increases; if $k - m < 0$, then the population decreases. (d) The expression refers to the *total* number of immigrants after $t$ years. (e) 106.09 million.

## 2.3

5. (a) $x(t) = \frac{4}{5}e^{-t} + \frac{2}{5}\sin 2t + \frac{1}{5}\cos 2t$; (b) $x(t) = e^{-t}(t + 1)$; (c) $x(t) = \frac{3}{2}e^{-t} - \frac{1}{2}e^t$; (d) $x(t) = 2 - e^{-t}$; (e) $x(t) = t^2 - 2t + 2 - e^{-t}$.

7. (a) $x(t) = (x_0 + m/4)e^{-2t} + mt/2 - m/4$. (c) The solution approaches a straight line with slope $m/2$.

10. (b) 0.418 mg.

## 2.4

1. (a) 11.55 yr; (b) 3.9 percent; (c) yes.

3. (a) $T_d = (\ln 2)/k$; (c) 10 percent.

5. (a) 13.9 h earlier; (b) about 4:30 the previous afternoon; (c) about 9:30 p.m.

7. (c) No.

9. (a) $x(t) = (1 + x_0)e^{-t} + t - 1$.

11. (a) $x(t) = mt - m + (m + 1)e^{-t}$.

13. (a) $x(t) = \frac{(\omega^2 + \omega + 1)e^{-t} + \sin \omega t - \omega \cos \omega t}{1 + \omega^2}$. (d) There are difficulties when the "true" solution becomes 0. A better characterization of the error is to consider the phase shift between the true and the approximate solutions, as well as the amplitudes of the two solutions.

15. (a) The half-life is 6.58 h. (b) The measurement uncertainties lead to estimates of the half-life between 5.48 h and 8.23 h.

## *Chapter 3*

### 3.1

1. (a) $x(t) = \frac{1}{\frac{1}{2} - t}$, $(-\infty, 1/2)$. (b) $x(t) = \frac{-1}{t + \frac{1}{2}}$, $(1/2, \infty)$.

3. (a) Hint: Use a step size of 1/1000 or smaller. (b) Hint: What is the sign of the second derivative of the true solution? (c) $x(t) = \frac{1}{\frac{34}{35} - t}$.

5. (a) $x(t) = \frac{1}{2 - kt}$, $(-\infty, 2/k)$. (b) $x(t) = -\frac{1}{0.5 + kt}$, $(-1/(2k), \infty)$. (c) The intervals of existence are $(1/(2k), \infty)$ and $(-\infty, -1/(2k))$, respectively.

7. (a) $x(t) = \frac{1}{\sqrt{-2t + 1}}$, $(-\infty, 1/2)$. (b) $x(t) = -\frac{1}{\sqrt{-2t + 1}}$, $(-\infty, 1/2)$. (c) $x(t) = \frac{1}{\sqrt{-2t + 1/x_0^2}}$, $\left(-\infty, 1/\left(2x_0^2\right)\right)$. (d) $x(t) = -\frac{1}{\sqrt{-2t + 1/x_0^2}}$, $\left(-\infty, 1/\left(2x_0^2\right)\right)$. (e) $x_0 \neq 0$.

11. (a) $x(t) = (1 + (1 - p)t)^{1/(1-p)}$, $(-\infty, 1/(p - 1))$. (b) $x(t) = \left(x_0^{1-p} + (1 - p)t\right)^{1/(1-p)}$, $\left(-\infty, x_0^{1-p}/(p - 1)\right)$.

13. (a) $x(t) = ((p + 1)t + 1)^{1/(p+1)}$, $(-1/(p + 1), \infty)$; (b) $x(t) = \left((p + 1)t + x_0^{p+1}\right)^{1/(p+1)}$, $\left(-x_0^{p+1}/(p + 1), \infty\right)$.

### 3.2

1. (a), (b), (d) All initial conditions. (c) $x(t_0) \neq 0$.

3. (c) Hint: $f$ and $\partial f/\partial x$ are defined and continuous everywhere in $S$ if $S$ is a sufficiently small box around $(0, 1)$.

5. $f(t, x) = x^2$ and $\partial f/\partial x = 2x$ are both continuous functions for all $x$. Hence, by Theorem 3.3, we have a unique solution through any initial point. The interval of existence of this solution has nothing to do with the question of uniqueness or existence of the solution.

7. (a) Hint: Rewrite the equation as $x' = f(t, x) = -x/t$.

## 3.3

1. (a) $K = 12$; (b) $K = 1$; (c) $K = 1$; (d) $K = 2$; (e) $K = 2$; (f) $K = e$; (g) $K = 1$; (h) $K = (n + 1)^2$.

3. (a) $|x(t) - \hat{x}(t)| \le |x(t_0) - \hat{x}(t_0)| e^{8(t-t_0)}$, valid for as long as the graphs of both functions lie in $S$.
   (b) Same bound as in (a).

7. (c) The short term.

9. (c) Nothing.

11. (a) 2 is a Lipschitz constant for any rectangle $S$. (d) $0.986 \le \hat{x}_0 \le 1.013$.

12. (b) and (d) have the greatest potential sensitivity to errors in initial conditions.

13. (a) $K = 11$. (d) Hint: Are solutions sensitive to errors in initial conditions if they approach a stable equilibrium?

# Chapter 4

## 4.1

1. Hint: Consider $f(x) = f(x \cdot 1)$ and $f(x - x)$.

3. Only (b) is linear.

5. Figures (b) and (d) are graphs of functions $y = f(x)$. Figures (c) and (d) are graphs of functions $x = g(y)$. Figure (a) is not the graph of a function.

9. (a) $(17, 39)$; (b) $(-20, 2)$; (c) $(10, -4)$; (d) $(0, 0)$.

11. (a) One parametric representation is $\mathbf{g}(t) = (0, 2) + t(1, 3)$; (b) $\mathbf{h}(t) = (1, 1) + t(3, 2)$; (c) no.

13. (a) Yes; (b) no.

15. (a) $\mathbf{0}$; (b) no.

17. $\mathbf{p}$ is a nonlinear function of $t$ that is not one-to-one, though its domain includes all real numbers.

19. The relative loudness of the instruments should be preserved as the volume increases; otherwise, distortion results.

## 4.2

1. No.

3. (a) $D(f) = g'h + gh'$; (b) $D(f) = 0$; (c) $D(f) = g''$; (d) $D(f) = g'(h(t))h'(t)$.

5. (a) $1/3$; (b) $4/3$; (c) $e - 1$; (d) $e$.

7. (a) $U(1) = U(2t) = 1$, so $U$ is not one-to-one.

11. No.

13. (a) $T(x, y) = (x, -y)$. (b) Hint: Show that $\mathbf{T}$ can be expressed as a matrix-vector product.

15. Yes.

17. $\mathcal{L}(\cosh kx) = s/(s^2 - k^2)$; $\mathcal{L}(\sinh kx) = k/(s^2 - k^2)$.

## 4.3

1. (a) $x' + 2x = t$; (b) $x' - x = -10$; (c) $x' + e^x = 5$. All are homogeneous except (c) and (h).

3. (a) $x' - a(t)x = b(t)$; (b) $x' - a(x)x = b(t)$; (c) $x'' - 4x' - 3x = \sin t$. Each operator is homogeneous.

6. Suggestion: Show that $\text{Op}(x) = (d^n x/dt^n) x^m$ is nonlinear.

## 4.4

1. The general solutions are $x(t) = x_0 e^{t^2/2}$, $x(t) = x_0 e^{1 - \cos t}$, and $x(t) = 1/(1/x_0 - t)$, respectively.

5. (b) Let $L$ be the corresponding operator. Since $L$ is linear, we have $L(x_1 + x_2) = L(x_1) + L(x_2)$; therefore, $x_1 + x_2$ is a solution if $x_1$ and $x_2$ are solutions.

11. Hint: Look for a linear combination of a suitable sine and cosine function.

13. (a) $L(x) = x'' + 3x' + 2x$; (b) $L(e^{kt}) = (k^2 + 3k + 2)e^{kt}$. (c) Hint: We must have $k^2 + 3k + 2 = 0$. (Why?)

## 4.5

1. (a) $L(x) = x' + x$; (c) $x_h(t) = e^{-t}$; (e) $L(Cx_h(t) + x_p(t)) = CL(x_h(t)) + L(x_p(t)) = 0 + t$.

3. (a) $L(x) = x' - x/t$; (b) $x_h(t) = t$; (f) $x(t) = Ct + 10t^2$; (h) no.

5. (b) $x_h = c_1 \cos t + c_2 \sin t$; (c) $A = 1/2$.

11. (a) \$27,125; (b) \$24,000 contributed and \$3125 in interest.

## Chapter 5

### 5.1

1. 98 N/m.

3. $|\phi| \leq 0.244$ rad $= 14.0°$; $|\phi| \leq 0.538$ rad $= 30.8°$; $|\phi| \leq 0.749$ rad $= 42.9°$.

6. 9800 N/m.

### 5.2

1. (a) About 0.248 m. (b) The length would vary between 24.70 and 24.95 cm.

2. $x(t) = \frac{1}{5} \sin \frac{1}{2} t$.

3. (a) $x(t) = c_1 \sin 2t + c_2 \cos 2t$; period $\pi$. (c) $x(t) = c_1 \sin 4t + c_2 \cos 4t$; period $\pi/2$. (e) $x(t) = c_1 \sin 3t + c_2 \cos 3t$; period $2\pi/3$.

4. (a) $x(t) = \sqrt{8} \cos (t - \pi/4)$, crosses the equilibrium at $t = 3\pi/4$, maximum speed $\sqrt{8}$ m/s. (b) $x(t) = \sqrt{2} \cos (3t + \pi/4)$, crosses the equilibrium at $t \approx 0.262$, maximum speed $\sqrt{2}$ m/s. (c) $x(t) \approx \frac{5}{2} \cos (2t - 2.498)$, crosses the equilibrium at $t \approx 0.464$, maximum speed $5/2$ m/s. (d) $x(t) = \sqrt{2} \cos (\sqrt{2} t + \pi/4)$, crosses the equilibrium at $t \approx 0.555$, maximum speed $\sqrt{2}$ m/s.

8. Make either the capacitance or the inductance $1/4$ as large and leave the other component alone.

10. $0.0241 \ \mu F \leq C \leq 0.0294 \ \mu F$.

12. (a) About 0.67 rad. (b) About 2.8 s. (c) About 1.24 m on either side of the swing.

14. (a) $2x'' + 50x = 0$. (b) $x(t) = 0.004 \sin 5t$. (c) At $t = \pi/5$; at $t = 2\pi/5$. (d) $A = 0.004$ m and $T = 2\pi/5$ s. (e) 0.5 m/s or $-0.5$ m/s.

16. Partial answer: If $x_0 > 0$ and $v_0 > 0$, then $\cos \delta > 0$ and $\sin \delta > 0$, so $0 < \delta < \pi/2$. If $x_0 > 0$ and $v_0 < 0$, then $\cos \delta > 0$ and $\sin \delta < 0$, so $-\pi/2 < \delta < 0$.

17. $x(t) = \sin t = \cos (t - \pi/2)$.

### 5.3

1. (a) $x(t) = -\frac{3}{2} e^{-3t} + \frac{5}{2} e^{-t}$; never crosses the equilibrium; maximum distance from the origin is approximately 1.242 units, which occurs when $t = \ln (3/\sqrt{5})$.

2. (a) No.

5. (a) $x(t) = c_1 e^t + c_2 e^{2t}$; (b) $x(t) = 4e^t - 3e^{2t}$. As $t \to \infty$, $x \to -\infty$; (c) $\lim_{t \to \infty} |x(t)| = 0$ if and only if $x(0) = x'(0) = 0$.

7. (a) $x(t) = c_1 e^{3t} + c_2 e^{-t/3}$. (b) Unless the initial conditions are chosen so that $c_1 = 0$, solutions grow exponentially with time. (c) We have $c_1 = (3v_0 + x_0)/10$; if $x_0 = -3v_0$, then $|x(t)| \to 0$ as $t \to \infty$.

11. $c \approx 2.99$.

## 5.4

1. (a) $\lambda = (-1 \pm \sqrt{23})/6$. (b) $\lambda = (-3 \pm i\sqrt{11})/2$. (c) The exponential function cannot be a solution, because if we substitute $x(t) = e^{\lambda t}$ into the equation, we obtain the relation $\lambda^2 e^{\lambda t} + e^{2\lambda t} = 0$, which cannot be satisfied for an interval of $t$ values when $\lambda$ is a constant.

3. Hint: Euler's formula implies $e^{2it} = \cos 2t + i \sin 2t$ and $e^{-2it} = \cos 2t - i \sin 2t$.

5. (a) $x(t) = e^{-t} \sqrt{\frac{17}{8}} \cos(\sqrt{8}\, t - 0.8148)$. (b) $e^{-t} \sqrt{17/8}$. (c) $\pi/\sqrt{2}$. (d) $t_0 \approx 0.834$, $x'(t_0) \approx -1.77$, $x''(t_0) \approx 3.55$; (e) $t_1 \approx 0.168$, $x(t_1) \approx 1.16$, $x''(t_1) \approx -10.5$. (f) Approximately 0.382 unit at $t \approx 1.28$. (g) Use the envelope equation—the amplitude of the oscillations is less than 0.001 after about 7.3 s.

7. (a) $x(t) = \left(\frac{1}{2} - i\right) e^{(-2+2i)t} + \left(\frac{1}{2} + i\right) e^{(-2-2i)t} = e^{-2t}(\cos 2t + 2 \sin 2t) = \sqrt{5} e^{-2t} \cos(2t - \delta)$, where $\delta = \tan^{-1} 2 \approx 1.107$. The mass crosses the equilibrium at intervals of $\pi/2$ s and crosses the first time at $t \approx 1.34$ s. The displacement never exceeds 0.01 unit for $t > \frac{1}{2} \ln(100\sqrt{5}) \approx 2.7$ s.
(b) $x(t) = \left(\frac{1}{2} - \frac{1}{3}i\right) e^{(-1+3i)t} + \left(\frac{1}{2} + \frac{1}{3}i\right) e^{(-1-3i)t} = e^{-t}\left(\cos 3t + \frac{2}{3} \sin 3t\right) = \frac{1}{3}\sqrt{13} e^{-t} \cos(3t - \delta)$, where $\delta = \tan^{-1} \frac{2}{3} \approx 0.588$. The mass crosses the equilibrium at intervals of $\pi/3$ s and crosses for the first time when $t \approx 0.720$ s. The displacement never exceeds 0.01 unit from the equilibrium for $t > \ln(100\sqrt{13}/3) \approx 4.8$ s. (c) $x(t) = \left(-\frac{1}{2} + \frac{1}{4}i\sqrt{3}\right) e^{(-1/2+i\sqrt{3})t} - \left(\frac{1}{2} + \frac{1}{4}i\sqrt{3}\right) e^{-(1/2+i\sqrt{3})t} = -e^{-t/2}\left(\cos \sqrt{3}\, t + \frac{1}{2}\sqrt{3} \sin \sqrt{3}\, t\right) = \frac{1}{2}\sqrt{7} e^{-t/2} \cos\left(\sqrt{3}\, t - \delta\right)$, where $\delta = \tan^{-1} \frac{1}{2}\sqrt{3} \approx -2.428$. The mass crosses the equilibrium at intervals of $\pi/\sqrt{3} \approx 1.81$ s and crosses for the first time when $\sqrt{3}\, t - \delta = 3\pi/2$ or $t \approx 1.32$ s. The displacement never exceeds 0.01 unit from the equilibrium for $t > 2 \ln(50\sqrt{7}) \approx 9.8$ s.

9. (a) $\phi(t) = e^{-t/(20m)}(c_1 \cos \omega t + c_2 \sin \omega t)$, where $c_1 = 1/2$, $c_2 = 1/(40m\omega)$ and

$$\omega = \sqrt{10 - \frac{1}{400m^2}}.$$

(b) Approximately 0.025 rad. (c) Approximately 1.86 kg. (d) Yes. (Suggestion: Plot the envelope equation for a fixed value of $t$ as a function of $m$.)

13. (a) As the capacitance increases, the pseudoperiod increases. If the capacitance becomes large enough, the solution no longer oscillates. (b) The pseudoperiod increases as $L$ increases. (c) The pseudoperiod increases as the resistance increases. If the resistance becomes large enough, the solution no longer oscillates—it becomes critically damped or overdamped.

15. (b) $c \approx 0.195$; (c) $k \approx 2.48$.

## 5.5

1. (a) $x(t) = e^{-t}(2 - t)$; crosses the equilibrium at $t = 2$ with velocity $-e^{-2} \approx -0.135$; does not cross the equilibrium if $x'(0) = -3/2$. (b) $x(t) = 2e^{-2t}$; never crosses the equilibrium for $x'(0) = -4$ and $x'(0) = -2$. (c) $x(t) = e^{-t/2}(2 - 3t)$ and crosses the equilibrium at $t = 2/3$ with velocity $-3e^{-1/3} \approx -2.15$; also crosses the equilibrium if $x'(0) = -2$.

3. (a) $c^2 = 4mk$.

5. (a) $x(t) = e^{-t}[x_0 + (x_0 + v_0)t]$. (c) $x_0 > 0$ and $v_0 < -x_0$, or $x_0 < 0$ and $v_0 > -x_0$.

8. (a) $Q'' + 2000Q' + 10^6 Q = 0$, $Q(0) = 0$, $Q'(0) = 1$. (b) $Q(t) = te^{-1000t}$. (f) Suggestion: Look at the absolute difference between the solutions as a function of time.

## 5.6

1. $x(t) = 2 - e^{-t} - 2te^{-t}$.

3. (a) $x(t) = \cos t + \frac{1}{2} \sin 2t$. (c) $x(t) = -2e^{-2t} + 6e^{-t} + \sin t - 3 \cos t$. (e) Not applicable because the equation is nonlinear.

5. $x(t) = \frac{3}{2}e^{-t} + \frac{1}{2} \cos t + \frac{1}{2} \sin t$.

## 5.7

1. (a)–(b) $x(t) = \sin 5t - \sin 6t = -2 \cos \frac{11}{2}t \sin \frac{1}{2}t$.

3. (a) $x(t) = \frac{1}{9}E + c_1 \cos 3t + c_2 \sin 3t$; period $2\pi/3$. (b) $x(0) = E/9$, $x'(0) = 0$. (c) Suggestion: What if there is any error in the initial condition?

5. (a) $x(t) = c_1 \cos 3t + c_2 \sin 3t - \frac{1}{16} \cos 5t$, period $2\pi$. (b) $x(t) = c_1 \cos \sqrt{5}\,t + c_2 \sin \sqrt{5}\,t - \frac{1}{76} \cos 9t$, not periodic. (c) $x(t) = c_1 \cos 2t + c_2 \sin 2t + \frac{1}{3} \cos t$, period $2\pi$.

7. (a) $x(t) = t \sin 4t + \cos 4t$, not periodic, exhibits resonance. (b) $x(t) = \frac{1}{2}(1 - \cos 2t + \sin 2t)$, $\pi$-periodic. (c) $x(t) = \sin \sqrt{5}\,t - \sqrt{5}\,t \cos \sqrt{5}\,t$, not periodic, exhibits resonance.

9. (a) $x(t) = \sin \sqrt{5}\,t + \cos 2t$. (c) The solution is not periodic.

11. (a) $x(t) = [(\pi/2) \sin 2t - \sin \pi t]/(\pi^2 - 4)$, which is not periodic because $\pi$ and 2 are not rationally related.

13. (a) $\mathrm{s}^{-2}$ for $\omega^2$ and $C/s^2$ for $V/L$; (b) the period is $2\pi/\omega$; (c) yes; (d) yes; (e) no.

15. (a) $f(t) = 1$; (b) $f(t) = \sin t$; (c) $f(t) = \sin \frac{1}{2}t$; (d) $f(t) = 2 \sin 3t$.

## 5.8

1. (a) $x(t) = 2(e^{-3t} - e^{-t}) + 4 \sin t + 2 \cos t$. The transient term is $2(e^{-3t} - e^{-t})$, and the steady state is $4 \sin t + 2 \cos t$. The steady state is periodic with period $2\pi$. The amplitude is $\sqrt{20}$, and the gain is $1/\sqrt{20}$.

2. (a) $x(t) = -e^{-t/2}(6 \sin 2t + 36 \cos 2t) + 4 \sin 2t - 32 \cos 2t$. The transient term is $e^{-t/2}(6 \sin 2t + 36 \cos 2t)$, and the steady state is $4 \sin 2t - 32 \cos 2t$. The steady state is periodic with period $\pi$. The amplitude is $4\sqrt{65}$, and the gain is $4/\sqrt{65}$.

3. (a) $x(t) = e^{-t/4}(\sin 2t + 16 \cos 2t) + \sin 2t - 16 \cos 2t$. The transient term is $e^{-t/4}(\sin 2t + 16 \cos 2t)$, and the steady state is $\sin 2t - 16 \cos 2t$. The steady state is periodic with period $\pi$. The amplitude is $\sqrt{257}$, and the gain is $16/\sqrt{257}$.

6. (a) $A < 2$ if $0 < \alpha < 2.93$ (approximately) or $\alpha > 3.07$ (approximately). (b) $A > 3$ if $2.97 < \alpha < 3.02$ (approximately).

8. (a) $x_p = e^{-kt}/(k^2 - ck + \omega^2)$. (b) $x(t) = x_p + e^{-ct/2}\left(b_1 \sin \frac{1}{2}\sqrt{4\omega^2 - c^2}\,t + b_2 \cos \frac{1}{2}\sqrt{4\omega^2 - c^2}\,t\right)$. (c) The general solution oscillates with decreasing amplitude and tends to 0 as $t \to \infty$. (d) The general solution is the same if $k < 0$. The oscillations increase exponentially, and $|x(t)| \to \infty$ as $t \to \infty$.

13. $c \approx 1, \omega \approx 1$.

15. $c \approx 0.74, \omega \approx 3.03$.

## 5.9

1. Theorem 5.9 does not apply to (a) or (e). (Why?) Otherwise, it guarantees unique solutions for (b)–(d) on an interval containing $t = 0$.

3. (a) and (b) are linearly independent pairs; (c) and (d) are not.

5. They are linearly independent and are solutions of the equation $x'' - x = 0$, for instance. The time $t_0$ at which the initial conditions are specified does not affect their linear independence.

8. The solutions can have relative extrema at the same point $t_0$ if $a(t_0) = b(t_0) = 0$.

# Chapter 6

## 6.1

3. (a) $x(t) = c_1 \cos \omega t + c_2 \sin \omega t$; (b) $x'_1 = x_2, x'_2 = -\omega^2 x_1$.

5. $B = -14$.

7. (a), (b), and (c) are autonomous systems.

9. (a) $\left(\frac{1}{2}\pi + 2n\pi, 1\right)$ and $\left(\frac{1}{2}\pi + (2n + 1)\pi, -1\right)$, where $n$ is any integer; (b) $(a, a)$ for all $a$.

11. $y' = a_2 y - k_2 xy - my^2$.

## 6.2

1. (a) $(0, 0)$, $(0.75, 0)$, $(-0.75, 0)$. (b)–(d) $(0, 0)$.

3. (a) Most trajectories approach the origin initially but leave it eventually. (b), (d) Trajectories approach the origin. (c) Trajectories move away from the origin.

5. (a) $x = 3$; (b) $x' = -5$; (c) $x' = 2.5$; (d) $x = 0.8$; (e) after about 3 oscillations.

11. (a) At each point, the slope lines have slope $2/(1 - 2x)$. (b) At each point, the slope lines have slope $2y + 2$. (c) Hint: Show that all solutions move into the first quadrant.

## 6.3

1. (a) Saddle; (b) spiral sink; (c) spiral source; (d) sink.

4. $(0, 10)$ and $(0, -10)$.

6. Center; trajectories spiral anticlockwise.

8. (a) $(0, 0)$ and $(4, 10)$; (b) saddle at the origin, center at $(4, 10)$.

## 6.4

1. (a) $\mathbf{x}(0.04) \approx (1.0376, 0.8377)$; (b) $\mathbf{x}(0.04) \approx (0.9212, -1.9200)$.

3. (a) $x(t) = \cos 2t + \frac{1}{2}\sin 2t$; (b) $\mathbf{x}(t) = \begin{pmatrix} \cos 2t + \frac{1}{2}\sin 2t \\ -2\sin 2t + \cos 2t \end{pmatrix}$.

## Chapter 7

### 7.1

1. (a) $(2, 11)$; (b) $(3, -2)$; (c) $(16, 5)$; (d)–(f) undefined.

3. (a) $\mathbf{AB} = \begin{pmatrix} 1 & 20 \\ 3 & 40 \end{pmatrix}$, $\mathbf{BA} = \begin{pmatrix} 1 & 2 \\ 30 & 40 \end{pmatrix}$; (b) $\mathbf{AB} = \mathbf{BA} = \mathbf{I}$.

6. $\mathbf{B} = \begin{pmatrix} -1 & 4 \\ 2 & 6 \end{pmatrix}$.

8. $\begin{pmatrix} 0 & a \\ 0 & 0 \end{pmatrix}\begin{pmatrix} 0 & b \\ 0 & 0 \end{pmatrix} = \mathbf{0}$ for all $a, b$. $\begin{pmatrix} -a & a \\ 0 & 0 \end{pmatrix}\begin{pmatrix} b & 0 \\ b & 0 \end{pmatrix} = \mathbf{0}$.

10. (a) $a_{11}a_{22} \neq 0$; (b) $\mathbf{D}^{-1} = \begin{pmatrix} 1/a_{11} & 0 \\ 0 & 1/a_{22} \end{pmatrix}$; (d) $\mathbf{D}_1 = \begin{pmatrix} 3/2 & 0 \\ 0 & -2 \end{pmatrix}$.

### 7.2

2. (a) Spiral source; (b) source.

3. Hint: $\mathbf{p}$ is a fixed point if $\mathbf{Ap} = \mathbf{0}$. Show that the only solution is $\mathbf{p} = \mathbf{0}$.

5. (a) $x_1(t) = c_1 e^t$, $x_2(t) = c_2 e^t$.

### 7.3

1. (a) $\lambda_1 = -2$ with $\mathbf{p}_1 = (3, 1)$, $\lambda_2 = 1$ with $\mathbf{p}_2 = (2, 1)$. (b) $\lambda_1 = -3$ with $\mathbf{p}_1 = (-1, 1)$, $\lambda_2 = -1$ with $\mathbf{p}_2 = (1, 1)$. (c) $\lambda_1 = \lambda_2 = 1$ with $\mathbf{p} = (1, 2)$. (d) $\lambda_1 = -2 + i$ with $\mathbf{p}_1 = (1, 3i)$, $\lambda_2 = -2 - i$ with $\mathbf{p}_2 = (1, -3i)$.

4. (a) They are both $a$; (b) $(1, 0)$ and $(0, 1)$.

8. (a) Show that the eigenvalues of the associated matrix are both negative.

### 7.4

1. (a) $\mathbf{x}(t) = 2e^{-2t}(3, 1) - e^t(2, 1)$. (b) $\mathbf{x}(t) = -\frac{3}{2}e^{-3t}(-1, 1) + \frac{5}{2}e^{-t}(1, 1)$. (c) $\mathbf{x}(t) = -e^{2t}(1, 1) + e^{3t}(2, 1)$. (d) $\mathbf{x}(t) = 7e^{-3t}(1, 2) - 4e^{-2t}(1, 3)$.

2. (a) $\mathbf{x}(t) = e^t(\sin t + \cos t, \cos t - \sin t)$, spiral source. (b) $\mathbf{x}(t) = 5e^{-t}(\cos t - \sin t, \cos t - 3\sin t)$, spiral sink.

5. (a) Complex form: $\mathbf{x}(t) = (1-i)e^{(-2+2i)t}\begin{pmatrix} 1 \\ 2i \end{pmatrix} + (1+i)e^{(-2-2i)t}\begin{pmatrix} 1 \\ -2i \end{pmatrix}$; real form $\mathbf{x}(t) =$
$e^{-2t}\begin{pmatrix} 2\cos 2t + 2\sin 2t \\ 4\cos 2t - 4\sin 2t \end{pmatrix}$. (b) Complex form: $\mathbf{x}(t) = (1+3i)e^{(1+3i)t}\begin{pmatrix} 1 \\ 2+i \end{pmatrix} + (1-3i)e^{(1-3i)t}\begin{pmatrix} 1 \\ 2-i \end{pmatrix}$;
real form: $\mathbf{x}(t) = e^t\begin{pmatrix} 2\cos 3t - 6\sin 3t \\ -2\cos 3t - 14\sin 3t \end{pmatrix}$.

8. Suggestion: Use the properties of the Wronskian established in Exercise 6.

10. (a) $\mathbf{x}(t) = \begin{pmatrix} e^{2t} + te^{2t} \\ 2e^{2t} + te^{2t} \end{pmatrix}$, source; (b) $\mathbf{x}(t) = \begin{pmatrix} -2e^t + 7te^t \\ 3e^t + 14te^t \end{pmatrix}$, source.

13. (a) The equation is critically damped if $b = (a/2)^2$. (c) It is the same. (d) $b = (a/2)^2$.

## 7.5

1. (a) $-1 \pm i$; (d) the absolute error tends to 0 as $t \to \infty$.

3. (a) $-3$ and $-1$; (d) the absolute error approaches 0 as $t \to \infty$.

4. (a) Suggestion: Show that the eigenvalues are independent of $b$ and that each term in the solution tends to 0 as $t \to \infty$.

11. (a) Show that the solution contains terms of the form $e^{\lambda t}$ and $te^{\lambda t}$, and argue that all of them tend to 0 as $t \to \infty$ if $\lambda < 0$.

12. Suggestion: Write each component of the general solution in phase-amplitude form and show that $\|\mathbf{x}(t)\| \le Me^{at}$ for an appropriate constant $M$.

13. (a) $\mathbf{x}(t) = -\frac{1}{2}e^t\begin{pmatrix} 1 \\ 1 \end{pmatrix} - \frac{3}{2}e^{-t}\begin{pmatrix} 1 \\ -1 \end{pmatrix}$. (b) $\hat{\mathbf{x}}(t) = -\frac{2}{5}e^t\begin{pmatrix} 1 \\ 1 \end{pmatrix} - \frac{17}{10}e^{-t}\begin{pmatrix} 1 \\ -1 \end{pmatrix}$.

15. $\mathbf{x}_1(t) = \begin{pmatrix} e^{2t} \\ e^{2t} \end{pmatrix}$ and $\mathbf{x}_2(t) = \begin{pmatrix} \frac{1}{5}e^{3t} + \frac{9}{10}e^{2t} \\ \frac{1}{10}e^{3t} + \frac{9}{10}e^{2t} \end{pmatrix}$.

17. (a) The origin is a source with double eigenvalue 1. (b) $\mathbf{x}(t) = e^t\begin{pmatrix} 2 - 3t \\ 1 - 6t \end{pmatrix}$ and $\hat{\mathbf{x}}(t) = e^t\begin{pmatrix} 2.1 - 3.2t \\ 1 - 6.4t \end{pmatrix}$.

## 7.6

1. (a) $\mathbf{x}(t) = \begin{pmatrix} 6e^{2t} - 3e^{-2t} - 2e^{-t} \\ 6e^{2t} - 4e^{-2t} - 3e^{-t} \\ 6e^{2t} - 3e^{-2t} - e^{-t} \end{pmatrix}$, saddle. (c) $\mathbf{x}(t) = \begin{pmatrix} -43e^t + 12e^{-2t} + 32e^{-t} \\ -6e^{-2t} + 8e^{-t} \\ -43e^t + 36e^{-2t} + 8e^{-t} \end{pmatrix}$, saddle.

3. (a) $\mathbf{x}(t) = \begin{pmatrix} -2e^{-t} + 3e^t \\ e^t(1 + 3t) \\ 3e^{-t} + e^t(3t - 2) \end{pmatrix}$, saddle. (b) $\mathbf{x}(t) = e^{-2t}\begin{pmatrix} -t - t^2/2 \\ 1 - t^2/2 \\ 1 - t - t^2/2 \end{pmatrix}$, sink. (c) $\mathbf{x}(t)$
$= e^{-t}\begin{pmatrix} 1 - 4t - t^2 \\ -10t - 2t^2 \\ 1 - 12t - 3t^2 \end{pmatrix}$, sink.

4. (a) $\mathbf{x}(t) = (x_0e^{-t}, y_0e^{0.1t}, z_0e^t)$; (c) along the $x_3$ axis.

5. (a) saddle; (b) $\mathbf{x}(t) = \begin{pmatrix} -9.8e^t + 19.8e^{-t} \\ -9.8e^t - 9.9e^{-2t} + 19.8e^{-t} \\ -9.8e^t + 9.9e^{-2t} \end{pmatrix}$.

10. The eigenvalues of an upper triangular $3 \times 3$ matrix are the diagonal entries.

11. (a) The eigenvalues and associated eigenvectors are $\lambda_1 = 2$, $\mathbf{p}_1 = (1, 1, 1, 0)$; $\lambda_2 = 1$, $\mathbf{p}_2 = (1, 1, 0, 0)$; $\lambda_3 = -2$, $\mathbf{p}_3 = (0, 1, 1, 0)$; $\lambda_4 = -1$, $\mathbf{p}_4 = (1, 0, 1, 1)$. The general solution is

$$\mathbf{x}(t) = c_1 e^{2t}(1, 1, 1, 0) + c_2 e^t(1, 1, 0, 0) + c_3 e^{-2t}(0, 1, 1, 0) + c_4 e^{-t}(1, 0, 1, 1).$$

(c) The eigenvalues and associated eigenvectors are $\lambda_1 = -2$, $\mathbf{p}_1 = (-1, 0, 2, 1)$; $\lambda_2 = 1$, $\mathbf{p}_2 = (1, 1, -1, 1)$; $\lambda_3 = 3$, $\mathbf{p}_3 = (0, 2, 1, 4)$; $\lambda_4 = -1$, $\mathbf{p}_4 = (0, 1, 0, 1)$. The general solution is

$$\mathbf{x}(t) = c_1 e^{-2t}(-1, 0, 2, 1) + c_2 e^t(1, 1, -1, 1) + c_3 e^{3t}(0, 2, 1, 4) + c_4 e^{-t}(0, 1, 0, 1).$$

## 7.7

1. (a) $\mathbf{\Psi}(t) = \mathbf{P}e^{\mathbf{\Lambda}t}\mathbf{P}^{-1} = \begin{pmatrix} e^{-2t} & 3e^{-t} - 3e^{-2t} \\ 0 & e^{-t} \end{pmatrix}$. (b) $\mathbf{\Psi}(t) = e^{-3t}\begin{pmatrix} \cos t & -\sin t \\ \sin t & \cos t \end{pmatrix}$.

3. Hint: Consider $e^{\mathbf{\Lambda}(t-t)}$.

10. (c) No.

15. (a) Suggestion: Determine the components of $\mathbf{C}\mathbf{f}(t)$ and use the linearity of differentiation. (b) Suggestion: Use the definition of matrix-vector multiplication and apply the result in (a).

## 7.8

1. (a) $\mathbf{x}(t) = \begin{pmatrix} -\frac{14}{3}e^{-2t} + \frac{9}{2}e^{-t} + \frac{7}{6}e^t - 2te^{-t} \\ 10e^{-t} - \frac{35}{3}e^{-2t} + \frac{5}{3}e^t - 4te^{-t} \end{pmatrix}$.

2. (a) $\mathbf{x}(t) = \begin{pmatrix} \frac{9}{2}e^{-t} - 4e^{-2t} + \frac{3}{2}\cos t + \frac{3}{2}\sin t \\ -10e^{-2t} + 9e^{-t} + 2\cos t + 3\sin t \end{pmatrix}$.

3. (a) $\mathbf{x}(t) = \begin{pmatrix} \frac{1}{2}e^{-t} + 2te^{-t} - 4t + \frac{3}{2}e^t \\ \frac{1}{2}e^{-t} - 1 + \frac{3}{2}e^t + te^{-t} - 3t \end{pmatrix}$.

4. (a) $\mathbf{x}(t) = \begin{pmatrix} 9e^{2t} - 6e^t - 4\cos t - 8\sin t \\ -4e^t + 9e^{2t} - 3\sin t - 4\cos t \end{pmatrix}$.

6. Suggestion: Use the definition of matrix-vector multiplication and the product rule for scalar functions.

## *Chapter 8*

## 8.1

1. Unique solutions are guaranteed for (a) and (c) for all initial conditions and for (d) if $t \neq 0$.

## 8.2

1. (a) $y = 2(1 - x)$ and $x = 0$; (b) $y = 1 - 1.8x$ and $y = 0$.

3. Hint: The fixed point is a saddle fixed point.

5. (a) $x$ preys on $y$. (b) $(0, 0)$, $(0, 1)$, $(1, 0)$ and $\left(\frac{1+a}{1+ba}, \frac{1-b}{1+ba}\right)$. (c) The coexistence equilibrium is stable, and all other ones are unstable. (d) Competitive coexistence. (e) $(0, 0)$ and $(0, 1)$ are unstable fixed points, and $(1, 0)$ is a stable fixed point. (f) $y$ dies out.

7. (a) The prey grows exponentially. (b) The predator population goes to zero exponentially. (c) $x' = (\lambda - ay - cx)x$.

## 8.3

1. (a) $x' = y$, $y' = -x + (1 - x^2)y$. (b) $x' = y$, $y' = -x + y$. (d) The linear system is a spiral source. The van der Pol system approaches a periodic solution. (e) No.

3. (c) $\mathbf{f}(\mathbf{x}) = (y, -x - (\alpha + x^2)y)$, which is defined and continuous everywhere. The Jacobian matrix is

$$\mathbf{D_x f} = \begin{pmatrix} 0 & 1 \\ -1 - 2xy & -(\alpha + x^2) \end{pmatrix},$$

which is defined and continuous everywhere. Solutions are unique. (d) They do not touch.

## 8.4

1. (a) $x(t) = C \cos t$. The amplitude is $|C|$, and the period is $2\pi$. (b) $A = 1$, $T = 6.7$. (c) $A = 2$, $T = 8.4$. (d) $A = 3$, $T = 16.1$.

3. (b) Periodic; (c) yes.

5. (a) Periodic; (b) $x(100) \approx 0$, $x(120) \approx -1.4$.

7. (a) All solutions go to the origin. (b) The initial condition for which $x'(0) = 3$ implies that the pendulum flips over once before it approaches the straight-down position. (c) $2.45 < x'(0) < 3.5$, approximately. (d) The pendulum flips over more times initially, but it eventually settles down to the straight-down equilibrium in typical cases.

9. (a) Not periodic. (b) $\omega = 1$: steady-state periodic with $T \approx 6.3$ and an amplitude of 4.2; $\omega = 2$: steady-state periodic with $T \approx 3.1$ and an amplitude of 1.24.

## 8.5

4. (a) $x' = x - \pi$. (e) The solution of Eq. (8.47) is $x(t) = 2 \tan^{-1}(e^{-t} \tan(3/2))$. The solution of the linearized equation is $x(t) = e^t(3 - \pi) + \pi$. The relative difference between the solution is less than 10 percent for $0 \leq t < 2.3$ or so.

5. (a) $x_n = \pm n\pi$ for all integers $n$. The corresponding linearization is $x' = x - x_n$ for $n$ even and $x' = -(x - x_n)$ for $n$ odd. The fixed points are unstable if $n$ is even and are stable if $n$ is odd. (b) $y_n = \left(\frac{1}{2} \pm n\right)\pi$ for all integers $n$. The corresponding linearization is $y' = y - y_n$ at all fixed points of the form $y_n = \left(\frac{3}{2} \pm 2n\right)\pi$. These fixed points are unstable. The linearization at the remaining fixed points is $y' = -(y - y_n)$, which implies that they are stable. (c) $x_f = 1$, linearization $x' = -(x - 1)$, stable; $x_f = 2$, linearization $x' = x - 2$, unstable. (d) Fixed point $y_f = 1$, linearization $y' = y - 1$, unstable; fixed point $y_f = 2$, linearization $y' = -(y - 2)$, stable.

7. (b) $\mathbf{x} = \frac{1}{2}\begin{pmatrix} (a+b-\pi)e^t + (a-b-\pi)e^{-t} + 2\pi \\ (a+b-\pi)e^t - (a-b-\pi)e^{-t} \end{pmatrix}$. (c) $\mathbf{x} = \begin{pmatrix} ce^t + \pi \\ ce^t \end{pmatrix}$. (d) $\mathbf{x} = \begin{pmatrix} -ce^{-t} + \pi \\ ce^{-t} \end{pmatrix}$.

10. (a) The fixed points are $\mathbf{x}_f = (0,0)$, $(1,0)$, $(0,1)$, and $(5/6, 1/3)$. All but $(5/6, 1/3)$ are unstable.

12. Solution trajectories of the nonlinear system that start on the axes remain on the axes, but those of the linearized system do not.

14. (a) $x_f = 0$ and $x_f = -\lambda$. (b) If $\lambda > 0$, then the origin is unstable and $x_f = -\lambda$ is stable; and conversely if $\lambda < 0$.

16. (a) The fixed points are $x_f = \pm\sqrt{-\lambda}$. (b) The positive equilibrium is unstable, and the negative one is stable.

18. (a) $x_1' = x_2$, $x_2' = -0.2x_2 - \sin x_1$, with fixed points $\mathbf{x}_f = (\pm n\pi, 0)$ for all integers $n$. (c) The fixed points are stable for $n$ even and unstable for $n$ odd. (d) All solutions approach the origin.

20. (b) The fixed points are $(0,0)$, which is a saddle; $(4,0)$, saddle; and $(3, 0.5)$, sink. (c) The fixed points are $(0,0)$, saddle; and $(2,0)$, sink.

22. (a) The Jacobian matrix is $\mathbf{D_x}(0) = \begin{pmatrix} 0 & -1 \\ 1 & 0 \end{pmatrix}$, whose eigenvalues are $\pm i$. (b) $r' < 0$ for $r > 0$, and $r' > 0$ for $r < 0$.

## 8.6

1. (a) $x' = \lambda x + b$, where $\lambda$ and $b$ are constants. (b) The only fixed point is $x_f = -b/\lambda$. (c) Bifurcation at $\lambda = 0$, where $x_f$ changes from a sink (for $\lambda < 0$) to a source (for $\lambda > 0$).

3. (b) When $\lambda = 1$, two fixed points coalesce to one at $x = 0$. (c) A saddle-node bifurcation.

4. (b) A saddle-node bifurcation.

8. (a) A saddle-node bifurcation.

10. (a) $ab < 0$. (b) $ab > 0$. (c) For appropriate values of $a$ and $b$, $x_f^{\pm} = \pm\sqrt{-a\lambda/b}$. (d) $x_f^-$ is stable if $b > 0$, and $x_f^+$ is stable if $b < 0$. (e) Yes. (f) A saddle-node bifurcation.

12. (a) The fixed points are $x_f = 0$ and $x_f = \lambda - 1$. (c) A transcritical bifurcation.

15. (b) The number of fixed points changes from one (when $\lambda < 0$) to three (when $\lambda > 0$). The origin is stable for $\lambda < 0$ and unstable for $\lambda > 0$. The remaining fixed points are stable.

## 8.7

3. (a) If $c > 0$ then $\phi(t)$ increases, so solution curves traverse the phase plane in an anticlockwise direction. If $c < 0$, then $\phi(t)$ decreases, so solution curves traverse the phase plane in a clockwise direction. (b) Limit cycles exist whenever $a\lambda/b < 0$. (c) If $b > 0$, then the limit cycle is stable; if $b < 0$, then it is unstable. (d) For example, $a = -1$, $b = 1$, $c = -1$, and $\lambda = 1$.

5. (a) $(\pi/6, 0)$, a spiral sink and $(5\pi/6, 0)$, a saddle. (c) At $\lambda = 1$, the two fixed points coalesce into one fixed point at $(\pi/2, 0)$, and the eigenvalues of the corresponding Jacobian matrix are $0$ and $-1$. (d) There are no fixed points for $\lambda > 1$ and two fixed points for $\lambda < 1$. There appears to be a periodic solution for $\lambda > 1$.

## 8.8

1. (a) $x_0 = 0.31416$, $x_1 = 0.1416$, $x_2 = 0.416$, $x_3 = 0.16$, $x_4 = 0.6$, and $x_n = 0$ for all $n > 4$. (b) $x_0 = 0.3141592$, $x_1 = 0.141592$, $x_2 = 0.41592$, $x_3 = 0.1592$, $x_4 = 0.592$, $x_5 = 0.92$, $x_6 = 0.2$, and $x_n = 0$ for all $n > 6$. (c) Differences grow by a factor of 10 each iteration. (d) 13. (e) $k + 3$.

3. (a) Constant; (b) increasing; (c) decreasing; (d) decreasing and alternating in sign; (e) periodic.

5. (a) $x = 0$, stable for $a < 1$, and $x = a - 1$, stable for $1 < a < 3$; (c) transcritical.

7. (a) $x = 0$, unstable; $x = \frac{1}{2}(3 - \sqrt{5})$, stable; $x = \frac{1}{2}(3 + \sqrt{5})$, unstable; (b) no fixed points; (c) $x = 0$ is a stable fixed point.

9. (a) Period-4 orbit; (b) period-8 orbit; (c) period-16 orbit; (d) period-32 orbit.

# *Chapter 9*

## 9.1

1. The global error for $h = 10^{-2}$ is approximately 0.0047. The results for $h = 10^{-1}$ and $h = 1$ suggest that the global error does decrease linearly with $h$.

3. (b) Yes. The global error decreases approximately by a factor of 10 as the step size is decreased by 10.

8. (a) $x(t) = \left(\frac{10}{9} - t\right)^{-1}$. (c) Yes. Among the suggested step sizes, there is a linear dependence on $h$ for $h$ between $10^{-3}$ and $10^{-5}$.

## 9.2

2. (a) $k < 0$; (b) $|a| < 1$; (c) $a = 1$; (d) $k = 0$.

3. (a) Saddle; (b) source; (c) source.

4. A saddle if $0 < |c| < 2$ and a source if $|c| > 2$.

## 9.3

2. (a) $0 < h < 1$; (b) $0 < h < 100$; (c) $0 < h < 1/100$.

5. (a) $x_1(t) = c_1 e^{t/10} + c_2 e^{-10t}$, $x_2(t) = 10.1 c_1 e^{t/10}$, a saddle fixed point. (b) $0 < h < 1/5$. (c) $0 < h < 1/10$. (d) No.

7. (a) $x_1(t) = c_1 e^{-2t} - c_2 e^{-120t}$, $x_2(t) = c_1 e^{-2t} + c_2 e^{-120t}$, sink. (b) $0 < h < 1/60$. (c) $0 < h < 1/120$. (d) Yes.

9. (a) $x_1(t) = c_1 - c_2 e^{-10t}$, $x_2(t) = c_1 + c_2 e^{-10t}$. (b) $0 < h < 1/5$. (c) No.

11. (a) $0$ is unstable and $\pm\sqrt{a/b}$ are stable; (c) $0 < h < 1/a$; (d) $0 < h < 1/(2a)$.

13. (a) A center. (b) $\mathbf{x}_{n+1} = \mathbf{B}\mathbf{x}_n = (\mathbf{I} + h\mathbf{A})\mathbf{x}_n$, where $\mathbf{B} = \mathbf{I} + h\mathbf{A} = \begin{pmatrix} 1 & h \\ -9h & 1 \end{pmatrix}$. (c) $1 \pm 3ih$. (d) A source for all values of $h$.

## 9.4

1. (a) Heun's method for $h = 0.1$ gives $\hat{x}(1) = 21.9139$ and for $h = 0.01$ gives $\hat{x}(1) = 22.1643$. The absolute error in the solutions at $t = 1$ is 0.25 and 0.0029, respectively. The fourth-order Runge-Kutta method with $h = 0.1$ gives $\hat{x}(1) = 22.1667$ and with $h = 0.01$ gives $\hat{x}(1) = 22.1672$. The absolute error in the solutions is $5.0 \times 10^{-4}$ and $5.8 \times 10^{-8}$, respectively.

5. $x' = \left(\frac{3x}{\sqrt{2}}\right)^{2/3}$.

7. The fourth-order Taylor series expansion approximates $x(t + h)$ as $x(t + h) = x(t) + h[x(t)]^2 + h^2[x(t)]^3 + h^3[x(t)]^4 + h^4[x(t)]^5$. The Taylor expansion of the true solution about $t = 0$ to fourth order is $x(t) = 1 + t + t^2 + t^3 + t^4$.

9. $0 < h < 2/k$.

## *Chapter 10*

## 10.1

1. (a)–(b) Let $M = k = 1$. (c) Let $M = 4$ and $k = 3$.

3. (a) $1/(s - 3)$; (b) $2/(s + 1) + 1/s$; (c) $1/(s - \ln 2) + 6/s$; (d) $1/(s + \ln 3)$.

5. (a) Hint: If $f$ is bounded, then there is a positive constant $M$ such that $|f(t)| \leq M$ for all $t$. (b) No.

7. Suggestion: Determine whether there is a function for which the stated limit holds but that tends to $\pm\infty$ as $t \to 0$.

9. $\mathcal{L}(1/t)$ does not exist, but $\mathcal{L}(1/(t + 1))$ does exist.

11. We have $\mathcal{L}(e^{it}) = 1/(s - i) = (s + i)/(s^2 + 1)$, which implies $\mathcal{L}(\cos t) = s/(s^2 + 1)$ and $\mathcal{L}(\sin t) = 1/(s^2 + 1)$.

13. Yes.

15. (a) Hint: The jump discontinuity in $f$ does not affect the value of the integral, so $\mathcal{L}(f) = 1/s^2$. Notice also that a finite number of jump discontinuities also does not affect the value of the integral. (c)–(d) In contrast to (a), where $f(t) \neq t$ at only an isolated point, we have $f(t) \neq t$ over an interval. In this case, $\mathcal{L}(f) \neq 1/s^2$.

## 10.2

1. (a) $x(t) = 10e^{kt}$; (b) $x(t) = 1000e^{kt} + 100(1 - e^{kt})/k$; (c) $x(t) = 100 - 97e^{-t/10}$.

3. (a) $x(t) = \cos t + \sin t + \frac{1}{2}t \sin t$; (b) $x(t) = t - \frac{2}{3}\sin 3t$.

5. (a) $x(t) = \frac{9}{2}e^{-t} - 4e^{-2t} + \frac{1}{2}e^{-3t}$; (b) $x(t) = \frac{3}{10}e^{-3t} - \frac{2}{5}e^{-2t} + \frac{1}{10}(\cos t + \sin t)$.

## 10.3

1. (a) $e^t + 5te^t$; (b) $e^t + 2te^t + \frac{5}{2}t^2e^t$; (c) $e^{-2t}\sin t$; (d) $\frac{1}{2}e^{3t}\sin 2t$.

2. (a) $1 - e^t + te^t$; (b) $2 + t - 2e^t + te^t$; (c) $e^{-t} + 2te^{-t}$; (d) $4t^2\sin 2t - 2t\cos 2t + \sin 2t$.

3. (a) $x(t) = e^{-t}(2\sin t + \cos t)$; (b) $x(t) = 5te^{-t} + 5e^{-t} + 2\sin t - 4\cos t$.

5. (a) $x(t) = e^{-t}\cos t - e^{-t}\sin t$; (b) $x(t) = 2e^{-2t}\cos 2t + 4e^{-2t}\sin 2t$; (c) $x(t) = e^{3t}\cos t - 2e^{3t}\sin t$.

7. $x_1(t) = 2e^{5t} - e^{-t}$, $x_2(t) = 2e^{5t} + e^{-t}$.

10. $x_1(t) = 4te^{-t} + e^{-t} + 2te^{-2t}$, $x_2(t) = 4te^{-2t} + 6te^{-t} - e^{-2t}$.

15. (a) $X(s) = (sx(0) + x'(0) + ax(0))/(s^2 + as + b)$.

19. (a) $x(t) \to 0$ as $t \to \infty$, and it does not oscillate. (b) $x(t) \to 0$ as $t \to \infty$, and it does not oscillate.

## 10.4

1. (a) $f(t) = 4[H(t-3) - H(t-1)]$; (c) $f(t) = t/2 + (1-t/2)H(t-2)$; (e) $f(t) = [H(t-2\pi) - 1)]\sin t$.

5. (a) $x(t) = \frac{3}{10}(t - \sin t) - \frac{3}{10}H(t-10)[(t-10) - \sin(t-10)]$.

7. (a) $x(t) = \begin{cases} 0, & 0 \le t \le \pi, \\ -\frac{2}{3}\sin t - \frac{1}{3}\sin 2t, & \pi \le t \le 2\pi, \\ -\frac{4}{3}\sin t, & t > 2\pi. \end{cases}$

9. (a) $x(t) = \frac{1}{3}\left[1 - \cos 3t - H\left(t - \frac{2}{3}\pi\right)(1 - \cos 3t)\right]$.

## 10.5

2. $x(t) = H(t - t_0)$.

3. (a) The mass is struck instantaneously in such a way that its momentum increases by one unit at $t = 0$.

4. (a) $k = e$; (b) $t_h = \ln 2$; (c) $k = e - 1$.

5. (a) $x(t) = \cos 3t + \frac{2}{3} \sin 3t + \frac{1}{3} H(t - 3) \sin 3(t - 3)$; (b) yes; (c) no.

7. (a) $x(t) = 2\cos 2t + \frac{1}{4} t \sin 2t - \frac{5}{2} \sin 2t - 2H(t-5) \sin 2(t-5)$; (b) continuous at $t = 5$; (c) not continuous at $t = 5$.

8. (a) $q(t) = \left( e^{-150t} \sin \sqrt{77500}\,t \right) \Big/ \sqrt{77500}$.

# INDEX